Polyphenols

Polyphenols

Food, Nutraceutical, and Nanotherapeutic Applications

Edited by

Mithun Rudrapal
Department of Pharmaceutical Sciences
School of Biotechnology and Pharmaceutical Sciences
Vignan's Foundation for Science, Technology & Research
Guntur, India

WILEY

Published by John Wiley & Sons, Inc., Hoboken, New Jersey.
Published simultaneously in Canada.

For general information on our other products and services or for technical support, please contact our Customer Care Department within the United States at (800) 762-2974, outside the United States at (317) 572-3993 or fax (317) 572-4002.

Wiley also publishes its books in a variety of electronic formats. Some content that appears in print may not be available in electronic formats. For more information about Wiley products, visit our web site at www.wiley.com.

Library of Congress Cataloging-in-Publication Data
Names: Rudrapal, Mithun, editor.
Title: Polyphenols: food, nutraceutical, and nanotherapeutic applications / edited by Mithun Rudrapal, Department of Pharmaceutical Sciences, School of Biotechnology and Pharmaceutical Sciences, Vignan's Foundation for Science, Technology & Research, Guntur, India.
Description: Hoboken, New Jersey : John Wiley & Sons, [2024] | Includes bibliographical references and index.
Identifiers: LCCN 2023026387 (print) | LCCN 2023026388 (ebook) | ISBN 9781394188833 (hardback) | ISBN 9781394188857 (epdf) | ISBN 9781394188840 (epub) | ISBN 9781394188864 (ebook)
Subjects: LCSH: Polyphenols. | Polyphenols--Physiological effect. | Polyphenols--Therapeutic use. | Functional foods.
Classification: LCC QP671.F52 P658 2024 (print) | LCC QP671.F52 (ebook) | DDC 572/.2--dc23/eng/20230621
LC record available at https://lccn.loc.gov/2023026387
LC ebook record available at https://lccn.loc.gov/2023026388

Cover Image: © altmodern/Getty Images
Cover Design: Wiley

Set in 9.5/12.5pt STIXTwoText by Integra Software Services Pvt. Ltd, Pondicherry, India

Contents

List of Contributors

Raghu Ram Achar
Division of Biochemistry, School of Life Sciences
JSS Academy of Higher Education and Research
Mysuru
Karnataka
India

Kamoru A. Adedokun
Roswell Park Comprehensive Cancer Center
University at Buffalo
Buffalo
NY
USA

Mujidat T. Adeyemo-Imodoye
Department of Medical Laboratory Science
Somolu General Hospital
Lagos Health Service Commission
Lagos State
Nigeria

Jyotirmoi Aich
School of Biotechnology and Bioinformatics
D Y Patil Deemed to be University
Navi Mumbai
Maharashtra
India

Tanisha Ambawat
School of Biotechnology and Bioinformatics
D Y Patil Deemed to be University
Navi Mumbai
Maharashtra, India

Anuthilakesh T.
Division of Biochemistry, School of Life Sciences
JSS Academy of Higher Education and Research
Mysuru
Karnataka
India

Anyanwu O. Gabriel
Department of Biochemistry
Bingham University
Karu
Nasarawa
Nigeria

Apoorva M.R.
Division of Biochemistry, School of Life Sciences
JSS Academy of Higher Education and
Research
Mysuru
Karnataka
India

Merin Babu
Synthite Industries Pvt. Ltd.
Synthite Valley
Kadayiruppu
Kerala
India

Ibrahim O. Bello
Department of Biological Sciences
Southern Illinois University Edwardsville
Edwardsville
IL
USA

Atul R. Bendale
Mahavir Institute of Pharmacy
Nashik
Maharashtra
India

Bhargavi K.
Division of Biochemistry, School of Life Sciences
JSS Academy of Higher Education, and
Research
Mysuru
Karnataka
India

Mustansir Bhori
School of Biotechnology and Bioinformatics
D Y Patil Deemed to be University
Navi Mumbai
Maharashtra
India

Baby Kumaranthara Chacko
C.V.J. Creative Centre
Bio Ingredients, Synthite Industries Pvt. Ltd.
Synthite Valley
Kadayiruppu
Kerala
India

Shreya Das
School of Biotechnology and Bioinformatics
D Y Patil Deemed to be University
Navi Mumbai
Maharashtra
India

Sweta Das
B. K. Birla College
Birla College Campus Rd
Kalyan
Maharashtra
India

Ameyota De
School of Biotechnology and Bioinformatics
D Y Patil Deemed to be University
Navi Mumbai
Maharashtra
India

Prashanta Kumar Deb
Department of Pharmaceutical Sciences and
Technology
Birla Institute of Technology
Ranchi
Jharkhand
India

Shine Devarajan
School of Biotechnology and Bioinformatics
D Y Patil Deemed to be University
Belapur
Navi Mumbai
Maharashtra
India

Sneha Dokhale
B. K. Birla College
Birla College Campus Rd
Kalyan
Maharashtra
India

Priaunsh Dutt
School of Biotechnology and
Bioinformatics
D Y Patil Deemed to be University
Navi Mumbai
Maharashtra
India

Samiksha Garse
School of Biotechnology and
Bioinformatics
D Y Patil Deemed to be University
Navi Mumbai
Maharashtra, India

Sachin Goel
Department of Biological Sciences and
Engineering
Netaji Subhas University of Technology
New Delhi
India

Madhusmita Gogoi
Department of Pharmaceutical Sciences
Faculty of Science and Engineering
Dibrugarh University
Dibrugarh
Assam
India

Sikiru O. Imodoye
Huntsman Cancer Institute
University of Utah
Salt Lake City
UT
USA

Viju Jacob
Nutraceuticals Division, Bio Ingredients
Synthite Industries Pvt. Ltd.
Synthite Valley
Kadayiruppu
Ernakulam
Kerala
India

C.V.J. Creative Centre
Bio Ingredients, Synthite Industries Pvt. Ltd.
Synthite Valley
Kadayiruppu
Kerala
India

Gourav Jain
Department of Pharmacy
Shri Govindram Seksariya Institute of
Technology & Science (SGSITS)
Indore
Madhya Pradesh
India

Aayushi Kadam
Department of Food and Human Nutritional
Sciences
University of Manitoba
Canada

Ramut T. Kamorudeen
Department of Public Health
University of South
Wales, Pontypridd
UK

Koyel Kar
Department of Pharmaceutical Chemistry
BCDA College of Pharmacy & Technology
Hridaypur
West Bengal
India

Monalisa Kesh
Amity Institute of Biotechnology
Amity University Uttar Pradesh
Noida
India

Swapnil Devidas Khamborkar
Synthite Industries Pvt. Ltd.
Synthite Valley
Kadayiruppu
Kerala
India

Manohar M.V.
Centre of excellence in Molecular Biology &
Regenerative Medicine
Department of Biochemistry
JSS Medical College, JSS-AHER
Mysuru
Karnataka, India

Akhil Nagar
Department of Pharmaceutical Chemistry
R. C. Patel Institute of Pharmaceutical
Education and Research
Shirpur
Maharashtra
India

Neelesh Kumar Nema
Nutraceuticals Division
Bio Ingredients, Synthite Industries Pvt. Ltd.
Synthite Valley
Kadayiruppu
Ernakulam
Kerala
India

Nithish G.S.
Indian Institute of Science Education and Research Tirupati
C/O Sree Rama Engineering College (Transit Campus Mangalam)
Tirupati
Andhra Pradesh
India

Ogah F. Chinenye
Department of Biochemistry
Bingham University
Karu
Nigeria

André M. Oliveira
Federal Centre of Technological Education of Minas Gerais (CEFET-MG)
Alameda das Perdizes
Contagem
MG
Brazil

Ifeoluwa T. Oyeyemi
Department of Biosciences and Biotechnology
University of Medical Sciences
Ondo
Nigeria
Department of Chemistry and Biochemistry
Old Dominion University
Hampton Boulevard
Norfolk
VA
USA

Amogha G. Paladhi
Department of Botany
CHRIST (Deemed to be) University
Bengaluru
Karnataka
India

Santwana Palai
Department of Veterinary Pharmacology & Toxicology
College of Veterinary Science and Animal Husbandry
Odisha University of Agriculture & Technology
Bhubaneswar
Odisha
India

Sharad P. Pandey
Department of Pharmacy
Shri Govindram Seksariya Institute of Technology & Science (SGSITS)
Indore
Madhya Pradesh
India

Arpita Paul
Department of Pharmaceutical Sciences
Faculty of Science and Engineering
Dibrugarh University
Dibrugarh
Assam
India

Aneena Peter
Synthite Industries Pvt. Ltd.
Synthite Valley
Kadayiruppu
Kerala
India

Pruthvi G.R.
Division of Biochemistry
School of Life Sciences
JSS Academy of Higher Education and Research
Mysuru
Karnataka
India

Nayana Rajan
Synthite Industries Pvt. Ltd.
Synthite Valley
Kadayiruppu
Kerala
India

Mithun Rudrapal
Department of Pharmaceutical Sciences
School of Biotechnology and Pharmaceutical Sciences, Vignan's Foundation for Science, Technology & Research
Guntur
India

Sachithra Sabu
Synthite Industries Pvt. Ltd.
Synthite Valley
Kadayiruppu
Kerala
India

Rakesh Sagar
Department of Pharmacy
Shri Govindram Seksariya Institute of Technology & Science (SGSITS)
Indore
Madhya Pradesh
India

Abhishek K. Sah
Department of Pharmacy
Shri Govindram Seksariya Institute of Technology & Science (SGSITS)
Indore
Madhya Pradesh
India

Linson Cheruveettil Sajan
Synthite Industries Pvt. Ltd.
Synthite Valley
Kadayiruppu
Kerala
India

Puneet K. Samaiya
Department of Pharmacy
Shri Govindram Seksariya Institute of Technology & Science (SGSITS)
Indore
Madhya Pradesh
India

Malik A. Sanusi
Department of Biochemistry and Molecular Biology
Obafemi Awolowo University
Ile-Ife
Osun State
Nigeria

Biswatrish Sarkar
Department of Pharmaceutical Sciences and Technology
Birla Institute of Technology
Ranchi
Jharkhand
India

Smitha Sarojam
Synthite Industries Pvt. Ltd.
Synthite Valley
Kadayiruppu
Kerala
India

Yashika Shadija
School of Biotechnology and Bioinformatics
D Y Patil Deemed to be University
Navi Mumbai
Maharashtra
India

Kanchanlata Tungare
School of Biotechnology and Bioinformatics
D Y Patil Deemed to be University
Navi Mumbai
Maharashtra
India

Sugumari Vallinayagam
Department of Biotechnology
Vel Tech Rangarajan Dr. Sagunthala R and D
Institute of Science and Technology
Chennai
Tamil Nadu

Zwanden S. Yahaya
Department of Pharmaceutics and Industrial
Pharmacy
Faculty of Pharmaceutical Sciences
Kaduna State University,
Kaduna
Nigeria

Aamena Yusuf
School of Biotechnology and Bioinformatics
D Y Patil Deemed to be University
Navi Mumbai
Maharashtra
India

Kamaruz Zaman
Department of Pharmaceutical Sciences
Faculty of Science and Engineering
Dibrugarh University
Dibrugarh
Assam
India

Preface

The book titled ***"Polyphenols: Food, Nutraceutical, and Nanotherapeutic Applications"*** provides us with detailed information on the potential health benefits of food polyphenols against human diseases and their therapeutic interventions as nutraceuticals and nanoformulations. Plant-based foods, including nutraceuticals, functional foods, and dietary supplements, have gradually become an area of increasing interest because of the potential health implications of polyphenol-based food products available in the global market. Long-term consumption of foods rich in polyphenols has been reported to improve conditions in diabetes, osteoporosis, cancer, neurological disorders, and cardiovascular diseases. The market demand for polyphenol-based food products and nutraceutical preparations is increasing rapidly with social and economic developments and the improvement in people's health awareness. The beneficial effects of polyphenol-based nutraceuticals and their nanoformulations attract more consumers. However, many polyphenols have low oral bioavailability, which limits the application of polyphenols as nutraceuticals and in the form of nanomedicine. This book would be a useful resource for researchers working in the areas of dietary polyphenols, nanodelivery, and nutraceutical applications. The book chapters provide up-to-date information relating to the ongoing research on food polyphenols along with their health and therapeutic implications as antioxidants. The chapters are described in a lucid and consistent manner in order to aid flow, continuity, and technical presentation. In this book, the health benefits of plant polyphenols in the prevention of various major diseases, such as cancer, neurological disorders, central nervous system disorders, cardiovascular diseases, bacterial infections, and inflammatory disorders, are summarized. This book particularly delineates food polyphenols and their antioxidant potentials with special emphasis on their nutraceutical applications and nanodeliveries. The bioavailability and pharmacokinetic issues and toxicity are also demonstrated. This book will be useful to drug discovery scientists (R&D), pharmaceutical scientists, biomedical scientists, food scientists, healthcare professionals, phytochemists, biochemists, clinicians, pharmacologists, dietician, nutritionists, research students, professors, and other researchers working in the field of polyphenols, nutraceuticals, and nanomedicine research.

Mithun Rudrapal
June, 2023

1

Food Polyphenols

Antioxidant Properties and Health Benefits

Pruthvi G.R.[1], *Apoorva M.R.*[1], *Anuthilakesh T.*[1], *Bhargavi K.*[1], *Nithish G.S.*[2], *and Raghu Ram Achar*[1,*]

[1] *Division of Biochemistry, School of Life Sciences, JSS Academy of Higher Education and Research, Mysuru, Karnataka, India*
[2] *Indian Institute of Science Education and Research Tirupati, C/o Sree Rama Engineering College (Transit Campus), Tirupati, Andhra Pradesh, India*
* *Corresponding author*

1.1 Introduction

1.1.1 Dietary Polyphenols

Polyphenols are a diverse group of naturally occurring compounds that are widely distributed in plant-based foods such as fruits, vegetables, nuts, grains, and spices. They are characterized by their multiple phenolic ring structures and they play a significant role in the color, flavor, and nutritional properties of foods. Over the past few decades, research has shown that polyphenols have various health-promoting benefits, including antioxidant, anti-inflammatory, anticancer, and cardioprotective effects. As a result, there has been growing interest in the role of polyphenols in human health, and they are now recognized as essential bioactive compounds in a diet. This introduction provides a glimpse into the fascinating world of food polyphenols and their potential health benefits.

Academic literature has documented approximately 50,000 distinct types of polyphenols, which are secondary metabolites of plants produced via the shikimate-derived phenyl propanoic and/or the polyketide pathways, in plants [1].

However, polyphenols can be broadly classified into four major groups: phenolic acids, flavonoids, phenolic amides, and other polyphenols [2].

Flavonoids

Flavonoids are a significant class of low molecular weight polyphenolic substances found widely in the plant kingdom. Their presence is found in almost all parts of plants. These compounds have a structural makeup of two aromatic rings with fifteen carbon atoms [3]. A three-carbon atom chain connects the aromatic rings depending on the C ring that the B ring is linked to, its conformation, and the degree of oxidation [4]. Flavonoids can be classified into different subclasses such as chalcones, flavones, flavonols, and isoflavones [5]. Flavonoids provide the characteristic color, flavor, and aroma of fruits and flowers, helping to attract pollinators [6].

Polyphenols: Food, Nutraceutical, and Nanotherapeutic Applications, First Edition. Edited by Mithun Rudrapal.

Flavonoids possess a variety of biological properties favoring human health and agriculture including anti-allergic, anti-inflammatory, antimicrobial, antioxidative, anticarcinogenic, and antiviral properties [5].

Phenolic Acids

Phenolic acids are aromatic rings with one carboxylic group comprising phenolic acids (–COOH). Although they are similar to alcohols, they become weak acids because of the presence of an aromatic ring and the hydrogen atom of the phenolic hydroxyl group [7]. They form a major group of polyphenols produced by plants and play a vital role as a human dietary component. Phenolic acids have a high antioxidant activity. Thus, a diet rich in phenolic acids helps the body cope with oxidative stress that occurs, thereby reducing the risk of diseases related to oxidative stress. Structurally, phenolics are present as bound forms such as amides, esters, or glycosides [8]. Phenolics are classified as hydroxybenzoic acid and hydroxycinnamic acid. Hydroxycinnamic acids are chlorogenic, combining both caffeic and quinic acids, whereas hydroxybenzoic acids contain a soluble C6–C1 structure that is derived from benzoic acids [9]. Like flavonoids, phenolic compounds also possess biological properties.

Polyphenol Amides

Some polyphenols have N-containing functional substituents. These are classified as polyphenolic amides. Capsaicinoids in chili and avenanthramides in oats are some of the major components of common foods [10].

Other Polyphenols

The polyphenols that do not fall into the above classes are grouped under "other" and are designated as non-flavonoid polyphenols. This category includes stilbenes, lignans, and other polyphenols, as shown in Figure 1.1 [10]. The classes and subclasses are listed in Table 1.1.

1.1.2 FDA-approved Antioxidant Polyphenols

FDA-approved antioxidant polyphenols are epigallocatechin, epigallocatechin-3-gallate (EGCG), hydroxytyrosol, catechin, apigenin, anthocyanins, hydroxy-cinnamic acids, proanthocyanidin b4, curcumin, kaempferol-3-ogalactoside, quercetin, kaempferol morin, apigenin, daidzein, ellagic acid, gallic acid, corilagin, dihydrocaffeic acid, caffeic acid (+)-catechin, and resveratrol [11].

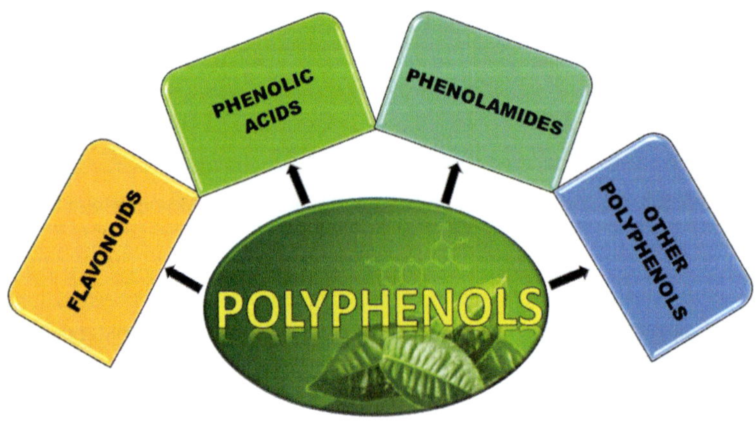

Figure 1.1 Classification of dietary polyphenols.

Table 1.1 Classes and subclasses of dietary polyphenols.

Classes	Subclasses	Examples
Flavonoids	• Flavones • Flavonols • Isoflavones • Flavanones • Anthocyanins	• Luteolin, apigenin • Quercetin, catechin • Genistein, daidzein • Hesperetin, naringenin • Malvidin, cyanidin
Phenolic amides	• Avenanthramides • Capsaicinoids	• Avenanthramide a: R = H; avenanthramide b: R = OCH_3; avenanthramide c: R = OH • Capsaicin, dihydrocapsaicin
Phenolic acids	• Hydroxybenzoic acid • Hydroxycinnamic acids	• Salicylic acid, gallic acid, vanillic acid, benzoic acid, ellagic acid • Caffeic acid, ferulic acid, cinnamic acid, coumaric acid, sinapinic acid
Other polyphenols	• Stilbenes • Lignans • Others including tannins, lignins, xanthones, chromones, anthraquinones	• Resveratrol, piceatannol • Sesamol, pinoresinol, enterodiol

1.1.3 Polyphenols and Their Antioxidant Properties

Polyphenols are a group of organic compounds that occur naturally and are identified by the presence of multiple phenol units and their associated functional groups. The term "polyphenol" is derived from the Greek words "polus", which means "many" and "phenol". Phenol refers to the chemical structure comprising an aromatic benzene ring with a hydroxyl (–OH) group, similar to that found in alcohols (therefore ending with the suffix -ol) [12].

The term "polyphenol" has been in existence since 1894. Although there is no clear definition of polyphenols, it is widely accepted that they are natural compounds with a polyphenolic structure, characterized by the presence of multiple hydroxyl (–OH) groups or other functional rings. Polyphenols can be classified into four main classes: phenolic acids, flavonoids, stilbenes, and lignans [13].

Condensed tannins, the most common type of polyphenols, can be found in nearly all plant families. Larger polyphenols play a vital role in forest ecology by aiding in the decomposition of forest litter and nutrient cycles, and they are often concentrated in leaf tissue, epidermis, bark layers, flowers, and fruits. The total amount of natural phenols and polyphenols, expressed as a percentage of the dry green leaf mass, varies between 1%–25%. The absolute quantity of total phenols in plant tissues varies significantly depending on the assay used and the type of polyphenol, as well as the literature source.

1.1.4 Polyphenols in Functional Foods and Nutraceuticals

Functional foods are food products that provide health benefits beyond basic nutrition, often referred to as "rainbow color foods". These foods can be categorized into modified or fortified functional foods, and can be easily incorporated into a balanced diet. They can help bridge any gaps in your diet, preventing nutrient deficiencies while also improving your overall health by increasing your intake of essential nutrients such as vitamins, minerals, fiber, heart-healthy fats, and probiotics.

Conventional functional foods include a variety of fruits, such as berries, kiwi, pears, peaches, apples, oranges, bananas, and pomegranates, as well as vegetables such as broccoli, cauliflower, kale, zucchini, and spinach. Nuts, seeds, legumes, whole grains, seafood, fermented foods, herbs, and spices are also included in this category. Alternatively, modified functional foods include fortified juices, dairy products, milk alternatives, grains, cereals, granola, eggs, and poultry. Supplements of vitamins, minerals, botanicals, herbs, botanical compounds, amino acids, and probiotics are also considered modified functional foods. By including these functional foods in our diets, we can fill in any nutritional gaps and prevent deficiencies while enhancing our overall health.

1.2 Classification, Sources, and Functions

1.2.1 Hydroxybenzoic Acids, Hydroxycinnamic Acids, and Stilbenes

Polyphenols	Structure	Functions	Sources	References
Hydroxybenszoic acid	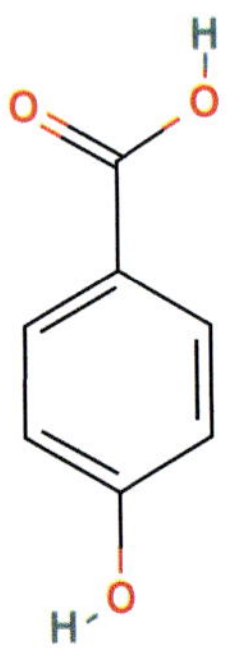	These compounds have demonstrated potential in reducing oxidative stress and inflammation, promoting better lipid profiles, and exhibiting anti-atherosclerotic, anti-inflammatory, analgesic, antibacterial, hepatoprotective, antiviral, and antineoplastic effects in both in vivo and in vitro studies. It is suggested that the compound may play a role in reversing biochemical changes related to cardiac dysfunction, diabetes, and metabolic disorders because of its antioxidant, anti-inflammatory, and antineoplastic properties. The therapeutic potential of the compound has been noted in various disorders including gastrointestinal, neuropsychological, metabolic, and cardiovascular systems.	Blackberries, cranberries, grapefruit, grapes, mangos, pomegranate, raspberries, rhubarb, strawberries, juices made from these fruits, tea, red and white wines, chestnuts, peanuts, pecans, walnuts, wheat, select herbs and spices	[14–17]
Protocatechuic acid	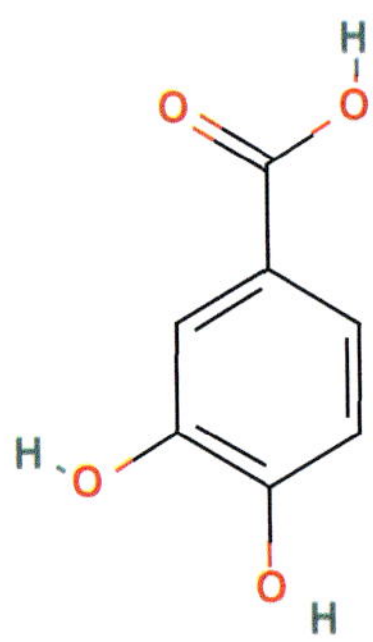			
Gallic acid	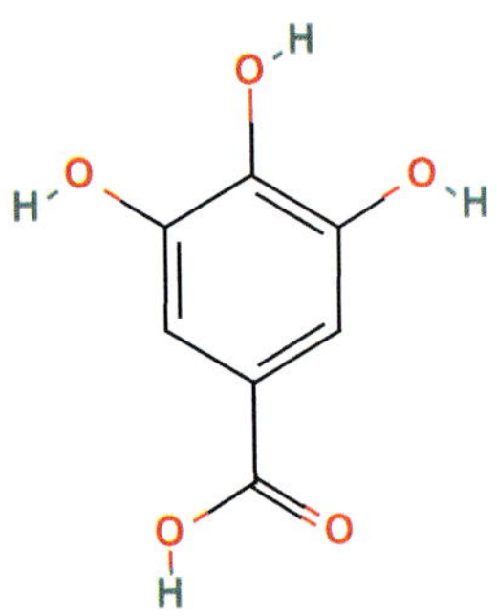			

(Continued)

Polyphenols	Structure	Functions	Sources	References
Hydroxycinnamic acids	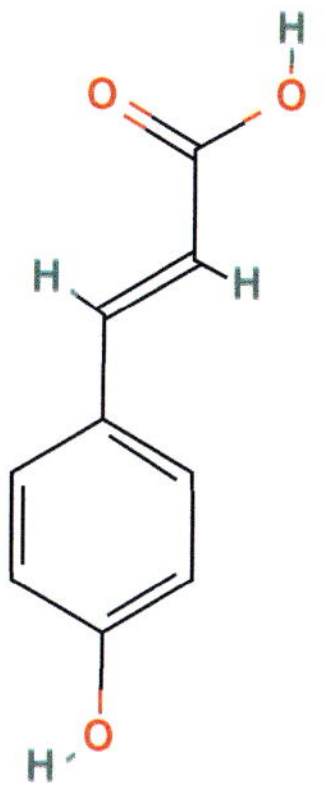	These compounds possess antioxidant, anti-inflammatory, anticollagenase, antityrosinase activities, UV protective, and anticarcinogenic properties that make it a promising agent for antiaging and anti-inflammatory purposes, as well as for use as a preservative and hyperpigmentation correcting ingredient. Additional benefits include antibacterial, antiviral, anti-atherosclerotic, immunostimulatory, antidiabetic, cardioprotective, antiproliferative, hepatoprotective, anticancer, and anti-hepatocellular carcinoma activity.	Blueberry, kiwi, cherry, plum, aubergine, apple, pear, chicory, artichoke, potato, corn flour, wheat flour, rice flour, oat, cider, coffee, wine, papaya, onion, strawberry, raspberry, sugarcane, grape, cherry, orange, grapefruit, watermelon, blackberries, peach, banana	[18, 19]
Caffeic acid	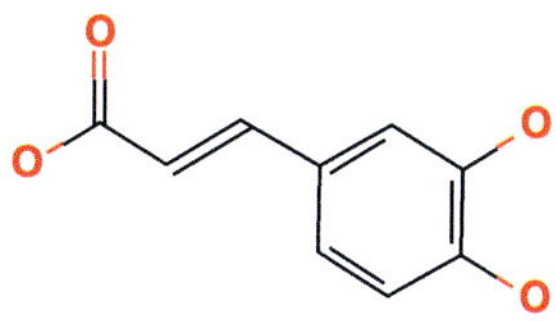			
Chlorogenic acid	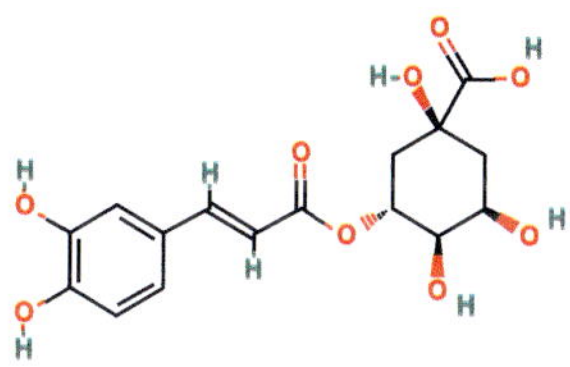			
O-Coumaric acid	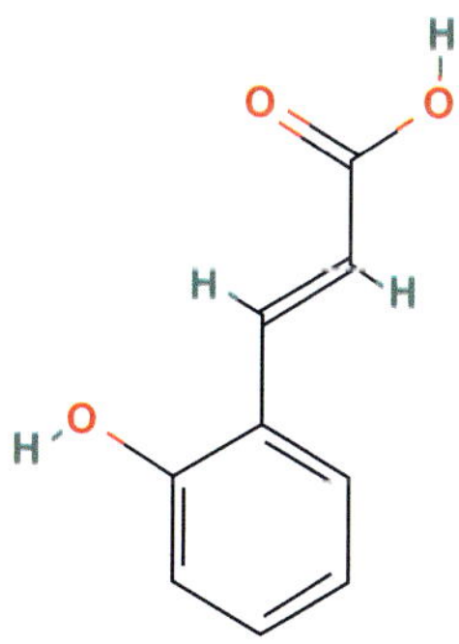			
Ferulic acid	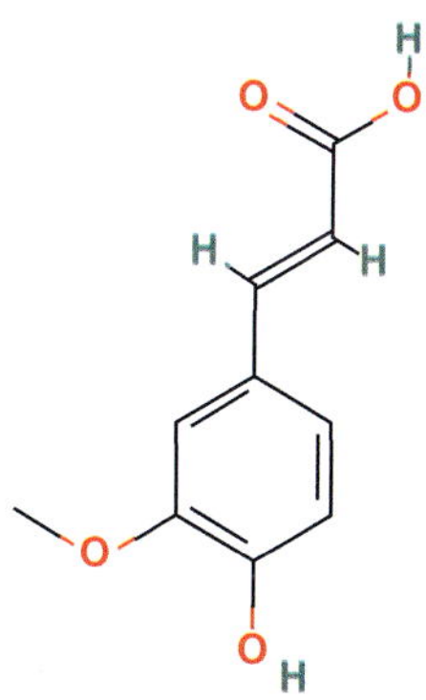			

1.2.2 Flavonoids, Flavonols, Flavones, Isoflavones, Flavanones, and Flavanols

Flavonols	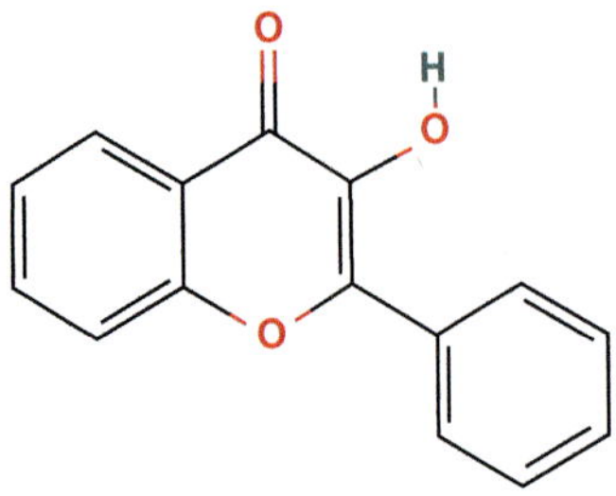	Flavonoids reduce the risk of cardiovascular diseases, metabolic disorders, and certain types of cancer by reducing oxidative stress, inhibiting low-density lipoprotein oxidation and platelet aggregation, and acting as vasodilators in blood vessels. • Kaempferol inhibits cancer cell growth and angiogenesis, induces cancer cell apoptosis, and preserves normal cell viability, exerting a protective effect. • Myricetin has neuroprotective action, antidiabetic, anticancer, immunomodulatory, cardiovascular, analgesic, and antihypertensive properties, which have been demonstrated in preclinical studies on Alzheimer's, Parkinson's and Huntington's diseases, as well as amyotrophic lateral sclerosis.	Apples, asparagus, broccoli, chili pepper, Chinese cabbage, kale, leeks, lettuce, onions, spinach, chives, dill, fennel leaves, oregano, blueberry, cherry, cranberry, wild leeks (whole), black tea, red wine, cranberry, dock sweet potato leaves, chard, swiss, broad beans, immature seeds, rutabagas, garlic blueberry, peppers, hot chili, green, blackberry, lotus root, lemon	[20–24]
Quercetin	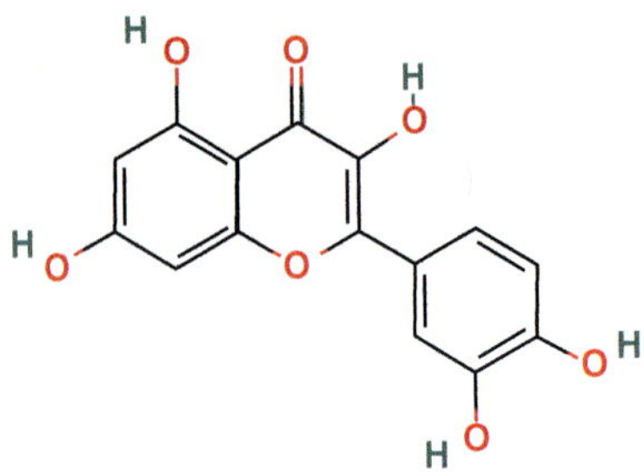			
Kaempferol	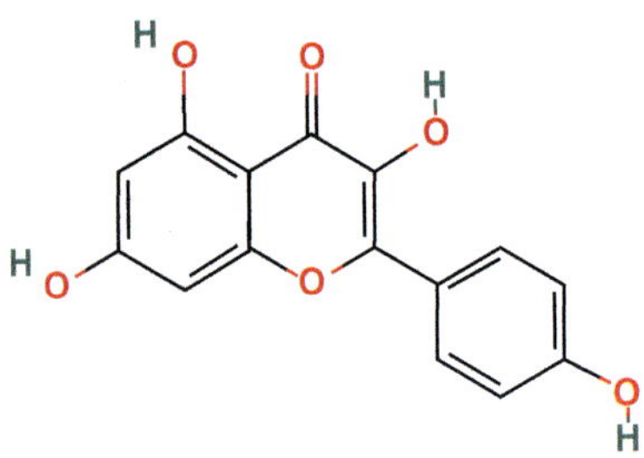			
Myricetin	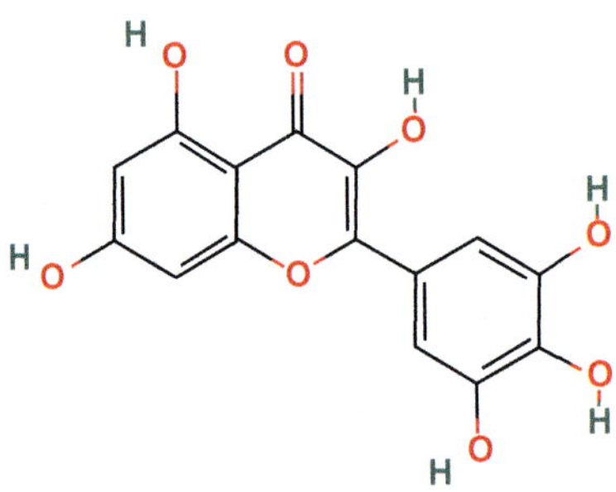			
Flavones	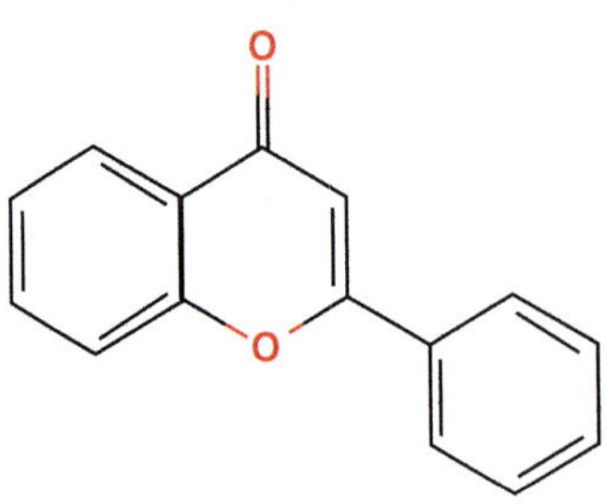	• Offers protection against various cancer types, including breast, uterine, colon, lung, ovarian-prostate, skin, liver, and stomach cancer. • Demonstrates neuroprotective properties	Parsley, chamomile, celery, vine-spinach, artichokes, oregano, celery, parsley, broccoli, onion leaves, carrots, peppers, cabbages, apple skins, chrysanthemum flowers	[25, 26]

(Continued)

Apigenin	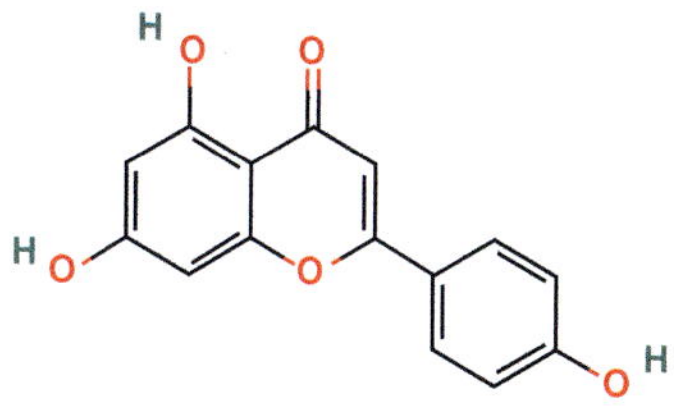			
Luteolin	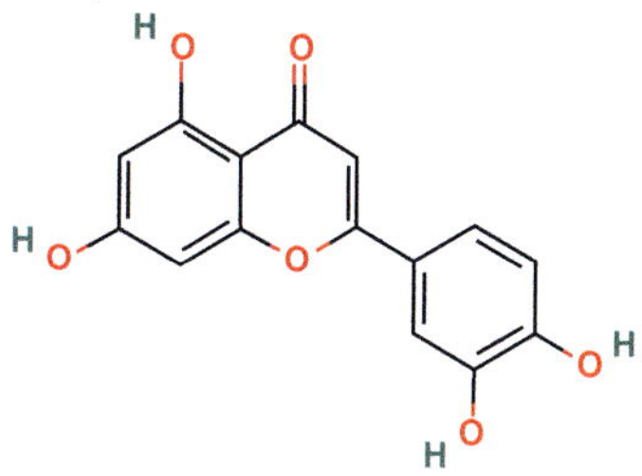			
Flavanones	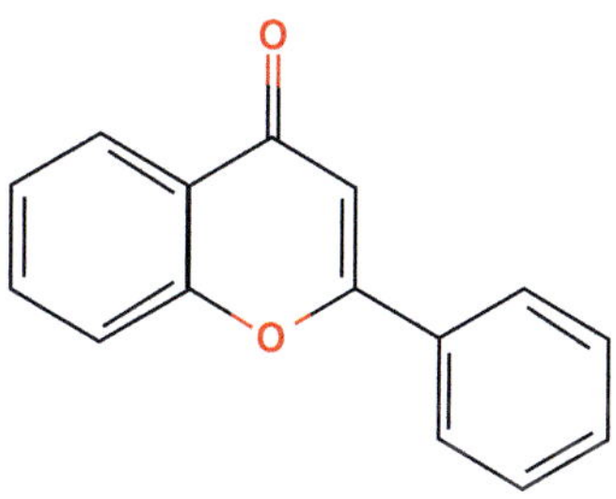	Hesperidin has shown benefits for cardiovascular function, type II diabetes, and cutaneous functions. • For skin, hesperidin has been found to have wound healing, UV protection, anti-inflammatory, antimicrobial, anti-skin cancer, and skin lightening properties. • Hesperidin promotes carbohydrate metabolism, increases antioxidant defenses, modulates immune system activity, and has anti-atherogenic and anti-inflammatory effects by decreasing lipid peroxidation biomarkers and protein carbonylation.	Orange juice, grapefruit juice, lemon juice	[27–29]
Hesperetin	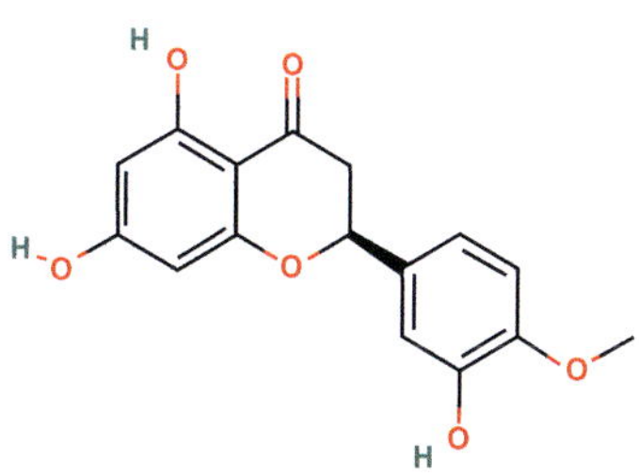			

(Continued)

(Continued)

Naringenin	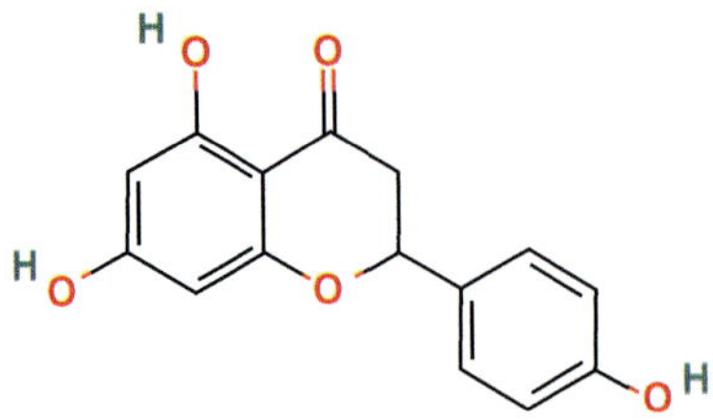			
Eriodictyol	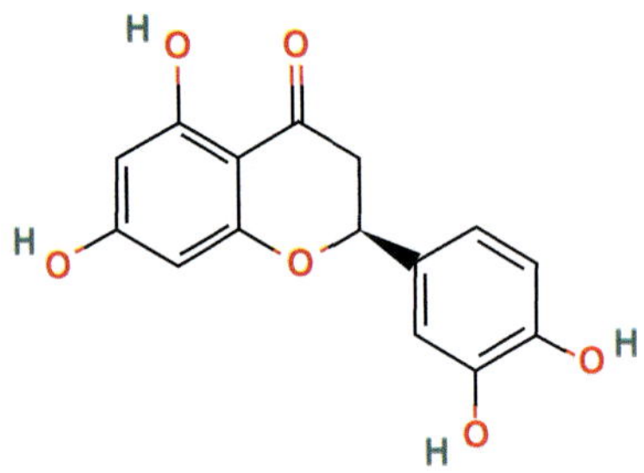			
Isoflavones	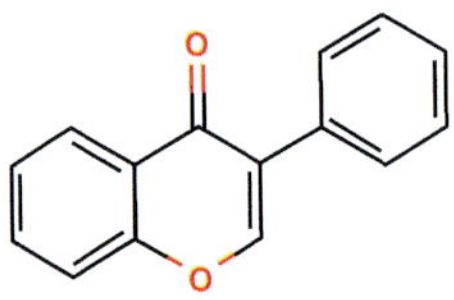	• Offers protection against age-related diseases including cardiovascular disease, osteoporosis, hormone-dependent cancer, and loss of cognitive function through weak oestrogenic action and antioxidant activity. • Protects against chronic diseases like cancer, osteoporosis, and ischemic heart disease and inhibits carcinogenesis and photo-carcinogenesis via antioxidant and antiproliferative effects. • Effective in reducing erythema formation induced by ultraviolet radiation. • Genistein reduces the risk of cardiovascular disease, osteoporosis, and postmenopausal symptoms such as hot flashes and vaginal dryness.	Soy beans, green beans, mung beans, soy flour, tofu, miso, tempeh, soy milk	[30, 31]

(Continued)

Daidzein	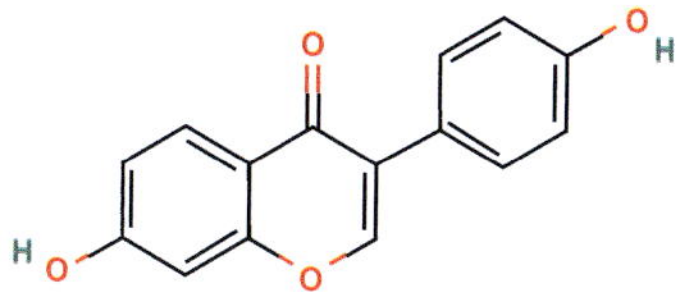			
Genistein	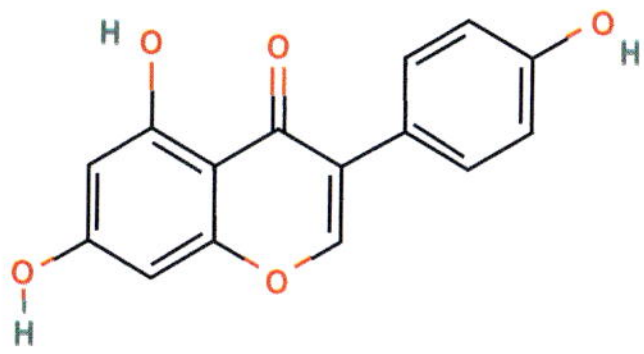			
Glycitein	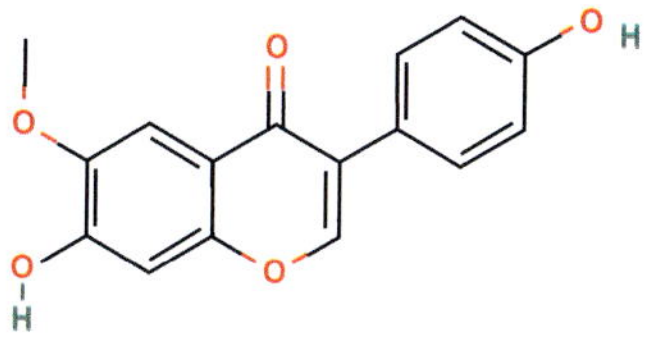			
Monomeric flavanols				
Catechin	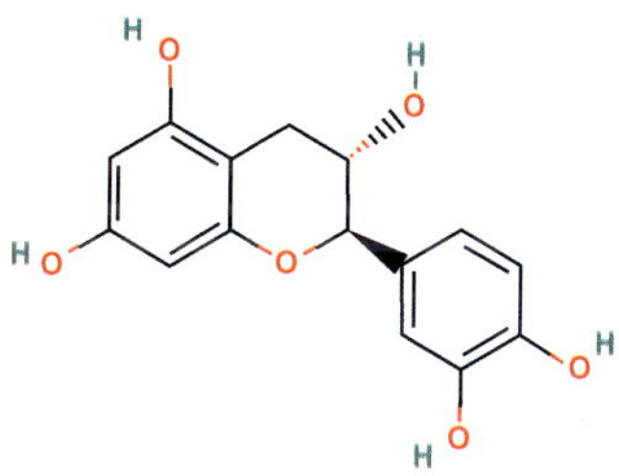	• Provides benefits to inflammatory bowel disease by regulating the infiltration and proliferation of immune-related cells including neutrophils, colonic epithelial cells, macrophages, and T lymphocytes. • Promotes gut health and improves plasma antioxidant activity, brachial artery dilation, fat oxidation, and resistance of low-density lipoproteins to oxidation.	Chocolate, Beans, Apricot, cherry, grape, peach, blackberry, apple, green tea, black tea, red wine, cider	[32]
Epicatechin	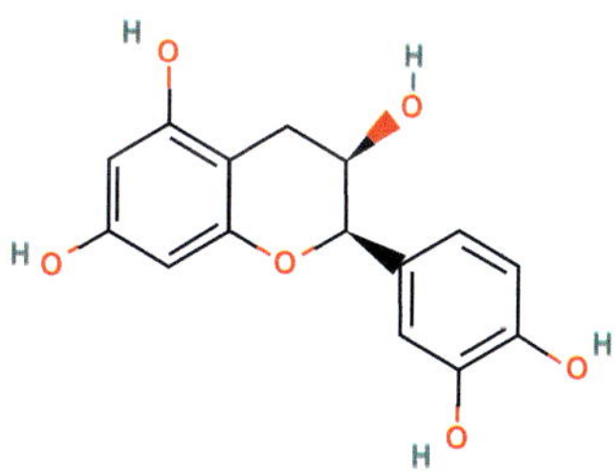			

1.2.3 Anthocyanidins, Lignans, Hydrolysable Tannins, and Condensed Tannins

Anthocyanins	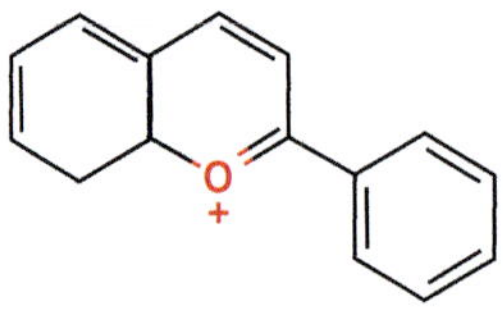	• Has antidiabetic, anticancer, anti-inflammatory, antimicrobial, and anti-obesity effects, as well as neuroprotective and cardiovascular disease revention properties.	Blackberry, black currant, blueberry, black grape, cherry, rhubarb, strawberry, red wine, plum red, red cabbage, cranberry, raspberry, apples, peaches plums; vegetables, such as red cabbage and red onions, kidney beans, pomegranates, eggplant	[33–36]
Cyanidin	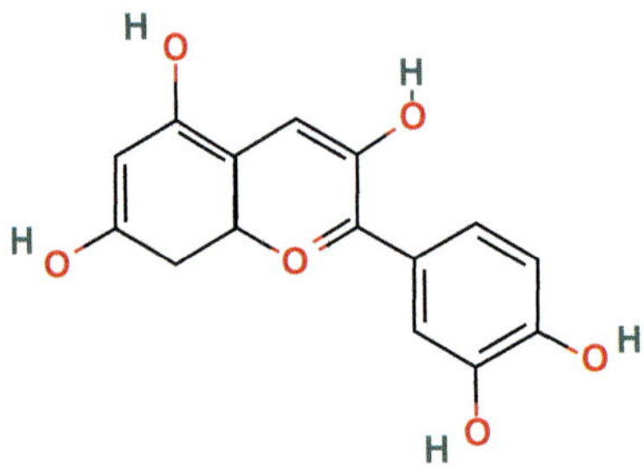			
Pelargonidin	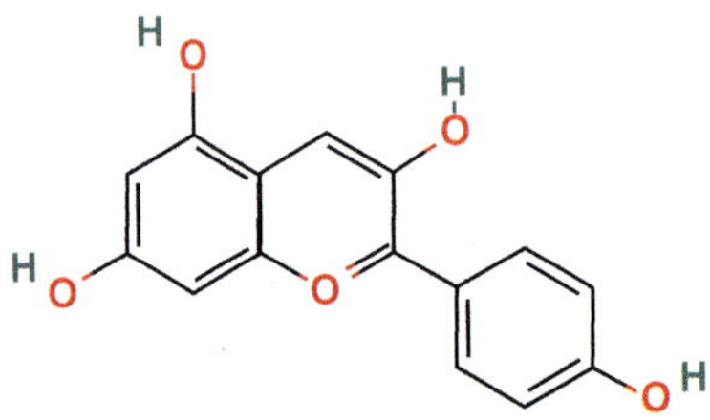			

1.3 Health Benefits and Antioxidant Properties of Polyphenols

1.3.1 Mechanism of Action for Health Benefits

Polyphenols, particularly dietary polyphenols, are plant-derived molecules that have various biological activities in humans. These activities include antioxidant and radical scavenging, modulation of inflammation and human enzymes, and binding to nuclear receptors. [37]

The health benefits of dietary polyphenols are mainly anti-inflammatory, antioxidant, and anti-proliferative, or they follow epigenetic regulation. The mechanisms of action of polyphenols involve their ability to bind to enzymes and modulate them, acting as free radical scavengers that prevent the formation of reactive oxygen species (ROS) or remove them before they damaging vital components of the cell [23].

Cardiovascular Disease

Grape skins and seeds are rich in antioxidants including resveratrol, quercetin, catechin, and proanthocyanidin, which can prevent platelet aggregation and shield low-density lipoproteins from oxidation [38]. Wine and alcohol polyphenols can be used in the treatment of myocardial ischemia reperfusion injury, although their respective mechanisms of cardio protection work in different ways.

If absorbed into the blood, flavonols may prevent cardiovascular disease as they are powerful antioxidants with the potential to shield biological macromolecules from oxidative damage in vivo [39]. In addition to polyphenols increasing high density lipoprotein and enhancing endothelial function, polyphenols may also have antioxidant, antiplatelet, and anti-inflammatory effects that protect against cardiovascular diseases [40]. By preferentially inhibiting the activity of cyclooxygenase 1 (COX 1), which produces thromboxane A2, an inducer of platelet aggregation and vasoconstrictor, resveratrol, a wine polyphenol, reduces platelet aggregation [40].

Resveratrol, a phytochemical, inhibits the platelet aggregation by decreasing the interaction of platelets with collagen and thrombin, as studied with different in vitro models [41]. Clearly, resveratrol influences a variety of molecular functions to influence the cardiovascular system. The ability of phytoalexins to prevent apoptosis is one of their most intriguing low concentration actions. Red wine contains a polyphenol fraction that has a cardioprotective effect of resveratrol, and an ischemia reperfusion injury acts as an efficient scavenger of oxyen free radicals [42].

Diabetes Milletus

Diabetes and cardiovascular disease are both at risk because of insulin resistance, a defining characteristic of metabolic disorders. A similar study revealed that polyphenols like resveratrol and epigallocatechin-3-gallate improved energy metabolism in conditions like diet-induced obesity and insulin resistance in vivo [43].

Berberine (BBR) is a naturally occurring substance with numerous pharmacological properties that has been isolated from plants, including *Coptis chinensis* and *Hydrastis canadensis* [44]. Both antioxidant and anti-inflammatory properties are present. The mechanisms underlying BBR's antioxidant and anti-inflammatory effects were intricate and involved numerous cellular kinases and signaling pathways, including the nuclear factor-B (NF-B) pathway and AMP-activated protein kinase (AMPK) [45].

Mechanism of Action

BBR in the intestines may block the action of alpha-glucosidase in order to decrease intestinal glucose absorption and thereby lower blood sugar levels [46]. Additionally, BBR has a role in the treatment of metabolic syndromes and it is a promising molecule for antihyperglycaemia, insulin sensitization, and antiobesity treatments by stimulating glycolysis, suppression of adipogenesis, and activation of the AMPK pathway [47].

Antiaging and Antioxidation

Curcumin plays a vital role in antiaging by changing the levels of the protein taking part in the process of aging [48]. Additionally, tannic acid is a naturally occurring plant polyphenol with an effective antiphotoaging effect because of its antioxidant potential. Since polyphenols have phenolic hydroxyl groups on their molecules, polyphenols can directly scavenge ROS and regulate the production and activity of endogenous antioxidants and oxidase enzymes to exert antioxidant action. Two intracellular enzymes, sodium oxide dismutase 1 (SOD1) in the cytosol and SOD2 in the mitochondrial matrix, rapidly convert superoxide into hydrogen peroxide as a key method of neutralizing oxidants [49]. Catalase (CAT) and glutathione peroxidases (GSH-Px) further deactivate hydrogen peroxide to produce oxygen and water [50]. By inhibiting the synthesis or deactivating the active species and precursors of free radicals, polyphenols slow the creation of free radicals and hence lower the rate of oxidation. They commonly serve as a direct scavenging free radical of the chain processes of lipid peroxidation (chain breakers). Chain breakers remove an electron from the free radical, neutralize it, and make the radicals themselves less reactive, stopping the chain reactions [10].

Polyphenols and Neurodegeneration

Polyphenols have a strong antioxidative tendency, and therefore, consuming them may offer protection from neurological illnesses [40]. The consumption of green tea may have a beneficial effect in reducing the risk of Parkinson's disease. The ability of catechins to chelate iron adds to their therapeutic role in Parkinson's disease. This property aids in their antioxidant action by preventing the generation of free radicals by redox-active transition metals [44].

Quercetin is a polyphenol abundant in nature and one of the most frequently occurring polyphenols. Numerous plant-based products contain quercetin. The highest level of quercetin has been found in onions. As a result of the molecule's lipophilicity and ease with which it can pass the blood–brain barrier (BBB), it offers protection from neurodegenerative disorders [51].

Quercetin is a potential metal-chelating agent, and this property established the basis for reducing lead (Pb)-induced toxicity in the brain by reducing Pb concentrations in both the brain and blood. Pb poisoning has negative effects such as disruption of the BBB, impairment of memory and learning, and growth retardation [52]. Quercetin has a protective impact on the central nervous system through a variety of mechanisms, including decreasing lipid peroxidation, and increasing superoxide dismutase and CAT activity [50].

Mechanism of Action

Quercetin may work as a direct antioxidant and by energizing cellular defenses against oxidative stress [53]. It reduces oxidative stress-induced neuronal damage by scavenging oxygen radicals and performing metal-chelating procedures. Attenuating nitric oxide (NO) synthase and xanthine oxidase are the next steps in the scavenging mechanism of quercetin [54].

1.3.2 Antioxidant Properties

Polyphenols exhibit potent antioxidant properties that can be attributed to their ability to donate an electron or hydrogen atom and neutralize free radicals. The presence of a highly conjugated system and specific hydroxylation patterns, such as the 3-hydroxy group in flavonoids, play significant roles in their antioxidant activity [10].

Polyphenols have been recognized for their potential role in mitigating aging mechanisms because of their antioxidant properties. The free radical theory of aging, also known as the oxidative stress theory of aging, suggests that age-related functional declines are caused by the accumulation of structural damage to macromolecules, such as lipids, DNA, and proteins, resulting from oxidative damage. Polyphenols can act as antioxidants by neutralizing free radicals through the donation of an electron or hydrogen atom, with certain hydroxylation patterns, such as the 3-hydroxy group in flavonoids, that are considered crucial for their antioxidant activities [55].

Oxidative stress would also result in cellular senescence, functional alterations, and pathological conditions. Antioxidant properties of polyphenols counteract the oxidative stress and provide health benefits through their antiaging mechanisms. The antioxidant properties of polyphenols occur through the following four mechanisms.

1) **Hydroxyl groups**: Polyphenols can scavenge ROS by directly donating an electron or hydrogen atom. The scavenging activity of polyphenols is determined by the presence of phenolic hydroxyl groups on their molecules, which can be influenced by the number, position, substituent patterns, and glycosylation of phytochemical molecules.

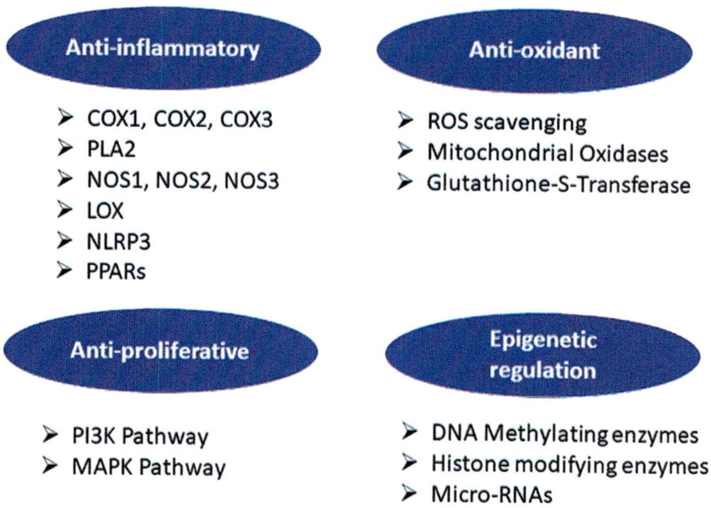

For example, the antioxidant activity of kaempferol derivatives varies based on their hydroxy substitutions, with kaempferol having the highest activity that is 2.7 times that of Trolox [56].

2) **Oxidase enzyme production**: Polyphenols can regulate the production and activity of oxidase enzymes, which helps neutralize oxidants. The main enzymes responsible for this mechanism are SOD1 and SOD2, which convert superoxide to hydrogen peroxide. Hydrogen peroxide is then deactivated to water and oxygen by CAT or GSH-Px. Studies have shown that oral administration of epimedium flavonoids can improve CAT and GSH-Px activities in *Drosophila melanogaster* [29].
3) **Nrf2 mediated pathway**: Polyphenols can enhance cellular antioxidant activity by regulating the Nrf2 pathway, a transcriptional factor that controls the expression of detoxifying enzymes, such as SOD, GPx1, NQO1, GST, and HO-1, by binding to antioxidant response elements (AREs) in the promoter regions of enzyme genes [57]. Resveratrol may disrupt the Nrf2-Keap complex in the cytosol, allowing Nrf2 to translocate to the nucleus and bind to the ARE-containing 5′-promoter region of NQO1, leading to its transcriptional activation [58].
4) **Regulating micro RNA**: Micro RNAs (miRNAs) are short, single-stranded, non-coding RNA molecules with a length of 19–22 nucleotides that bind to mRNA's 3′UTR and regulate various biological pathways and processes such as cell proliferation, cell death, cancer, aging, and other human diseases. Approximately 60% of human transcripts are predicted to be regulated by miRNAs [36].

Polyphenols, such as quercetin, hesperidin, naringenin, anthocyanin, catechin, and curcumin, have been found to reverse the changes of specific miRNAs induced by ApoE mutation. These miRNAs, including mmu-miR-291b-5p, mmu-miR-296-5p, mmu-miR-30c-1, mmu-miR-467b, and mmu-miR-374, collectively regulate 34 pathways, including the GSH metabolism pathway [59]. Curcumin has been found to downregulate the expression of certain miRNAs such as miR-17-5p, miR-20A, and miR-27a. These miRNAs have been shown to modulate the production of ROS, which are known to contribute to oxidative stress and various diseases. By downregulating these miRNAs, curcumin may have a protective effect against oxidative stress and related diseases [60].

1.3.3 Critical View of Their Mode of Action

Polyphenols exhibit antioxidant properties through several known approaches, as mentioned earlier. Besides their antioxidant effects, they have also been found to possess anticancer properties and play a crucial role in reducing the risk of cardiovascular and neurodegenerative diseases. Polyphenols can exert their effects either directly, such as in the case of antioxidant activity, or indirectly by modulating cellular signaling pathways and machinery that regulate various cellular functions under both normal and pathological conditions. Therefore, polyphenols have the potential to be used as therapeutic agents in the prevention and treatment of various diseases.

The impact of polyphenols on transcription factors can either activate or inhibit certain pathways, ultimately affecting the development of pathological conditions. For instance, polyphenols can inhibit the JNK, ASK1, and p38 pathways, resulting in reduced inflammatory reactions and apoptosis of neurons. This can also lead to a decrease in iNOS expression and NO release. Alternatively, polyphenols can activate proteins, such as eNOS, that regulate NO and affect the risk of cardiovascular disease. Therefore, polyphenols have the potential to impact various signaling pathways, leading to positive and negative outcomes [61].

1.3.4 Crosstalk

Polyphenols interact with multiple signaling pathways and modulate their activity, leading to a range of biological effects [62]. This activity is achieved through crosstalk between signaling pathways, where polyphenols can influence multiple pathways simultaneously. For example, polyphenols can inhibit the production of ROS and activate the Nrf2 pathway, leading to the activation of antioxidant enzymes [63]. Polyphenols can also inhibit NF-kB signaling, which leads to a reduction in the production of pro-inflammatory cytokines. Additionally, polyphenols can modulate the activity of AMPK, which regulates cellular metabolism and energy balance. Crosstalk between these pathways can lead to a range of effects such as improved insulin sensitivity, reduced oxidative stress, and improved cardiovascular health. Understanding the crosstalk between these pathways is crucial for understanding the complex effects of polyphenols on human health [64].

Some common pathways involved in the cross talk of food polyphenols include:

- **Antioxidant pathway**: The antioxidant pathway refers to a complex series of cellular processes that help to neutralize and remove harmful ROS from the body. ROS are highly reactive molecules that are produced as by-products of the cellular metabolism, and can damage cellular components such as DNA, proteins, and lipids. The antioxidant pathway involves a variety of enzymatic and non-enzymatic mechanisms that work together to protect the body from oxidative stress. Some of the key components of the antioxidant pathway include:

Enzymatic antioxidants. These are enzymes that help break down ROS and convert them into less harmful substances. Examples include SOD, CAT, and GSH-Px.

Non-enzymatic antioxidants. These are molecules that can neutralize ROS directly, without the need for enzymatic activity. Examples include vitamins C and E, beta-carotene, and flavonoids.

Cellular repair mechanisms. When cellular components are damaged by ROS, the cell can activate repair mechanisms to fix the damage and prevent further harm.

Overall, the antioxidant pathway is a complex and dynamic system that plays a critical role in maintaining cellular health and preventing disease. It is important to consume a balanced diet rich in antioxidants, and to engage in healthy lifestyle behaviors, such as regular exercise and stress reduction, to support the antioxidant pathway and reduce oxidative stress.

Inflammation pathway: The inflammation pathway is a complex biological response that occurs in response to tissue injury or infection. It is an essential part of the immune system's defense mechanism that aims to eliminate harmful stimuli and initiate the healing process.

The pathway begins with the release of chemical mediators, such as cytokines, histamines, and prostaglandins, which trigger the dilation and increased permeability of blood vessels. This causes an influx of white blood cells, particularly neutrophils, into the affected tissue.

Neutrophils then release enzymes that break down the damaged tissue and engulf and destroy any pathogens present. This process creates an inflammatory exudate, which contains dead cells, cellular debris, and fluid, resulting in characteristic signs of inflammation such as redness, swelling, heat, and pain.

As the tissue begins to heal, macrophages, a type of white blood cell, remove the debris and initiate the repair process. The inflammatory response eventually resolves, and the tissue returns to its normal state.

However, if the inflammatory response is prolonged or excessive, it can result in chronic inflammation, which can lead to tissue damage and disease. Examples of conditions associated with chronic inflammation include rheumatoid arthritis, inflammatory bowel disease, and atherosclerosis.

Metabolic pathway: A metabolic pathway is a series of chemical reactions that occur within a cell to convert a starting molecule, known as a substrate, into a final product. These reactions are often catalyzed by enzymes, which are specialized proteins that facilitate special chemical reactions.

Metabolic pathways can be grouped into two main categories: catabolic pathways and anabolic pathways. Catabolic pathways break down larger molecules into smaller molecules, releasing energy in the process. Alternatively, anabolic pathways build larger molecules from smaller molecules, consuming energy in the process.

One well known metabolic pathway is cellular respiration, which is a catabolic pathway that breaks down glucose molecules to release energy in the form of adenosine triphosphate, the primary energy source for cells. Another example of a metabolic pathway is the biosynthesis of amino acids, which is an anabolic pathway that builds larger molecules from smaller molecules to create the building blocks of proteins.

Signaling pathway: A signaling pathway refers to a series of chemical reactions that occur within a cell in response to a specific stimulus or signal. These pathways are responsible for coordinating cellular activities such as growth, differentiation, metabolism, and response to environmental stimuli.

Signaling pathways typically involve a series of molecular interactions between different proteins, enzymes, and other signaling molecules. The process begins when a signaling molecule, such as a hormone, neurotransmitter, or growth factor, binds to a specific receptor on the surface of a cell. This binding event triggers a series of downstream reactions that activate a cascade of proteins and enzymes, ultimately resulting in a specific cellular response.

There are several different types of signaling pathways, including receptor tyrosine kinase pathways, G protein-coupled receptor pathways, and intracellular pathways. These pathways are critical to maintaining homeostasis within the body and for responding to changes in the environment. Dysregulation of signaling pathways is implicated in a wide range of diseases, including cancer, diabetes, and neurological disorders.

Microbiome pathway: The microbiome pathway refers to the interactions between microorganisms and their hosts, particularly in the context of the human microbiome. The human microbiome comprises trillions of microorganisms, including bacteria, viruses, fungi, and other microbes, that live on and within the human body. These microorganisms play a critical role in human health, helping to regulate the immune system, digest food, and produce essential vitamins and other compounds.

The microbiome pathway can be disrupted by a variety of factors, including diet, antibiotics, and other medications, as well as environmental factors such as pollution and stress. When the microbiome is disrupted, it can lead to a range of health problems, including obesity, diabetes, autoimmune diseases, and even mental health disorders.

Research into the microbiome pathway is ongoing, and scientists are working to better understand how these microorganisms interact with their hosts and how these interactions can be manipulated to improve health outcomes. This includes exploring the potential of probiotics and other microbiome-targeted therapies to treat and prevent disease.

References

1 Zhang, L., Han, Z., and Granato, D. (2021). Polyphenols in foods: classification, methods of identification, and nutritional aspects in human health. In: *Advances in Food and Nutrition Research*, 98. 1–33. Elsevier. https://doi.org/10.1016/bs.afnr.2021.02.004.

2 Prabhu, S., Molath, A., Choksi, H. et al. (2021). Classifications of polyphenols and their potential application in human health and diseases. *International Journal of Physiology, Nutrition and Physical Education* 6 (1): 293–301. https://doi.org/10.22271/journalofsport.2021.v6.i1e.2236.

3 Mahajanakatti, A.B., Deepak, T.S., Achar, R.R. et al. (2022). Nanoconjugate synthesis of elaeocarpus ganitrus and the assessment of its antimicrobial and antiproliferative properties. *Molecules* 27 (8): 2442. https://doi.org/10.3390/molecules27082442.

4 Harbone, J.B. (2017). *The Flavonoids Advances in Research since 1986*. Routledge. https://doi.org/10.1201/9780203736692.

5 Panche, A.N., Diwan, A.D., and Chandra, S.R. (2016). Flavonoids: an overview. *Journal of Nutritional Science* 5: e47. https://doi.org/10.1017/jns.2016.41.

6 Janick, J. (ed.) (2005). *Plant Breeding Reviews: Janick/Plant Breeding Reviews, Volume 25*. John Wiley & Sons, Inc. https://doi.org/10.1002/9780470650301.

7 Kumar, N. and Goel, N. (2019). Phenolic acids: natural versatile molecules with promising therapeutic applications. *Biotechnology Reports* 24: e00370. https://doi.org/10.1016/j.btre.2019.e00370.

8 Pereira, D., Valentão, P., Pereira, J., and Andrade, P. (2009). Phenolics: from chemistry to biology. *Molecules* 14 (6): 2202–2211. https://doi.org/10.3390/molecules14062202.

9 Khoddami, A., Wilkes, M.A., and Roberts, T.H. (2013). Techniques for analysis of plant phenolic compounds. *Molecules* 18 (2): 2328–2375. https://doi.org/10.3390/molecules18022328.

10 Tsao, R. (2010). Chemistry and biochemistry of dietary polyphenols. *Nutrients* 2 (12): 1231–1246. https://doi.org/10.3390/nu2121231.

11 Han, X., Shen, T., and Lou, H. (2007). Dietary polyphenols and their biological significance. *International Journal of Molecular Sciences* 8 (9): 950–988. https://doi.org/10.3390/i8090950.

12 Murali, M., Gowtham, H.G., Shilpa, N. et al. (2022). Exploration of anti-HIV phytocompounds against SARS-CoV-2 main protease: structure-based screening, molecular simulation, ADME analysis and conceptual DFT studies. *Molecules* 27 (23): 8288. https://doi.org/10.3390/molecules27238288.

13 Aishwarya Rajendra, P., Ravi, M.B., Raghavendra, S.K.N. et al. (2022). Lemongrass oil disrupts the biofilm of Candida albicans MTCC 1637T on soft denture reliners at lower concentrations compared to thyme and tea tree oils. *Journal of Applied Biology and Biotechnology* Published online February 15, https://doi.org/10.7324/JABB.2022.100214.

14 Juurlink, B.H., Azouz, H.J., Aldalati, A.M. et al. (2014). Hydroxybenzoic acid isomers and the cardiovascular system. *The Journal of Nutrition* 13 (1): 63. https://doi.org/10.1186/1475-2891-13-63.

15 Bento-Silva, A., Koistinen, V.M., Mena, P. et al. (2020). Factors affecting intake, metabolism and health benefits of phenolic acids: do we understand individual variability? *European Journal of Nutrition* 59 (4): 1275–1293. https://doi.org/10.1007/s00394-019-01987-6.

16 Zhang, S., Gai, Z., Gui, T. et al. (2021). Antioxidant effects of protocatechuic acid and protocatechuic aldehyde: old wine in a new bottle. In: *Evid Based Complement Alternat Med* (ed. F.S. Youssef), 1–19. https://doi.org/10.1155/2021/6139308.

17 Kahkeshani, N., Farzaei, F., Fotouhi, M. et al. (2019). Pharmacological effects of gallic acid in health and disease: a mechanistic review. *The Iranian Journal of Basic Medical Sciences* 22 (3): https://doi.org/10.22038/ijbms.2019.32806.7897.

18 Taofiq, O., González-Paramás, A., Barreiro, M., and Ferreira, I. (2017). Hydroxycinnamic acids and their derivatives: cosmeceutical significance, challenges and future perspectives, a review. *Molecules* 22 (2): 281. https://doi.org/10.3390/molecules22020281.

19 Espíndola, K.M.M., Ferreira, R.G., Narvaez, L.E.M. et al. (2019). Chemical and pharmacological aspects of caffeic acid and its activity in hepatocarcinoma. *Frontiers in Oncology* 9: 541. https://doi.org/10.3389/fonc.2019.00541.

20 Waheed Janabi, A.H., Kamboh, A.A., Saeed, M. et al. (2019). Flavonoid-rich foods (FRF): a promising nutraceutical approach against lifespan-shortening diseases. *The Iranian Journal of Basic Medical Sciences* (Online First). https://doi.org/10.22038/ijbms.2019.35125.8353.

21 Li, Y., Yao, J., Han, C. et al. (2016). Quercetin, inflammation and immunity. *Nutrients* 8 (3): 167. https://doi.org/10.3390/nu8030167.

22 Anand David, A., Arulmoli, R., and Parasuraman, S. (2016). Overviews of biological importance of quercetin: a bioactive flavonoid. *Pharmacognosy Reviews* 10 (20): 84. https://doi.org/10.4103/0973-7847.194044.

23 Silva Dos Santos, J., Gonçalves Cirino, J.P., de Oliveira Carvalho, P., and Ortega, M.M. (2021). The pharmacological action of kaempferol in central nervous system diseases: a review. *Frontiers in Pharmacology* 11: 565700. https://doi.org/10.3389/fphar.2020.565700.

24 Taheri, Y., Suleria, H.A.R., Martins, N. et al. (2020). Myricetin bioactive effects: moving from preclinical evidence to potential clinical applications. *BMC Complementary Medicine Therapy* 20 (1): 241. https://doi.org/10.1186/s12906-020-03033-z.

25 Shukla, S. and Gupta, S. (2010). Apigenin: a promising molecule for cancer prevention. *Pharmaceutical Research* 27 (6): 962–978. https://doi.org/10.1007/s11095-010-0089-7.

26 Lin, Y., Shi, R., Wang, X., and Shen, H.M. (2008). Luteolin, a flavonoid with potential for cancer prevention and therapy. *Current Cancer Drug Targets* 8 (7): 634–646. https://doi.org/10.2174/156800908786241050.

27 Kim, J., Wie, M.B., Ahn, M. et al. (2019). Benefits of hesperidin in central nervous system disorders: a review. *Anatomy and Cell Biology* 52 (4): 369. https://doi.org/10.5115/acb.19.119.

28 Salehi, B., Fokou, P., Sharifi-Rad, M. et al. (2019). The therapeutic potential of naringenin: a review of clinical trials. *Pharmaceuticals* 12 (1): 11. https://doi.org/10.3390/ph12010011.

29 Lee, S., Yang, H., Son, G. et al. (2015). Eriodictyol protects endothelial cells against oxidative stress-induced cell death through modulating ERK/Nrf2/ARE-dependent heme oxygenase-1 expression. *International Journal of Molecular Sciences* 16 (12): 14526–14539. https://doi,org/10.3390/ijms160714526.

30 Liggins, J., Bluck, L.J.C., Runswick, S. et al. (2000). Daidzein and genistein content of fruits and nuts. *The Journal of Nutritional Biochemistry* 11 (6): 326–331. https://doi.org/10.1016/S0955-2863(00)00085-1.

31 Thangavel, P., Puga-Olguín, A., Rodríguez-Landa, J.F., and Zepeda, R.C. (2019). Genistein as potential therapeutic candidate for menopausal symptoms and other related diseases. *Molecules* 24 (21): 3892. https://doi.org/10.3390/molecules24213892.

32 Fan, F.Y., Sang, L.X., and Jiang, M. (2017). Catechins and their therapeutic benefits to inflammatory bowel disease. *Molecules* 22 (3): 484. https://doi.org/10.3390/molecules22030484.

33 Khoo, H.E., Azlan, A., Tang, S.T., and Lim, S.M. (2017). Anthocyanidins and anthocyanins: colored pigments as food, pharmaceutical ingredients, and the potential health benefits. *Food and Nutrition Research* 61 (1): 1361779. https://doi.org/10.1080/16546628.2017.1361779.

34 Li, D., Wang, P., Luo, Y. et al. (2017). Health benefits of anthocyanins and molecular mechanisms: update from recent decade. *Critical Reviews in Food Science and Nutrition* 57 (8): 1729–1741. https://doi.org/10.1080/10408398.2015.1030064.

35 Liang, Z., Liang, H., Guo, Y., and Yang, D. (2021). Cyanidin 3-O-galactoside: a natural compound with multiple health benefits. *International Journal of Molecular Sciences* 22 (5): 2261. https://doi.org/10.3390/ijms22052261.

36 Amini, A.M., Muzs, K., Spencer, J.P.E., and Yaqoob, P. (2017). Pelargonidin-3- O -glucoside and its metabolites have modest anti-inflammatory effects in human whole blood cultures. *Nutrition Research* 46: 88–95. https://doi.org/10.1016/j.nutres.2017.09.006.

37 Vivarelli, S., Costa, C., Teodoro, M. et al. (2023). Polyphenols: a route from bioavailability to bioactivity addressing potential health benefits to tackle human chronic diseases. *Archives of Toxicology* 97 (1): 3–38. https://doi.org/10.1007/s00204-022-03391-2.

38 Kondratyuk, T.P. and Pezzuto, J.M. (2004). Natural product polyphenols of relevance to human health. *Pharmaceutical Biology* 42 (sup1): 46–63. https://doi.org/10.3109/13880200490893519.

39 Gee, J.M., DuPont, M.S., Rhodes, M.J.C., and Johnson, I.T. (1998). Quercetin glucosides interact with the intestinal glucose transport pathway 11This work was supported by a UK biotechnology and biological sciences research council competitive strategic grant. *Free Radical Biology and Medicine* 25 (1): 19–25. https://doi.org/10.1016/S0891-5849(98)00020-3.

40 Pandey, K.B. and Rizvi, S.I. (2009). Plant polyphenols as dietary antioxidants in human health and disease. *Oxidative Medicine and Cellular Longevity* 2 (5): 270–278. https://doi.org/10.4161/oxim.2.5.9498.

41 Borriello, A., Cucciolla, V., Della Ragione, F., and Galletti, P. (2010). Dietary polyphenols: focus on resveratrol, a promising agent in the prevention of cardiovascular diseases and control of glucose homeostasis. *Nutrition, Metabolism and Cardiovascular Diseases* 20 (8): 618–625. https://doi.org/10.1016/j.numecd.2010.07.004.

42 Ray, P.S., Maulik, G., Cordis, G.A. et al. (1999). The red wine antioxidant resveratrol protects isolated rat hearts from ischemia reperfusion injury. *Free Radical Biology and Medicine* 27 (1–2): 160–169. https://doi.org/10.1016/S0891-5849(99)00063-5.

43 Zhang, J. and Liu, Q. (2015). Cholesterol metabolism and homeostasis in the brain. *Protein Cell* 6 (4): 254–264. https://doi.org/10.1007/s13238-014-0131-3.

44 Maradesha, T., Martiz, R.M., Patil, S.M. et al. (2023). Integrated network pharmacology and molecular modeling approach for the discovery of novel potential MAPK3 inhibitors from whole green jackfruit flour targeting obesity-linked diabetes mellitus. Devarajan S, ed. *PLOS ONE* 18 (1): e0280847. https://doi.org/10.1371/journal.pone.0280847.

45 Li, Z., Geng, Y.N., Jiang, J.D., and Kong, W.J. (2014). Antioxidant and anti-inflammatory activities of berberine in the treatment of diabetes mellitus. *Evidence-Based Complementary and Alternative Medicine* 1–12. https://doi.org/10.1155/2014/289264.

46 Pang, B., Zhao, L.H., Zhou, Q. et al. (2015). Application of berberine on treating type 2 diabetes mellitus. *International Journal of Endocrinology* 1–12. https://doi.org/10.1155/2015/905749.

47 Yin, J., Zhang, H., and Ye, J. (2008). Traditional chinese medicine in treatment of metabolic syndrome. *Endocrine, Metab Immune Disord-Drug Targets* 8 (2): 99–111. https://doi.org/10.2174/187153008784534330.

48 Queen, B. and Tollefsbol, T. (2010). Polyphenols and aging. *Current Aging Science* 3 (1): 34–42. https://doi.org/10.2174/1874609811003010034.

49 Çelebi, H., Şimşek, İ., Bahadir, T., and Tulun, Ş. (2023). Use of banana peel for the removal of boron from aqueous solutions in the batch adsorption system. *International Journal of Environmental Science and Technology* 20 (1): 161–176. https://doi.org/10.1007/s13762-022-04566-1.

50 Luo, J., Si, H., Jia, Z., and Liu, D. (2021). Dietary anti-aging polyphenols and potential mechanisms. *Antioxidants* 10 (2): 283. https://doi.org/10.3390/antiox10020283.

51 Deepika and Maurya, P.K. (2022). Health benefits of quercetin in age-related diseases. *Molecules* 27 (8): 2498. https://doi.org/10.3390/molecules27082498.

52 Elumalai, P. and Lakshmi, S. (2016). Role of quercetin benefits in neurodegeneration. In: *The Benefits of Natural Products for Neurodegenerative Diseases*, 12. (ed. M.M. Essa, M. Akbar, and G. Guillemin), 229–245. Advances in Neurobiology. Springer International Publishing. https://doi.org/10.1007/978-3-319-28383-8_12.

53 Khan, H., Ullah, H., Aschner, M. et al. (2019). Neuroprotective effects of quercetin in Alzheimer's disease. *Biomolecules* 10 (1): 59. https://doi.org/10.3390/biom10010059.

54 Pradeep, S., Prabhuswaminath, S.C., Reddy, P. et al. (2022). Anticholinesterase activity of areca catechu: in vitro and in silico green synthesis approach in search for therapeutic agents against Alzheimer's disease. *Frontiers in Pharmacology* 13: 1044248. https://doi.org/10.3389/fphar.2022.1044248.

55 Liguori, I., Russo, G., Curcio, F. et al. (2018). Oxidative stress, aging, and diseases. *Clinical Interventions in Aging* 13: 757–772. https://doi.org/10.2147/CIA.S158513.

56 Cao, G., Sofic, E., and Prior, R.L. (1997). Antioxidant and prooxidant behavior of flavonoids: structure-activity relationships. *Free Radical Biology and Medicine* 22 (5): 749–760. :https://doi.org/10.1016/S0891-5849(96)00351-6.

57 Kobayashi, M. and Yamamoto, M. (2005). Molecular mechanisms activating the Nrf2-keap1 pathway of antioxidant gene regulation. *Antioxid Redox Signal* 7 (3–4): 385–394. https://doi.org/10.1089/ars.2005.7.385.

58 Zhai, Y., Liu, J., Zhou, L. et al. (2018). Detection of Candida species in pregnant Chinese women with a molecular beacon method. *Journal of Medical Microbiology* 67 (6): 783–789. https://doi.org/10.1099/jmm.0.000740.

59 Shreevatsa, B., Dharmashekara, C., Swamy, V.H. et al. (2021). Virtual screening for potential phytobioactives as therapeutic leads to inhibit NQO1 for selective anticancer therapy. *Molecules* 26 (22): 6863. https://doi.org/10.3390/molecules26226863.

60 Kumari, V.B.C., Huligere, S.S., Shbeer, A.M. et al. (2022). Probiotic potential lacticaseibacillus casei and limosilactobacillus fermentum strains isolated from dosa batter inhibit α-glucosidase and α-amylase enzymes. *Microorganisms* 10 (6): 1195. https://doi.org/10.3390/microorganisms10061195.

61 Xu, W., Zhou, Y., Lin, L. et al. (2022). Hypoglycemic effects of black brick tea with fungal growth in hyperglycemic mice model. *Food Science and Human Wellness* 11 (3): 711–718. https://doi.org/10.1016/j.fshw.2021.12.028.

62 Sun, L.R., Zhou, W., Zhang, H.M. et al. (2019). Modulation of multiple signaling pathways of the plant-derived natural products in cancer. *Frontiers in Oncology* 9: https://www.frontiersin.org/articles/10.3389/fonc.2019.01153 (Accessed March 4, 2023).

63 Kanner, J. (2020). Polyphenols by generating H2O2, affect cell redox signaling, inhibit PTPs and activate Nrf2 axis for adaptation and cell surviving: in vitro, in vivo and human health. *Antioxidants* 9 (9): 797. https://doi.org/10.3390/antiox9090797.

64 Burgos-Morón, E., Abad-Jiménez, Z., Martínez de Marañón, A. et al. (2019). Relationship between oxidative stress, ER stress, and inflammation in type 2 diabetes: the battle continues. *Journal of Clinical Medicine* 8 (9): 1385. https://doi.org/10.3390/jcm8091385.

2

Plant Polyphenols as Nutraceuticals and Their Antioxidant Potentials

Neelesh Kumar Nema[1,], Nayana Rajan[1], Merin Babu[1], Sachithra Sabu[1], Swapnil Devidas Khamborkar[1], Smitha Sarojam[1], Linson Cheruveettil Sajan[1], Aneena Peter[1], Baby Kumaranthara Chacko[1], and Viju Jacob[1,2]*

[1] *Synthite Industries Pvt. Ltd, Synthite Valley, Kadayiruppu, Kerala, India*
[2] *C.V.J. Creative Centre, Bio Ingredients, Synthite Industries Pvt. Ltd., Synthite Valley, Kadayiruppu-682311, Kerala, India*
* *Corresponding author*

2.1 Introduction

2.1.1 Plant Polyphenols: Important Secondary Metabolites

Polyphenols, which are widely distributed in fruits, vegetables, spices cereals, and beverages in the form of glycosides or aglycones, are structurally aromatic compounds with more than one phenolic group immediately next to the pyran ring or C-skeleton [1]. Soluble glycosides are generally found in vacuoles, whereas insoluble metabolites are found in the cell wall, providing mechanical strength to the plant [2]. Additionally, glycosides are secondary metabolites generated through the intrinsic biosynthetic pathways of plants, i.e., shikimic acid/phenylpropanoid and/or the polyketide pathways. First, a shikimic acid pathway generates the precursor amino acids known as phenylalanine and tyrosine, which further enter the phenylpropanoid metabolic pathway where their end metabolites combine with malonyl-CoA to form P-coumaroyl-CoA. This structure is the backbone for many polyphenols, including flavonoids [3–5]. Although the fundamental structure remains the same, critical enzymes, e.g., lyase, synthase, isomerase, and reductase, contribute to generating various polyphenols in the phenylpropanoid pathway by changing the position of the methoxyl and hydroxyl groups in the parent structure [6].

2.1.2 Medicinal Uses

Polysaccharides serve an important role in protecting from various ailments, such as diabetes, cardiovascular complications, arthritis, Alzheimer's, and cancer, as well as promoting human health [7, 8].

2.1.3 Polyphenols in the Nutraceutical Market

Plant parts are frequently employed, along with vitamins and minerals, in the nutraceutical sector because of their health benefits, which help to keep the market attractive and profitable. Multifaceted

Polyphenols: Food, Nutraceutical, and Nanotherapeutic Applications, First Edition. Edited by Mithun Rudrapal.

functions, and notably, management of aging ailments, such as memory loss, immunological disorders, atherosclerosis (cardiovascular disease (CVD)), diabetes, cancer, arthritis, and osteoporosis, are driving the demand for polyphenols in the healthcare market. The need for herbal products, the accessibility and affordability of ingredients, the results of product-focused research, the reduction of processing costs, the simplification of the product development process, and clear regulations are other major driving forces that are currently allowing new businesses to enter the nutraceuticals and healthcare markets in response to rising demand. Foods, including tea, wine, chocolates, fruits, vegetables, and olive oil, contain more than 8,000 polyphenols [9]. According to one market research study, the polyphenol and associated market would develop at a 7.4% compound annual growth rate (CAGR) from 2021 ($1.6 billion) to 2030, reaching $2.7 billion. Polyphenols are now available in specially designed and produced forms, i.e., functional beverages, which are the most popular functional foods for providing overall nutrition and well-being to consumers [10]. With a 5.9% CAGR between 2021 and 2030, the market for functional beverages, which was valued at $110,148.9 million in 2020, is expected to reach $200,080.3 million by 2030. Market figures are increasing day by day and now attract many manufacturers, including pharmaceuticals, nutraceuticals, health, and food supplements, for investment [9] and business.

2.2 Types of Polyphenols and Phenolic Acids

2.2.1 Nature and Chemistry of Polyphenols

Numerous phytochemicals have been linked to health benefits, including sulfur-containing compounds, alkaloids, terpenoids, steroids, stilbenoids, polysaccharides, and phenolics (flavonoids and non-flavonoids, e.g., anthocyanins, flavones, flavanols, isoflavones, stilbenes, tannins, etc.) [11, 12]. Polyphenols are the most common and widely distributed secondary metabolites that are largely consumed in the human diet for health benefits and medicinal effects [13]. According to White–Bate–Smith–Swain–Haslam, polyphenols are water-soluble 5–7 aromatic ring compounds with molecular weights ranging from 500 Dalton to 4000 Dalton and at least 12 phenolic hydroxyl groups. A 2-phenyl-benzo-pyrane nucleus comprising two benzene rings (A and B) joined by a heterocyclic pyran ring is the primary structural component of flavonoids (C).

2.2.2 Classification of Polyphenols

Harborne and Simmonds classified phenolic and polyphenol compounds (phenolic aglycones) in 1964 [14] on the basis of the carbon atoms associated with the structure and total number of phenol rings: phenolic acids and related chemicals have a C6 structure, phenolic acids and associated derivatives have a C6–C1 structure, and flavonoids and anthocyanins have a C15 structure [15]. Because flavonoids have a variety of structural types and are widely used by humans for health and medical reasons, they are grouped as flavones, flavanones, flavonols, flavanonols, flavanols (Flavans), anthocyanin, etc. [13], as mentioned in Figure 2.1.

2.3 Polyphenols in the Human Body

2.3.1 Absorption and Metabolism

Following food consumption, dietary flavonoids or polyphenols are processed by digestive fluids in the gastrointestinal (GI) system and, lastly, by colon bacteria [16]. Intended health benefits from

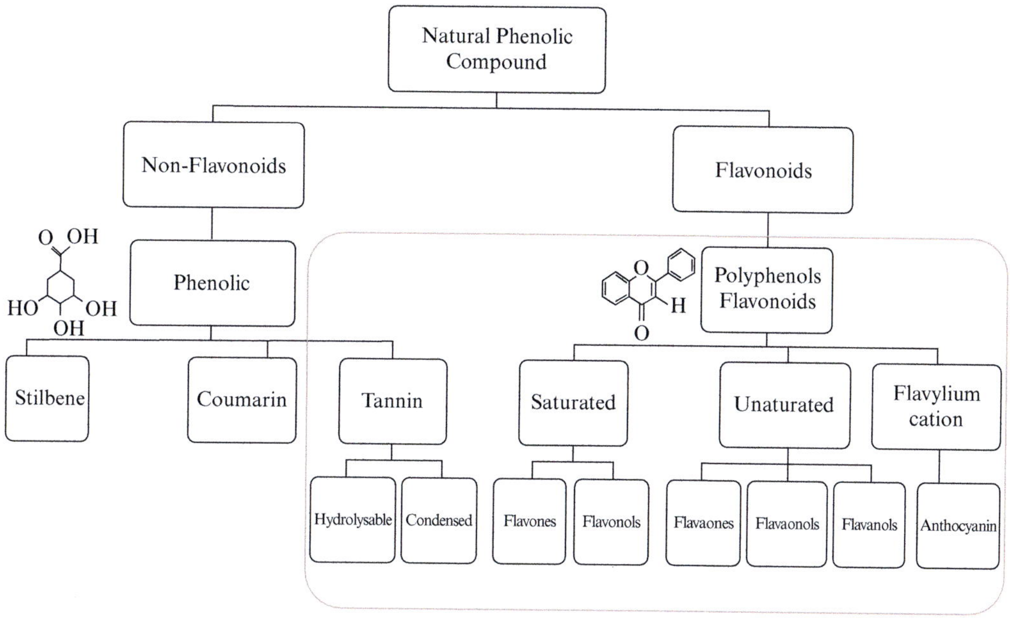

Figure 2.1 Classification and types of polyphenols.

ingesting polyphenols and associated metabolites depend greatly on the processes of absorption, transportation, bioavailability, and bioactivity [17]. Following their ingestion, some polyphenols pass through the stomach on their way to the small intestine where they are then absorbed. Research indicates that glycosides are less absorbed in the GI tract and excreted in unchanged form. The enzyme lactase phloridzin hydrolase (LPH), which is especially found in the epithelial cells of the small intestine, splits flavonoid-O-β-glycosides into their aglycone and glycine halves. Aglycone passively diffuses over the membrane of the epithelial cell because of its high lipophilic nature. Cytosolic β-glucosidase is another enzyme that hydrolyzes some of the phenolic–glycosides (CBG). Upon absorption, polyphenol bioavailability is influenced by its capacity to pass a membrane and by the structural integrity [18]. Aglycones have more antioxidant action than glycosides. The loss of the antioxidant activity of flavonoids following glycosylation may be attributed to conjugated glycosides removing hydroxyl groups, preventing them from scavenging reactive oxygen species (ROS) or chelating transition metals. Furthermore, because glycosylation enlarges the molecule, transit across membranes may be reduced, resulting in diminished antioxidant action. However, glycosylation enhances the water solubility of the compound and subsequently intensifies its absorption from the GI tract [19].

2.3.2 Bioavailability and Bioefficacy

Bioavailability is mostly influenced by the polyphenol structural diversity. Flavonoids are rapidly and extensively degraded in the intestine and liver cells; therefore, they are likely to present in circulation and urine as metabolites (e.g., phase II metabolites). Tmax (time to peak concentrations) ranged between 0.5 hours and 9 hours. When compared with its glucosides and rutinosides, the peak concentration of quercetin (Cmax) in plasma was 20 times greater. Similarly, Tmax is also faster (10 times) than the other variants. Following their release from sugars, flavonoid aglycones undergo conjugation events, including glucuronidation and sulfation with or without methylation, which occurs mostly in enterocytes and the liver. Conjugation promotes their excretion and hence shortens their plasma half-life; consequently, their biologic characteristics change. The evidence available on flavonoids' bioavailability and plasma kinetics are quite varied. Gallic acid and isoflavones have the highest rate of absorption among flavonoids, followed by catechins, flavanones, and quercetin glycosides.

To ensure the bioavailability of these compounds and aid in the resolution of food processing and ingestion concerns, microencapsulation technologies have emerged as a highly successful and promising approach. The metabolic pathways of these molecules are still the subject of a few studies and research efforts, which makes is easy to understand and determine how they work. As a result, further study is needed to develop better methods for capitalizing on the health-promoting properties of these chemicals.

2.4 Polyphenols and Their Antioxidant Potential

2.4.1 Oxidative Stress and Human Diseases/Disorders

Oxidative stress (OS) occurs when the ratio of pro-oxidants to antioxidants in the biological system is skewed toward the former. The excessive production of free radicals, such as ROS and nitrogen species (pro-oxidants), as well as a deficiency in endogenous defense mechanisms, such as enzyme- and non-enzyme-based antioxidants, are the main causes of OS. ROS and RNS, which are formed as by-products of cellular metabolism and in response to some environmental stimuli, are both potentially reactive molecules containing an unpaired electron in their outermost electron shell.

At low to moderate concentrations, these reactive species modulate some vital physiological cell processes; however, at higher concentrations, they harm important cellular biomolecules and are responsible for chronic pathological conditions and diseases in humans such as cancer, neurological disorders, cardiovascular problems, Type-2 diabetes, chronic obstructive pulmonary disease (COPD), and reproductive system dysfunction.

2.4.1.1 Impact on Decellular Biomolecules: DNA, Protein, and Lipids

Elevated concentration of ROS/RNS can affect the major cellular structures, including lipid membrane, protein, and DNA, which leads to potential detrimental effects on the human body. ROS initiates DNA damage starting from the depletion of bases, cleavage of single or double-strands, alteration of a nitrogen base, sugar adaptation, cross-linking of proteins, mutations, etc. These changes are responsible for the transcription process (to make new molecules of messenger RNA (mRNA)) and ultimately lead to carcinogenesis, aging, and neurodegenerative diseases [20].

Higher oxidant proportions can generate some significant compounds, called protein carbonyls, by peptide bond cleavage and amino acid oxidation or mutation. At extreme conditions, irreversible binding of carbonyl groups to the peptide backbone results in high molecular weight aggregates and are insusceptible to usual proteasomal degradation, which may be directed to aging related diseases, i.e., Alzheimer's disease (AD), CVD, inflammatory disease (ID), diabetes (Type-1&2), and arthritis [21].

Lipids are the next susceptible target of ROS and they undergo peroxidation to produce lipid hydroperoxides (LOOH) as primary products and many secondary products such as malondialdehyde (MDA) and 4-hydroxynonenal (4-HNE) [22]. MDA and 4-HNE are two extensively investigated OS markers. MDA is highly mutagenic and 4- HNE is toxic, introducing damage to the lipid membrane permeability, fluidity, and cellular integrity, ultimately leading to cancer, neurodegenerative diseases, and aging [23].

2.4.1.2 Oxidative Stress-induced Cardiovascular Diseases (CVD)

Higher ROS levels can promote cardiovascular risk factors such as metabolic syndromes, cardiovascular complications, and dyslipidemia. Atherosclerosis is a medical disorder caused by ROS that can cause progressive hardening and constriction of arteries, reducing the flow and delivery of blood and oxygen throughout the body and eventually leading to plaque development and progression [24].

2.4.1.3 Rheumatoid Arthritis (RA)

Highly reactive oxygen free radicals are thought to contribute to the onset of the disease. The development of rheumatoid arthritis (RA), a chronic inflammatory disease, is significantly influenced by changes in the lipid metabolism brought on by OS. The OS can be identified by the infiltration of macrophages and T lymphocytes that have been activated into the tissues around the joints. Free radicals at the site of inflammation are involved in both the onset and development of this condition [25].

2.4.1.4 Oxidative Stress and Cancer

Carcinogenesis is a multistep process that begins with mutation and ends with the selective clonal proliferation of the altered cell. ROS play a crucial role in the progression of cancers, such as those of the breast, lung, liver, colon, prostate, ovary, and brain, by initiating and/or controlling the specific multistep process. The overproduction of ROS and defective antioxidant and/or DNA repair systems can result in oxidative damage to cellular macromolecules. Furthermore, a ROS alters important transcription processes by stimulating signal transduction pathways, i.e., Nrf2 and NF-κB. ROS-induced changes in gene expression patterns contribute to the carcinogenesis process [1, 26–28].

2.4.1.5 Oxidative Stress-induced Respiratory Disorders

The respiratory system has a large potential to produce excess reactive species either directly or indirectly because it serves as the major entry point for many substances that are inhaled. Pathological conditions that can be encountered as a result of OS are adult respiratory distress syndrome (ARDS), pulmonary emphysema, bronchial asthma, mucus hyper-secretion silicosis, and vascular barrier dysfunction chronic obstructive pulmonary disease (COPD) [29, 30].

2.4.1.6 Oxidative Stress-induced Reproductive System Dysfunctions

OS and redox imbalances modulate several reproductive functions, including congenital abnormalities, reproductive tract infections, male infertility, sperm maturation, sperm count, asthenospermia or asthenozoospermia, DNA damage, genetic diseases, spermatogenesis, testis function, folliculogenesis, oocyte maturation, ovum quality, embryogenesis, fetal growth, and sexual dysfunction, and even play a major role in sperm DNA damage/fragmentation and developing decisive conditions that can affect the fertility status of both male and female. ROS can facilitate spermatozoa maturation, motility, spermatocyte fusion, acrosome responses, and sperm capacitation at moderate levels by activating tyrosine phosphorylation. The sperm plasma membrane will suffer damage from ROS-induced lipid peroxidation, which may result in sperm death [31, 32]. OS has an impact on several essential pathophysiology in females, including polycystic ovarian syndrome (PCOS), endometriosis, recurrent miscarriage, spontaneous abortion, preeclampsia, infertility, and intrauterine growth restriction [33].

2.4.1.7 Oxidative Stress and Neurodegenerative Disorders

OS is responsible for various neurological disorders like Alzheimer's, Parkinson's, Amyotrophic, and Huntington's diseases, along with lateral multiple sclerosis, depression, and dementia. AD is characterized by a progressive loss of cognitive function and behavioral deterioration that impair daily and routine activities because of the development of extracellular amyloid plaques, protein aggregate accumulation, intra-neuronal neurofibrillary tangles, and loss of synaptic connections in specific brain regions. OS plays a pivotal role in mediating these decisive effects by damaging DNA, lipids, and proteins. Additionally, Parkinson's disease is caused by ROS-induced selective neuronal dysfunction in dopaminergic neurons in the substantia nigra pars compacta. The nigrostriatal dopaminergic pathway, which is one of the causes of Parkinson's disease, also negatively modulates dopaminergic levels in the brain [34]. The gradual degradation of upper and lower motor neurons in the spinal cord, cortex, and brain stem characterizes ALS, a fatal neurodegenerative condition [35]. Overall, OS-mediated lipid peroxidation, DNA and RNA damage, and protein loss have all been observed in patients with Parkinson's disease and ALS. Mutations in the Huntingtin (HTT) gene can cause HD, which causes neuronal degeneration in the striatum, followed by degeneration of the cerebral cortex and thalamus. The key indicators observed in HD patients include OS-induced biomolecule damage as well as protein carbonyls [36, 37].

2.4.1.8 Oxidative Stress and Aging

Mitochondrial oxidative damage is marked as the major root cause of cell aging. OS-directed damage of cellular biomolecules is also considered a risk factor for aging and related diseases [38].

2.4.2 Role of Antioxidants

Antioxidants are a topic of research for scientists, academics and industrialists. One of the Elsevier Scopus database studies on the keyword "antioxidant" filtered 105,440 hits. When 200 highly cited papers out of 327,657 between 1976 and 2020were examined using bibliometric indicators with the VOS viewer and biblioshiny tools, it was discovered that flavonoids and polyphenol derivatives are mostly linked to antioxidant potential [39].

Antioxidants are substances that, when present in low concentrations relative to those of an oxidizable substrate, considerably delay or prevent its oxidation. Antioxidants, in the end, reduce the occurrence of degenerative consequences by decreasing OS, DNA/lipid/protein damage, malignant transformations, and other parameters that might cause cell impairment. Antioxidants can be generally categorized as either natural or synthetic antioxidants. Endogenous and exogenous systems are included in the natural antioxidant category. Endogenous defense systems operate through both their enzymatic and non-enzymatic equivalents. Enzymatic antioxidants are further categorized as primary and secondary defense systems. Primary antioxidants (e.g., catalase, superoxide dismutase (SOD), and glutathione peroxidase) are involved in chain-breaking activities by inhibiting chain initiation and by interfering with chain propagation. Secondary antioxidants (e.g., glucose-6-phosphate dehydrogenase and glutathione reductase) can trigger singlet oxygen quenching, peroxide decomposition, chelation of pro-oxidative metal ions, UV absorption, and hinder oxidative enzymes. Secondary antioxidants and primary antioxidants can work together synergistically in certain situations. For example, secondary antioxidants can stabilize primary antioxidants by fostering an acidic environment, regenerating primary antioxidants by donating hydrogen, chelating pro-oxidative transition metal cations, and quenching molecular oxygen. Alternatively, non-enzymatic antioxidants have a low molecular weight, lipophilic entities (e.g., lipoic acid, ubiquinol, and plasmalogen), hydrophilic entities (e.g., uric acid, glutathione, bilirubin, melatonin, and amino acids), and sometimes metal-binding proteins (e.g., ferritin, transferrin, lactoferrin, ceruloplasmin, metallothioneins, and albumin). Antioxidants, both enzymatic and non-enzymatic, are carefully segregated throughout the cytoplasm and organelles to counteract the negative effects of ROS and provide maximum protection. Exogenous antioxidants, including vitamins (e.g., ascorbic acid/ascorbate, tocopherols, tocotrienols, and retinol), trace elements (e.g., selenium, zinc, manganese, copper, and iron), and carotenoids (e.g., α, β-carotene, lutein, zeaxanthin, lycopene, and β-cryptoxanthin) (e.g., phenolic and polyphenols) [40], can be analyzed by various in vitro and in vivo methods as mentioned in Figure 2.2.

2.4.3 Mechanism of Action and Signaling Pathways

The majority of cellular ROS are produced by mitochondrial oxidative metabolism or by cellular reactions to pathogen invasion, xenobiotics, and cytokines, which disrupt signal transduction pathways established for cellular defense systems. O_2 responsible for ROS is primarily produced by Complexes I and III in the respiratory chain and can be quickly converted into H_2O_2 by the enzyme superoxide dismutase (SOD) before being reduced to water by catalase or glutathione peroxidase. Extrinsic factors, such as tumor necrosis factor (TNF-α), epidermal growth factor (EGF), interleukin-1 (IL-1), hypoxia, and radiation, all contribute to increased ROS production [41]. Polyphenols modulate RO-induced signaling pathways, such as MAPK, Akt, PI3K, and PKC, to promote cell survival via various factors, i.e., NOX, NADPH oxidases, TNF-α, EGF, and IL-1β [42]. Some ROS-induced signaling pathways, secondary messenger, and antioxidant potential of polyphenols at specific sites are mentioned in Figure 2.3.

All important cellular processes, e.g., proliferation, differentiation gene expression, mitosis, apoptosis (natural cell death), stress responses, and immune defense, are regulated by one of the protein kinase pathways known as mitogen-activated protein kinase (MAPK) pathways that belong to the CDK/MAPK/GSK3/CLK (CMGC) kinase family. A crucial signaling mechanism called MAPK is activated by the simultaneous phosphorylation of tyrosine and threonine residues. It is involved in the conversion of extracellular signals into cellular responses against growth factors (EGF, PDGF, NGF, and VEGF), hormones (prostaglandin, PGE2), and cytokines (IL-17, IL-6, and IL-2). MAPK is found generally in three types of families of extracellular signal-regulated kinases (ERKs); stress-activated protein kinase/c-Jun N terminal kinase (SAPK/JNKs) that are activated in response to stresses such as heat, high osmolarity, UV irradiation, and also pro-inflammatory cytokines, e.g., TNF-α and IL-1; and the p38 mitogen-activated protein kinases (p38s) subfamily that is stimulated by lipopolysaccharides, inflammatory cytokines, and environmental stress. Growth factors and mitogens often activate ERKs, whereas cellular stressors and inflammatory cytokines, such as IL-6, IL-8, TNF-α, and others, stimulate JNKs and p38s [43, 44]. Polyphenols, especially flavonoids interact with Toll-like receptors (TLR) and manage the MAPK signaling pathway [45] with the help of the subfamilies of the ERK, JNK, and p38 pathways.

One of the most potent intracellular pro-survival signaling systems that interact with polyphenols and their metabolites to alter cell growth and survival is the phosphoinositide-3-kinase-Akt (PI3K/Akt) route, which is in addition to the MAPK pathway. Through the Akt/PKB pathways, PI3K activation promotes cell survival and inhibits apoptosis, whereas the PI3K pathway inhibition hastens the process [43]. Growth factors, including EGF, PDGF, NGF, and VEGF, as well as

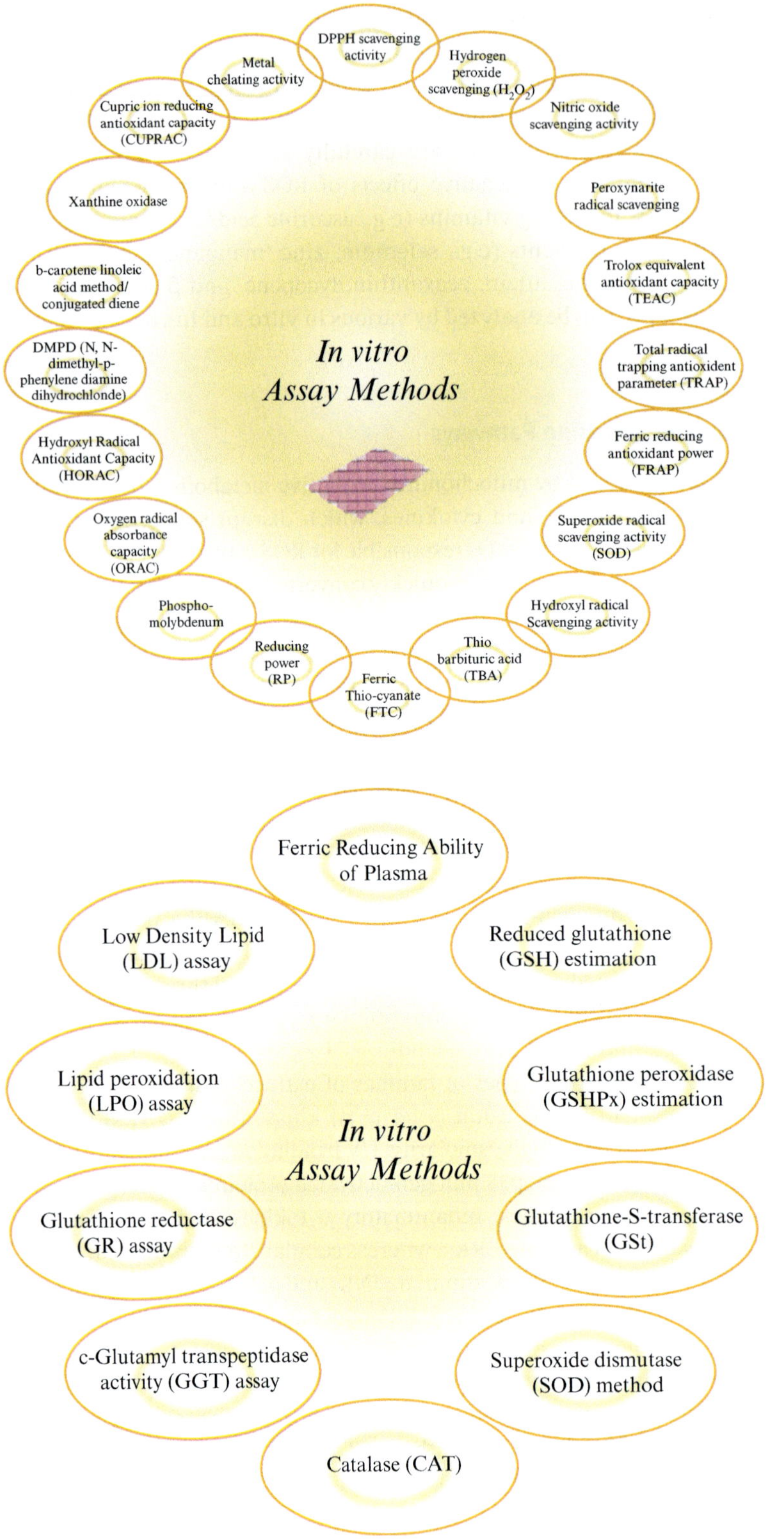

Figure 2.2 In vitro and in vivo methods for antioxidant evaluation.

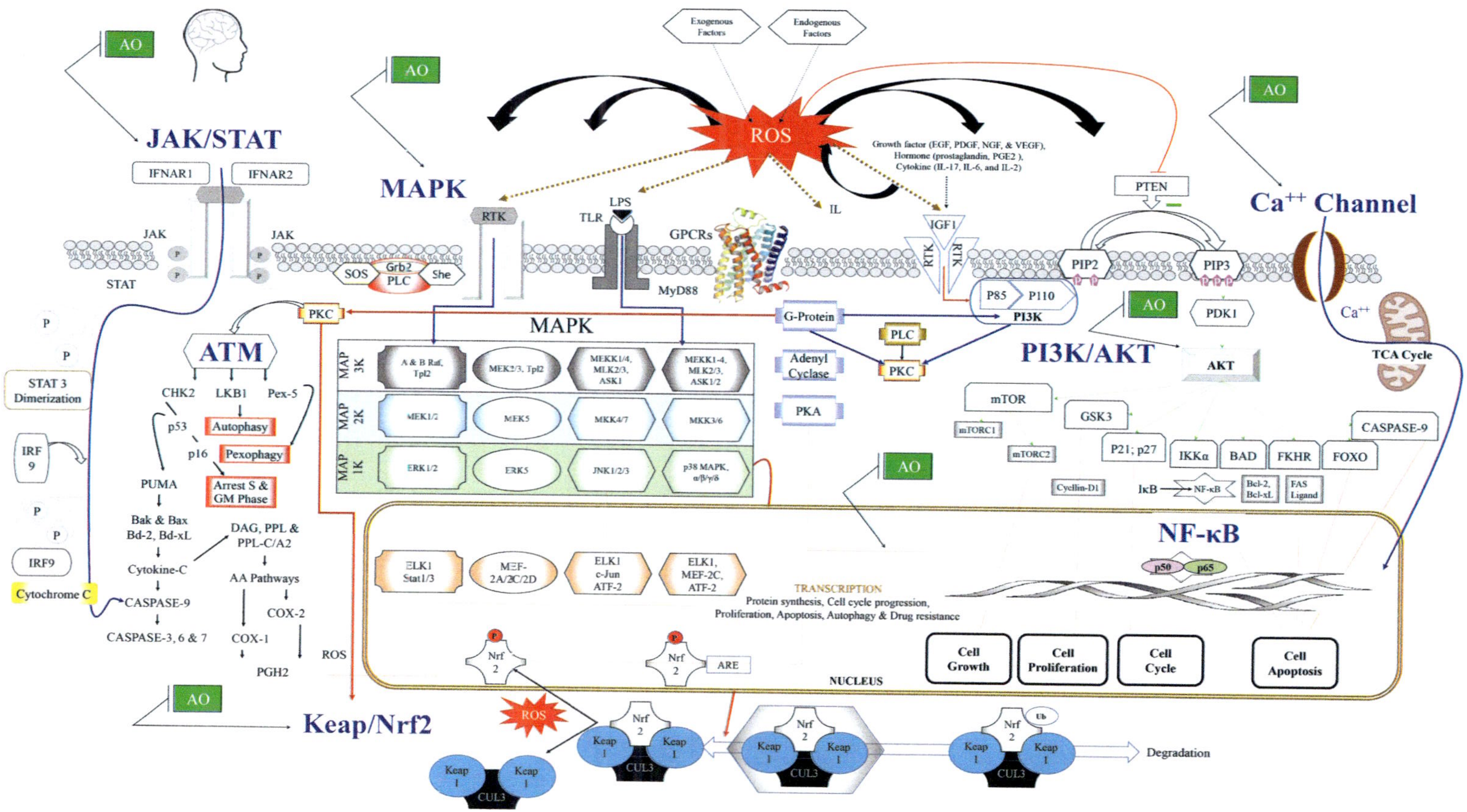

Figure 2.3 ROS-induced signaling pathways, secondary messengers, and antioxidant potential of polyphenols at specific sites in cells.

IGF-1R hormones like PGE2 and cytokines like IL-17, IL-6, and IL-2, activate different receptors, such as tyrosine kinases (RTKs), and dimerize Class I PI3K subunit p85 and p85 to control the AKT pathway. The second messenger phosphatidylinositol 3, 4-bisphosphate (PIP2) is produced when PIP2 is active, and PIP3 subsequently activates the Akt/PKB pathway to facilitate growth factor-mediated cell survival. The PI3K pathway is inhibited, which accelerates apoptosis. Akt/PKB is adversely regulated through the tumor suppressor protein/lipid PTEN [41, 46]. Different polyphenols alter PI3K/Akt activity, prevent a variety of human illnesses, and have positive influences on human health [41]. Kaempferol is a flavonoid that inhibits apoptosis via the Akt signaling pathway. Myricetin therapy prevents H_2O_2-induced apoptosis from killing cells. Genistein promotes the activity of endothelial nitric oxide synthase (eNOS) and increases phosphorylation of eNOS through the PI3K/Akt pathway [47, 48].

Another enzyme, protein kinase C (PKC), is a member of the protein kinase family (lipid-dependent kinases comprising11 isozymes) that participates in signal-transduction cascades by interfering with the phosphorylation of serine/threonine. This family plays a distinct role in cell signaling via three classified groups: conventional PKCs (cPKCs; α, β, and γ) activated through DAG, Ca^{2+}, and phospholipid, novel PKCs (nPKCs; ε, η, δ, and θ) activated only through DAG, and atypical PKCs (λ and ζ) activated by neither Ca^{2+} nor DAG channels [49]. Curcumin inhibits PKC, which downstream affects NF-κB and activator protein-1 (AP-1). Downstream of PKC also inhibits UGT, the ERK pathway, and EGFR, which modulates the transcription factors for physiologic changes. In response to oncogenic stimuli, one more polyphenol, resveratrol, inhibits the activity of kinases including PKC, MAPK, and IB kinase (IKK), as well as transcription factors, including hypoxia-inducible factor-1 (HIF-1), STAT3, NF-κB, and AP-1, and acts as an anticancer drug. By activating PKC and altering cell survival/cell cycle genes, epigallocatechin gallate (EGCG) can prevent OS-induced cell death. At lower doses, quercetin stimulates PKC activity; however, at greater concentrations, it inhibits calcium- and phospholipid-dependent PKCs. Similarly, tannic acid, chlorogenic acid, apigenin, and other polyphenol extracts from several plants and PKC activity from fruits take part in cell signaling and transcription, regulating numerous diseases states [50].

Keap1-Nrf2-ARE has a crucial function in regulating cellular metabolism and redox equilibrium as well as eliciting an adaptive response to OS, which can otherwise result in a variety of inflammatory disorders such as diabetes, AD, Parkinson's disease, and cancer [51]. To promote the separation of Nrf2 and Keap1, ROS either oxidizes critical reactive cysteine residues that control Keap1 activity (Cys273, Cys288, and Cys151) or activates kinases like PKC, MAPK, phosphatidylinositide 3-kinases (PI3Ks), and protein kinase-like endoplasmic reticulum kinases (PERK). When under OS, Nrf2 is liberated from its binding to keap1, which is normally broken down by proteasomes, and is then transported to the nucleus where it attaches to the sMaf protein and ARE to start transcription. Resveratrol is downstream Keap1 proteins and modulates Nrf2 nuclear accumulation. Similarly, polydatin from *Polygonum cuspidatum's* downstream Keap1 protein promotes Nrf2 transcriptional activity and the ARE-binding capability [52]. Numerous natural polyphenols have important roles in disease prevention by interacting with these mechanisms in both diet and medication.

2.4.4 Antioxidant Potential of Polyphenols

Characteristic structural attributes make polyphenols a strong antagonist for oxidants. Hydroxyl groups in the polyphenol structure can stabilize the free radicals by establishing hydrogen bonds. Polyphenols are capable of metal chelation and thereby prevent metal-catalyzed free radical formation. The hydrophobic benzoic ring, as well as hydroxyl groups in polyphenols, can beneficially interact with membrane proteins via hydrogen bonds, leading to low chances of OS and associated anomalies. Phenolic ligands can suppress the strength of binding interactions of some pro-inflammatory mediators. Interleukin-6 (IL-6), a major inflammatory mediator and low-density lipoprotein (LDL), affects dietary flavones from tea or cocoa, shows a beneficial effect on cardiovascular health, and retards OS-induced injuries, cellular NO production, and platelet cluster formation. Procyanidin, an active phenolic compound, contributes to anti-inflammatory activities by reducing the concentration of RONS and prostaglandin E_2, which regulates the integrity of several immune cells. Phenolic substances can up- or down-regulate transcriptional components engaged in antioxidant pathways [53]. Flavone, the isoflavones daidzein and genistein, the flavonols isorhamnetin, kaempferol, and quercetin, the flavanone naringenin, and the anthocyanin pelargonidin all reduce the expression of the iNOS protein and miRNA as well as the production of NO in a dose-dependent manner, promoting anti-inflammatory activities in the human body [54]. Dietary polyphenols can play a pivotal role in regulating neurological disorders like AD. AD patients observed a diminished level of the neurotransmitter acetylcholine, which in turn is responsible for the gradual memory loss. The phenolic compounds have several hydroxyl groups that can establish hydrogen bonds, hydrophobic contacts, and π–π interactions with the amino acid residues that make up the active site of acetylcholine. This strategy thus ensures the inhibition of acetylcholine hydrolysis by acetylcholinesterase (AChE). Caffeine, cinnamic acid, resveratrol, curcumin, and quercetin are involved in AChE inhibition [55]. Citrus flavanones, such as hesperidin, hesperetin, and neohesperidin, as well as the polyphenols derived from the gut bacterial metabolism of flavan-3-ols, are characterized by the smooth blood–brain barrier permeability and exhibit a beneficial effect against OS-induced neurodegenerative injuries. Polyphenols demonstrate chemopreventive action in the human body. Apoptosis, metastasis, and replication are just a few of the cellular mechanisms that might change during cancer development (the process of programmed cell death that occurs in multicellular organisms), and inefficient activity of repair mechanisms that proceeds through complex stages like initiation, promotion, and progression. Polyphenols can interfere either at these stages or proceed by arresting the cell cycle, induction of apoptosis, modulating signaling pathways, or influencing epigenetic alterations in cellular biomolecules [56].

Quercetin has inhibitory effects on the aryl hydrocarbon receptor that modulates the cytochrome p450 enzyme, which is enough to activate some carcinogenic species, and a transcription factor that can be activated by polycyclic aromatic hydrocarbons involved in lung carcinogenicity [57]. Polyphenols, including EGCG, lycopene, and curcumin, operate by the down regulation of some signal transduction pathways to reduce the effect of OS and have beneficial effects on prostate cancer. The risk of developing breast cancer as well as bladder cancer is minimized by the antiproliferative activity of polyphenols carnosol, resveratrol, EGCG, luteolin, and curcumin to a certain extent [56]. Curcumin has some favorable impact on the healthy function of the female reproductive system because it can eliminate/retard the chances of OS by accelerating heme oxygenase-1 (HO-1), nuclear factor erythroid 2-related factor 2 (Nrf2), and SOD expression levels. Because it has the power to cause apoptosis and prevent angiogenesis, curcumin may be employed in the treatment of cancer (the physiological process through which new blood vessels form from pre-existing vessels, formed in the earlier stage of vasculogenesis). Curcumin can also affect insulin levels as well as obesity, which has a direct influence on PCOS [58]. One curcumin enriched formulation, Re-party^{TM}, recovers alcohol-induced adverse effects developed by OS [59]. Curcumin is also beneficial to several neurodegenerative disorders working through modulating various molecular pathways, viz. NF-κB, Nrf2-ARE, PI3K/Akt, ERK1/2, MAPK, JNK, and iNOS/NO pathways [51]. Some polyphenols, including epicatechins, can regulate OS, cellular integrity, and aging-related pathways via mitochondrial RNA [60, 61]. Resveratrol also reduces aging-associated cognitive decline [62]. Polyphenols also take an active role in ensuring renal health. The characteristic stilbene structure enables resveratrol to reduce elevated ROS levels in the HK-2 (human kidney 2) renal cell line. The major flavanol EGCG can trap iron through chelation and prevents the situation of OS via RNS. Ellagic acid is also involved in renal health protection by reducing OS [63].

2.4.5 Overview of the Health Benefits of Nutraceuticals

Polyphenols, particularly flavonoids, have been named "nature's biological response regulators" because of extensive observations of their inherent capacity to affect the body's reaction against seasonal and vulnerable change [42]. The extensive distribution, diversity, and low toxicity of polyphenols/flavonoids compared with other bioactive/plant metabolites have led to significant human and animal consumption. Flavonoids that are mostly found in fruits, vegetables, spices, and plant-derived drinks, such as tea, coffee, and red wine, activate various signaling pathways, such as ERK, PkB/Akt, PI3K, and PKC, to improve cell survival.

At doses of 100–300 nmol/L, the flavonoid (–)-epicatechin (EC), which is mostly present in green tea, causes a fast increase in CREB phosphorylation that is ERK- and PI3K-dependent. EC also stimulates ERK and Akt phosphorylation. The polyphenolic component (–)-epicatechin-3-gallate (ECG), which is widely present in green tea, has been demonstrated to protect keratinocytes against UV-B-induced photodamage and H_2O_2-induced OS by inhibiting p38 and ERK1/2 pathways.

Polyphenols from *Curcuma longa* may be potential molecules and a good option for inhibiting oxidative-induced NLRP-3 pathways [64]. Some of the polyphenols that work against various OS-induced signaling pathways are highlighted in Table 2.1.

2.5 Revolutionary Approaches

2.5.1 Antioxidant Prospects of Unexplored Polyphenols

Polyphenols, triterpenes, alkaloids, and other natural matrices have given rise to a range of extraction strategies for natural antioxidants. In addition to standard extraction techniques, green chemistry extraction methods, such as microwave- or ultrasound-assisted extraction techniques, and nanotechnologies are in demand [92, 93]. Because of their selectivity, consistency, and productivity, these environmentally friendly technologies are widely used in industrial settings and are currently recognized as one of the most efficient energy-saving techniques. These techniques may be used to extract polyphenols that have recently attracted attention because of their effectiveness in a variety of non-communicable disease prevention and treatment applications. Several polyphenols, including nepetin, hispidulin, patuletin, tiliroside, spinacetin, tambulin, ombuin, syringetin, laricitrin, petunidin, peonidin, and morin, have been the focus of extensive recent research as potential therapeutic candidates for cellular dysfunctions. These studies have not only focused on technologies. Nepetin is an essential flavone that guards membrane protection and cell life by preventing peptide oligomerization and aggregation. An expanding field of study has focused on the same substance with powerful anti-inflammatory capabilities [94, 95]. Several *Artemisia*, *Salvia*, and *Scoparia* plants contain the unexplored flavonoid hispidulin, which has broader medicinal applications because of its anti-adipogenesis, antithrombotic, and antioxidant activities [96]. Because of its fundamental quercetin-like structure, patuletin is an important flavonoid that may be explored extensively as it is thought to have an antiproliferative function [97]. Tiliroside is a glycoside flavonoid that is present in *Amaranthaceae* plants and it has antioxidant and anti-inflammatory properties [98]. One of the spinach flavonoids, spinacetin, requires special attention

Table 2.1 Polyphenols for inhibiting oxidative stress by various signaling pathways and second messengers.

S. No.	Polyphenol Class & Structure	Parent Structure	Food Source	Signaling Pathways And Second Messenger	References
1.	Flavones	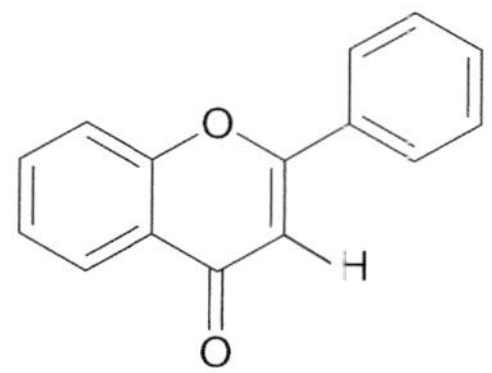	Apigenin	↓ROS, ↓MAPK/ERK, ↓PI3K/Akt, ↓H_2O_2-induced apoptosis, ↓AP-1, ↓JNK, ↓PKC, ↓COX-2, ↓NF-κB, ↑Caspase-3	[65]
			Chrysin	↓NF-κB, STAT3, Notch1, microRNA, ↓PI3K	[66]
			Luteolin	↓ROS, ↓MAPK/ERK, ↓PI3K/Akt, ↓H_2O_2-induced apoptosis	
2.	Flavonols	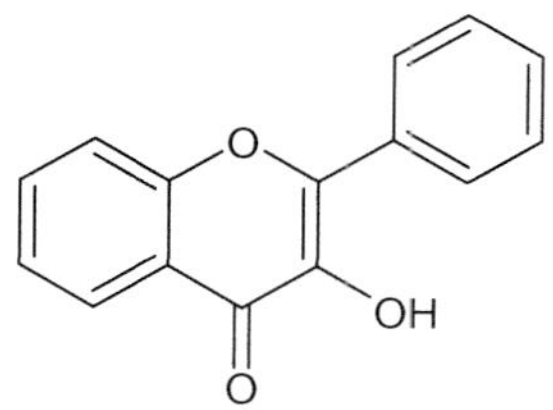	Kaempferol	↓ROS, ↓MAPK/ERK, ↓PI3K/↑↓Akt, ↓Nrf-2 ↓SOD; ↓H_2O_2-induced apoptosis, ↓QR	[65, 67]
			Quercetin	↓ROS, ↓MAPK/ERK, ↓PI3K/Akt, ↓Nrf-2 ↑P53	[68, 69]
			Rutin	↓ROS, ↓MAPK/ERK, ↓PI3K/Akt, ↓MDA	[70–72]
			Myricetin	↓CYP1A1, ↑QR	[73, 74]
3.	Flavanones	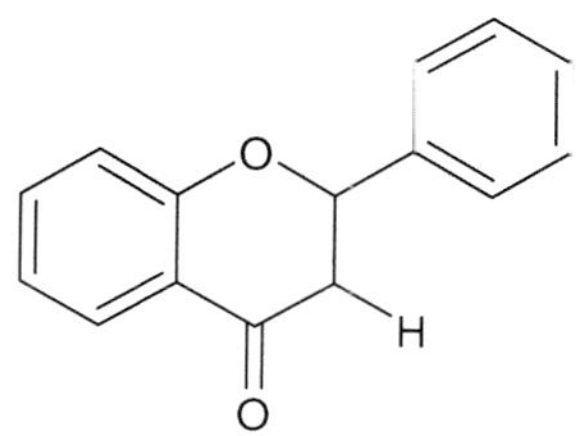	Eriodictyol	↓ROS, ↓MAPK (ERK/JNK/p38), ↓PI3K/Akt, ↓mTOR, ↓Nrf-2/ARE	[75, 76]
			Hesperetin	↓H_2O_2-induced apoptosis, ↓cyclinD1, cyclinE1, cyclin-dependent kinase 2	[77]
			Naringenin	↓ROS, ↓MAPK/ERK, ↓PI3K/Akt, ↓Nrf-2/ARE	[78]

(Continued)

Table 2.1 (Continued)

S. No.	Polyphenol Class & Structure	Parent Structure	Food Source	Signaling Pathways And Second Messenger	References
4.	Flavanonols	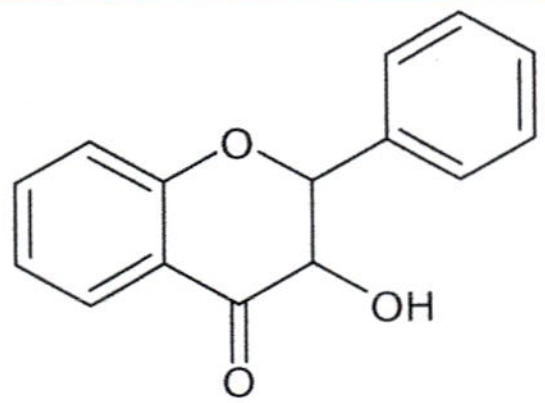	Dihydro-quercetin (Taxifolin)	Downstream Nrf2/ARE	[79, 80]
5.	Flavanols	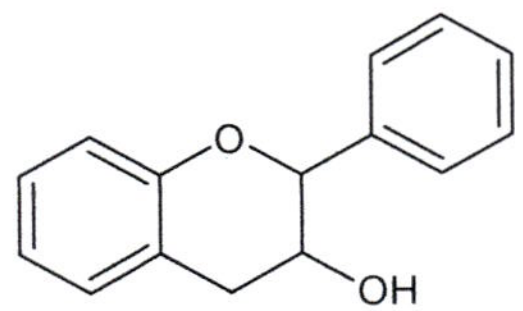	(+)/(−)-Catechin (C) (+)/(−)-Epicatechin (EC) (+)/(−)-Gallocatechin (GC) (+)/(−)-Epigallo-catechin (EGC) (+)/(−)-Epicatechin-3-gallate (EGCG) (+)/(−)-Epigallocatechin-3-gallate (EGCG)	↓ROS; ↑↓PI3K/Akt, ↓H_2O_2-induced apoptosis; ↑GSH; ↓Nrf-2, ↓JNK	[65]
6.	Iso flavonoids	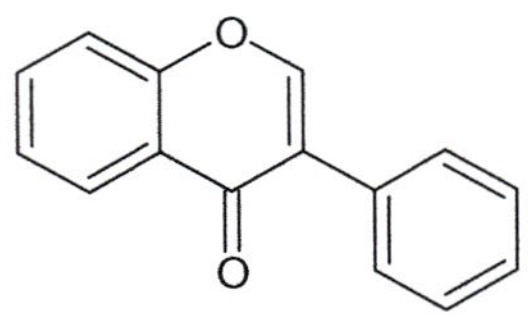	Genistein	↓ROS, ↓COX-2, ↓MAPK/ERK, ↓PI3K/Akt, ↓AP-1, ↓JNK, ↓PKC, ↓NF-κB, ↑Caspase-3, ↑P38	[81]
			Daidzein	↑Caspase 3/7 activity, Interfere Bax/Bcl2	[82, 83]
			Glycitein	↓JNK, ↓H_2O_2-induced apoptosis	[84]

7.	Neo-flavonoids	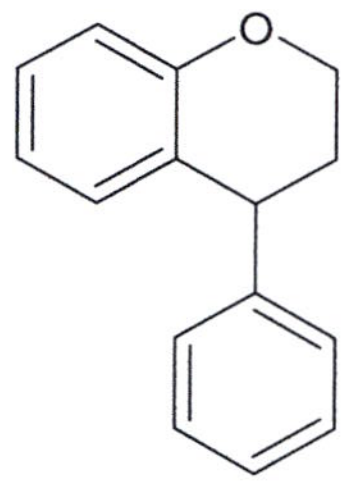	Dalbergiones	Inhibited the production of NO, ↓Nrf-2/ARE	[85, 86]
8.	Anthocyanin	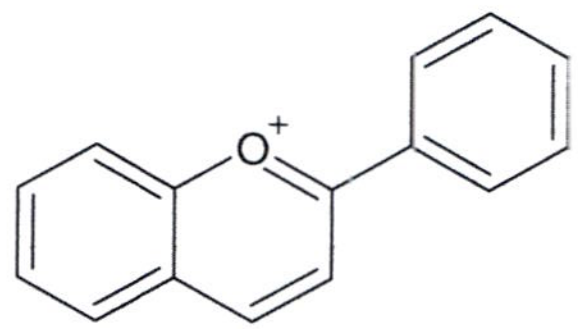	Aurantinidin Capensinidin Cyanidin Delphinidin Europinidin Hirsutidin Malvidin Pelargonidin Peonidin Petunidin Pulchellidin Rosinidin	↓ROS, ↓COX-2, ↓MAPK (ERK/JNK/p38), ↓JNK, ↓Nrf-2/ARE	[87, 88]

(Continued)

Table 2.1 (Continued)

S. No.	Polyphenol Class & Structure	Parent Structure	Food Source	Signaling Pathways And Second Messenger	References
9.	Hydrolysable Tannins	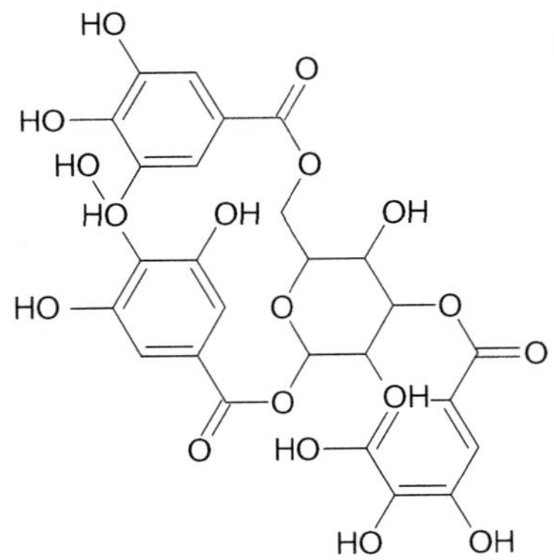	Gallotannins Ellagitannins	↓ROS, ↓MAPK/ERK, ↓ROS, ↓MAPK/ERK, ↓JAK/STAT3, ↓PI3K/Akt/mTOR	[89] [90]
10.	Condensed Tannins	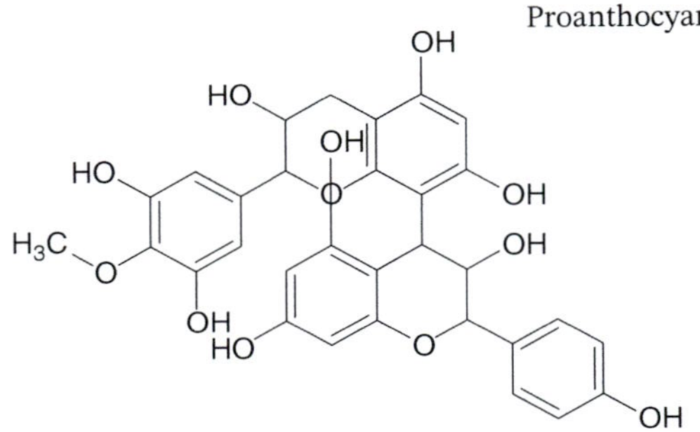	Proanthocyanidins	↓ROS, ↓MAPK/ERK, ↓JAK/STAT3, ↓PI3K/Akt/mTOR, ↓NF-κB	[91]

because it may have substantial antioxidant properties [99, 100]. Tambulin and ombuin are two noble polyphenols that require extensive derivatization because of their pursuit of both antibacterial and antiproliferative properties [101]. Berries have been shown to include anthocyanins, such as petunidin and peonidin, as well as polyphenols, including syringetin and laricitrin, which require more extensive characterization [102]. Because of its complementary effect in boosting anticancer drugs, morin is another bioflavonoid of pharmaceutical and nutraceutical relevance [103]. These are a few understudied polyphenols with antioxidant properties that can be researched further to aid in the development of conventional medical practices.

2.5.2 Global Market Value and Commercial Significance

The dietary supplements segment is immense globally. The Asia–Pacific Region, USA, Canada, Japan, China, and Europe are driving the market with a 5.6% CAGR. The Asia–Pacific Region is the world's largest functional food business area dominating the market with a majority of the market share because of the rising nutritional enrichment consciousness in China. India is expected to continue to have a beneficial influence on regional market growth [104]. Customers choose items with added health advantages as they are now more health conscious. In the dietary supplement industry, polyphenols are the primary option. The rising usage of polyphenols in health foods is predicted to fuel growth in the dietary supplement industries, which is expected to reach $2.9 billion by 2030, representing a 7.4% CAGR from the existing market of 1.67 billion USD [104]. Analysts predict that the Asia–Pacific Region followed by Latin America is growing at a 7.3% annual rate, especially in the polyphenol business [9]. In 2027, the polyphenols market in the USA is expected to be worth $272.7 million USD. China, the world's second-biggest economy, is expected to reach a projected market size of $362.4 million USD by 2027, with a 10.9% CAGR from 2020 to 2027. From 2020 to 2027, Japan and Canada are two more important geographic markets that will grow at rates of 3.8% and 6.4%, respectively. In Europe, Germany is predicted to grow at a CAGR of roughly 4.5% [105]. "Polyphenols" is a popular search term in Asia and the Pacific. The keyword "polyphenols" was frequently searched by Singaporeans on Google trends. New Zealand's polyphenol search was 75% that of Singapore's. With 67%, the United States was second, followed by Australia (55%), Canada (54%), and China (53%). Ireland ranked fifth in the world for polyphenol searches, with 51% of the top five zones [106]. The current Google trend also suggests that the Asia–Pacific Region is leading the polyphenol and antioxidant business, which would provide new opportunities for industrialists.

2.6 Conclusion

Human requirements are the beginning point for all innovations. Recently, the world of science recognized the importance and requirement of polyphenols in human life. Polyphenols are the most abundant secondary metabolites present in the plant kingdom. Their activities on the biological functions are endless. Many studies have demonstrated that polyphenols are a strong, naturally occurring antioxidant and they can be used in many functional foods. Polyphenol consumption or a polyphenol-rich diet can significantly slow the progression of several chronic pathological conditions, including cancer, diabetes, cardiovascular disease, and aging etc. Our approaches should be focuses on gaining a better understanding of the biological effects of polyphenols and their significant impact on human health. Scientific research on polyphenols clarified epidemiological information and bioavailability characteristics, including the rates of absorption, accumulation, and elimination, and the mechanism of action for many biological effects. Polyphenols that control intracellular signaling pathways through some important pathways, such as MAPK, interleukins (ILs-1 and 6), TNF-α, NF-κB, and JNK, and also that play a significant role in the body's natural antioxidant defenses mechanism should be explored much. Research on the impact of polyphenols on human health is still in its early phases; however, it appears to hold promise for the prevention of chronic human diseases, illnesses, and infirmities in the future. According to the current state of the global market, there are numerous opportunities to investigate unexplored polyphenols. More research is required to enhance polyphenol bioavailability and its mode of action in the bodies. Following the confirmation of undiscovered polyphenols' capabilities, the food, nutraceuticals, health supplements, and pharmaceutical industries can use them to combat a range of noncommunicable disorders.

Acknowledgments

The authors acknowledge the Management of Synthite Industries Pvt. Ltd., India for their support and inspiration.

Abbreviations

AD, Alzheimer's disease; PkB/Akt, Protein kinase B (Ak strain transforming); AP-1 Activator protein-1, ARE, Antioxidant-responsive element; Bax, Bcl-2 associated X protein; Bcl, B-cell lymphoma protein; Bcl-2, B-cell lymphoma protein 2; CBG, Cytosolic β-glucosidase; CNS, Central nervous system; COX, cyclooxygenase; CVD, Cardiovascular diseases; eNOS, Endothelial nitric oxide synthase; ER, endoplasmic reticulum; ERK, Extracellular signal-regulated protein kinase; Epac1, Exchange protein directly activated by cAMP1; H_2O_2, Hydrogen peroxide; HD, Huntington's disease; IFN, Interferon; IL, Interleukin; iNOS, Inducible nitric oxide synthase; IRF, interferon regulatory factor; JAK, Janus kinases; JNK, c-Jun N-terminal kinase; Keap, Kelch-like ECH-associated protein; LOX, Lipo-oxygenase; LPH, Lactase phloridzin hydrolase; LPS, Lipopolysaccharide; MAPK, Mitogen-activated protein kinases; MDA, Malondialdehyde; MMP, Matrix metalloproteinases; mTOR, Mechanistic target of rapamycin; MyD88, myeloid differentiation primary response 88; NADH, Nicotinamide adenine dinucleotide hydrogen; NF-κB, Nuclear factor kappa B; NLRP3, NLR family pyrin domain containing 3; NO, Nitric oxide; Nrf2, Nuclear factor erythroid 2-related factor 2; OS, Oxidative stress; PD, Parkinson's disease; PGE2, Prostaglandin E2; PI3K/Akt, Phosphatidylinositol 3-kinase/Ak strain transforming; PPAR, Peroxisome proliferator-activated receptor; PPAR-γ, Peroxisome proliferator activated receptor; RA, Rheumatoid arthritis; ROS, Reactive oxygen species; SAPK/JNKs Stress-activated Protein Kinase/c-Jun N terminal kinase; STAT, Signal transducer and activator of transcription protein; SOD, Superoxide dismutase; TGF, Transforming growth factor; TGF-β, Tumor growth factor- β; TLR, toll-like receptor; TNF-α, Tumor necrosis factor-α.

References

1 Rudrapal, M., Khairnar, S., Khan, J. et al. (2022). Dietary polyphenols and their role in oxidative stress-induced human diseases: insights into protective effects, antioxidant potentials and mechanism(s) of action. *Frontiers in Pharmacology* 13 (February): 1–15. https://doi.org/10.3389/fphar.2022.806470.

2 Wink, M. (1997). Compartmentation of secondary metabolites and xenobiotics in plant vacuoles. *Advances in Botanical Research* 25 (C): 141–169. https://doi.org/10.1016/S0065-2296(08)60151-2.

3 Petrussa, E., Braidot, E., Zancani, M. et al. (2013). Plant flavonoids-biosynthesis, transport and involvement in stress responses. *International Journal of Molecular Sciences* 14 (7): 14950–14973. https://doi.org/10.3390/ijms140714950.

4 Liu, W., Feng, Y., Yu, S. et al. (2021). The flavonoid biosynthesis network in plants. *International Journal of Molecular Sciences* 22 (23): 1–18. https://doi.org/10.3390/ijms222312824.

5 Casañal, A., Zander, U., Munoz, C. et al. (2013). The strawberry pathogenesis-related 10 (PR-10) Fra a proteins control flavonoid biosynthesis by binding to metabolic intermediates. *The Journal of Biological Chemistry* 288 (49): 35322–35332. https://doi.org/10.1074/jbc.M113.501528.

6 Ahmad, R.S., Hameed, A., Sameen, A. et al. (2019). Polyphenols; biosynthesis, classification and pharmacological utilization. In: *Food and Nutrition: 2nd Edition Chapter* no. October. 1–31. [Online]. Available. https://www.researchgate.net/publication/343472098_Chapter_Polyphenols_Biosynthesis_Classification_and_Pharmacological_Utilization.

7 Viuda-Martos, M., Ruiz-Navajas, Y., Fernández-López, J. et al. (2010 December). Spices as Functional Foods. *Critical Reviews in Food Science and Nutrition* 51 (1): 13–28. https://doi.org/10.1080/10408390903044271.

8 Leja, K.B. and Czaczyk, K. (2016 December). The industrial potential of herbs and spices - A mini review. *Acta Scientiarum Polonorum. Technologia Alimentaria* 15 (4): 353–365. https://doi.org/10.17306/J.AFS.2016.4.34.

9 Chauhan, S. and Deshmukh, R. (2022). Polyphenol market-global opportunity analysis and industry forecast, 2021-2030. *Allied Market Research*, Allied Market Reserach, Portland, United States, 2021–2030. [Online]. Available: www.alliedmarketresearch.com. (accessed Jun. 21, 2022).

10 Pérez-Gregorio, R. (2021). Phenolic compounds and functional beverages. *Beverages* 7 (4): 2020–2022. https://doi.org/10.3390/beverages7040071.

11 Pandey, K.B. and Rizvi, S.I. (2009). Plant polyphenols as dietary antioxidants in human health and disease. *Oxidative Medicine and Cellular Longevity* 2 (5): 270–278. https://doi.org/10.4161/oxim.2.5.9498.

12 Rauter, A.P., Ennis, M., Hellwich, K-H. et al. (2018). *Nomenclature of flavonoids (IUPAC Recommendations 2017). Purc and Applied Chemistry* 90 (9): https://doi.org/10.1515/pac-2013-0919.

13 Kumar, S. and Pandey, A.K. (2013). Chemistry and biological activities of flavonoids: an overview. *The Scientific World Journal* 2013 (4): 1–16. https://doi.org/10.1155/2013/162750.

14 Ollis, W.D. (1965). Biochemistry of phenolic compounds. *Journal of the American Chemical Society* 87 (5): 1157–1158. https://doi.org/10.1021/ja01083a067.

15 Vermerris, W. and Nicholson, R. (2008). Families of phenolic compounds and means of classification. In: *Phenolic Compound Biochemistry* (ed. Vermerris, W. and Ralph, N.), 1–34. Dordrecht: Springer. https://doi.org/10.1007/978-1-4020-5164-7_1.

16 Hollman, P.C.H. (2004). Absorption, bioavailability, and metabolism of flavonoids. *Pharmaceutical Biology* 42: 74–83. https://doi.org/10.1080/13880200490893492.

17 Adinarayana, D. and Syamasundar, K.V. (1982). Phytochemicals from Pterocarpus marsupium. *Pbyrocbmistry* 21: 1083.

18 Chen, L., Cao, H., and Xiao, J. (2018). *Polyphenols: Absorption, Bioavailability, and Metabolomics*, 10. Elsevier Inc. https://doi.org/10.1016/B978-0-12-813572-3.00002-6.

19 Makarova, M.N. (2011). Bioavailability and metabolism of flavonoids. *Eksperimental'naya I Klinicheskaya Farmakologiya* 74 (6): 33–40.

20 Barnham, K.J., Masters, C.L., and Bush, A.I. (2004). Neurodegenerative diseases and oxidatives stress. *Nature Reviews. Drug Discovery* 3 (3): 205–214. https://doi.org/10.1038/nrd1330.

21 Kawa, I.A., Masood, A., Amin, S. et al. (2019). Clinical perspective of posttranslational modifications. In: *Protein Modificomics* (ed. Dar, T.A. and Singh, L.R.), 37–68. Elsevier. https://doi.org/10.1016/B978-0-12-811913-6.00002-3.

22 Ayala, A., Muñoz, M.F., and Argüelles, S. (2014). Lipid peroxidation: production, metabolism, and signaling mechanisms of malondialdehyde and 4-hydroxy-2-nonenal. *Oxidative Medicine and Cellular Longevity* 2014: https://doi.org/10.1155/2014/360438.

23 Li, Y., Zhao, T., Li, J. et al. (2022). Oxidative stress and 4-hydroxy-2-nonenal (4-HNE): implications in the pathogenesis and treatment of aging-related diseases. *Journal of Immunology Research* 2022: https://doi.org/10.1155/2022/2233906.

24 Singh, A., Kukreti, R., Saso, L., and Kukreti, S. (2019). Oxidative stress: a key modulator in neurodegenerative diseases. *Molecules* 24 (8): 1–20. https://doi.org/10.3390/molecules24081583.

25 Mateen, S., Moin, S., Khan, A.Q. et al. (2016). Increased reactive oxygen species formation and oxidative stress in rheumatoid arthritis. *PLoS One* 11 (4): 1–15. https://doi.org/10.1371/journal.pone.0152925.

26 Arfin, S., Jha, N.K., Jha, S.K. et al. (2021). Oxidative stress in cancer cell metabolism. *Antioxidants* 10 (5): 1–28. https://doi.org/10.3390/antiox10050642.

27 Zahra, K.F., Lefter, R., Ali, A. et al. (2021). The involvement of the oxidative stress status in cancer pathology: a double view on the role of the antioxidants. *Oxidative Medicine and Cellular Longevity* 2021: https://doi.org/10.1155/2021/9965916.

28 Slika, H., Mansour, H., Wehbe, N. et al. (2022). Therapeutic potential of flavonoids in cancer: ROS-mediated mechanisms. *Biomedicine & Pharmacotherapy* 146: 112442. https://doi.org/10.1016/j.biopha.2021.112442.

29 Cannavò, L., Perrone, S., Viola, V. et al. (2021). Oxidative stress and respiratory diseases in preterm newborns. *International Journal of Molecular Sciences* 22 (22): 1–12. https://doi.org/10.3390/ijms222212504.

30 Victoni, T., Barreto, E., Lagente, V. et al. (2021 February). Oxidative Imbalance as a crucial factor in inflammatory lung diseases: could antioxidant treatment constitute a new therapeutic strategy? *Oxidative Medicine and Cellular Longevity* 2021: 1–11. https://doi.org/10.1155/2021/6646923.

31 Mannucci, A., Argento, F.R., Fini, E. et al. (2022). The impact of oxidative stress in male infertility. *Frontiers in Molecular Biosciences* 8 (January): 1–9. https://doi.org/10.3389/fmolb.2021.799294.

32 Aitken, R.J. and Curry, B.J. (2011). Redox regulation of human sperm function: from the physiological control of sperm capacitation to the etiology of infertility and DNA damage in the germ line. *Antioxidants & Redox Signaling* 14 (3): 367–381. https://doi.org/10.1089/ars.2010.3186.

33 Mohammadi, M. (2019). Oxidative stress and polycystic ovary syndrome: a brief review. *International Journal of Preventive Medicine* 10 (1): 86. https://doi.org/10.4103/ijpvm.IJPVM_576_17.

34 Dias, V., Junn, E., and Mouradian, M.M. (2013). The Role of oxidative stress in parkinson's disease. *Journal Profile Parkinson's Disease* 3 (4): 461–491. https://doi.org/10.3233/JPD-130230.

35 Zarei, S., Carr, K., Reiley, L. et al. (2015). A comprehensive review of amyotrophic lateral sclerosis. *Surgical Neurology International* 6 (1): 171. https://doi.org/10.4103/2152-7806.169561.

36 Islam, M.T. (2017 January). Oxidative stress and mitochondrial dysfunction-linked neurodegenerative disorders. *Neurological Research* 39 (1): 73–82. https://doi.org/10.1080/01616412.2016.1251711.

37 Patel, M. (2016). Targeting oxidative stress in central nervous system disorders HHS public access. *Physiology and Behavior* 176 (1): 139–148. https://doi.org/10.1016/j.tips.2016.06.007.Targeting.

38 Cui, H., Kong, Y., and Zhang, H. (2012). Oxidative Stress, Mitochondrial Dysfunction, and Aging. *Journal of Signal Transduction* 2012 (646354): 1–13: https://doi.org/10.1155/2012/646354.

39 Mamdapur, G.M.N., Nema, N.K., Chacko, B.K., and Paul, J. (2021). Bibliometric analysis of the 200 most cited articles on antioxidant from 1976-2020. *Digit. @ Univ. Nebraska - Lincoln*.

40 Augustyniak, A., Bartosz, G., Cipak, A. et al. (2010). Natural and synthetic antioxidants: an updated overview. *Free Radical Research* 44 (10): 1216–1262. https://doi.org/10.3109/10715762.2010.508495.

41 Zhang, J., Wang, X., Vikash, V. et al. (2016). ROS and ROS-mediated cellular signaling. *Oxidative Medicine and Cellular Longevity* 2016: 1–18. https://doi.org/10.1155/2016/4350965.

42 Mansuri, M.L., Parihar, P., Solanki, I., and Parihar, M.S. (2014). Flavonoids in modulation of cell survival signalling pathways. *Genes and Nutrition* 9 (3): https://doi.org/10.1007/s12263-014-0400-z.

43 Soares-Silva, M., Diniz, F.F., Gomes, G.N., and Bahia, D. (2016). The mitogen-activated protein kinase (MAPK) pathway: role in immune evasion by trypanosomatids. *Frontiers in Microbiology* 7 (FEB): 1–9. https://doi.org/10.3389/fmicb.2016.00183.

44 Son, Y., Cheong, Y.-K., Kim, N.-H. et al. (2011). Mitogen-activated protein kinases and reactive oxygen species: how can ROS activate MAPK pathways? *Journal of Signal Transduction* 2011: 1–6. https://doi.org/10.1155/2011/792639.

45 Behl, T., Upadhyay, T., Singh, S. et al. (2021). Polyphenols targeting MAPK mediated oxidative stress and inflammation in Rheumatoid Arthritis. *Molecules* 26 (6570): 1–24. https://doi.org/10.3390/molecules26216570.

46 Yap, T.A., Garrett, M.D., Walton, M.I. et al. (2008 Aug). Targeting the PI3K–AKT–mTOR pathway: progress, pitfalls, and promises. *Current Opinion in Pharmacology* 8 (4): 393–412. https://doi.org/10.1016/j.coph.2008.08.004.

47 Si, H., Yu, J., Jiang, H. et al. (2012). Phytoestrogen genistein up-regulates endothelial nitric oxide synthase expression via activation of cAMP response element-binding protein in human aortic endothelial cells. *Endocrinology* 153 (7): 3190–3198. https://doi.org/10.1210/en.2012-1076.

48 Kang, K.A., Wang, Z.H., Zhang, R. et al. (2010). Myricetin protects cells against oxidative stress-induced apoptosis via regulation of PI3K/Akt and MAPK signaling pathways. *International Journal of Molecular Sciences* 11 (11): 4348–4360. https://doi.org/10.3390/ijms11114348.

49 Black, A.R., and Black, J.D. (2013). Protein kinase C signaling and cell cycle regulation. *Frontiers in Immunology* 3 (JAN): 1–17. https://doi.org/10.3389/fimmu.2012.00423.

50 Das, J., Ramani, R., and Suraju, M.O. (2016 October). Polyphenol compounds and PKC signaling. *Biochimica Et Biophysica Acta - General Subjects* 1860 (10): 2107–2121. https://doi.org/10.1016/j.bbagen.2016.06.022.

51 Joshi, P., Bisht, A., Joshi, S. et al. (2022). Ameliorating potential of curcumin and its analogue in central nervous system disorders and related conditions: a review of molecular pathways. *Phytotherapy Research* (May): 1–38. https://doi.org/10.1002/ptr.7522.

52 Zhou, Y., Jiang, Z., Lu, H. et al. (2019). Recent advances of natural polyphenols activators for keap1-nrf2 signaling pathway. *Chemistry and Biodiversity* 16 (11): https://doi.org/10.1002/cbdv.201900400.

53 Rahman, M.M., Rahaman, M.S. Islam, M.R. et al. (2021). Role of phenolic compounds in human disease: current knowledge and future prospects. *Molecules* 27 (1): 233. https://doi.org/10.3390/molecules27010233.

54 Hämäläinen, M., Nieminen, R., Vuorela, P. et al. (2007). Anti-inflammatory effects of flavonoids: genistein, kaempferol, quercetin, and daidzein inhibit STAT-1 and NF-κB activations, whereas flavone, isorhamnetin, naringenin, and pelargonidin inhibit only NF-κB activation along with their inhibitory effect on i. *Mediators of Inflammation* 2007: https://doi.org/10.1155/2007/45673.

55 Reddy, V.P., Aryal, P., Robinson, S. et al. (2020 Jan). Polyphenols in Alzheimer's disease and in the gut–brain axis. *Microorganisms* 8 (2): 199. https://doi.org/10.3390/microorganisms8020199.

56 Briguglio, G., Costa, C., Pollicino, M. et al. (2020). Polyphenols in cancer prevention: new insights (review). *International Journal of Functional Nutrition* 1 (2): 1–11. https://doi.org/10.3892/ijfn.2020.9.

57 Androutsopoulos, V.P., Tsatsakis, A.M., and Spandidos, D.A. (2009). Cytochrome P450 CYP1A1: wider roles in cancer progression and prevention. *BMC Cancer* 9: 1–17. https://doi.org/10.1186/1471-2407-9-187.

58 Kamal, D.A.M., Salamt, N., Yusuf, A.N.M. et al. (2021). Potential health benefits of curcumin on female reproductive disorders: a review. *Nutrients* 13 (9): 1–20. https://doi.org/10.3390/nu13093126.

59 Nema, N.K, Sarojam, S., Khamborkar, S.D. et al. (2021). Novel health supplement for recovery of alcohol induced adverse effects and associated disorders: hepG2 cells study and safety dose profiling on rats. *Nutrition and Food Science* 7 (5): 1–9. https://doi.org/10.24966/FSN-1076/100121.

60 Luo, J., Si, H., Jia, Z., and Liu, D. (2021). Dietary anti-aging polyphenols and potential mechanisms. *Antioxidants* 10 (2): 1–20. https://doi.org/10.3390/antiox10020283.

61 Mukherjee, P.K., Maity, N., Nema, N.K., and Sarkar, B.K. (2011). Bioactive compounds from natural resources against skin aging. *Phytomedicine* 19 (1): 64–73. https://doi.org/10.1016/j.phymed.2011.10.003.

62 Vacca, R.A., Valenti, D., Caccamese, S. et al. (2016). Plant polyphenols as natural drugs for the management of down syndrome and related disorders. *Neuroscience and Biobehavioral Reviews* 71: 865–877. https://doi.org/10.1016/j.neubiorev.2016.10.023.

63 Guerreiro, Í., Ferreira-Pêgo, C., Carregosa, D. et al. (2022). Polyphenols and their metabolites in renal diseases: an overview. *Foods* 11 (7): 1–22. https://doi.org/10.3390/foods11071060.

64 Nema, N.K, Mamdapur, G.M.N., Sarojam, S. et al. (2021). Preventive medicinal plants and their phytoconstituents against SARS-CoV-2/COVID-19. *Pharmacognosy Reviews* 13 (4): 173–191. https://doi.org/10.5530/pres.13.4.10.

65 Ramos, S. (2008). Cancer chemoprevention and chemotherapy: dietary polyphenols and signalling pathways. *Molecular Nutrition & Food Research* 52 (5): 507–526. https://doi.org/10.1002/mnfr.200700326.

66 Talebi, M., Talebi, M., Farkhondeh, T. et al. (2021). An updated review on the versatile role of chrysin in neurological diseases: chemistry, pharmacology, and drug delivery approaches. *Biomedicine & Pharmacotherapy* 141: 111906. https://doi.org/10.1016/j.biopha.2021.111906.

67 Yang, C., Yang, W., He, Z. et al. (2021). Kaempferol alleviates oxidative stress and apoptosis through mitochondria-dependent pathway during lung ischemia-reperfusion injury. *Frontiers in Pharmacology* 12 (March): 1–13. https://doi.org/10.3389/fphar.2021.624402.

68 Figueira, I., Menezes, R., Macedo, D. et al. (2016). Polyphenols beyond barriers: a glimpse into the brain. *Current Neuropharmacology* 15 (4): 562–594. https://doi.org/10.2174/1570159×14666161026151545.

69 Erythroid, F. and Signaling, F. (2021). *Quercetin alleviates oxidative damage by activating nuclear enterocytes.*

70 Koval'skii, I.V., Krasnyuk, I.I., Krasnyuk, I.I. et al. (2014). Mechanisms of rutin pharmacological action (review). *Pharmaceutical Chemistry Journal* 48 (2): 73–76. https://doi.org/10.1007/s11094-014-1050-6.

71 Stroke, E. (2021). Interventional cardiology pharmacological properties of rutin and its potential uses for Alzheimer ' s disease. *Journal of Experimental Stroke and Translational Medicine Interview* 13: 1–12.

72 Enogieru, A.B., Haylett, W., Hiss, D.C. et al. (2018). Rutin as a potent antioxidant: implications for neurodegenerative disorders. *Oxidative Medicine and Cellular Longevity* 2018: https://doi.org/10.1155/2018/6241017.

73 Agraharam, G., Girigoswami, A., and Girigoswami, K. (2022). Myricetin: a multifunctional flavonol in biomedicine. *Current Pharmacology Reports* 8 (1): 48–61. https://doi.org/10.1007/s40495-021-00269-2.

74 Park, K.S., Chong, Y., and Kim, M.K. (2016). Myricetin: biological activity related to human health. *Applied Biological Chemistry* 59 (2): 259–269. https://doi.org/10.1007/s13765-016-0150-2.

75 Lee, S.E., Yang, H., Son, G.W. et al. (2015). Eriodictyol protects endothelial cells against oxidative stress-induced cell death through modulating ERK/Nrf2/ARE-dependent heme oxygenase-1 expression. *International Journal of Molecular Sciences* 16 (7): 14526–14539. https://doi.org/10.3390/ijms160714526.

76 Islam, A., Islam, M.S., Rahman, M.K. et al. (2020). The pharmacological and biological roles of eriodictyol. *Archives of Pharmacal Research* 43 (6): 582–592. https://doi.org/10.1007/s12272-020-01243-0.

77 Estruel-Amades, S., Massot-Cladera, M., Garcia-Cerdà, P. et al. (2019). Protective effect of hesperidin on the oxidative stress induced by an exhausting exercise in intensively trained rats. *Nutrients* 11 (4): 1–17. https://doi.org/10.3390/nu11040783.

78 Wang, K., Chen, Z., Huang, L., et al. (2017). Naringenin reduces oxidative stress and improves mitochondrial dysfunction via activation of the Nrf2/ARE signaling pathway in neurons. *International Journal of Molecular Medicine* 40 (5): 1582–1590. https://doi.org/10.3892/ijmm.2017.3134.

79 Das, A., Baidya, R., Chakraborty, T. et al. (2021). Pharmacological basis and new insights of taxifolin: a comprehensive review. *Biomedicine & Pharmacotherapy* 142: 112004. https://doi.org/10.1016/j.biopha.2021.112004.

80 Cai, C., Liu, C., Zhao, L. et al. (2018). Effects of taxifolin on osteoclastogenesis in vitro and in vivo. *Frontiers in Pharmacology* 9: 1–10. https://doi.org/10.3389/fphar.2018.01286.

81 Goh, Y.X., Jalil, J., Lam, K.W. et al. (2022). Genistein: a review on its anti-inflammatory properties. *Frontiers in Pharmacology* 13 (January): 1–23. https://doi.org/10.3389/fphar.2022.820969.

82 Kumar, V. and Chauhan, S.S. (2021). Daidzein induces intrinsic pathway of apoptosis along with ER α/β ratio alteration and ROS production. *The Asian Pacific Journal of Cancer* 22 (2): 603–610. https://doi.org/10.31557/APJCP.2021.22.2.603.

83 Choi, E.J. and Kim, G.H. (2014). The antioxidant activity of daidzein metabolites, O-desmethylangolensin and equol, in HepG2 cells. *Molecular Medicine Reports* 9 (1): 328–332. https://doi.org/10.3892/mmr.2013.1752.

84 Kang, K.A., Zhang, R., Piao, M.J. et al. (2007). Inhibitory effects of glycitein on hydrogen peroxide induced cell damage by scavenging reactive oxygen species and inhibiting c-Jun N-terminal kinase. *Free Radical Research* 41 (6): 720–729. https://doi.org/10.1080/10715760701241618.

85 Swamy, M.K. (2021). Plant-derived bioactives: chemistry and mode of action. *Plant-Derived Bioactives Chemistry and Mode of Action* (May): 1–589. https://doi.org/10.1007/978-981-15-2361-8.

86 Shao, F., Panahipour, L., Omerbasic, A. et al. (2022). Dalbergiones lower the inflammatory response in oral cells in vitro. *Clinical Oral Investigations* (0123456789): https://doi.org/10.1007/s00784-022-04509-7.

87 Salehi, B., Sharifi-Rad, J., Cappellini, F. et al. (2020). The therapeutic potential of anthocyanins: current approaches based on their molecular mechanism of action. *Frontiers in Pharmacology* 11: 1–20. https://doi.org/10.3389/fphar.2020.01300.

88 Reis, J.F., Monteiro, V.V.S., Gomes, R.S. et al. (2016). Action mechanism and cardiovascular effect of anthocyanins: a systematic review of animal and human studies. *Journal of Translational Medicine* 14 (1): 1–16. https://doi.org/10.1186/s12967-016-1076-5.

89 Marcińczyk, N., Gromotowicz-Popławska, A., Tomczyk, M., and Chabielska, E. (2022). Tannins as hemostasis modulators. *Frontiers in Pharmacology* 12 (January): 1–21. https://doi.org/10.3389/fphar.2021.806891.

90 Xu, J., Cao, K., Liu, X. et al. (2022). Punicalagin regulates signaling pathways in inflammation-associated chronic diseases. *Antioxidants* 11 (1): 1–12. https://doi.org/10.3390/antiox11010029.

91 Yang, L., Xian, D., Xiong, X. et al. (2018). Proanthocyanidins against oxidative stress: from molecular mechanisms to clinical applications. *BioMed Research International* 2018: https://doi.org/10.1155/2018/8584136.

92 Hano, C. and Tungmunnithum, D. (2020). Plant polyphenols, more than just simple natural antioxidants: oxidative stress, aging and age-related diseases. *Medicines* 7 (5): 26. https://doi.org/10.3390/medicines7050026.

93 Nema, N.K., Rajan, N., Sabu, S. et al. (2022). Use of nanotechnology for the improvement of sensory attributes of foods. In: *Application of Nanotechnology in Food Science, Processing and Packaging* (ed. Egbuna, C., Jeevanandam, J., C. Patrick-Iwuanyanwu, K. et al.), 31–46. Cham: Springer. https://doi.org/10.1007/978-3-030-98820-3_3.

94 Xu, J., Zheng, T., Zhao, C. et al. (2021). Resistance of nepetin and its analogs on the fibril formation of human islet amyloid polypeptide. *International Journal of Biological Macromolecules* 166: 435–447. https://doi.org/10.1016/j.ijbiomac.2020.10.202.

95 Chen, X., Han R., Hao, P. et al. (2018). Nepetin inhibits IL-1β induced inflammation via NF-κB and MAPKs signaling pathways in ARPE-19 cells. *Biomedicine & Pharmacotherapy* 101: 87–93. https://doi.org/10.1016/j.biopha.2018.02.054.

96 Lee, S.G., Kim, J.S., Min, K. et al. (2018). Hispidulin inhibits adipogenesis in 3T3-L1 adipocytes through PPARγ pathway. *Chemico-Biological Interactions* 293: 89–93. https://doi.org/10.1016/j.cbi.2018.07.027.

97 Alvarado-Sansininea, J.J., Sánchez-Sánchez, L., López-Muñoz, H. et al. (2018). Quercetagetin and patuletin: antiproliferative, necrotic and apoptotic activity in tumor cell lines. *Molecules* 23 (10): https://doi.org/10.3390/molecules23102579.

98 Corrêa, W.R., Serain A.F., Leticia Aranha Netto L.A. et al. (2018). Anti-inflammatory and antioxidant properties of the extract, tiliroside, and patuletin 3-O-β-D-glucopyranoside from Pfaffia townsendii (Amaranthaceae). *Evidence-based Complementary and Alternative Medicine* 2018: https://doi.org/10.1155/2018/6057579.

99 Olennikov, D.N., Chirikova, N.K., and Kashchenko, N.I. (2015). Spinacetin, A new Caffeoylglycoside, and other phenolic compounds from Gnaphalium uliginosum. *Chemistry of Natural Compounds* 51 (6): 1085–1090. https://doi.org/10.1007/s10600-015-1498-y.

100 Riso, P., Brusamolino, A., Scalfi, L., and Porrini, M. (2004). Bioavailability of carotenoids from spinach and tomatoes. *Nutrition, Metabolism and Cardiovascular Diseases* 14 (3): 150–156. https://doi.org/10.1016/S0939-4753(04)80035-8.

101 Nooreen, Z., Singh, S., Singh, D.K. et al. (2017). Characterization and evaluation of bioactive polyphenolic constituents from Zanthoxylum armatum DC., a traditionally used plant. *Biomedicine & Pharmacotherapy* 89: 366–375. https://doi.org/10.1016/j.biopha.2017.02.040.

102 Stanoeva, J.P., Stefova, M., Andonovska, K.B. et al. (2017). Phenolics and mineral content in bilberry and bog bilberry from Macedonia. *International Journal of Food Properties* 20 (April): S863–S883. https://doi.org/10.1080/10942912.2017.1315592.

103 Solairaja, S., Andrabi, M.Q., Dunna, N.R., and Venkatabalasubramanian, S. (2021). Overview of morin and its complementary role as an adjuvant for anticancer agents. *Nutrition and Cancer* 73 (6): 927–942. https://doi.org/10.1080/01635581.2020.1778747.

104 DBMR (2022). Polyphenol market size, report, statistics, trends, scope, & forecast trends by 2029. *DATA BRIDGE MARKET RESEARCH*, https://www.databridgemarketresearch.com/reports/global-polyphenols-market (accessed Jun. 21, 2022).

105 Research and Market (2022). Polyphenols - global market trajectory & analytics. *Research and Market*, https://www.researchandmarkets.com/reports/5030379/polyphenols-global-market-trajectory-and (accessed Jun. 21, 2022).

106 Google Trend (2022). Polyphenols, dietary supplement, triterpene, gycosides, oleoresin - explore - google trends. *Google Trend*, https://trends.google.com/trends/explore?cat=45&q=Polyphenols, DietarySupplement,Triterpene,Gycosides,Oleoresin (accessed Jun. 21, 2022).

3

Polyphenols

Plant-based Nutraceuticals in Human Health

Monalisa Kesh[1,] and Sachin Goel[2]*

[1] Amity Institute of Biotechnology, Amity University Uttar Pradesh, Noida, India
[2] Department of Biological Sciences and Engineering, Netaji Subhas University of Technology, New Delhi, India
* Corresponding author

Abbreviations

ADME: Absorption, Distribution, Metabolism, and Excretion
BACE1: Beta-site amyloid precursor protein cleaving enzyme 1
MPTP: 1-methyl-4-phenyl-1,2,3,6-tetrahydropyridine
MFC: Minimum Fungicidal Concentration
MIC: Minimum Inhibitory Concentration
MBC: Minimum Bactericidal Concentration

3.1 Introduction

3.1.1 What Are Polyphenols?

Micronutrients, called polyphenols, are naturally found in plants [1, 2]. They can be easily added to human diets by consuming fruits, vegetables, drinks, and spices, and they are present in many supplements. All food products made from plants include polyphenols. When ingested as a part of a balanced diet, these metabolites have been linked to the health advantages humans receive from eating food originating from plants. Higher plants produce polyphenols advantageous to human health. These benefits include acting as anti-allergic, antioxidant, antihypertensive, anticancer, anti-inflammatory, and antimicrobial agents [3]. Because of the potential positive impact on human health, scientists are becoming more interested in polyphenols and other dietary phenolics [3]. Investigations have revealed the importance of polyphenolics in keeping cancer, diabetes, osteoporosis, and cardiovascular diseases under check [4].

3.1.1.1 Why We Need Polyphenols

The absence of polyphenols is not linked to any adverse effects. However, because of their ability to lower the risk of chronic diseases, they are recognized as “lifespan requirements”. According to

Polyphenols: Food, Nutraceutical, and Nanotherapeutic Applications, First Edition. Edited by Mithun Rudrapal.

studies, people who exceed 650 mg of polyphenols daily have a reduced risk of fatality compared with individuals who consume less than 500 mg. Our body is protected by polyphenols through:

3.1.1.1.1 Improving Heart Health

Polyphenols maintain healthy blood vessels and assist in controlling blood pressure, which improve circulation. Another risk factor of heart disease, chronic inflammation, is also decreased by polyphenols.

3.1.1.1.2 Lowering the Diabetes Risk

Our blood sugar levels can be reduced and better controlled by polyphenols. They also encourage the production of the hormone insulin, which tells our body to use sweets effectively. These outcomes can lessen insulin resistance, a state in which our body does not react to the hormone appropriately. Our risk of ailments, such as obesity and diabetes, is decreased by maintaining blood sugar levels and insulin resistance.

3.1.1.1.3 Anticancer Properties

The anti-inflammatory and antioxidant functions of polyphenols may reduce our risk of developing cancer. Polyphenols may even stop the development of tumors and eradicate living cancer cells.

3.1.1.1.4 Raising Immunity

Polyphenols can protect against immunity-related illnesses. Additionally, polyphenols inhibit dangerous bacteria in our stomach and encourage the growth of healthy bacteria. This effect promotes healthy digestion; however, a balanced bacterial ecosystem is also necessary for a robust immune system.

3.2 Types and Categories of Polyphenols

One of the most prevalent and extensively dispersed classes of organic products of natural flora is dietary phenolics, often known as polyphenols. More than 4,000 flavonoids have been discovered among approximately 8,000 phenolic structures. Despite being classified chemically as substances having phenolic structural characteristics, polyphenols are a wide class of natural products that includes many subgroups of phenolic substances. Polyphenols can be found in abundance in different fruits, wine, and green vegetables. There are numerous ways to classify polyphenols based on their diversity and widespread distribution in plants. Polyphenols have been categorized based on their place of origin, biological purpose, and chemical makeup. Additionally, most polyphenols appear as glycosides having various sugar units along with acylated sugars at various locations on the polyphenolic structure. Polyphenol categorization is discussed briefly in this chapter.

3.2.1 Phenolic Acids

Phenolic acids are types of non-flavonoid polyphenolic chemicals [5]. While polyphenolic compounds are typically present in cereals and seeds in their bound forms, they are also found in the free form in fruits and vegetables [5]. They can only be released by alkaline or acid hydrolysis.

3.2.2 Flavonoids

A few of the subclasses of flavonoids can be differentiated as anthocyanins, flavanones, flavones, flavonols, and flavan-3-ols. Plants also contain certain flavonoids, including neoflavonoids and isoflavones. Chalcones are nonetheless regarded as flavonoid family members. The basic structural formation of flavonoids includes aglycones; nevertheless, in plants, most of these substances appear as glycosides. The biological properties of these molecules, such as their antioxidant activity, are influenced by their structural diversity and glycosylation patterns.

3.2.3 Isoflavones, Neoflavonoids, and Chalcones

Ring B of isoflavones is joined to ring C at location C3. Isoflavones are primarily found among plants of the leguminous family. Isoflavones have a significant role in human health because many cultures rely heavily on beans, particularly soy beans, in their diet. Along with biochanin A, glycetein, and formononetin, the two primary isoflavones in red clovers and soy are genistein and daidzein. Food plants rarely contain neoflavonoids, although dalbergin has been considered the most prevalent and extensively dispersed neoflavonoid in the plant kingdom. Fruits, such as apples, and hops or beers, include open-ring chalcones.

3.2.4 Flavonols, Flavones, Flavanones, and Flavanonols

Exclusively found in plants, this group contains flavones, their 3-hydroxy descendants flavonols, in addition to their methoxides and glycosides. This group comprises the largest subcategory of polyphenols. There are at least 279 and 347 distinct glycosidic combinations in the two most popular flavonol aglycones, quercetin and kaempferol, respectively. In the past 15 years, there have been more flavanones identified along with their 3-hydroxy derivatives, i.e., flavanonols. Some flavanones, such as pyrano-flavanones, prenylated-flavanones, furano-flavanones, and benzylated-flavanones, exhibit distinctive patterns of substitution that result in this subgroup's considerable number of substituted derivatives.

3.3 Polyphenols: Food Sources

Polyphenols can be obtained from plant-based diets with whole grains, fruits, and vegetables [6–8], which are simple to consume to improve your health. However, some sources are more nutrient rich than others. Along with the other vital nutrients they include, these following eight foodstuffs have the highest per-serving polyphenol content.

3.3.1 Berries

Berries are a simple addition to any diet because they have few calories and are rich in polyphenols, fiber, and vitamin C. The largest quantities are found in elderberries and chokeberries, where a serving of half a cup contains 1,123 mg and 870 mg of polyphenols.

3.3.2 Herbs and Spices

We need to look no further than our spice cabinet to add polyphenols to our dinner. In addition to their polyphenol content, a variety of minerals, including calcium, magnesium, and potassium, are frequently present in dried herbs and spices.

3.3.3 Cocoa Powder

With 516 mg of polyphenols per tablespoon, cocoa powder is a rich source even though we should limit our sugar intake. However, the cocoa content can be decreased by heating and processing to form chocolate goods. For instance, milk chocolate contains only 35 milligrams per tablespoon compared with 249 milligrams in dark chocolate.

3.3.4 Nuts

Nuts are a quick and simple method to add fiber, protein, and important fatty acids to our daily diet; however, portions should be considered because they are heavy in calories. Chestnuts, which include approximately three nuts, have the highest concentration of polyphenols among all nuts (347 mg per ounce). Other healthy options include almonds with 53 mg per ounce serving and hazelnuts and pecans with 140 mg.

3.3.5 Flaxseeds

Flaxseeds are occasionally used to ease constipation and enhance digestion. They include a lot of fiber, and each tablespoon also contains 229 mg of polyphenols. Flaxseeds can be baked into cookies and breads or added to cereal, sandwiches, and salads.

3.3.6 Vegetables

We should consume 2.5–3 cups of vegetables daily. Most vegetables contain polyphenols; as a result, sufficient vegetables in our diet can help us reap the health benefits of these antioxidants.

3.3.7 Olives

Olives are abundant in polyphenols, fatty acids, and vitamin E. Five black olives, or 20 grams, have 113 mg of polyphenols compared with 70 mg in the same amount of green olives.

3.3.8 Coffee and Tea

We consume polyphenols if we drink a cup of coffee or tea to start the day. Thirty-five milligrams of polyphenols are present in approximately 20 g of coffee, enough to produce one cup. Black, green, and ginger teas are generally consumed in lesser quantities; however, even one cup can provide some polyphenols to our diet.

3.4 Polyphenols in Food

3.4.1 Factors Influencing Food Phenolic Content

Our diet is rich in phenolic compounds, which are secondary metabolites. These substances have significant effects on color, flavor, and astringency, and may affect the sensory qualities of food. Sufficient intake of these phenolic compounds can have health advantages. Polyphenols can be found in various grape bunch components such as the stems, peels, and pulp. As explained by Pastrana-Bonilla et al., the average total phenolic quantity in grape varieties is approximately 23.8 mg/g of gallic acid, comparable in pulp, and approximately 374.6 mg/g of gallic acid in skins [9]. Furthermore, numerous scholars documented both high amounts of overall and individual phenolic chemicals in table grapes [10]. Numerous studies have noted substantial quantities of phenolic chemicals in grape raisins [11]. The phenolic makeup of grapes is influenced by variety, environment, and viticultural techniques. The content of grape phenolics will therefore depend on all of these individual or combination factors, with the grape variety playing a key role. Phenolic biosynthesis and accumulation during the maturity and development of grape fruits are significantly influenced by genotypic variations. However, the overall phenolic content is significantly influenced by the genotype, ecological, and control techniques. Numerous studies claim that a region's geology and soil conditions, vineyard altitude, sunlight exposure, climate, and solar radiation are crucial environmental elements that affect the phenolic composition of the grapes. Other elements, including the risk of disease from farming practices and grape ripeness, are also crucial [12–14]. Ripe cherries were shown to contain more phenolic compounds, according to many studies. The levels of bioactive chemicals can be greatly changed by different pre- and post-harvest factors, including cultivar, rootstock, climate, storage conditions, soil type, and processing. According to Gonçalves et al., greater temperatures and sun radiation encouraged the manufacture of phenolic acids, while reducing the amount of anthocyanins in the environment [15]. According to Stöhr et al., uses of many preharvest compounds, such as oxalic acid (2 mM), to enhance cherry quality have been researched [16]. Anthocyanins, flavone, neochlorogenic acids, and chlorogenic acids were all enhanced [17]. However, a late period of final maturity is when the phenolic content tends to peak. Several sweet cherry varieties had an increase in all phenolic components and antioxidant activity during cold storage. In addition to soil, drainage and climatic conditions, additional secondary metabolite concentrations in vegetables and fruits are also regulated by these factors. Phenolic chemicals found in fruits and vegetables showed sensitivity to environmental fluctuations, with warm temperatures showing a positive association with the quantities of phenolic components. Crops grown in soil may vary every year in terms of phytochemical content and overall production [18]. Different biological properties of dietary phenolics are of nutritional importance because they are linked to the potential benefits that promote human health by inhibiting many diseases. Owing to their pharmacological qualities, these substances may occasionally be employed for medicinal purposes. Because of their toxicity, many phenolics with a low molecular weight, such as thymol, are employed as antiseptics in medical science [19].

3.4.2 Ingestion and Absorption of Phenolic Compounds in Food

In the environment, food polyphenolics are primarily found as glycosides, esters, and sometimes in their polymeric forms, which require digestive enzymes or gut microbiota to be hydrolyzed in order to be absorbed. Thus, both their absorption and metabolism control the biological characteristics of such phenolic compounds that are undergoing this breakdown. Because of the functional tasks that plant polyphenols carry out in the human body, they have garnered a lot of attention. They have been linked to a detrimental influence on the digestive system in addition to their considerable advantages to our health. Because of the relatively low absorption of phenolics after a meal, improper digestion of the food may occur, further leading to stomach issues. While it may be helpful to observe the inhibitory activity of polyphenolics on the absorption of dietary factors high in calories (saccharides and fats), their inhibitory activity on the digestibility of proteins are undesirable because of the decreased utilization of amino acids. Both in vivo and in vitro techniques are used to measure how much of these compounds are digested and absorbed, with the former being the more dependable. As a result, numerous in vitro models have been developed to mimic how human digestion works in an effort to accurately represent actual conditions. Numerous models of intestinal epithelial cells have been used to analyze the procedures of polyphenolic absorption [20–22]. A few ex vivo studies suggested that gastric or intestinal portions of the body, including the jejunum and colon, were the sites of phenolic acid absorption. The movement and accessibility of polyphenols from the gastrointestinal (GI) lumen into the cytoplasm of enterocytes are influenced by specific chemical properties such as lipophilicity, molecular weight, stereochemistry, and the presence of groups capable of hydrogen bonding. Intestinal carriers, such as P-glycoprotein and SGLT1 cotransporters, as well as a passive diffusion process, are thought to be how phenolic chemicals are absorbed. These transporters carry the medications into the interior of the cell and are expressed on the cell membranes [23, 24]. In the GI system, the interaction of phenolic compounds with a variety of macronutrients, micronutrients, and host proteins, delay or reduce nutrient digestion and absorption. Their size, shape, molecular intricacy, charges, dietary matrix, and inclusion of other phenolics or medications all affect their absorption and pharmacokinetics properties. They can control the production of Phase I and II enzymes, and inhibit them such that they may result in possible drug-nutrient and drug-drug interactions. Understanding such interactions has sparked the creation of innovative biotechnological and nanotechnological methods that prevent undesirable interactions, shield phenolic compounds from severe chemical conditions, and enable tailored delivery. To supplement and provide a viable medicinal replacement for various disease treatments, it is vital to investigate the biomolecules that are capable of transporting and protecting these substances. Because most phenolic compounds have complicated structures that affect how they interact with other biomolecules, ions, vitamins, and enzymes on a molecular level, further research is needed. A better understanding of the biology and phases of dietary phenolic compounds is needed for successful implementation of their therapeutic potentials [25, 26].

3.4.3 Thermal Processing of Food and Its Phenolic Content

Most food processing methods use a series of procedures to transform the raw material in desired ways, with each action having an effect on the content of the food. Polyphenols can be divided into the following groups: stilbenes, proanthocyanidins, flavones, flavonols, flavanones, flavanols, isoflavones, anthocyanins, hydroxybenzoic acids, and hydroxycinnamic acids [27]. To exercise their beneficial health effects, bioactive substances and phenolics must first tolerate the conditions required for food processing, followed by releasing them and being accessible in the GI system before they can proceed through digestion and reach the desired tissue of interest. It is essential to investigate alterations in the phenolic compounds throughout the steps to determine the nutritional mark of the processed goods. Measuring alterations in the food phenolics while processing is important for more accurately determining the nutritional importance of the prepared goods [28]. According to research on the impact of processing on phenolics, the bioavailability and accessibility of food polyphenolics are significantly influenced by food processing. Oxygen, pH, temperature, metal ions, enzymes, light, and sugars are elements that affect the stability of anthocyanins. Anthocyanins come in four varying forms depending on the pH. The cell structure is disturbed during the processing operation steps, including the steps of juice extraction, followed by cutting and dicing, ultimately allowing the substrate and enzyme to combine. Enzyme hydrolysis occurs during the process, and the sugar cleaves off at specific locations. By applying blanching, a mild heating technique, the enzyme can be inactivated to reduce the rate of pigment deterioration; this method improves anthocyanin preservation and stability [28]. Oxygen is another significant contributor to the destruction of anthocyanins, either directly or indirectly through the activity of polyphenol oxidases. For phenolic acids, the varieties of food and the methods used to process them determine the changes that occur in the different forms, free or bounded, throughout processing. Furthermore, the stability of phenolic acids is significantly impacted by temperature, air, and enzymes. The bulk of phenolic acids is found bound in grains. However, during the thermal processing of those grains, an increase in 200%–300% of the total syringic, vanillic, and ferulic acids was observed, indicating that when grains are hydrothermally processed, phenolic acids and related derived products may be released from the cell wall [29]. Processing generally tends to affect phenolic acids in foods in both favorable and unfavorable ways. Therefore, it is crucial to identify the best processing conditions that increase the product's shelf life and prevent the bioactive components from degrading. The phenolic compounds go through considerable structural changes throughout the processing of flavonols, flavanone, flavan-3-ols, and isoflavones, which may affect their biological activity, similar to how black tea is processed. Additionally, catechin and its byproducts give black tea its flavor and astringent quality, similar to how polyphenols experience oxidation when fermenting cocoa beans [30, 31]. The effects of processing on the flavonol amount of onions and asparagus (blanching, chopping, maceration, and boiling) were evaluated; chopping significantly reduced the rutin level in asparagus. The largest reduction in total flavonols was observed after 60 minutes of boiling, proving that flavonols in these plant foods are sensitive to thermal heating. Similar amounts of quercetin derivatives were lost when broccoli was boiled or fried; however, more quercetin derivatives were retained when broccoli was steamed [32]. In conclusion, fermentation does seem to positively affect the structural composition of these food phenolics, which may therefore indicate an effect on their bioavailability, bioefficacy, and other characteristics, whereas heat processing has a variety of effects.

3.4.4 Problems Encountered in Polyphenol Digestion and Metabolism

Although numerous in vitro studies show that plant polyphenols have strong biological effects, the real in vivo effects may only be minimal. Given that the total antioxidant value in plasma is greater than 103 mol/l, a specified amount of 20–50 mol/l of additional antioxidants from dietary sources would be required to considerably boost the systemic antioxidant capacity. This is unlikely to happen as unconjugated serum levels of approximately 1 mol/l are usually reached, even with high dietary polyphenol consumption. However, more dramatic effects could occur if dietary phenols come into direct contact with the mucosa of the digestive tract. Numerous pieces of data imply that luminal levels of polyphenols may be significantly greater than serum levels. Large amounts of polyphenol chemicals are transported to the colon via poor intestinal absorption, where many of them are extensively metabolized by colonic flora. High dietary polyphenol concentrations change the flora in the colon. The intestinal mucosa may be directly impacted by unmodified flavonoids and the microbial metabolic byproducts of dietary polyphenols, which may have antioxidant effects towards dietary components of the reactive oxygen species (ROS). In that vein, phenolic compound amounts in human fecal water have been tested. Fecal water polyphenol levels are presumably more significant markers of possible bioactivity because fecal water interacts with the intestinal mucosa more frequently than stool solids [33, 34]. The polyphenols in green tea have demonstrated positive preclinical outcomes in inflammatory bowel disease (IBD) animal models. Another important factor in the pathophysiology of IBD is oxidative stress [35]. Green tea polyphenols were also tested in conjunction with two other strong antioxidants, S-adenosylmethionine and 2(R,S)-n-propylthiazolidine-4(R) carboxylic acid, in the dextran sodium sulphate mouse model of non-T-cell mediated colitis. All three antioxidants improved the colon length, body weight, hematocrit, diarrhea, and colon histology when compared with untreated mice in the same model [36]. A common viral agent, *Helicobacter pylori*, can seriously harm the digestive system, leading to various diseases. Mucosal damage is largely attributed to the considerable oxidative stress caused by an *H. pylori* infection [37]. Polyphenols also help manage non-inflammatory diarrheal conditions. An apple polyphenol extract had inhibition activity on cholera toxin-induced diarrhea in a dose-dependent way [38]. The bioactive molecules in numerous other substances that are used as antidiarrheals in conventional medicine have been identified as polyphenol components [39].

3.5 Health Beneficial Effects

3.5.1 Antioxidant Effects of Dietary Polyphenols

Dietary polyphenols have a major impact on our health. A high intake of vegetables, fruits, wine, and whole grain products high in polyphenols are beneficial for keeping several diseases in check. Instead of vitamin C, phytochemicals, particularly polyphenols, make up most of the antioxidant properties of fruits [40]. By preventing production or by neutralizing the free radicals, polyphenols help to decrease the oxidation rate. They can frequently function as effective radical scavengers. These chain-breakers can neutralize the radicals by interrupting the continuous reactions [41, 42]. It is well known that foodstuff and beverages high in polyphenols may boost the plasma's antioxidant capacity. The effects of limiting polyphenolics in plasma, along with their metabolic products and their consequence on the concentration levels of other reductants, may all be contributing factors to the increase in plasma's antioxidative capacity, which can further decrease lymphocyte DNA oxidative damage. Similar findings have been achieved with foods and beverages high in polyphenols. The *Salvia africana lutea* extract is considered one notable terpene source such as 3-beta-acetoxy-7-alpha-methoxyrosmanol and triterpenes including beta-amyrin, ursolic acids, and oleanolic acids. The methanolic extract demonstrated certain biological effects [43]. According to Dienaite et al. [44], results on extracts obtained from leaves and rootstock of *Paeonia officinalis* implicated that they may be used as antidiabetic drugs. Another plant with polyphenols that combines antioxidant and antidiabetic effects is the raspberry, as studied by Wu et al. In order to propose potential mechanisms explaining how the polyphenols from raspberries can inhibit digestive enzymes, the authors also employed docking analysis [45]. Metal chelators are another well-known property of polyphenols. Chelation of metallic elements, such as Fe^{2+}, can halt the oxidation caused by highly reactive hydroxyl radicals by directly slowing the Fenton reaction [46]. To study the total antioxidant function, numerous antioxidant models have been designed. Collectively, these techniques may effectively depict how polyphenols operate as antioxidants and, in doing so, can provide an idea on the actual purpose of food polyphenolics in our health and nutrition, despite their limited relation to the principles governing antioxidant effects in an organism [47, 48].

3.5.2 Glucose Lowering Effect of Dietary Polyphenols

Numerous research studies have discussed the antidiabetic properties of polyphenols. In this regard, the glucose lowering potential of tea catechins has been studied and published in 2001 and 2005 by Rizvi et al. These studies discussed how these tea polyphenolic compounds are present, as well as their significant biological properties, including antidiabetic effects [49, 50]. The border of the small intestine is brushed by various enzymes that can contribute to the hypoglycemic impact of dietary polyphenolic substances. By modifying intracellular signaling, polyphenols may aid in lessening the release of liver glucose and further enhance glucose absorption in the peripheral organs. Antioxidant-active polyphenols can prevent the production of advanced glycation end products [51, 52]. Dietary polyphenols from fruits, berries, and vegetables, such as strawberries, blueberries, raspberries, blackcurrants, pumpkin, beans, legumes, maize, black rice, eggplant, green tea, and black tea, have been demonstrated to inhibit several digestive enzymes [53–58]. Although there is significant in vivo and in vitro evidence, there are few clinical studies that can show how polyphenols are an effective treatment for type-2 diabetes (T2D). A clinical study using double blinded procedures has demonstrated that everyday administration of 1,500 mg curcumin will reduce fasting blood sugar levels and body weight while easing diabetes symptoms in T2D patients [59, 60]. In a study published in 2012, where the activity of beta cells over a nine-month period of time were enhanced, curcumin slowed the progression of prediabetes to T2D. In approximately 118 T2D patients, supplemental curcumin with piperine in a 100:1 ratio administered for three months reduced inflammation, elevated levels of adiponectin, and reduced levels of leptin [61]. HbA1c was reduced in a randomized controlled experiment with 60 T2D patients who received green tea powder each day for eight weeks; however, no alterations in body pressure, weight, or body lipid levels were observed [62]. There is more evidence that shows certain dietary polyphenols may affect blood glucose levels and may assist, manage, and avoid complications from diabetes. More clinical trials are required to learn the effects of foods high in polyphenols, their effective dosage, and processes behind their actions in treating diabetes.

3.5.3 Anti-obesity Effect of Dietary Polyphenols

Over the last three decades, the incidence of obesity has progressively increased throughout the world. Numerous therapies have been created with the aim of reducing the difficulties associated with obesity. Functional food consumption and bioactive ingredients are now regarded as innovative techniques for the inhibition and nursing of this illness. Polyphenols may be regarded as nutraceutical and nutritional supplements suggested for various syndromes because of their biological features. Different study findings suggest that they provide anti-obesity effects via a variety of methods. Some of the mechanisms by which food polyphenolics confer anti-obesity properties, either singly or in particular combinations, include the inhibitory activity of enzymes, excitability of calorie consumption, suppression of hunger, and regulation of glucose metabolism [63–65]. One important method by which these substances act is thought to be enzyme inhibition. Amylase, glucosidase, and lipase, three important digestive enzymes, are known to be inhibited by polyphenols. As a result, the digestion of carbs and lipids is slowed, which reduces calorie intake. In the small intestine, glucosidase hydrolyzes oligosaccharides to create glucose, which is then transported into the blood by a glucose transporter [66]. The team isolated polyphenolic extract from the peels and fruit of *Vitis rotundifloia*. They found interesting results showing how methanolic extract inhibited pancreatic lipase. It was also discovered that ellagitannin-rich and proanthocyanidin-rich polyphenolic extracts from several berries significantly changed the in vitro steapsin activity. Cloudberry polyphenols can also exhibit inhibitory action against pancreatic lipase or steapsin [67, 68]. Walnut polyphenolic extract has inhibited porcine pancreatic lipase. Similarly, a flavonol glycoside, galangin, showed steapsin inhibition. Galangin was given orally to female rats for six weeks. A dosage of 50 mg/kg was used, and this caused a roughly 40% decline in body mass compared with rats in the control category. Rahim and team talked about how the steapsin inhibitory action of gallic acid, epigallocatechin, and epigallocatechin gallate is highly potent [69–71]. Plant phenolics can decrease hunger by inhibiting appetite sensors, moderating MCH receptors, or delaying the release of hormones that stimulate hunger. Flavonoid-rich extracts from a variety of plants, including Indian cactus, bitter orange, and tea (*Camellia sinensis*), have been proven to have appetite-suppressing properties [72–75]. Effective thermogenic stabilizers and amplifiers of energy expenditure can also be produced by polyphenolic substances. In the future, it is anticipated that more thorough research, especially in human trials, is needed to develop evidence-based and more efficient methods for using natural phenolics as functional supplements or foods to inhibit and treat human obesity.

3.5.4 Antimicrobial Effect of Dietary Polyphenols

Fungus, gram(+) bacteria, and gram(−) bacteria are all susceptible to the antibacterial effects of polyphenols. Of the more than 10,000 chemicals that can considered, numerous classifications have been published. Despite the fact that *Staphylococcus aureus* has been the subject of most of the studies on polyphenolic antibacterial activity, very few mention clinically isolated samples. Depending on the study, microdilution or disc diffusion testing is done to determine their minimum fungal concentration (MFC), minimum inhibitory concentration (MIC), and minimum bactericidal concentration (MBC). Miyasaki and team analyzed three extracts (*Rosa rugosa, Scutellaria baicalensis, Terminalia chebula*) using MS+UV study statistics as their foundation because they demonstrated interesting results. Norwogonin with 128 mg/mL MIC and 256 mg/mL MBC was shown to be the more effective than clinically significant strains of *Acinetobacter baumannii*. The three extracts contained a combination of different acids. In this investigation, *Scutellaria baicalensis* extracts in dimethyl sulphoxide were used to purify norwogonin. Heated water was used to purify ellagic acid from rugosa and the same warm water was used to purify *Terminalia chebula* extracts. These treatments were tested against certain *Acinetobacter* isolates, which were significantly the first account of norwogonin's antibacterial action against *A. baumannii* [76]. Theaflavin's antibiotic function against *Acinetobacter* has been demonstrated by other authors and it was boosted when coupled with epicatechin at a ratio of 2:1. The study implies theaflavin's putative antibiotic functions, which may prove useful for treating resistant microbes clinically. Additionally, this study details the theaflavin–epicatechin combination to combat significant healthcare-associated pathogens. In another study, ethanol extract showed antibacterial activity against *A. baumannii* [77]. Blood and sputum samples were used to isolate the strains for this investigation. Similar to the first study, another study was unable to show any synergy between the polyphenolic extracts and antibiotics against *A. baumannii*. Gallic acid equivalents/mL were used to express the overall amount of polyphenols in the extract. It is also noteworthy that the extract of *Amaranthus retroflexus* inflorescences had the greatest polyphenol concentration. This extract was also the most effective against the studied microbial strains, indicating a connection between the antibacterial activity and polyphenol concentration [78]. Betts and team examined 15 *Pseudomonas aeruginosa* multidrug resistant strains. The above-mentioned bacteria is vulnerable towards antibiotics and polyphenols when used together, proving their synergistic effects both in vitro and in vivo [79]. *Polygonum cuspidatum* crude ethanol extracts, as well as ethanolic extracts made from the floral and leafy parts of *Amaranthusare*, are both effective towards *P. aeruginosa* [77, 78]. Khan's investigation discussed how *Acacia nilotica* had the strongest antibacterial activity. Literature suggests that tannins, flavonoids, and other phenolic chemicals are primarily accountable for antibiotic functions [80]. In a different analysis, among eight methanolic floral isolations that contain polyphenols like tannins and flavonoids (*Sophora secundiflora, Periploca laevigata, Sphaeralcea ambigua, Optunia ficus-indica, Marrubium vulgare, Sesbania drummondii*, and *Guitierrezia microcephala*), *P. laevigata* was the extract with the greatest efficacy against the tested enterobacteria. However, a *Nothoscordum bivalve* bulb exhibited no action against any of the examined microorganisms [81]. Further research is required to demonstrate the importance underlying polyphenol consumption in the prevention, treatment, and control of certain illnesses and infectious diseases, even though in vitro investigations appear encouraging. The emergence of these diseases might be prevented by using disinfection solutions based on polyphenols.

3.5.5 Anti-inflammatory Effect of Dietary Polyphenols

Chronic inflammation has been identified as significant to numerous human disorders such as arthritis, T2D, neurological disorders, and cardiovascular conditions. The significance of polyphenols as therapeutic agents in a variety of acute and chronic illnesses is highlighted by their anti-inflammatory action [82, 83]. Numerous studies have looked at the immunological and anti-inflammatory function of food polyphenolics. The regulation of inflammatory signaling is aided by various factors including expression of certain genes and ROS scavenging. By scavenging free radicals, polyphenols can exert anti-inflammatory effects. Inhibition of pro-inflammatory enzymes, such as COX-2, LOX, and iNOS, as well as NF-κB and the activating protein-1 (AP-1), activation of the phase-II antioxidant detoxifying enzymes, and the process of activating protein kinase-C and nuclear factor erythroid-2 related factor, are significant factors in this aspect [84]. Curcumin downregulates TNF and IL-1, and also blocks multiple inflammatory biological fluids, including the enzymes IKK and MAPK as well as the cyclooxygenases COX and LOX. Curcumin resists the activity of STAT3, TLR-2, and TLR-4 [85]. In another study, the anti-inflammatory qualities of resveratrol were primarily responsible for its cardioprotective effects. Resveratrol inhibits COX, inactivates PPAR-gamma, and stimulates eNOS in mouse and rat macrophages, according to in vivo and in vitro research [86]. Numerous scientific research studies on plant chemicals and extracts demonstrate that polyphenols can help prevent and slow the onset of several chronic diseases, as well as other ailments.

3.6 Polyphenols in Disease Management

As demonstrated in Figure 3.1, polyphenols can aid in disease control.

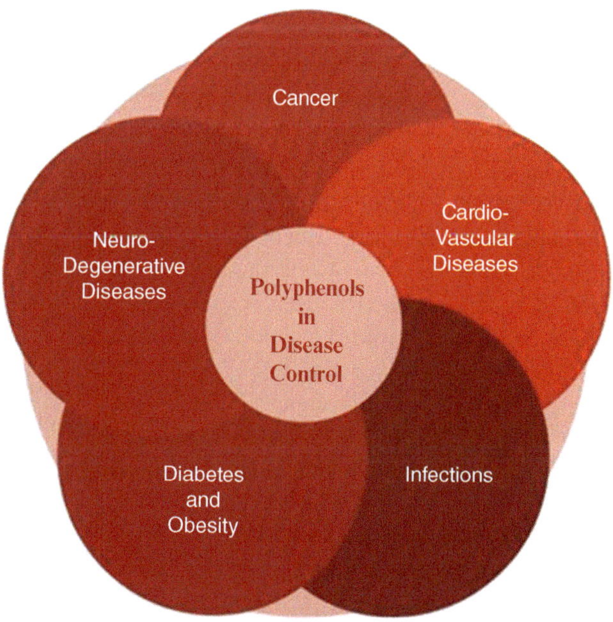

Figure 3.1 Polyphenols in disease management.

3.6.1 Polyphenols and Cancer

A set of ailments connected to a change in regulating cell growth and metabolism leads to cancer [87]. Any substance that can block the differentiation of malignant cells, can be employed as a chemopreventive agent. There are numerous cancers, including in the breasts, lung, colon, and prostatitis, which are also responsible for more than half of all cancer cases that are only now being discovered. Consuming a lot of green foods daily will help stop the growth and spread of cancer. In recent decades, the ingestion of fruits, berries, and vegetables have been reported to resist the advancement of various cancer types, including prostate and colorectal cancer [87]. These epidemiological relationships have also been verified by data collected from different investigations. However, there is a discrepancy between some studies that claimed there was no decrease in stomach, pancreas, or bladder cancer incidence with similar diets. On the contrary, experimental evidence is available that discusses polyphenol-rich vegetables and other substances that are beneficial in preventing the onset of colon cancer [87]. Activated carcinogens, reactive metabolites, nutrients, and mutagens are a few examples of how they may interact. Additionally, it has the power to alter numerous gene expressions linked to cancer. In human intervention trials and on human cell lines, green tea flavanols have demonstrated significant benefits and anticancer characteristics. Additionally, green tea consumption significantly lowers the risk of developing colon, breast, bladder, and bile duct cancers [87]. Flavanol epigallocatechin gallate is thought to be a primary mediator of several anticarcinogenic functions correlated to the consumption of green tea. High amounts of olive oil also contain phenolic alcohols, lignans, and secoiridoids, in addition to flavonoids, and are also thought to have anticancer properties. Various biological factors playing important roles in colorectal carcinogenesis can counteract these effects [87]. The removal of cell cycle progression, cancer cell signaling carcinogenic agents, stimulation, apoptosis, and modification of enzyme activity are just a few of the many ways that these polyphenols can prevent cancer. For instance, enhancing the activity of the P450 enzyme, catalase, quinone oxidoreductase, glutathione peroxidase, and/or NADPH can aid in the detoxification of carcinogens. Given its key role in cell growth regulation and its critical role in the MAPK signaling pathway, it is a desirable avenue for anticancer treatments [87].

3.6.2 Polyphenols and Neurodegenerative Diseases

Our aging society is being increasingly burdened by neurodegenerative illnesses [87]. Without accounting for other forms of dementia brought on by ischemia injury, Alzheimer's disease affects 15% of people over 65, while Parkinson's disease affects 1% of the population. Antioxidants may help avoid such disorders because they are reliant on oxidative stress, which particularly harms brain tissues. Aging rats' cognitive abilities and neural signal transduction were enhanced when their diets included aqueous extracts of spinach, strawberries, or blueberries that are high in polyphenols. Anthocyanin-rich blueberries were particularly effective. Because a brain's lack of vitamins E and C do not account for these effects, the potential role of polyphenols as antioxidants is postulated [87]. Epicatechin or catechin intravenously injected into mice reduced the memory impairment introduced by cerebral ischemia. Additionally, polyphenols defend against several neurotoxic medications whose toxicity is connected to an increase in oxidative stress in laboratory animals. Grape polyphenols were added to the diet to increase synaptic function as determined on isolated synaptosomes and to lessen the neurodegenerative effects of chronic ethanol consumption. In rats injected with MPTP, a medication designed to simulate a Parkinson's illness, the oral treatment of EGCG restored dopaminergic neurotransmission and stopped the increase in SOD and catalase caused by this agent. In another study, curcumin showed comparable protective properties [87]. When challenged by oxidized LDL, cultured neuronal cells respond better to catechins in vitro. Epigallocatechin gallate appears pro-oxidant and harmful at higher levels (30 μM). Therefore, preventing neurodegenerative disorders would be more successful in environments with low polyphenol content. With naringin or quercetin, subsequent research with polyphenols verified their poor blood–brain barrier permeability. Epicatechin's glucuronide conjugate failed to shield cortical neurons from the oxidative stress introduced by H_2O_2. However, it is possible that they are merely aglycones, as was demonstrated for genistein [87]. According to three studies conducted in France, Denmark, and Canada, moderate wine drinking was connected negatively with the risk of dementia. In the Danish cohort, no such connection was observed for beer or alcoholic beverages. This suggests that polyphenols may help to prevent dementia in some way. In an Italian cohort, cognitive impairment was less likely in modest drinkers in comparison with those that abstained; however, it was more likely in heavy drinkers. Additionally, a French cohort found a negative correlation between the consumption of flavones and Alzheimer's disease [87].

An example of protein-small molecule (ligand) docking is discussed in Figure 3.2, Figure 3.3, Tables 3.1, and 3.2.

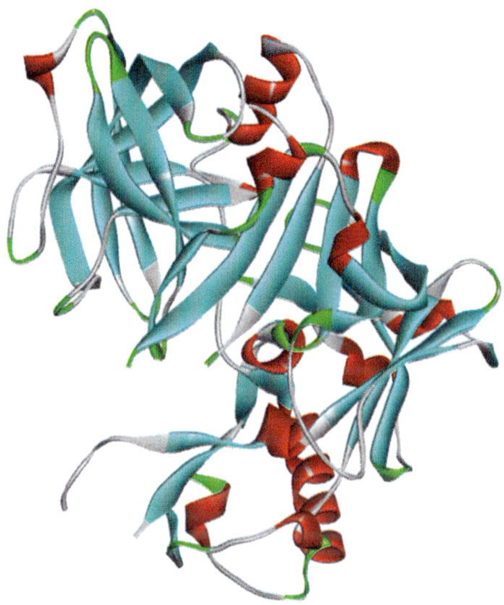

Figure 3.2 3D structure: BACE 1. Adapted from Johansson, P., 2018.

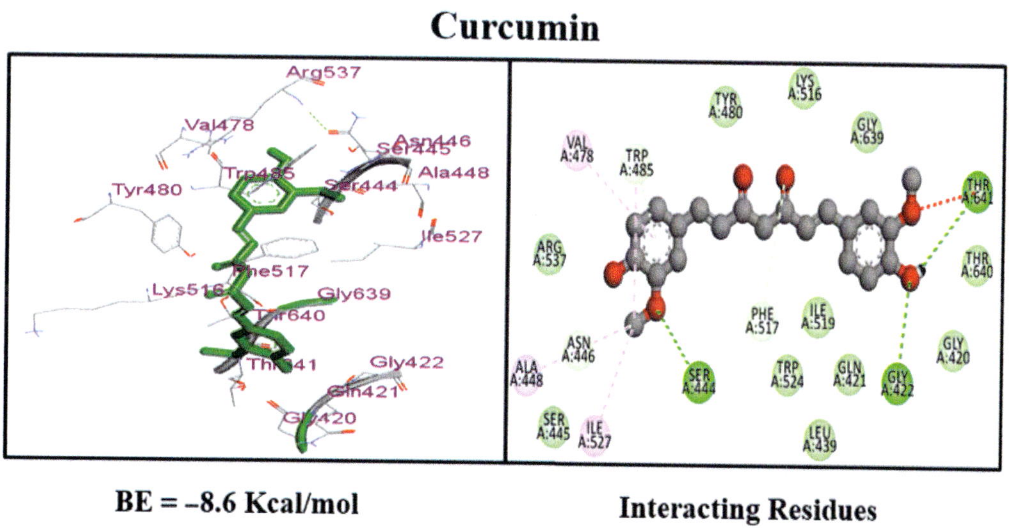

Figure 3.3 2D interaction of curcumin with BACE 1.

Table 3.1 Molecular docking analysis of an anticancer compound against beta-site amyloid precursor protein cleaving enzyme 1 (BACE 1).

S. No.	Compound	2d Structure	Binding Energy (Kcal/Mol)	Molecular Interactions
1.	Curcumin	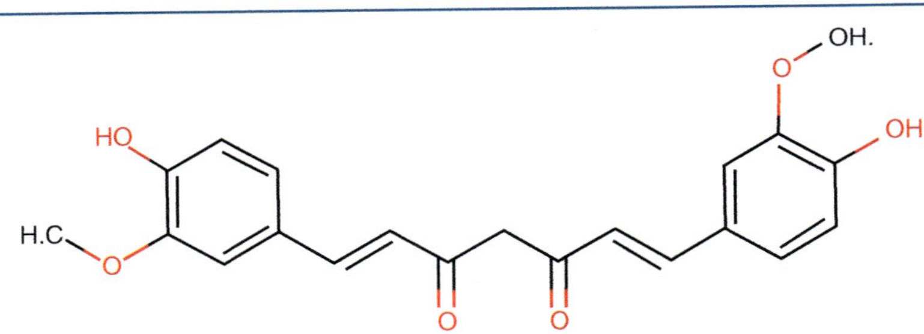	−8.6	Conventional hydrogen bond SER444, THR641 Van der Waals GLY420, GLN421, LEU439, SER445, TYR480, ILE519, TRP524, ILE527, ARG537, GLY639, THR640

Table 3.2 Absorption, distribution, metabolism, and excretion (ADME) properties of a flavonoid compound (curcumin).

S. No.	Compound	Molecular Formula	Adme Properties (Lipinski's Rule Of Five)		Structure	DRUG LIKELINESS
			Properties	Values		
1.	Curcumin	$C_{21}H_{20}O_6$	Molecular weight (≤ 500 Da)	368.4	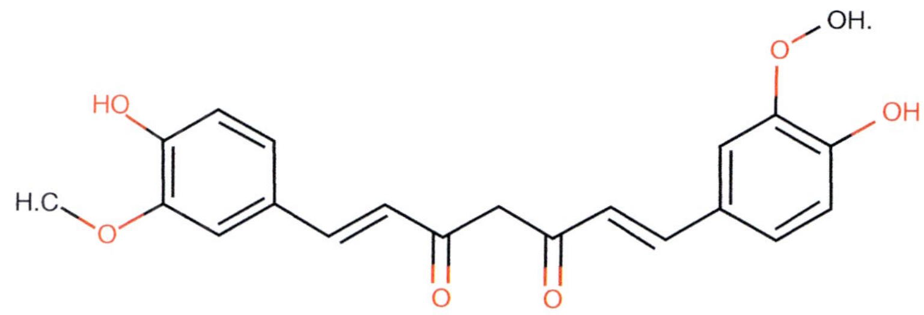 	Yes
			Log P (≤ 5)	3.2		
			H-bond donor (≤ 5)	2		
			H-Bond acceptor (≤ 10)	6		
			Violations	0		

3.6.3 Polyphenols and Cardiovascular Diseases

Strokes and coronary heart disease are two major contributors to mortality in modern nations. Environmental and genetic variables influence how cardiovascular disease starts, spreads, and develops. Physical activity, smoking, consumption of saturated fats, and other factors can lead to heart diseases. Finding the lone cause of these diseases among this complicated mix of factors is challenging. Consuming foods high in polyphenols, such as tea, green vegetables, fruits, cocoa, and berries, increases the chances of heart safety. While some flavonoids, like those in soy and cocoa, have beneficial effects on cardiovascular disease, others are less effective. A diet high in flavanols, which are found in cocoa, decreases the body's pressure level, and lowers the risk of heart problems. According to reports, drinking black tea more frequently can aid this cause. Tea, cocoa, and purple grapes all have a positive impact on heart health. The potential of polyphenols to modify an enzyme's activity is one mode of action for their beneficial effects on cardiac health. According to investigations, polyphenol consumption has been linked to endothelium-dependent relaxation. Inhibition of platelet aggregation and activation, whether chronic or acute, has been linked to coffee, grape juice, cocoa, and black tea, which may reduce age-related damage.

Case Study 1

SARS-CoV-2, which caused the global COVID pandemic in 2019, had resulted in unprecedented levels of morbidity and mortality [88, 89]. Four structural proteins, as well as the significant non-structural and adjacent proteins, make up the 29 proteins forming SARS-CoV-2. An encoded protease acts on a polyprotein to produce 16 of these proteins [90].

Molecular Docking: In the current study, SARS-Cov-2 proteins, spike glycoproteins (PDB ID: 6VYB), nucleocapsid phosphoprotein (PDB ID: 6VYO), membrane glycoprotein (PDB ID: 6M17), nsp10 (PDB ID: 6W4H), and RNA-dependent RNA polymerase (PDB ID: 6M71) structures were used to assess the binding affinities of 14 drug candidates. Despite multiple studies suggesting that food supplements and nutraceuticals may prove to be effective long-term strategies for preventing growing infectious illnesses, turmeric has been widely used and has extensive medical uses.

Using AutoDock 4.2, thorough docking investigations were conducted on 14 medicinal compounds having antiviral characteristics that are being investigated for clinical trials alongside curcumin. Ivermectin demonstrated the highest degree of affinity for the examined and desired proteins, with remarkable binding capacity.

Conclusion and Summary

1) In conclusion, we suggest curcumin as a therapy for the creation of COVID medications.
2) Curcumin was found to display a considerable reaction on the nucleocapsid and nsp10 of the coronavirus proteins.
3) Studies showed that the success of some detailed biological studies might be improved by strong binding with residues. As a result of our research, curcumin is now being considered as part of a medication combination therapy for SARS-CoV-2. Knowing how curcumin works in this study is fascinating and may be significant to future studies.
4) The current results are encouraging and supported by the possibility that curcumin is a naturally occurring and harmless substance with bioactivity.

Case Study 2

In a study analyzing dietary polyphenols, a high dietary consumption of polyphenols was linked to longevity. Professor Cristina Andrés Lacueva, Director of the Urinary Biomarker's Biomarkers and

Nutritional & Food Metabolomics Research Group, explains that the use of nutritional biomarkers has made it possible to estimate intake with greater accuracy and objectivity because it is not dependent solely on participant memory when filling out questionnaires. Bioavailability and individual variances are considered with nutritional biomarkers. The methodology provides a more trustworthy and precise assessment in this aspect. The study's findings demonstrate that participants with high polyphenol diets (> 650 mg/day) had a 30% lower overall mortality rate than those with low polyphenol intakes (500 mg/day). The study's principal author, Ral Zamora Ros, emphasizes that the findings support existing research that suggests how the ingestion of fruits, berries, and vegetables can resist the advancement of various diseases, such as cancer and other chronic illnesses, and can reduce overall mortality. Additionally, the study emphasizes the significance of assessing food consumption using nutritional biomarkers, as opposed to just meal frequency questionnaires, whenever possible [91].

3.7 Conclusion and Future Perspectives

Recent decades have seen an increase in interest in nutrition related to polyphenols, which are chemical compounds that are widely present in plants. Numerous studies imply that polyphenols can control metabolism, weight, appetite, several types of chronic disease, cell proliferation, and cell differentiation. There are already more than 8,000 polyphenols known. Numerous polyphenols contain antioxidant and anti-inflammatory characteristics, according to studies on animals, people, and epidemiology, which may help prevent or treat diseases like obesity, cancer, diabetes, neurodegenerative diseases, and cardiovascular disease. Overconsumption, however, has raised concerns, particularly when compounds are ingested alone rather than as part of a dietary matrix. It is difficult to fully understand the health impacts of polyphenols because there are so many of them, each with unique structures, metabolic pathways, and physiological functions. Consumer knowledge of the possible advantages and hazards will increase along with the scientific understanding of polyphenols and their marketing initiatives. Regulatory bodies should consider keeping up with scientific research to provide advice for polyphenol supplementation and intake. Suggestions for the consumption of fruits and vegetables ought to be integrated with current initiatives in nutrition education and advice for promoting a balanced diet. The present state of our knowledge of the mechanisms of disease treatment, dose needs, and potential adverse effects creates constraints on our capacity to take advantage of the knowledge. Additional human research is required to confirm the molecular mechanisms and public health consequences of polyphenols, as well as potential adverse outcomes for specific subgroups. The concentration at which polyphenols can be consumed safely and advantageously is unknown because studies in vitro and in vivo have employed amounts that are significantly greater than those frequently found in human diets. More research is required to determine whether, and in what manner, polyphenols can be ingested.

References

1 Belščak-Cvitanović, A., Durgo, K., Huđek, A. et al. (2018). Overview of polyphenols and their properties. In: *Polyphenols: Properties, Recovery, and Applications.*

2 Abbas, M., Saeed, F., Anjum, F.M. et al. (2017). Natural polyphenols: an overview. *International Journal of Food Properties* Taylor & Francis. 20: 1689–1699.

3 Tomás-Barberán, F.A. and Andrés-Lacueva, C. (2012). Polyphenols and health: current state and progress. *Journal of Agricultural and Food Chemistry* 60: 8773–8775.

4 Rasouli, H., Farzaei, M.H., and Khodarahmi, R. (2017). Polyphenols and their benefits: a review. *International Journal of Food Properties* 20: 1700–1741.

5 Tsao, R. (2010). Chemistry and biochemistry of dietary polyphenols. *Nutrition* 2: 1231–1246.

6 Manach, C., Scalbert, A., Morand, C. et al. (2004). Polyphenols: food sources and bioavailability. *The American Journal of Clinical Nutrition* 79: 727–747.

7 Williamson, G. (2017). The role of polyphenols in modern nutrition. *Nutrition Bulletin* 42: 226–235.

8 El Gharras, H. (2009). Polyphenols: food sources, properties and applications - A review. *The Journal of Food Science and Technology* (44): 2512–2518.

9 Pastrana-Bonilla, E., Akoh, C.C., Sellappan, S., and Krewer, G. (2003). Phenolic content and antioxidant capacity of muscadine grapes. *Journal of Agricultural and Food Chemistry* American Chemical Socety. 51: 5497–5503.

10 Eyduran, S.P., Akin, M., Ercisli, S. et al. (2015). Sugars, organic acids, and phenolic compounds of ancient grape cultivars (Vitis Vinifera L.) from Igdir province of Eastern Turkey. *Biological Research Society of Biology of Chile* 48: 1–8.

11 Meng, J., Fang, Y., Zhang, A. et al. (2011). Phenolic content and antioxidant capacity of Chinese raisins produced in Xinjiang Province. *Food Research International Elsevier* 44: 2830–2836.

12 Bergqvist, J., Dokoozlian, N., and Ebisuda, N. (2001). Sunlight exposure and temperature effects on berry growth and composition of cabernet sauvignon and grenache in the central San Joaquin valley of California. *American Journal of Enology and Viticulture* 52: 1–7.

13 Josep, M.U., Sort, X., Zayas, A., and Rosa, M.P. (2010). *Effects of soil and climatic conditions on grape ripening and wine quality of Cabernet Sauvignon.* Routledge 21: 1–17.

14 Bruno, G. and Sparapano, L. (2007). Effects of three esca-associated fungi on vitis vinifera L.: v. Changes in the chemical and biological profile of xylem sap from diseased cv. Sangiovese vines. *Physiological and Molecular Plant Pathology* Academic Press. 71: 210–229.

15 Gonçalves, B., Landbo, A.K., Knudsen, D. et al. (2004). Effect of ripeness and postharvest storage on the phenolic profiles of cherries (Prunus avium L.). *Journal of Agricultural and Food Chemistry* American Chemical Society. 52: 523–530.

16 Stöhr, H., Mosel, H.D., and Herrmann, K. (1975). The phenolics of fruits. VII. The phenolics of cherries and plums and the changes in catechins and hydroxycinnamic acid derivatives during the development of fruits. *Z. Lebensm. Unters. Forsch* 159: 85–91.

17 Martínez-Esplá, A., Zapata, P.J., Valero, D. et al. (2014). Preharvest application of oxalic acid increased fruit size, bioactive compounds, and antioxidant capacity in sweet cherry cultivars (prunus avium L.). *Journal of Agricultural and Food Chemistry*American Chemical Society. 62: 3432–3437.

18 Bourgaud, F., Gravot, A., Milesi, S., and Gontier, E. (2001). Production of plant secondary metabolites: a historical perspective. *Plant Science* Elsevier. 161: 839–851.

19 Lampe, J.W. (1999). Health effects of vegetables and fruit: assessing mechanisms of action in human experimental studies. *The American Journal of Clinical Nutrition* Oxford Academic. 70: 475s–490s.

20 Tarko, T., Duda-Chodak, A., and Zając, N. (2013). Digestion and absorption of phenolic compounds assessed by in vitro simulation methods. A review. *Rocz. PanstwowegoZakladuHig* 64 (2): 79–84.

21 CirkovicVelickovic, T.D. and Stanic-Vucinic, D.J. (2018). The role of dietary phenolic compounds in protein digestion and processing technologies to improve their antinutritive properties. *Comprehensive Reviews in Food Science and Food Safety* John Wiley & Sons Ltd. 17: 82–103

22 Zhang, X., Song, J., Shi, X. et al. (2013). Absorption and metabolism characteristics of rutin in caco-2 cells. *Science World Journal* 2013.

23 Zhao, Z. and Moghadasian, M.H. (2010). Bioavailability of hydroxycinnamates: a brief review of in vivo and in vitro studies. *Phytochemistry Reviews* Springer. 9: 133–145.

24 Lewandowska, U., Szewczyk, K., Hrabec, E. et al. (2013). Overview of metabolism and bioavailability enhancement of polyphenols. *Journal of Agricultural and Food Chemistry* American Chemical Society. 61: 12183–12199.

25 Swallah, M.S., Fu, H., Sun, H. et al. (2020). The impact of polyphenol on general nutrient metabolism in the monogastric gastrointestinal tract. *Journal of Food Quality* Hindawi Limited. 2020.

26 Domínguez-Avila, J.A., Wall-Medrano, A., Velderrain-Rodríguez, G.R. et al. (2017). Gastrointestinal interactions, absorption, splanchnic metabolism and pharmacokinetics of orally ingested phenolic compounds. *Food and Function* Royal Society of Chemistry. 8: 15–38.

27 Manach, C., Williamson, G., Morand, C. et al. (2005). Bioavailability and bioefficacy of polyphenols in humans. I. Review of 97 bioavailability studies. *The American Journal of Clinical Nutrition* Oxford Academic. 81: 230S–242S.

28 Skrede, G., Wrolstad, R.E., and Durst, R.W. (2000). Changes in anthocyanins and polyphenolics during juice processing of highbush blueberries (Vaccinium corymbosum L.). *The Journal of Food Science and Technology* John Wiley & Sons, Ltd. 65: 357–364.

29 Zielinski, H., Kozlowska, H., and Lewczuk, B. (2001). Bioactive compounds in the cereal grains before and after hydrothermal processing. *Innovative Food Science and Emerging Technologies* Elsevier. 2: 159–169.

30 Curtis, P.J., Dhatariya, K., Sampson, M. et al. (2012). Chronic ingestion of Flavan-3-ols and isoflavones improves insulin sensitivity and lipoprotein status and attenuates estimated 10-year CVD risk in medicated postmenopausal women with type 2 diabetesA 1-year, double-blind, randomized, controlled trial. *Diabetes Care* American Diabetes Association. 35: 226–232.

31 Muthumani, T. and Kumar, R.S.S. (2007). Influence of fermentation time on the development of compounds responsible for quality in black tea. *Food Chemistry* Elsevier. 101: 98–102.

32 Rothwell, J.A., Medina-Remón, A., Pérez-Jiménez, J. et al. (2015). Effects of food processing on polyphenol contents: a systematic analysis using phenol-explorer data. *Molecular Nutrition and Food Research* John Wiley & Sons, Ltd. 59: 160–170.

33 Halliwell, B., Rafter, J., and Jenner, A. (2005). Health promotion by flavonoids, tocopherols, tocotrienols, and other phenols: direct or indirect effects? Antioxidant or not? *The American Journal of Clinical Nutrition* Oxford Academic. 81: 268S–276S.

34 Gee, J.M. and Johnson, I.T. (2012). Polyphenolic compounds: interactions with the gut and implications for human health. *Current Medicinal Chemistry* Bentham Science Publishers. 8: 1245–1255.

35 Dryden, G.W., Deaciuc, I., Arteel, G., and McClain, C.J. (2005). Clinical implications of oxidative stress and antioxidant therapy. *Current Gastroenterology Reports* Springer. 7: 308–316.

36 Oz, H.S., Chen, T.S., McClain, C.J., and De Villiers, W.J.S. (2005). Antioxidants as novel therapy in a murine model of colitis. *The Journal of Nutritional Biochemistry* Elsevier. 16: 297–304.

37 Lee, K.M., Yeo, M., Choue, J.S. et al. (2004). Protective mechanism of epigallocatechin-3-gallate against helicobacter pylori-induced gastric epithelial cytotoxicity via the blockage of TLR-4 signaling. *Helicobacter* John Wiley & Sons, Ltd. 9: 632–642.

38 Saito, T., Miyake, M., Toba, M. et al. (2002). Inhibition by apple polyphenols of ADP-ribosyltransferase activity of cholera toxin and toxin-induced fluid accumulation in mice. *Microbiology and Immunology* John Wiley & Sons, Ltd. 46: 249–255.

39 Agbor, G.A., Léopold, T., and Jeanne, N.Y. (2004). The antidiarrhoeal activity of Alchornea cordifolia leaf extract. *Phytotherapy Research* John Wiley & Sons, Ltd. 18: 873–876.

40 Wang, H., Cao, G., and Prior, R.L. (1996). Total antioxidant capacity of fruits. *Journal of Agricultural and Food Chemistry* American Chemical Society. 44: 701–705.

41 Pietta, P.G. (2000). Flavonoids as Antioxidants. *Journal of Natural Products* American Chemical Society. 63: 1035–1042.

42 Rice-Evans, C.A., Miller, N.J., and Paganga, G. (1996). Structure-antioxidant activity relationships of flavonoids and phenolic acids. *Free Radical Biology and Medicine* Pergamon. 20: 933–956.

43 Etsassala, N.G.E.R., Badmus, J.A., Waryo, T.T. et al. (2019). Alpha-glucosidase and alpha-amylase inhibitory activities of novel abietane diterpenes from Salvia africana-lutea. *Antioxidants* Multidisciplinary Digital Publishing Institute. 8: 421.

44 Dienaitė, L., Pukalskienė, M., Pukalskas, A. et al. (2019). Isolation of strong antioxidants from paeonia officinalis roots and leaves and evaluation of their bioactivities. *Antioxidants* Multidisciplinary Digital Publishing Institute. 8: 249.

45 Wu, L., Liu, Y., Qin, Y. et al. (2019). HPLC-ESI-qTOF-MS/MS characterization, antioxidant activities and inhibitory ability of digestive enzymes with molecular docking analysis of various parts of raspberry (Rubus ideaus L.). *Antioxidants* Multidisciplinary Digital Publishing Institute. 8: 274.

46 Perron, N.R. and Brumaghim, J.L. (2009). A review of the antioxidant mechanisms of polyphenol compounds related to iron binding. *Cell Biochemistry and Biophysics* Springer. 53: 75–100.

47 Prior, R.L., Wu, X., and Schaich, K. (2005). Standardized methods for the determination of antioxidant capacity and phenolics in foods and dietary supplements. *Journal of Agricultural and Food Chemistry* American Chemical Society. 53: 4290–4302.

48 Halliwell, B. (2008). Are polyphenols antioxidants or pro-oxidants? What do we learn from cell culture and in vivo studies? *Archives of Biochemistry and Biophysics* Academic Press. 476: 107–112.

49 Ibrahim Rizvi, S., Abu Zaid, M., Anis, R., and Mishra, N. (2005). Protective role of tea catechins against oxidation-induced damage of type 2 diabetic erythrocytes. *Clinical and Experimental Pharmacology and Physiology* 32: 70–75.

50 Rizvi, S.I. and Zaid, M.A. (2001). Insulin-like effect of (−)epicatechin on erythrocyte membrane acetylcholinesterase activity in type 2 diabetes mellitus. *Clinical and Experimental Pharmacology and Physiology* John Wiley & Sons, Ltd. 28: 776–778.

51 Xiao, J.B. and Hogger, P. (2014). *Dietary Polyphenols and Type 2 Diabetes: Current Insights and Future Perspectives*. Bentham Science Publishers.

52 Hanhineva, K., Törrönen, R., Bondia-Pons, I. et al. (2010). Impact of dietary polyphenols on carbohydrate metabolism. *International Journal of Molecular Sciences*Molecular Diversity Preservation International. 11: 1365–1402.

53 Kwon, Y.I., Apostolidis, E., Kim, Y.C., and Shetty, K. (2007). Health benefits of traditional corn, beans, and pumpkin: in vitro studies for hyperglycemia and hypertension management. *Journal of Medicinal Food* 10: 266–275.

54 McDougall, G.J., Shpiro, F., Dobson, P. et al. (2005). Different polyphenolic components of soft fruits inhibit α-amylase and α-glucosidase. *Journal of Agricultural and Food Chemistry* American Chemical Society. 53: 2760–2766.

55 Ademiluyi, A.O. and Oboh, G. (2012). Phenolic-rich extracts from selected tropical underutilized legumes inhibit α-amylase, α-glucosidase, and angiotensin I converting enzyme in vitro. *The Journal of Basic and Clinical Physiology and Pharmacology* 23: 17–25.

56 Lee, W.K., Wong, L.L., Loo, Y.Y. et al. (2009). Evaluation of different teas against starch digestibility by mammalian glycosidases. *Journal of Agricultural and Food Chemistry* American Chemical Society. 58: 148–154.

57 Welsch, C.A., Lachance, P.A., and Wasserman, B.P. (1989). Dietary phenolic compounds: inhibition of Na+-dependent D-glucose uptake in rat intestinal brush border membrane vesicles. *The Journal of Nutrition* Oxford Academic. 119: 1698–1704.

58 Johnston, K., Sharp, P., Clifford, M., and Morgan, L. (2005). Dietary polyphenols decrease glucose uptake by human intestinal Caco-2 cells. *FEBS Letters* 579: 1653–1657.

59 Adibian, M., Hodaei, H., Nikpayam, O. et al. (2019). The effects of curcumin supplementation on high-sensitivity C-reactive protein, serum adiponectin, and lipid profile in patients with type 2 diabetes: a randomized, double-blind, placebo-controlled trial. *Phytotherapy Research* John Wiley & Sons, Ltd. 33: 1374–1383.

60 Hodaei, H., Adibian, M., Nikpayam, O. et al. (2019). The effect of curcumin supplementation on anthropometric indices, insulin resistance and oxidative stress in patients with type 2 diabetes: a randomized, double-blind clinical trial. *Diabetology and Metabolic Syndrome* 11: 1–8.

61 Chuengsamarn, S., Rattanamongkolgul, S., Luechapudiporn, R. et al. (2012). Curcumin extract for prevention of type 2 diabetes. *Diabetes Care* 35: 2121–2127.

62 Fukino, Y., Ikeda, A., Maruyama, K. et al. (2007). Randomized controlled trial for an effect of green tea-extract powder supplementation on glucose abnormalities. *European Journal of Clinical Nutrition* 62: 953–960.

63 Rodríguez-Pérez, C., Segura-Carretero, A., and Del Mar Contreras, M. (2017). Phenolic compounds as natural and multifunctional anti-obesity agents: a review. *Critical Reviews in Food Science and Nutrition* 59: 1212–1229.

64 Cory, H., Passarelli, S., Szeto, J. et al. (2018). The role of polyphenols in human health and food systems: a mini-review. *Frontiers in Nutrition*S.A. 5: 87.

65 Liu, J., He, Z., Ma, N., and Chen, Z.Y. (2020). Beneficial effects of dietary polyphenols on high-fat diet-induced obesity linking with modulation of gut microbiota. *Journal of Agricultural and Food Chemistry* 68: 33–47.

66 Nyambe-Silavwe, H., Villa-Rodriguez, J.A., Ifie, I. et al. (2015). Inhibition of human α-amylase by dietary polyphenols. *Journal of Functional Foods* 19: 723–732.

67 McDougall, G.J., Kulkarni, N.N., and Stewart, D. (2009). Berry polyphenols inhibit pancreatic lipase activity in vitro. *Food Chemistry* 115: 193–199.

68 You, Q., Chen, F., Wang, X. et al. (2011). Inhibitory effects of muscadine anthocyanins on α glucosidase and pancreatic lipase activities. *Journal of Agricultural and Food Chemistry* 59: 9506–9511.

69 Kumar, S. and Alagawadi, K.R. (2013). Anti-obesity effects of galangin, a pancreatic lipase inhibitor in cafeteria diet fed female rats. *Pharmaceutical Biology* 51: 607–613.

70 Shi, D., Chen, C., Zhao, S. et al. (2014). Walnut polyphenols inhibit pancreatic lipase activity in vitro and have hypolipidemic effect on high-fat diet-induced obese mice. *Journal of Food and Nutrition Research* 2: 757–763.

71 Rahim, A.T.M.A., Takahashi, Y., and Yamaki, K. (2015). Mode of pancreatic lipase inhibition activity in vitro by some flavonoids and non-flavonoid polyphenols. *International Food Research* 75: 289–294.

72 Klontz, K.C., Timbo, B.B., and Street, D. (2016). Consumption of dietary supplements containing citrus aurantium (Bitter Orange)—2004 californiabehavioral risk factor surveillance survey (BRFSS). *Annals of Pharmacotherapy* 40: 1747–1751.

73 Kuriyan, R., Raj, T., Srinivas, S.K. et al. (2007). Effect of CarallumaFimbriata extract on appetite, food intake and anthropometry in adult Indian men and women. *Appetite* 48: 338–344.

74 Moon, H.S., Lee, H.G., Choi, Y.J. et al. (2007). Proposed mechanisms of (−)-epigallocatechin-3-gallate for anti-obesity. *Chemico-Biological Interactions* 167: 85–98.

75 Geoffroy, P., Ressault, B., Marchioni, E., and Miesch, M. (2011). Synthesis of Hoodigogenin A, aglycone of natural appetite suppressant glycosteroids extracted from Hoodia gordonii. *Steroids* Elsevier. 76: 702–708.

76 Miyasaki, Y., Rabenstein, J.D., Rhea, J. et al. (2013). Isolation and characterization of antimicrobial compounds in plant extracts against multidrug-resistant acinetobacter baumannii. *PLoS One* Public Library of Science. 8: e61594.

77 Su, P.W., Yang, C.H., Yang, J.F. et al. (2015). Antibacterial activities and antibacterial mechanism of polygonum cuspidatum extracts against nosocomial drug-resistant pathogens. *Molecules* Multidisciplinary Digital Publishing Institute. 20: 11119–11130.

78 Marinaş, I.C.R., Chifiriuc, C., Oprea, E., and Lazăr, V. (2014). Antimicrobial and antioxidant activities of alcoholic extracts obtained from vegetative organs of A. retroflexus. *Roumanian Archives of Microbiology and Immunology* 73: 35–42.

79 Betts, J.W., Hornsey, M., Higgins, P.G. et al. (2019). Restoring the activity of the antibiotic aztreonam using the polyphenol epigallocatechin gallate (EGCG) against multidrug-resistant clinical isolates of pseudomonas aeruginosa. *Journal of Medical Microbiology* Microbiology Society. 68: 1552–1559.

80 Khan, R., Islam, B., Akram, M. et al. (2009). Antimicrobial activity of five herbal extracts against multi drug resistant (MDR) strains of bacteria and fungus of clinical origin. *Molecules* Molecular Diversity Preservation International. 14: 586–597.

81 Sánchez, E., Rivas Morales, C., Castillo, S. et al. (2016). Antibacterial and antibiofilm activity of methanolic plant extracts against nosocomial microorganisms. evidence-based complement. *Alternative Medicine* Hindawi Limited. 2016.

82 Elliott Middleton E. Jr., Kandaswami C., and Theoharides T.C. (2000). The effects of plant flavonoids on mammalian cells: implications for inflammation, heart disease, and cancer. *Pharmacological Reviews* 52 (4): 673–751.

83 Visioli, F. and Galli, C. (1998). The effect of minor constituents of olive oil on cardiovascular disease: new findings. *Nutrition Reviews* Oxford Academic. 56: 142–147.

84 Santangelo, C., Varì, R., Scazzocchio, B., and Filesi, C. (2014). Polyphenols, intracellular signalling and inflammation. *Annalidell'IstitutoSuperiore di Sanita* 43 (4): 394–405.

85 Siddiqui, A.M., Cui, X., Wu, R. et al. (2006). The anti-inflammatory effect of curcumin in an experimental model of sepsis is mediated by up-regulation of peroxisome proliferator-activated receptor-γ. *Critical Care Medicine* 34: 1874–1882.

86 Speciale, A., Chirafisi, J., Saija, A., and Cimino, F. (2011). Nutritional antioxidants and adaptive cell responses: an update. *Current Molecular Medicine* Bentham Science Publishers. 11: 770–789.

87 Scalbert, A., Manach, C., Morand, C. et al. (2005). Dietary polyphenols and the prevention of diseases. *Critical Reviews in Food Science and Nutrition* 45: 287–306.

88 Satyam, R., Bhardwaj, T., Goel, S. et al. (2020). miRNAs in SARS-CoV 2: a spoke in the wheel of pathogenesis. *Current Pharmaceutical Design* Bentham Science Publishers. 27: 1628–1641.

89 Choudhury, P.R., Saha, T., Goel, S. et al. (2022). Cross-species virus transmission and its pandemic potential. *Bulletin of the National Research Centre* Springer Berlin Heidelberg. 46.

90 Suravajhala, R., Parashar, A., Choudhir, G. et al. (2021). Molecular docking and dynamics studies of curcumin with COVID-19 proteins. *Network Modeling Analysis in Health Informatics and Bioinformatics* Springer Vienna. 10: 1–10.

91 Zamora-Ros, R., Rabassa, M., Cherubini, A. et al. (2013). High concentrations of a urinary biomarker of polyphenol intake are associated with decreased mortality in older adults. *The Journal of Nutrition* Oxford Academic. 143: 1445–1450.

4

Nanotechnological Approach in Nutraceuticals

Manohar M.V.[1], *Amogha G. Paladhi*[2], *Sugumari Vallinayagam*[3], *and Mithun Rudrapal*[4,*]

[1] *Centre of Excellence in Molecular Biology & Regenerative Medicine, Department of Biochemistry, JSS Medical College, JSS-AHER, Mysuru, Karnataka, India*
[2] *Department of Botany, CHRIST (Deemed to be) University, Bengaluru, Karnataka, India*
[3] *Department of Biotechnology, Vel Tech Rangarajan Dr. Sagunthala R and D Institute of Science and Technology, Chennai, India*
[4] *Department of Pharmaceutical Sciences, School of Biotechnology and Pharmaceutical Sciences, Vignan's Foundation for Science, Technology & Research, Guntur, India*
[*] *Corresponding author*

4.1 Introduction

Nanonutraceuticals are a fabrication process for extending the food quality and shelf life using nanocomposites for the protection of nutrition supplements in food, which acts as encapsulation against the factors causing spoilage. The process of using nanotechnology for the above is termed nanoencapsulation, and it comprises the use of an edible coating material that helps in the detection of pathogenicity, toxification, quality, thermal levels, and oxygen levels, thus helping monitor the packaged food materials to ensure the delivery of quality nutrition in the form of supplementation. The food packaging industry faces numerous issues regarding food storage, processing, and packaging, which includes the maintenance of quality, shelf life, and texture of the food for longer periods of time. Thus, the adaptation of applied nanotechnologies is an effective way to address these issues. A new phase in industry has been implemented with the use of nanotechnology. Industries, such as drug processors and nutraceuticals, have found a greater use of nanotechnology in the form of nanoencapsulation, which introduces a better way of addressing the above issues. Nanotechnology in food industries has a higher rate of implementation over a wide range because of its emulsion causing properties. Better implementation of nanotechnologies is possible with an implemented bioengineering technique to resolve the toxicity and high metal release properties of nanotechnologies. Thus, there is a need for the characterization of already featured nanoformulations such as cantilevers, nanofibers, nanofillers, and other nanocompounds. The multifunctional advantages of nanotechnology have made it possible for industries and consumers to better rely on packed food materials. It is important to have a longer shelf life in packed food materials because it serves the purpose in transport to military and extra-terrestrial space experiments. Various experimental approaches have been made with nanomaterials of different shapes, functions, and formulations with the purpose of encapsulation, biocompatibility, and products for delivery systems. With respect to the environment, nanotechnology for food packaging and processing has advantages as it can make use of inorganic constituents to form foils or films that are more biodegradable than native packing materials. The advantages of nanotechnology in food processing, packaging, and post packaging are further discussed. The use of nanoparticles in the food processing industries is broadly classified into three phases: production and processing (prepackaging), packaging and supplementation, and detection (post-packaging) [1].

Polyphenols: Food, Nutraceutical, and Nanotherapeutic Applications, First Edition. Edited by Mithun Rudrapal.

4.2 Nanotechnology in Food Processing

The use of nanomaterials as ingredients, packaging materials, and for processing packed food imparts better taste, texture, and consistency. Nanotechnology helps maintain a better shelf life in various food materials, which significantly lowers the spoilage and wasting of food materials by microbial activity, and the nanoprocessing helps to maintain a suitable environment for a longer shelf life [2]. Certain food industries use nanomaterials as delivery systems for better absorption because they are capable of sustained release without disturbing their morphology or structure. The absorption of nutrients depends on the particle size, which affects the rate of absorption. Thus, the use of nanoencapsulations will help the formation of emulsions and result in better site-directed absorption of biocompounds, which noticeably increases the efficiency of common delivery systems. Nanotechnology is also used in drug delivery systems where efficient absorption is one of the core goals for an effective drug response [3]. As nanotechnology has expanded its applications, it is also extended to the production of packing materials, such as nanobiopolymers, which are also reusable and biodegradable, as a replacement of currently used materials. Because of their ability to bind or encapsulate certain receptors, nanosensors are also used in the detection of mycotoxins, microorganisms, and other contaminants in food [4–6].

4.3 Texture, Nutrition, and Consistency of Foods

The use of nanotechnology introduces a whole new strategy for maintaining the packed food shelf life; thus, it helps for marketing and lasts longer. Nanotechnology improves the flavor, taste, and has better delivery of culinary balance [7]. The use of materials other than nanoparticles for the same purpose might compromise properties such as bioavailability, nutraceutical delivery, spoilage, increased microbial activity, and shelf life. Thus, the use of nanoencapsulation or nanoprocessing has better or improved properties [8].

SiO_2 nanomaterials, which range in size, are anticaking agents that help maintain quality sauces, creams, and spreads [9].

4.4 Delivery Systems

Studies have addressed the use of nanotechnology for nanoencapsulation of bioactive compounds, such as minerals, vitamins, colorants, drugs, antimicrobials, antioxidants, probiotic microbes, flavors, and micronutrients, for better availability [10, 11]. Encapsulation in food packaging helps maintain the taste and odor by maintaining the permeability of packaging, food texture, and the matrix, while regulating the specific release of active agents at a specific rate and time. Thus, a nanopackaging delivery system helps maintain the moisture and temperature [12–14, 6]. The advantage of using a nano-based delivery system is its chemical or biological degradation and its compatibility with other biocompounds in the system [15].

Various delivery system technologies, such as emulsions, simple solutions, association colloids, and biopolymer matrices, have been used to maintain the sustained release of active compounds for prolonged periods [16]. The delivery systems make use of nanotechnologies for all of the above approaches, which reduce the toxicity and increase the efficiency of distribution [17, 18]. Use of nanoparticles for encapsulation has advantages over native or traditional encapsulation systems [19], and they can be efficiently increased in the delivery systems by dendrimers, which are a class of polymers that are used as coating particles. Dendrimers are highly branched structures that can be used as sensors, catalysts, or agents for gene therapy and the delivery of drugs [20]. The properties of dendrimers, such as non-toxicity, biodegradability, and non-immunogenicity, make them useful in delivery systems [21, 22].

Nanocapsulation-based deliveries in the field of nutraceuticals are used to preserve aroma, flavor, and other components of the packed food materials [23]. Nanoformulations or nanocompositions in food packing industries are applied for efficient delivery and sustained or controlled release of nutraceuticals [24]. Nanocapsules, nanoparticles, nanospheres, and liposomal nanovesicles are the types of nanoparticles used for the controlled release and delivery of nutrients and functional ingredients [25]. These nanonutraceutical delivery systems are used as facilitators for carrying essential oils, minerals, vitamins, and phytochemicals to efficiently increase their bioavailability. Nanocochleates are multilayered, cigar-shaped, spiral, solid lipid bilayer structures developed by Bioral™ for the delivery of nutrients [26]. By mechanically manipulating the physical properties, nanotechnologies can improve the mode of delivery of bioactive compounds by increasing sustainability, solubility, shelf life, and sustained release in the gastrointestinal (GI) tract, which will facilitate efficient absorption of the delivered nutrients [27]. The rate of sustained release of these bioactive compounds can be increased by altering the molecular structures during digestion, the substitute particle size, and the solubility to increase the ratio of surface area:volume [28]. Thus, molecular transformations can be controlled to maximize bioaccessibility and absorption [9].

4.5 Nanotechnology in Food Packaging

Food packaging involves necessary preventions against mechanical, chemical, thermal, and microbial effects on the packed food material. Thus, the use of foils in earlier methods was re-engineered using nanocomposites because they can be used as coatings, release devices, barriers, and packaging materials. The use of certain nanocomposites provides a heat resistant property to the food packing material. Certain coating materials developed using nanotechnology also have a dirt-repellent capacity, can signal microbial activity, and chemically change in the environment of the system. The food packaging also provides necessary nutritional information for the consumers and helps improve food packing. The use of exfoliated nanocomposites enhances the tensile strength of film or foil used in the packaging [29]. The use of nanoparticles can impart better shelf life by inhibiting the entry or exit of water vapors, thus maintaining permeability [30].

Packaging materials are intended to be biodegradable, which can be enhanced by using inorganic particles like clay. The silica or silicates in clay create a biopolymeric matrix that can be used as surfactants that aid degradation. The introduction of inorganic particles will enhance the function of biopolymers as delivery systems, which are helpful in the preparation of capsules with fragile micronutrients that are edible [29, 31, 32].

Nanolaminates comprise two or more constituting layers of nanomaterials that are physically or chemically bound to each other, and they can be used in food industries as a foaming agent or edible coatings and films. Such nanolaminates have been applied to fruit, vegetables, meat, and the packaging of fried and bakery products. Nanolaminates have certain structural properties that do not allow for the entry or exit of gas, lipids, or moisture, thus acting as a barrier for functional agents like color, flavor, nutrients, antioxidants, and antimicrobials [33–37]. Nanolaminates are used as coating materials in the food industry because of their extremely thin structure, which is fragile.

Nanotubes are characterized by their hexagonal shape that forms a hollow carbon tube, which are exploited for their thermal resistance and as flexible agents in certain medical, sports, alumina, and industrial food processing equipment [30].

Nanofillers comprise nanofibers, nanowhiskers, or nanotubes, and they also can be made of solid layered silicate structures that are inorganic in nature. Among these nanofillers, because of its biodegradability, solid layered silicate structures have found use in the food packaging industry. They are also inexpensive and easy to process; some packagers also use a nylon and clay-based hybrid to produce nanoscale fillers [38].

Other approved food additives that can be used as nanoencapsulating agents are carrageenan, gelatin, chitosan, polylactic acid, polyglycolic acid, and alginate [39, 40, 18].

Antimicrobial food packaging systems also use oxides of nanometals such as silver, copper, titanium, carbon, and magnesium. Metals in the form of nanotubes are also used [26, 41, 42].

Active packaging has been designed as an antimicrobial wrapping system in which opened food packaging that is rewrapped by the customer will also prevent microbial activity [43]. Antimicrobial packaging also uses zinc nanoformulations, such as zinc oxide quantum dots, that are used in the form of amorphous powder. The dots are bound to a polystyrene film, termed zinc oxide-polystyrene, which can be suspended in a polyvinyl prolidone gel to act against microbes such as *Salmonella enteritidis*, *Escherichia coli*, and *Listeria monocytogenes* [44].

Active packaging is a nanotechnology method for tackling microbial activity. Microbial activity causes stress that causes the production of a reactive oxygen species (ROS), which is a core reason for the spoilage of food materials. This must be addressed without compromising the quality of food and its packaging; thus, active packaging is suggested to maintain the shelf life, antimicrobial activity, and the sustained release of nutrients [45].

Active nanopackaging technology has found great potential where the use of nanostructures imparts a high surface:volume ratio. The nanocomposites are structured as nanobiocomposites and electrospun nanofiber-based structures, which are used to enhance the desired properties and improve the effective functionalities with little nanofiller [46]. Studies have used engineered nanobioparticles for sustained, specific release of the active agents by the maintenance of the moisture and temperature to trigger particular mechanisms. These are termed second generation nanostructured materials [47]. The nanostructures are generated in such a way to have only one or two dimensions that are more active in the microscale, which results in reduced potential mobility, i.e., controlled activity. These nanostructured particles can be used for potential packaging of certain nutrients, where after delivery, the unwanted substances do not come in contact with the packed food for better safety [47].

Smart food packaging is a technique that has incorporated the use of nanosensors to aid in the detection of various microbial changes and biochemical changes during the preservation of packed nutraceutical compounds, allowing us to track the changes and release certain antimicrobial enzymes, flavors, or antioxidants that help extend the shelf life [48]. Intelligent food contact materials help to monitor the condition of packed food, such as rancidity, microbial activity, spoilage, browning, and the production of ROS, detecting the environment inside the food package [47].

Smart packaging with sensors detect the quality of food and quantity of substances that are converted into signals. The use of these sensors will facilitate identification of oxygen scavengers, moisture absorbers, and other biomarkers. This property can also be used to track the temperature, oxygen levels, and pathogens, thus aiding in determining the shelf life of products. Metal nanoparticles incorporated with enzymes will inhibit or detect microbes in the food products, and certain nano-based fluorophores indicate the spoilage of packed food products by the detection of gaseous amines [49].

Nanocantilevers are an outcome of biosensor technologies where the biological binding interaction based principle has the ability to detect antigens. The cantilevers are based on the interaction between an antigen and antibody or a substrate and enzyme or cofactor, receptor, and ligand that produces physical and/or electromechanical signals through which detection is possible [16]. The silicon-based nanocantilevers are engineered to bind to the antibodies, which results in modulation in the frequency of resonance based on the aggregate of antigens attached. Gfeller et al. used a nanocantilever to detect fecal pollution in water and food products where *E. coli* was used as an antigen and was aided by an agarose-based cantilever [50]. Nanocantilevers have also been used to decrease the effect of phenolic compounds that cause unwanted dark or brown colored pigmentation in the presence of oxygen in freshly cut fruits. This helps extend the storage and marketing of freshly cut fruits. Thus, nanomaterials have also found their direct use as anti-browning agents [51, 52, 53, 54].

Reinforced packaging is a technology of smart packaging that maintains better mechanical properties, thus producing better barrier capacity. The packaging ensures blocking of UV rays, thus avoiding the spoilage of the necessary biomes in the food materials such as *E. coli*. Reinforced packaging provides thermal stability that ensures the maintenance of useful enzymes and dissolved oxygen levels in the packed food. Thus, reinforced packaging when applied in a suitable manner will result in better quality and shelf life of the food materials [45].

4.6 Nanoencapsulation

Encapsulation shall be defined as the entrapment or capsulation of certain nutrient components in a dispersed medium such as emulsions. The goal of encapsulation is to emphasize, protect, immobilize, and sustain food materials. Nanocapsules of a specific structure and size can perform certain functions. Nanocapsules have a size less than 100 nm. Capsules that are sized on the micron scale are termed microcapsules. Thus, the process of encapsulation is also termed microencapsulation [55]. Encapsulation is adapted in food packaging to maintain taste, texture, or aroma, and nanocompounds are noble structures that do not mask the flavors and do not interact with the food molecules. Thus, encapsulation enhances the quality, shelf life, and ability of the constituents in the finished food products [56, 30]. In 2011, Markman et al. proposed nanoencapsulation using Maillard reaction based protein–polysaccharide conjugates to form clear beverages where the nanoencapsulations were based on an emulsion medium [57]. In 2009, Bromley showed vitamin D encapsulation using casein–maltodextrin and compared it with non-encapsulated vitamin D for the delivery of active compounds using the pre-emulsion compositions. These nanoencapsulations have been used in delivering conjugated fatty acids, omega 3, and omega 6 fatty acids as beverage ingredients, where they are diluted to form an emulsion that increases biosustainability, availability, and shelf life, and also creates very clear beverages [58]. Similar nanoencapsulations were applied to medicinal beverages and drug industries where vitamins, such as E and D, are supplemented. Certain nutrient contents, such as lycopene, beta-carotene, isoflavones, and phytosterols, are also formulated using nanoencapsulation for better supplementation and bioavailability to the consumer [48]. Nanoencapsulation and nanoemulsions are also used in the processing units of dairy products where they improve the texture and shelf life of fermented milk, yogurt, puddings, cheese, and other dairy-based products. Nanoencapsulation is also applied and developed in preparation of supplements that constitute a probiotic bacterial biome that is to be delivered to parts of the GI tract to interact with receptor specific regions [59]. The limitation of nanoencapsulation is its volatility, which should be addressed by using crystalline matrices during storage to reduce the activity of volatile compounds. As a result, essential oils can be stored for long periods and microbial activity can be avoided [60].

4.7 Nanotechnology for Food Preservation

4.7.1 Nano-based Food Additives

Nanotechnology has been used for its ability to maintain and deliver food in an appropriate condition by maintaining the antimicrobial behavior and having a low gas barrier condition. Nano-based food additives are bioengineered compounds that are structurally stable, which aids packaging and encapsulation of the nutrients in the food packaging industries. Nanoencapsulations are prepared in such a way that they do not affect the taste or appearance of the food packed. Certain nano-additives are emulsions used to achieve better supplementation by facilitating the encapsulated nutrients into the blood stream via absorption in the intestine. Nanoparticle emulsions are also an ideal approach to elevate the texture and uniformity in products such as ice creams and other spreads [1].

4.7.2 Nanotechnology-based Detection

4.7.2.1 Nanosensors

4.7.2.1.1 Pathogen Detection

Nanoelectro-mechanical systems (NEMS) are a specially designed technology that has found its use in detecting and preventing food spoilage. NEMS range from nanometer to millimeter scale, and they have the ability to move and analyze food. Thus, NEMS can be used to control the storage environment [51].

Catalyst-based ELISAs can be used to determine a pathogen present in food, which has led to the proposal of biosensors that are receptive transducers that act upon the target by changing the recognition of the organic compound into a quantifiable physical sign visible to the person who uses the packed food (Figure 4.1). The receptors present in this biosensor are equipped with a particular antibody, which when it reacts with the DNA of particular antigen, will produce a visible and potential modification that helps in the detection of microbial activity and food quality [61].

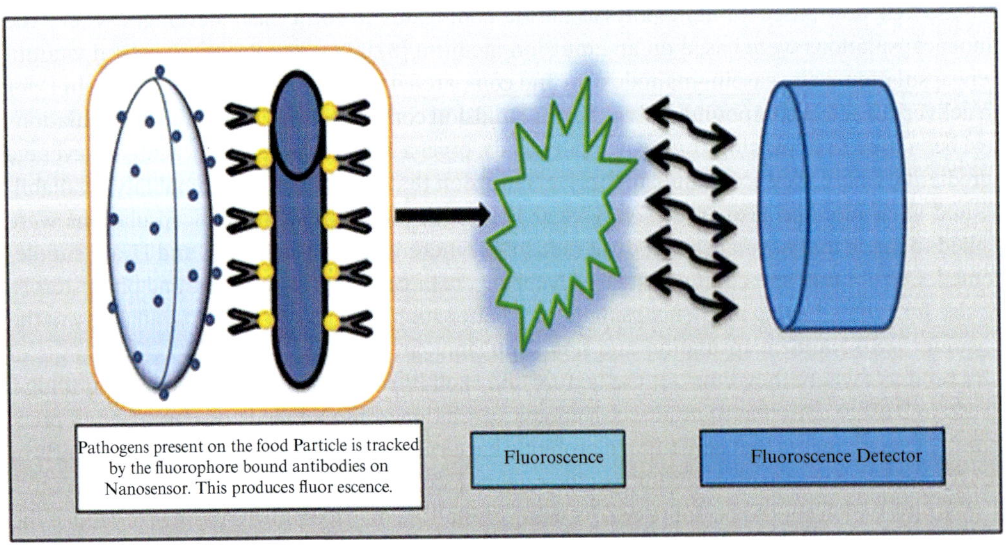

Figure 4.1 Mode of fluorophore activity as a functional detector of microbe detection.

4.7.2.1.2 *Wine Discrimination: An Electronic Nose*

Biosensors can be used in the production of wine to sense the quality of wine with respect to microbial activity. A surge of certain chemicals can lead to spoilage or a compromised quality of wine; thus, these electronic biosensors are termed "tongues" or "noses" in the field of wine production. Biosensors are microfluidic devices that are highly sensitive [62, 63].

4.7.2.1.3 *Lab-on-a-Chip: Microfluidic Devices*

This nanotechnology technique is termed lab-on-a-chip because of its capability to detect and sense a very small quantity of a compound of interest. Silicon chips used for tracking microliters or nanoliters of compounds are microfluidic systems. As a result, the lab-on-a-chip microfluidic device technology has expanded its application to medical, biomedical, and chemical analyses as well as new industries [64, 65, 66].

4.7.2.1.4 *Surface Plasmon Resonance-based Detection*

Surface plasmon resonance (SPR)-based detections are based on the property of gold nanoparticles that produce a specific characteristic signal in a sandwiched ELISA. The produced signal is restricted by SPR and the surface plasmon, elevating the SPR sensitivity [67].

4.7.2.1.5 *Nanotube Membranes*

Nanotube membranes are molecular level detections used to analyze and monitor the compartmentalization of various biomolecules such as DNA, RNA, proteins, antibodies, enzymes, vitamins, and minerals [68, 69].

4.7.3 Product Condition During Transportation

4.7.3.1 Natural Biopolymers

Biodegradable polymers are also termed biopolymers and they can be used to pack materials. Food packaging industries use biopolyesters as a biopolymer because of their compatibility to forming films and molds. Biopolymers are used in food packing industries because of their relatively good gas barrier properties that avoid the ROS that spoils the food [70].

4.7.3.2 Shelf Life

The shelf life of packed food depends on the condition in which they are packed, hygiene, materials used for packing, method of packing, and factors such as temperature and pH. Nanosensors have found great application in the field of food packaging because of their ability to provide a better shelf life (Figure 4.2). The maintenance of shelf life using nanoparticles is a necessary approach because of its multifunctional purpose, including detecting microbial activity, ROS, and pH change. The use of nanomaterials also contributes to consistency throughout the maintenance of food quality and when changes occur by environmental factors [47].

4.7.3.3 Threats and Legislation

Legislation regarding the use of nanomaterials as encapsulations and packaging was created by the EU in 2011 in the form of a non-binding legal document where engineered nanomaterials are defined under Reg.1169/2011 in the provision of information regarding food to the consumer. The terms and regulations for packaging of food materials require communication of the information regarding the packaging, constituents, and nanoencapsulations (if any), and the information

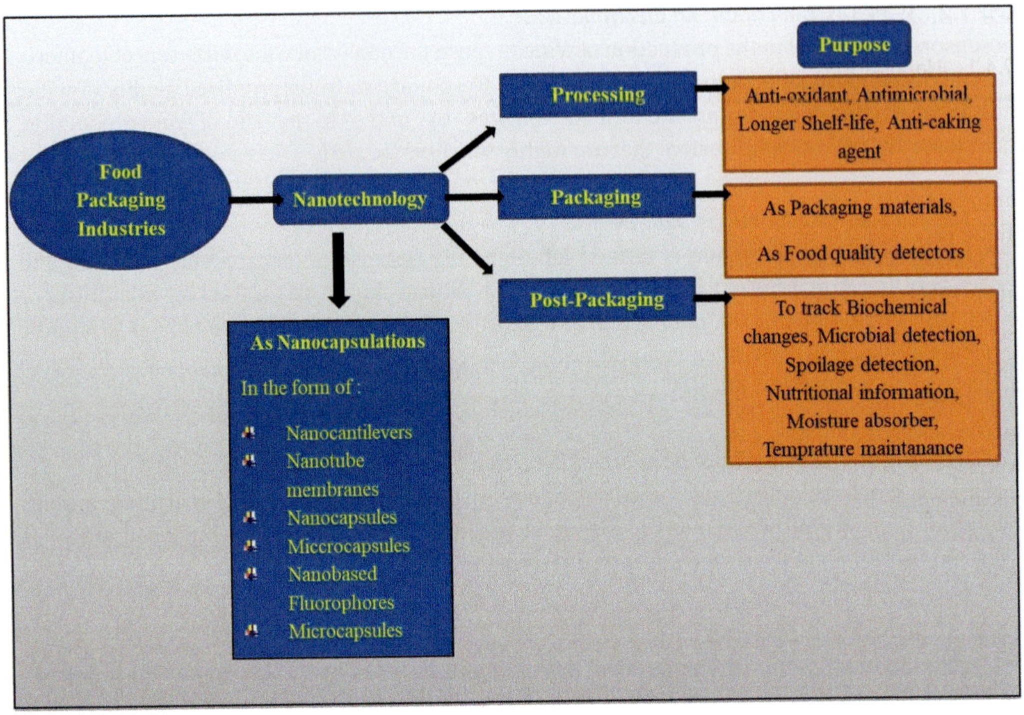

Figure 4.2 Uses of nanoencapsulation and its purpose in food packaging industries.

regarding packaging materials shall be mentioned in the EU catalogue. Only such products that contain the information are approved [71].

A legislative council that is innovative in forming a supportive testing method to identify the processing and food packaging with nanoformulations is led by National food legislations or communities [47].

4.7.3.4 Safety Issues

Another major concern in food packaging that involves nanotechnology for processing is when the packaging fails to meet the criteria, leading to foodborne sickness caused by bacterial contamination with pathogens such as *Listeria monocytogens*, *E. coli*, *Campylobacter* spp., and *Salmonella typhi* [72–74]. To avoid this, certain biopolymer nanocomposites that have effective anti-gas barrier forming capacities are used. Thus, the nanocomposites provide a low oxygen environment that is unsuitable for microbial activity [61].

4.7.3.5 Heavy Metal Release

There are possible adverse effects in the use of nanomaterials in the food industry. The use of nanomaterials that are smaller than that of typical metals may result in the release of heavy metals, such as zinc, silver, and copper oxides, which are reported to be heavy metal intoxicants. There are incidences where the use of nanomaterials has resulted in heavy metal toxicity in certain individuals, which is highly concerning because this might lead to the production of intracellular ROS, lung infections, and diseases related to CNS such as Alzheimer's disease [75–78].

4.7.4 Biological Adverse Effects

4.7.4.1 Nanoparticle-specific Allergies

Despite the advantages, nanoformulations can also cause adverse effects. The use of nanoparticles is known to cause certain allergies that are defined as nanoparticle-specific allergies. Because nanoparticles have heavy metal release, they can affect the immune system of certain individuals. Nanomaterials used for encapsulation might produce ROS in biological systems, which causes inflammation. The production of ROS affects the surfactants in the lung, causing asthma as an allergic response. ROS also induces epithelial inflammation in alveolar and intestinal epithelial cells, which causes inflammation in the GI tract. In adverse cases, this might also result in Crohn's disease [26, 79–81].

Nanoparticles that are smaller than 70 nm can enter the nucleus of a cell that contains the organisms' genetic material (DNA) and can cause detangling or unpairing of DNA and adversely block the transcription [26]. The entry of nanoparticles into epithelial cells may cause acute toxic responses of the skin, and when nanoparticles enter the brain they can cause damage to the skin cells and epithelial cells, leading to damaged skin [42] (Table 4.1). The negative effects of nanoparticles are paired with Parkinson's and Alzheimer's diseases [82, 42].

Table 4.1 Nanoparticles and their hazards to be addressed.

NANOPARTICLES	HAZARDS	REFERENCES
Au NPs	- Cause depolarization of a-tubulin, a major component of microtubules	[9, 1]
Ag NPs	- Adversely affects the cellular structure and its cytoskeleton - Capable of inducing nanoparticle-specific immune responses	
Carbon NPs—Fullerenes	- Damage to the brain, can be used as a benchmark to determine the rate of environmental poisoning	[83]
Si NPs	- Causes oxidative stress - Causes in vitro and in vivo cytotoxicity - Elevated lipid peroxidation, LPO, ROS - Decreased level of cellular glutathione	[1]
Carbon nanotubes	- Causes damage to lungs - Increase in lung inflammation - Increase in allergen-specific IgE levels in mice - Preexisting inflammation increases airway fibrosis in mice with allergic asthma	[84, 81]
SiO_2 NPs	- Induce allergen specific Th-2 type immune responses in vivo	[84]
Ovalbumin (OVA) + SiO_2 NPs	- Intransal exposure induced a relatively high rate of OVA-specific immunoglobulin G, IgE, IgG1 antibodies	[80]
Carbon NPs	- Allergic inflammation - Significantly enhanced IL-13 gene expression concurrent with the downregulation of the Th1-associated transcription factor Stat4.	[85, 84]

4.7.4.2 Regulations of Using Nanotechnology in Nutraceuticals

Certain aspects of the use of nanomaterials in food industries need to be regulated. Nanomaterials, regardless of the compositions of certain chemicals, should be assessed in final form because they have different chemical properties. The current safety methods for the use of nanoparticles for food are not suitable because the safety assessments are confidentially assessed by industries [26, 42].

4.8 Future Perspectives

Future investigations of nanoformulations and nanotechnology in the field of nutraceuticals should target decreasing or pacifying the toxic effects or hazards of nanomaterials in biological systems. There are several issues to be addressed to improve the current technology in the use of nanomaterials to produce relatively healthy and sustainable food products. More affordable nanotechnologies should also be evaluated. New approaches should be identified to standardize the procedures for using nanoencapsulations in delivery systems because they work inside a living entity. The impact of the use of nanomaterials should be studied under various circumstances to help understand nanomaterial use to deliver bioactive compounds [9, 86, 1, 23, 42].

4.9 Conclusion

Nanotechnology used in certain nutraceuticals, drug production, and packed food processing industries has played a valuable role in delivering better quality packed food from the industry to consumer and improving the shelf life. These improvements are in part created by the encapsulation of biocompounds in nanomaterials, which is a promising approach to maintain the activity of compounds and also release bioactive compounds in a sustained controllable manner [24]. Nanoencapsulation to deliver the bioactive compounds is a promising use of nanotechnology in the field of nutraceuticals. There is a need to create better awareness amongst the public sector about the use of nanotechnology for packing and processing of functional food and nutrients. Nanotechnology at its best will ensure the safety of food and enhanced nutritional quality and food provisions, which addresses one of the most discussed and "to be addressed" sustainable development goals provided by the World Health Organization (WHO).

References

1. Prasad, R.D., Sahoo, A.K., Shrivastav, O.P. et al. (2022). *ES food and agroforestry a review on aspects of nanotechnology in food science and animal nutrition.* 12–46.
2. Pradhan, N., Singh, S., Ojha, N. et al. (2015). Facets of nanotechnology as seen in food processing, packaging, and preservation industry. *BioMedical Research International* 2015: https://doi.org/10.1155/2015/365672.
3. Ezhilarasi, P.N., Karthik, P., Chhanwal, N., and Anandharamakrishnan, C. (2013). Nanoencapsulation techniques for food bioactive components: a review. *Food and Bioprocess Technology* 6 (3): 628–647. https://doi.org/10.1007/s11947-012-0944-0.
4. Bratovcic, A., Odobasic, A., Catic, S., and Sestan, I. (2015). Application of polymer nanocomposite materials in food packaging. *Croatian Journal of Food Science and Technology* 7 (2): 86–94. https://doi.org/10.17508/cjfst.2015.7.2.06

5 Lamprecht, A., Saumet, J.L., Roux, J., and Benoit, J.P. (2004). Lipid nanocarriers as drug delivery system for ibuprofen in pain treatment. *International Journal of Pharmaceutics* 278 (2): 407–414. https://doi.org/10.1016/j.ijpharm.2004.03.018.

6 Ubbink, J. and Krüger, J. (2006). Physical approaches for the delivery of active ingredients in foods. *Trends in Food Science and Technology* 17 (5): 244–254. https://doi.org/10.1016/j.tifs.2006.01.007.

7 Nakagawa, K. (2014). Nano- and microencapsulation of flavor in food systems. *Nano- and Microencapsulation for Foods* 9781118292: 249–271. https://doi.org/10.1002/9781118292327.ch10.

8 Liang, Y., Ma, T., Thakur, A. et al. (2014). Differentially expressed glycosylated patterns of α-1-antitrypsin as serum biomarkers for the diagnosis of lung cancer. *Glycobiology* 25 (3): 331–340. https://doi.org/10.1093/glycob/cwu115.

9 He, X. and Hwang, H.M. (2016). Nanotechnology in food science: functionality, applicability, and safety assessment. *Journal of Food and Drug Analysis* 24 (4): 671–681. https://doi.org/10.1016/j.jfda.2016.06.001.

10 Chen, S.Y., Lee, Y.R., Hsieh, M.C. et al. (2018). Enhancing the anticancer activity of antrodia cinnamomea in hepatocellular carcinoma cells via cocultivation with ginger: the impact on cancer cell survival pathways. *Frontiers in Pharmacology* 9 (JUL): 1–13. https://doi.org/10.3389/fphar.2018.00780.

11 Yang, J., Zhang, J., Zhu, L. et al. (2006). Synthesis of nano titania particles embedded in mesoporous SBA-15: characterization and photocatalytic activity. *Journal of Hazardous Materials* 137 (2): 952–958. https://doi.org/10.1016/j.jhazmat.2006.03.017.

12 Gibbs, B.F., Kermasha, S., Alli, I., and Mulligan, C.N. (1999). Encapsulation in the food industry: a review. *International Journal of Food Sciences and Nutrition* 50 (3): 213–224. https://doi.org/10.1080/096374899101256.

13 Pothakamury, U.R. and Barbosa-Cánovas, G.V. (1995). Fundamental aspects of controlled release in foods. *Trends in Food Science and Technology* 6 (12): 397–406. https://doi.org/10.1016/S0924-2244(00)89218-3.

14 Shahidi, F. and Han, X.Q. (1993). Encapsulation of food ingredients. *Critical Reviews in Food Science and Nutrition* 33 (6): 501–547. https://doi.org/10.1080/10408399309527645.

15 Weiss, J., Takhistov, P., and McClements, D.J. (2006). Functional materials in food nanotechnology. *Journal of Food Science* 71 (9): https://doi.org/10.1111/j.1750-3841.2006.00195.x.

16 Hall, R.H. (2002). Biosensor technologies for detecting microbiological foodborne hazards. *Microbes and Infection* 4 (4): 425–432. https://doi.org/10.1016/S1286-4579(02)01556-3.

17 Khosravi-Darani, K., Pardakhty, A., Honarpisheh, H. et al. (2007). The role of high-resolution imaging in the evaluation of nanosystems for bioactive encapsulation and targeted nanotherapy. *Micron* 38 (8): 804–818. https://doi.org/10.1016/j.micron.2007.06.009.

18 Ravi Kumar, M.N.V. (2000). A review of chitin and chitosan applications. *Reactive and Functional Polymers* 46 (1): 1–27. https://doi.org/10.1016/S1381-5148(00)00038-9.

19 Roy, K., Mao, H.Q., Huang, S.K., and Leong, K.W. (1999). Oral gene delivery with chitosan-DNA nanoparticles generates immunologic protection in a murine model of peanut allergy. *Nature Medicine* 5 (4): 387–391. https://doi.org/10.1038/7385.

20 Hughes, G.A. (2005). Nanostructure-mediated drug delivery. *Nanomedicine: Nanotechnology, Biology, and Medicine* 1 (1): 22–30. https://doi.org/10.1016/j.nano.2004.11.009.

21 Flanagan, J. and Singh, H. (2006). Microemulsions: a potential delivery system for bioactives in food. *Critical Reviews in Food Science and Nutrition* 46 (3): 221–237. https://doi.org/10.1080/10408690590956710.

22 Golding, M. and Sein, A. (2004). Surface rheology of aqueous casein-monoglyceride dispersions. *Food Hydrocolloids* 18 (3): 451–461. https://doi.org/10.1016/j.foodhyd.2003.08.003.

23 Singh, T., Shukla, S., Kumar, P. et al. (2017). Application of nanotechnology in food science: perception and overview. *Frontiers in Microbiology* 8 (AUG): 1–7. https://doi.org/10.3389/fmicb.2017.01501.

24 Sozer, N. and Kokini, J.L. (2009). Nanotechnology and its applications in the food sector. *Trends in Biotechnology* 27 (2): 82–89. https://doi.org/10.1016/j.tibtech.2008.10.010.

25 Wen, H.W., Decory, T.R., Borejsza-Wysocki, W., and Durst, R.A. (2006). Investigation of NeutrAvidin-tagged liposomal nanovesicles as universal detection reagents for bioanalytical assays. *Talanta* 68 (4): 1264–1272. https://doi.org/10.1016/j.talanta.2005.07.032.

26 Chaudhry, Q., Scotter, M., Blackburn, J. et al. (2008). Applications and implications of nanotechnologies for the food sector. *Food Additives and Contaminants - Part A Chemistry, Analysis, Control, Exposure and Risk Assessment* 25 (3): 241–258. https://doi.org/10.1080/02652030701744538.

27 Chen, L., Remondetto, G.E., and Subirade, M. (2006). Food protein-based materials as nutraceutical delivery systems. *Trends in Food Science and Technology* 17 (5): 272–283. https://doi.org/10.1016/j.tifs.2005.12.011.

28 Kommuru, T.R., Gurley, B., Khan, M.A., and Reddy, I.K. (2001). Self-emulsifying drug delivery systems (SEDDS) of coenzyme Q10: formulation development and bioavailability assessment. *International Journal of Pharmaceutics* 212 (2): 233–246. https://doi.org/10.1016/S0378-5173(00)00614-1.

29 Alexandre, M. and Dubois, P. (2000). Polymer-layered silicate nanocomposites: preparation, properties and uses of a new class of materials. *Materials Science and Engineering R: Reports* 28 (1): 1–63. https://doi.org/10.1016/S0927-796X(00)00012-7.

30 Rashidi, L. and Khosravi-Darani, K. (2011). The applications of nanotechnology in food industry. *Critical Reviews in Food Science and Nutrition* 51 (8): 723–730. https://doi.org/10.1080/10408391003785417.

31 Bharadwaj, R.K., Mehrabi, A.R., Hamilton, C. et al. (2002). Structure - property relationships in cross-linked polyester ± clay nanocomposites. *Polymer* 43: 3699–3705.

32 Sorrentino, A., Gorrasi, G., and Vittoria, V. (2007). Potential perspectives of bio-nanocomposites for food packaging applications. *Trends in Food Science and Technology* 18 (2): 84–95. https://doi.org/10.1016/j.tifs.2006.09.004.

33 Carrión-Granda, X., Fernández-Pan, I., and Maté, J.I. (2016). Antimicrobial edible films and coatings. *Edible Films and Coatings: Fundamentals and Applications* 67 (4): 281–295. https://doi.org/10.1201/9781315373713.

34 Cha, D.S. and Chinnan, M.S. (2004). Biopolymer-based antimicrobial packaging: a review. *Critical Reviews in Food Science and Nutrition* 44 (4): 223–237. https://doi.org/10.1080/10408690490464276.

35 Morillon, V., Debeaufort, F., Blond, G. et al. (2010). Factors affecting the moisture permeability of lipid-based edible films : a review factors affecting the moisture permeability of lipid-based edible films : a review. *Critical Reviews in Food Science and Nutrition* 42 (1): 2002.

36 Phan The, D., Debeaufort, F., Luu, D., and Voilley, A. (2008). Moisture barrier, wetting and mechanical properties of shellac/agar or shellac/cassava starch bilayer bio-membrane for food applications. *Journal of Membrane Science* 325 (1): 277–283. https://doi.org/10.1016/j.memsci.2008.07.052.

37 Ponce, A.G., Roura, S.I., Del Valle, C.E., and Moreira, M.R. (2008). Antimicrobial and antioxidant activities of edible coatings enriched with natural plant extracts: in vitro and in vivo studies. *Postharvest Biology and Technology* 49 (2): 294–300. https://doi.org/10.1016/j.postharvbio.2008.02.013.

38 Kawasumi, M. (2004). The discovery of polymer-clay hybrids. *Journal of Polymer Science* 42: 819–824.

39 Agulló, E., Rodríguez, M.S., Ramos, V., and Albertengo, L. (2003). Present and future role of chitin and chitosan in food. *Macromolecular Bioscience* 3 (10): 521–530. https://doi.org/10.1002/mabi.200300010.

40 Lopez-Rubio, A., Gavara, R., and Lagaron, J.M. (2006). Bioactive packaging: turning foods into healthier foods through biomaterials. *Trends in Food Science and Technology* 17 (10): 567–575. https://doi.org/10.1016/j.tifs.2006.04.012.

41 Doyle, M. (2006June). Nanotechnology: a brief literature review. *FRI Briefings*, 1–10. http://scholar.google.com/scholar?hl=en&btnG=Search&q=intitle:Nanotechnology+:+A+Brief+Literature+Review#0%5Cnhttp://scholar.google.com/scholar?hl=en&btnG=Search&q=intitle:Nanotechnology:+a+brief+literature+review#0. (Access date 26 July 2022)

42 Srinivas, P.R., Philbert, M., Vu, T.Q. et al. (2010). Nanotechnology research: applications in nutritional sciences. *Journal of Nutrition*140 (1): 119–124. https://doi.org/10.3945/jn.109.115048.

43 Moore, G. (2006). Nano and food packaging technologies converge. *Food Technology* 60 (3): 92–94.

44 Jin, T., Sun, D., Su, J.Y. et al. (2009). Antimicrobial efficacy of zinc oxide quantum dots against listeria monocytogenes, Salmonella Enteritidis, and Escherichia coli O157:H7. *Journal of Food Science* 74 (1): https://doi.org/10.1111/j.1750-3841.2008.01013.x.

45 Cerqueira, M.A., Vicente, A.A., and Pastrana, L.M. (2018). Nanotechnology in food packaging: opportunities and challenges. *Nanomaterials for Food Packaging: Materials, Processing Technologies, and Safety Issues* 1–11. https://doi.org/10.1016/B978-0-323-51271-8.00001-2.

46 Lagarón, J.M., Cabedo, L., Cava, D. et al. (2005). Improving packaged food quality and safety. part 2: nanocomposites. *Food Additives and Contaminants* 22 (10): 994–998. https://doi.org/10.1080/02652030500239656.

47 Dainelli, D., Gontard, N., Spyropoulos, D. et al. (2008). Active and intelligent food packaging: legal aspects and safety concerns. *Trends in Food Science and Technology* 19 (SUPPL. 1): S103–S112. https://doi.org/10.1016/j.tifs.2008.09.011.

48 Momin, J.K., Jayakumar, C., and Prajapati, J.B. (2013). Potential of nanotechnology in functional foods. *Emirates Journal of Food and Agriculture* 25 (1): 10–19. https://doi.org/10.9755/ejfa.v25i1.9368.

49 Chellaram, C., Murugaboopathi, G., John, A.A. et al. (2014). Significance of nanotechnology in food industry. *APCBEE Procedia* 8 (Caas 2013): 109–113. https://doi.org/10.1016/j.apcbee.2014.03.010.

50 Gfeller, K.Y., Nugaeva, N., and Hegner, M. (2005). Micromechanical oscillators as rapid biosensor for the detection of active growth of Escherichia coli. *Biosensors and Bioelectronics* 21 (3): 528–533. https://doi.org/10.1016/j.bios.2004.11.018.

51 Bodor, R., Žúborová, M., Ölvecká, E. et al. (2001). Isotachophoresis and isotachophoresis-zone electrophoresis of food additives on a chip with column-coupling separation channels. *Journal of Separation Science* 24 (9): 802–809. https://doi.org/10.1002/1615-9314(20010901)24:9<802::AID-JSSC802>3.0.CO;2-1.

52 Li, X., Li, W., Jiang, Y. et al. (2011). Effect of nano-ZnO-coated active packaging on quality of fresh-cut "Fuji" apple. *International Journal of Food Science and Technology* 46 (9): 1947–1955. https://doi.org/10.1111/j.1365-2621.2011.02706.x.

53 Canelle, L., Bousquet, J., Pionneau, C. et al. (2005). An efficient proteomics-based approach for the screening of autoantibodies. *Journal of Immunological Methods* 299 (1–2): 77–89. https://doi.org/10.1016/j.jim.2005.01.015.

54 Rojas-Graü, M.A., Soliva-Fortuny, R., and Martín-Belloso, O. (2009). Edible coatings to incorporate active ingredients to fresh-cut fruits: a review. *Trends in Food Science and Technology* 20 (10): 438–447. https://doi.org/10.1016/j.tifs.2009.05.002.

55 Mohanraj, V.J. and Chen, Y. (2006). Nanoparticles – a review. *Bioline International* 5 (June): 561–573.

56 Given, P.S. (2009). Encapsulation of flavors in emulsions for beverages. *Current Opinion in Colloid and Interface Science* 14 (1): 43–47. https://doi.org/10.1016/j.cocis.2008.01.007.

57 Markman, G. and Livney, Y.D. (2012). Maillard-conjugate based core-shell co-assemblies for nanoencapsulation of hydrophobic nutraceuticals in clear beverages. *Food and Function* 3 (3): 262–270. https://doi.org/10.1039/c1fo10220f.

58 Bromley, P.J. (2013). Nanotechnology and nonpolar active compounds in functional foods: an application note. In: *Bio-Nanotechnology: A Revolution in Food, Biomedical and Health Sciences*, 697–703. https://doi.org/10.1002/9781118451915.ch39.

59 Vidhyalakshmi, R., Bhakyaraj, R., and Subhasree, R.S. (2009). *Encapsulation "the future of probiotics"-A review. Advances in Biological Research* 3: 96–103.

60 Burt, S. (2004). Essential oils: their antibacterial properties and potential applications in foods - A review. *International Journal of Food Microbiology* 94 (3): 223–253. https://doi.org/10.1016/j.ijfoodmicro.2004.03.022.

61 Chung, I.M., Rajakumar, G., Gomathi, T. et al. (2017). Nanotechnology for human food: advances and perspective. *Frontiers in Life Science*10 (1): 63–72. https://doi.org/10.1080/21553769.2017.1365775.

62 Lange, D., Hagleitner, C., Hierlemann, A. et al. (2002). Complementary metal oxide semiconductor cantilever arrays on a single chip: mass-sensitive detection of volatile organic compounds. *Analytical Chemistry* 74 (13): 3084–3095. https://doi.org/10.1021/ac011269j.

63 García, M., Aleixandre, M., Gutiérrez, J., and Horrillo, M.C. (2006). Electronic nose for wine discrimination. *Sensors and Actuators, B: Chemical* 113 (2): 911–916. https://doi.org/10.1016/j.snb.2005.03.078.

64 Engelman, J.A., Luo, J., and Cantley, L.C. (2006). The evolution of phosphatidylinositol 3-kinases as regulators of growth and metabolism. *Nature Reviews Genetics* 7 (8): 606–619. https://doi.org/10.1038/nrg1879.

65 Lassen, S., Wiebring, A., Helmholz, H. et al. (2012). Isolation of a Na v channel blocking polypeptide from Cyanea capillata medusae - A neurotoxin contained in fishing tentacle isorhizas. *Toxicon* 59 (6): 610–616. https://doi.org/10.1016/j.toxicon.2012.02.004.

66 Vo-Dinh, T., Cullum, B.M., and Stokes, D.L. (2001). Nanosensors and biochips: frontiers in biomolecular diagnostics. *Sensors and Actuators, B: Chemical* 74 (1–3): 2–11. https://doi.org/10.1016/S0925-4005(00)00705-X.

67 Kwon, M.J., Lee, J., Wark, A.W., and Lee, H.J. (2012). Nanoparticle-enhanced surface plasmon resonance detection of proteins at attomolar concentrations: comparing different nanoparticle shapes and sizes. *Analytical Chemistry* 84 (3): 1702–1707. https://doi.org/10.1021/ac202957h.

68 Lee, S.B. and Martin, C.R. (2002). Electromodulated molecular transport in gold-nanotube membranes. *Journal of the American Chemical Society* 124 (40): 11850–11851. https://doi.org/10.1021/ja027494f.

69 Rouhi, M. (2002). Novel chiral separation tool. *Chemical & Engineering News Archive* 80 (25): 13. https://doi.org/10.1021/cen-v080n025.p013.

70 Tharanathan, R.N. (2003). Biodegradable films and composite coatings: past, present and future. *Trends in Food Science and Technology* 14 (3): 71–78. https://doi.org/10.1016/
[illegible]

71 Gallocchio, F., Belluco, S., and Ricci, A. (2015). Nanotechnology and food: brief overview of the current scenario. *Procedia Food Science* 5: 85–88. https://doi.org/10.1016/j.profoo.2015.09.022.

72 Hoffmann, S., Batz, M.B., and Morris, J.G. (2012). Annual cost of illness and quality-adjusted life year losses in the united states due to 14 foodborne pathogens. *Journal of Food Protection* 75 (7): 1292–1302. https://doi.org/10.4315/0362-028X.JFP-11-417.

73 Scallan, E., Hoekstra, R.M., Mahon, B.E. et al. (2015). An assessment of the human health impact of seven leading foodborne pathogens in the United States using disability adjusted life years. *Epidemiology and Infection* 143 (13): 2795–2804. https://doi.org/10.1017/S0950268814003185.

74 Scharff, R.L. (2012). Economic burden from health losses due to foodborne illness in the united states. *Journal of Food Protection* 75 (1): 123–131. https://doi.org/10.4315/0362-028X.JFP-11-058.

75 Fukui, H., Horie, M., Endoh, S. et al. (2012). Association of zinc ion release and oxidative stress induced by intratracheal instillation of ZnO nanoparticles to rat lung. *Chemico-Biological Interactions* 198 (1–3): 29–37. https://doi.org/10.1016/j.cbi.2012.04.007.

76 Gliga, A.R., Skoglund, S., Odnevall Wallinder, I. et al. (2014). Size-dependent cytotoxicity of silver nanoparticles in human lung cells: the role of cellular uptake, agglomeration and Ag release. *Particle and Fibre Toxicology* 11 (1): 1–17. https://doi.org/10.1186/1743-8977-11-11.

77 Karlsson, H.L., Cronholm, P., Hedberg, Y. et al. (2013). Cell membrane damage and protein interaction induced by copper containing nanoparticles-importance of the metal release process. *Toxicology* 313 (1): 59–69. https://doi.org/10.1016/j.tox.2013.07.012.

78 McShan, D., Ray, P.C., and Yu, H. (2014). Molecular toxicity mechanism of nanosilver. *Journal of Food and Drug Analysis* 22 (1): 116–127. https://doi.org/10.1016/j.jfda.2014.01.010.

79 Syed, S., Zubair, A., and Frieri, M. (2013). Immune response to nanomaterials: implications for medicine and literature review. *Current Allergy and Asthma Reports* 13 (1): 50–57. https://doi.org/10.1007/s11882-012-0302-3.

80 Yoshida, T., Yoshioka, Y., Fujimura, M. et al. (2011). Promotion of allergic immune responses by intranasally-administrated nanosilica particles in mice. *Nanoscale Research Letters* 6 (1): 6–11. https://doi.org/10.1186/1556-276X-6-195.

81 Ryman-Rasmussen, J.P., Tewksbury, E.W., Moss, O.R. et al. (2009). Inhaled multiwalled carbon nanotubes potentiate airway fibrosis in murine allergic asthma. *American Journal of Respiratory Cell and Molecular Biology* 40 (3): 349–358. https://doi.org/10.1165/rcmb.2008-0276OC.

82 Chau, C.F., Wu, S.H., and Yen, G.C. (2007). The development of regulations for food nanotechnology. *Trends in Food Science and Technology* 18 (5): 269–280. https://doi.org/10.1016/j.tifs.2007.01.007.

83 Yadav, T., Mungray, A.A., and Mungray, A.K. (2014). Fabricated nanoparticles: current status and potential phytotoxic threats. *Reviews of Environmental Contamination and Toxicology* 230: 83–110. https://doi.org/10.1007/978-3-319-04411-8_4.

84 Nygaard, U.C., Hansen, J.S., Samuelsen, M. et al. (2009). Single-walled and multi-walled carbon nanotubes promote allergic immune responses in mice. *Toxicological Sciences* 109 (1): 113–123. https://doi.org/10.1093/toxsci/kfp057.

85 Lefebvre, D.E., Pearce, B., Fine, J.H. et al. (2014). In vitro enhancement of mouse T helper 2 cell sensitization to ovalbumin allergen by carbon black nanoparticles. *Toxicological Sciences* 138 (2): 322–332. https://doi.org/10.1093/toxsci/kfu010.

86 Kalpana Sastry, R., Anshul, S., and Rao, N.H. (2013). Nanotechnology in food processing sector-An assessment of emerging trends. *Journal of Food Science and Technology* 50 (5): 831–841. https://doi.org/10.1007/s13197-012-0873-y.

5

Polyphenols

Nutraceutical and Nanotherapeutic Approaches

Mustansir Bhori[1], Ameyota De[1], Shreya Das[1], Aayushi Kadam[2,3], Jyotirmoi Aich[1], and Kanchanlata Tungare[1,]*

[1] *School of Biotechnology and Bioinformatics, D Y Patil Deemed to be University, Navi Mumbai, Maharashtra, India*
[2] *Department of Food and Human Nutritional Sciences, University of Manitoba, Winnipeg, Manitoba, Canada R3T 2N2*
[3] *Anatek Services Pvt. Ltd., Mumbai, India*
* *Corresponding author*

5.1 Introduction

Polyphenols are organic molecules that are only produced by plants. They have chemical characteristics with phenolic substances and have biological effects on inflammatory and oxidative stress, the digestion of macronutrients, and gut flora [1]. They are plant secondary metabolites that are typically used in defense against pathogens or UV radiation [2]. The bitterness, color, astringency, flavor, oxidative stability, and odor of foods can all be influenced by polyphenols. At the end of the 20th century, epidemiological research and meta-analyses strongly suggested that the long-term utilization of a diet high in plant polyphenols provided some defense against the growth of cancers, diabetes, cardiovascular diseases (CVDs), neurodegenerative diseases, and osteoporosis [3, 4]. These substances are widespread in foods, including nuts, fruits, seeds, vegetables, tree barks, and flowers, as well as popular drinks, such as beer, tea, and wine, making them essential components of the human diet. In addition, some can bind and precipitate macromolecules, including dietary proteins, digestive enzymes, and carbohydrates, which decreases the ability of food to be digested [5].

Phenolic compounds exhibit notable antioxidant and antibacterial properties; however, they are also highly unstable, prone to degradation, weakly soluble, and, in most cases, have a limited bioavailability. Thus, adding phenolic chemicals from plants to food can drastically change their stability, physicochemical characteristics, bioavailability, and solubility [6]. Because of their potential advantages in human health, polyphenols and other dietary phenolics are generating more scientific interest. As a result, these chemicals have gained prominence because of their widespread presence in plant-based foods and compelling evidence linking their ingestion to a number of ailments [7].

Polyphenols: Food, Nutraceutical, and Nanotherapeutic Applications, First Edition. Edited by Mithun Rudrapal.

5.2 Different Classes of Polyphenols

More than 8,000 polyphenolic chemicals have been found in various plant species. Phenylalanine or a nearby precursor, shikimic acid, serve as a common step for the synthesis of all plant phenolic compounds. Although there are also direct links between sugar (polysaccharide or monosaccharide) and an aromatic carbon, they primarily exist in conjugated forms, including one or more sugar residues attached to hydroxyl groups [8]. Polyphenols are divided into numerous classes depending on the number of phenol rings they contain, as well as the structural components that connect these rings together. The primary classes are flavonoids, lignans, phenolic acids, and stilbenes [9].

Ligans- The 2,3-dibenzylbutane structure of lignans, which are diphenolic substances, is created when two cinnamic acid residues dimerize. Secoisolariciresinol is one of many lignans that are thought to act as a phytoestrogen. The most abundant dietary source is linseed, which has low levels of matairesinol and secoisolariciresinol (up to 3.7 g/kg dry weight) [10].

Stilbenes- Two phenyl moieties are joined by a two-carbon methylene bridge in stilbene. Stilbene is rarely consumed in the human diet. The majority of stilbenes in plants function as phytoalexins, which are substances that are only produced in reaction to infection or damage. Resveratrol (3,4′,5-trihydroxystilbene), which is mostly found in grapes, is one of the most researched naturally occurring polyphenol stilbenes. Red wine, a grape-based drink, has a sizable amount of resveratrol.

Flavonoids- The group of polyphenols known as flavonoids has received the most research. The basic building blocks of this category are two aromatic rings joined by three carbon atoms to form an oxygenated heterocycle. Flavonoids come in more than 4,000 different forms, many of which influence the eye-catching hues of flowers, fruits, and leaves [11]. Flavonoids can be split into six subclasses according to the different types of heterocycles they contain: flavanones, flavonols, flavanols, flavones, isoflavones, and anthocyanins. Individual variances within each group are caused by differences in the quantity, arrangement, alkylation, and/or glycosylation levels of the hydroxyl groups [9]. Some of the most popular flavonoids include quercetin (QT), myricetin, and catechins.

Phenolic acids- The two families of phenolic acids, derivatives of benzoic acid and cinnamic acid, are widely present in foods. The concentration of hydroxybenzoic acid in food plants is normally low, with the exception of some black radishes, onions, and red fruits that can have concentrations of several tens of milligrams per kilogram of dry mass [12]. P-coumaric, caffeic, ferulic, and sinapic acids make up the majority of hydroxycinnamic acids, which are more prevalent than hydroxybenzoic acids.

5.3 Properties of Polyphenols

Polyphenols are compounds that have vast conjugated systems of pi-electron configurations in their aromatic structures, which give them their UV/V absorptive properties. They also possess auto-fluorescence characteristics, especially lignin and the phenolic portion of suberin. Polyphenols readily oxidize. Additionally, polyphenols have a high affinity for binding protein molecules. This property may result in the development of both soluble and insoluble protein polyphenol complexes [13].

Antioxidant property

Antioxidants, called polyphenols, include more than 4,000 different types of phenols found in nature. Many of these substances exhibit antioxidant action in vitro. In vivo, it is doubtful that they will play an antioxidant role [14]. However, they might have an impact on gene regulation, receptor sensitivity, inflammatory enzyme activity, and cell-to-cell communication [15]. Regulatory theory considers a polyphenol antioxidant's capacity to scavenge free radicals and stimulate few particular metal chelation reactions. To keep the cells' metabolic processes functioning normally, singlet oxygen, peroxynitrite, and hydrogen peroxide must be continuously eliminated. The benefits of dietary polyphenol consumption may be linked to positive effects in higher animal species. For example, there may be a decrease in inflammation, as seen in studies on endothelial cells, specifically in coronary artery disease, and via the modulation of oxidative low-density lipoprotein [16], and there may be a benefit to skin antiaging [17].

Antimicrobial property

Phytochemicals similar to flavonoids, including resveratrol, and tea catechins were initially created as protective compounds to deter animals from consuming plants. Researchers and physicians from all over the world have recently focused attention on phenolic acids and other molecular by-products of plants. Such phytochemicals have been shown to be excellent factors for boosting human health [18]. Eating fruits, vegetables, and other goods produced from plants is thought to provide some health benefits in part because these foods contain polyphenols. Fruits and other goods made from plants are popular because of their polyphenol content. Understanding how polyphenols are metabolized will help us better grasp how they affect our bodies in vivo. These substances undergo gut absorption and metabolism based on their interactions with intestinal bacteria. Food is the main source of polyphenols, which has sparked a discussion concerning the impact of phytochemicals obtained from plants on the gut flora. Researchers hypothesized that these phytochemicals might alter the biological activity and/or makeup of the gut's microbial population. Numerous studies demonstrate the importance of nutrients from plants in reducing the risk of inflammatory disorders and infectious diseases. For example, tea catechins, particularly epigallocatechin-3-gallate, were successful in curing significant nosocomial bacterial infections [19, 20]. Strong antibacterial effects are produced by flavonoids and certain phenolic acids, such as caffeic acid, quinic acid, chlorogenic acid, and gallic acid, against common microbial strains that impact the human respiratory system or urinary tract system, including the *Candida* species (108). By inhibiting the bacterial DNA-B helicase enzyme, the flavonoid galangin is able to prevent *Klebsiella pneumoniae* (a Gram-negative bacterium) from replicating [19].

Anticariogenic property

Both in vitro and in vivo research have been used to examine the effects of polyphenols. Additional research on the effectiveness of polyphenols against *Streptococci mutans* and in vivo tests on animals and people were conducted [21, 22]. Polyphenol chemicals found in tea, coffee, and chocolate have an antimicrobial effect and may help stop cariogenic processes. *Streptococcus mutans* and *Streptococcus sanguinis* considerably lessen the formation of biofilms and the production of acid according to studies by Milgrom et al. Similarly, the ability of *S. mutans* to adhere to saliva-coated hydroxyapatite beads is inhibited by trigonelline, caffeine, and chlorogenic acid found in roasted coffee and green tea. Studies on black, green, and oolong tea have shown that tea polyphenols have an anti-caries impact through an antibacterial mechanism of action. Additionally, the antibacterial properties of galloyl esters of (α)-epicatechin, (α)-epigallocatechin, and (α)-gallocatechin are increasingly being demonstrated.

5.4 Nutraceutical-based Polyphenols

5.4.1 Polyphenols Derived from Foods

Dietary polyphenols are a class of secondary metabolites that are abundant in a variety of foods, including fruits, vegetables, wine, tea, extra virgin olive oil, chocolate, and other items made with cocoa. Most of these polyphenols are isomers, derivatives, or isoflavones of flavones, catechins, and phenolic acids. Dietary polyphenols perform a wide range of biologically important tasks, including preventing oxidative stress and degenerative illnesses. The majority of these biological activities, according to experimental evidence, can be linked to their innate antioxidant capacities. By triggering endogenous defense mechanisms and altering cellular signaling processes, such as NF-κB activation, AP-1 DNA binding, glutathione biosynthesis, PI3-kinase/Akt pathway, MAPK proteins (ERK, JNK, and P38) activation, and Nrf2 translocation into the nucleus, dietary polyphenols may provide indirect protection [23, 24]. A few dietary phenols and their biological activities are listed in Table 5.1.

Table 5.1 Sources of dietary polyphenols and their biological activities.

Sr. No.	Dietary Polyphenol	Source	Biological Activity	References
1	ECG	Apples, peaches, pears, raisins, cherries, red wine, tea	Inhibiting cyclooxygenase and lipoxygenase, using the MAPK proteins (ERK, JNK, and p38) to activate ARE-mediated gene expression	[24, 25]
2	Catechin	Apples, apricots, plums, blueberries, blackberries, chocolate, wine	CAT,glutathione S-transferase (GST), and SOD activity enhancement, increasing the amount of GSH in cells.	[26, 27]
3	Curcumin	Turmeric	CYP1A2, CYP3A4, and CYP2C9 inhibition. Activating ARE and Nrf2 to boost GSTP1 expression. Increasing the expression of heat shock protein 70, CAT, and SOD.	[28–30]
4	Caffeic acid	Blueberry, pear, orange, lemon, spinach, lettuce	Inhibiting the oxidation of dopamine by peroxynitrite	[31, 32]
5	Resveratrol	Grapes, peanuts, red wine	Avoiding DNA oxidative damage. By catalyzing o'acetyl transferase and sulfotransferase, it can decrease the development of PhIP-DNA adducts. CYP 1A1/1A2 expression and activity inhibition.	[33, 34]

(Continued)

Table 5.1 (Continued)

Sr. No.	Dietary Polyphenol	Source	Biological Activity	References
6	Quercetin	Celery, onions, fennel, spinach, broccoli, plums, blackberries, red wine, tea	Increasing NADPH:quinone oxidoreductase-1 expression and activity (NQO1). Improving the transcriptional activity that is mediated by Nrf2 and ARE. Stabilizing and enhancing Nrf2.	[35, 36]
7	Gallic acid	Raspberry, pomegranate juice	Preventing tyrosinase, xanthine oxidase, and superoxide radical production	[37, 38]
8	Hydroxytyrosol	Virgin olive oil, wine	Preventing the synthesis of eicosanoids and platelet aggregation. Preventing the formation of thromboxane B2. Decreasing the adherence of monocytoid cells to activated endothelium. Decreasing the mRNA and protein of VCAM-1	[39, 40]

5.4.2 Management of Chronic Diseases

Epidemiological studies have consistently demonstrated a negative correlation between diets high in polyphenols and the likelihood of developing chronic diseases [4, 41]. By accepting an electron to create relatively stable phenoxy radicals, the phenolic groups in polyphenols are able to sabotage chain oxidation reactions in biological components [42]. It is generally known that diets and drinks high in polyphenols may boost the plasma's antioxidant capacity. Following consumption of foods high in polyphenols, plasma's increased antioxidative capacity may be attributed to the existence of reducing polyphenols, as well as their metabolites, the reduction of other reducing agents' concentrations (sparing effects of polyphenols on other endogenous antioxidants), or the absorption of pro-oxidative food components such as iron [41]. The amount of oxidative damage to lymphocyte DNA decreases with antioxidant use. The preventive properties of polyphenols have been demonstrated by similar observations with polyphenol-rich foods and beverages [42].

5.4.2.1 Cardiovascular Disease

The frequent consumption of foods high in polyphenols, such as fruits, vegetables, chocolate, tea, and wine, may have cardioprotective benefits according to numerous epidemiological and human intervention studies [43–45]. Studies have found a link between the consumption of flavones, flavanols, and flavonols, as well as a lower possibility of coronary artery disease [4]. Additionally, the consumption of anthocyanin and flavanone lowered the risk of CVD-related mortality [45]. Numerous studies on people, animals, and cells have suggested that polyphenols may have positive effects on the vascular system by triggering antioxidant defenses [46, 47], lowering blood pressure [48, 49], enhancing endothelial function [50–52], inhibiting platelet aggregation [53, 54], oxidizing low density lipoprotein [47, 55], and lowering inflammatory responses [56, 57]. By preventing the expression of metalloproteinase 1 (MMP1) and the disintegration of atherosclerotic plaques, QT, a polyphenol found in abundance in onions, has been demonstrated to be conversely linked with death from coronary heart disease [58]. The invasion and proliferation of smooth muscle cells in the artery wall have been demonstrated to be inhibited by tea catechins, which may help reduce the development of an atheromatous lesion [59]. With evidence of preventing platelet aggregation, polyphenols may have antithrombotic properties. By preferentially inhibiting the activity of cyclooxygenase 1 (COX 1) that produces thromboxane A2, which is an inducer for platelet aggregation and vasoconstrictor, resveratrol, a wine polyphenol, reduces platelet aggregation. Additionally, rat aortic rings and isolated arteries can also be relaxed by resveratrol. Other mechanisms by which resveratrol exhibits vasorelaxant activity include the capacity to stimulate Ca^{++} activated K^{+} channels and to improve nitric oxide (NO) signaling in endothelial cells [60]. The potential of polyphenols to regulate the activity and level of nitric oxide synthase (eNOS), and hence NO bioavailability to the endothelium, is a proposed mechanism for their influence on vascular function [61, 62]. Aortic ring tests employing physiological polyphenol concentrations have revealed that polyphenols cause endothelium-dependent relaxation [63, 64]. The capacity of polyphenols to link with kinase signaling pathways, such as the PI3-kinase/Akt pathway, as well as intracellular Ca^{+2} on eNOS phosphorylation, and following NO generation, are likely to be involved in this regulation of vascular NO [65].

5.4.2.2 Cancer

Polyphenols often have a protective effect on human cancer cell lines and result in a decrease in tumor occurrence or tumor progression [66]. These effects have been observed at different locations, including the mouth, stomach, duodenum, colon, liver, lung, mammary glands, and skin. Numerous polyphenols have been studied, including isoflavones, QT, lignans, catechins, ellagic acid, flavanones, red wine polyphenols, curcumin, and resveratrol; all of them demonstrated protective effects in various models despite having diverse modes of action [67]. It is commonly accepted that eating plenty of fruit and vegetables daily can assist in halting the development and spread of cancer. Case controlled studies conducted for more than 20 years have shown a reverse relationship between regular vegetable and fruit diets, as well as the occurrence of different forms of cancer [68]. The polyphenols may interact with reactive intermediates, activated carcinogens and mutagens, and mutagen-activated carcinogens. They may also modify the activity of important proteins that regulate cell cycle progression and have an impact on the expression of genes related to cancer [20]. The elimination of carcinogenic chemicals [17], the modification of cancer cell signaling [69] and cell cycle progression [70], the activation of apoptosis [71], and the modulation of enzyme activity [72] are a few of the possible methods by which they can perform these anticancer effects. Most significantly, the anticancer effects of green tea flavanols have been documented in animal models [73], human cell lines [74], and human intervention studies [75]. The metabolism of the pro-carcinogens is affected by altering the function of cytochrome P450 enzymes that are implicated in the activation of carcinogens. By boosting the activity of phase II conjugating enzymes, they may also speed up their excretion. It is possible that the toxicity of polyphenols is what caused this activation of phase II enzymes [41]. Consuming green tea has been shown to greatly lower the risk of biliary tract cancer [76], bladder cancer [77], breast cancer [78], and colon cancer [79]. The flavanol epigallocatechin gallate (EGCG) has been demonstrated to stimulate apoptosis and prevent cancer cell proliferation by changing the expression of cell cycle regulatory proteins, as well as the action of signaling proteins involved in metastasis, cell transformation, and proliferation, and is thought to be the primary mediator of a number of anticancer properties associated with green tea [80]. Resveratrol has been shown to be efficient against most cancers, including breast, lung, prostate, skin, colorectal, and gastric cancer, at all stages of the disease. Additionally, resveratrol has been demonstrated to inhibit metastasis and angiogenesis. Resveratrol has the ability to alter a number of pathways involved in cell proliferation, apoptosis, and inflammation according to extensive data from human cell cultures. Resveratrol inhibits hydroperoxidase, protein kinase C, cyclooxygenase, Akt, NF-B matrix metalloproteinase-9, focal adhesion kinase, cell cycle regulators, and Bcl-2 phosphorylation. These anticarcinogenic effects of resveratrol appear to be closely associated with its antioxidant activity [81]. Because of its ability to scavenge free radicals, QT is also found to have anticancer properties in mice opposed to benzo(a) pyrene-induced lung carcinogenesis [82].

5.4.2.3 Diabetes

Numerous research studies have evaluated the antidiabetic properties of polyphenols. Researchers have looked into the antidiabetic potential of tea catechins [83]. Polyphenols may influence blood sugar levels via a variety of methods such as by preventing the gut or peripheral tissues from absorbing glucose. Maltose was found to have the hypoglycemic effects of diacetylated anthocyanins at a 10 mg/kg diet intake, whereas sucrose or glucose did not [84]. This shows that the inhibition of α-glucosidase in the GI mucosa is what is causing these effects. A dose of at least 50 mg/kg or more of catechin was also observed to inhibit α-amylase and sucrase in rats. Studies have investigated the ability of polyphenols to inhibit intestinal glycosidases and the glucose transporter [85]. Individual polyphenols, including isoflavones from soy beans, epicatechin gallate, (−)epicatechin, (+)catechin, (−)epigallocatechin, glycyrrhizin from licorice root, saponins, chlorogenic acid, and tannic acid, can lessen S-glut-1-mediated intestinal glucose transfer. Additionally, the passage of glucose is slowed by saponins from the stomach to the small intestine [86]. It has additionally been claimed that resveratrol has antidiabetic properties. Modulation of SIRT1 is one of the many mechanisms introduced to explain the antidiabetic activity of this stilbene, which enhances insulin sensitivity and whole-body glucose homeostasis in diabetic rats [87]. Resveratrol dramatically reduces renal dysfunction and oxidative stress in diabetic mice and suppresses alterations in the kidney caused by diabetes (diabetic nephropathy). Resveratrol therapy also reduced insulin output and postponed the development of insulin resistance. It was proposed that a potential mechanism involved the beta cells' K^+ ATP and K^+ V channel blockage [88]. QT is known to have potent antidiabetic properties. According to a recent study, QT can prevent changes in diabetic individuals caused by oxidative stress. In diabetics, QT significantly defended the antioxidant system against lipid peroxidation. Protocatechuic acid, flavonoids, polyphenolic acids, and anthocyanins are all present in *Hibiscus sabdariffa* extract. According to a study by Lee et al. [89] the polyphenols in *H. sabdariffa* extracts reduce the pathophysiology, serum lipid profile, and oxidative indicators in the kidney associated with diabetic nephropathy.

5.4.2.4 Neurodegenerative Disorders

Neurodegenerative disorders, such as Alzheimer's disease [90, 91] and Parkinson's disease [92], constitute a growing challenge in our aging society because of the rising prevalence with advancing age. These, as well as other neurodegenerative conditions, seem to be brought on by a variety of factors such as neuroinflammation, glutamatergic excitotoxicity, an increase in oxidative stress, iron, and/or a decrease in endogenous antioxidants [93]. Average wine consumption might reduce the occurrence of several age-related neurological illnesses, such as Alzheimer's disease, according to epidemiological studies [48]. Consuming polyphenols may offer protection against neurological illnesses because of their potent antioxidative properties [94]. When compared with individuals who drank less or none at all, individuals who drank a few [23, 36] glasses of wine a day had about an 80% lower occurrence of dementia and Alzheimer's disease [95]. Additionally, regular consumption of foods and/or beverages high in flavonoids has been linked to an approximately 50% reduction in the chance of dementia [96], the preservation of cognitive function with age [94], a delay in the onset of Alzheimer's disease [97], and a decreased chance of developing Parkinson's disease. Flavonoids may operate to protect the brain in a variety of ways, such as by shielding weaker neurons, boosting current neuronal function, or encouraging neuronal regeneration [98]. For instance, polyphenols defend neurons from oxidative stress as well as Aβ-induced neuronal injury [99], and *Ginkgo biloba* extracts, rich in polyphenols, have also been demonstrated to be neuroprotective [100] by guarding against NO as well as beta-amyloid-induced neurotoxicity [101]. The citrus flavanone tangeretin has been shown to preserve nigro-striatal integrity and functionality in the context of Parkinson's disease following 6-hydroxydopamine lesioning, indicating it may function as a neuroprotective agent in opposition to the pathophysiology linked with Parkinson's disease [102]. Tyrospol and caffeic acid have both been demonstrated to protect against 5-S-cysteinyl-dopamine neurotoxicity as well as peroxynitrite neurotoxicity in vitro [103], in addition to the neuroprotection brought on by flavonoids. Resveratrol, which is abundant in wine, effectively scavenges OH• and O_2 in vitro, along with lipid hydroperoxyl free radicals. The antioxidant activity is likely responsible for the protective result of average red wine consumption against dementia in older people. In a model of Alzheimer's disease, resveratrol suppresses nuclear factor-κ B signaling, protecting against microglia-dependent amyloid toxicity. The activity is associated with the stimulation of SIRT-1 [104]. It has been shown that consuming vegetable and fruit juices with a polyphenol concentration at least three times a week may be crucial in postponing the start of Alzheimer's disease [97]. Green tea drinking has been related in nutritional studies to a lower risk of Parkinson's disease occurrence. EGCG has been demonstrated in animal models to exert a precautionary role opposed to the neurotoxin N-methyl-4-phenyl-1,2,3,6-tetrahydropyridine (MPTP), which is an inducer of a Parkinson's-like disease, that can work by either competitively inhibiting the drug's uptake because of molecular similarity or by scavenging MPTP-mediated radical formation [105]. The ability of catechins to chelate iron contributes to their therapeutic role in Parkinson's disease. This characteristic aids in their antioxidant action by inhibiting the generation of free radicals via redox-active transition metals. Additionally, because the brain is not fully equipped with an effective antioxidant defense system, the antioxidant function is also linked to the activation of the expression of antioxidants and detoxifying enzymes [106].

5.5 Polyphenol-based Nanotechnology

Bioactive natural products like polyphenols can be protected against degradation, improved in terms of absorption, retention time, and bioavailability, controlled in terms of administration, and enhanced in terms of intracellular penetration by using nanotechnology-based systems [107]. Liposomes, micelles, and nanoparticles (NPs) have been created as nanocarriers to distribute polyphenols. This has resulted in a notable improvement in the rate of dissolution and absorption, as well as bioavailability. Furthermore, compared with free polyphenols, some nanosystems resist undesirable liver metabolism, preserving therapeutic levels of polyphenols in blood circulation for longer [108]. Consequently, the restrictions related to in vivo use can be solved using polyphenol NPs.

Synthesis of Polyphenol Nanoparticles (NPs)

For the synthesis and improvement of the activity of polyphenol NPs, several approaches are being researched.

Coacervation Technique
One of the easiest methods for creating nanocarriers is coacervation. The coacervate phase of a polyelectrolyte mixture is generated, and active components are then deposited inside the newly created matrix. The intricacy of the conservation process depends on how many polymers are involved. The characteristics of numerous variables, including concentration, ionic strength, pH, and the types and ratios of the biopolymers, affect the complexity and force of interactions in biopolymers. The fundamental cause of coacervation is the electrostatic interaction between molecules with opposing charges. Additionally, complex coacervation is negatively influenced by hydrogen bonding and hydrophobic interactions. The biopolymer shell's chemical makeup and surface properties, which are closely correlated with the nanoencapsulation capabilities, determine how effectively the NPs work [109]. The drying method has a significant impact on the size range of the coacervation procedure, which is between 100 nm and 600 nm. Together with acacia gum and gelatin, oppositely charged polymers like chitosan and alginate are frequently used in coacervation [110]. Alginate, which is insoluble at low pH, inhibits the ability of chitosan to dissolve in an acidic media. Flavonoids, including chlorogenic acid, caffeoyl derivatives, QT, kaempferol, rutin, and dicaffeoyl quinic acid, are used to successfully encapsulate polyphenolic extract [111].

Solvent Evaporation
The first method used to create polymeric NPs from a premade polymer was solvent evaporation. To create nanospheres, this technique calls for the creation of an oil-in-water (o/w) emulsion [112, 113]. Prior to incorporating the active ingredient (polyphenol), the organic phase of the polymer is first prepared by dissolving it in a polar organic solvent. Because chloroform and dichloromethane are poisonous, ethyl acetate with a safer toxicological profile has been used for biomedical purposes [114]. A surfactant, such as the often-manufactured polyvinyl acetate, is a component of the aqueous phase. The organic solution is processed using ultrasonication or high-speed homogenization to emulsify in the presence of the surfactant in the aqueous medium, resulting in dispersion of the nanodroplets [115]. As the polymer solvent evaporates, NPs are generated in a suspension. Either low pressure or constant magnetic stirring are used during the evaporation process, which occurs at room temperature. Following solvent removal, the solidified NPs can be washed, centrifugally collected, then frozen and dried for long-term retention. This process enables the formation of nanospheres [116].

Spray drying method
Numerous fields have long used the well-established technology of spray drying (e.g., foods and chemicals). Spray drying was first employed at the turn of the 20th century in the pharmaceutical sector to dry blood. Since then, spray drying has been employed in numerous pharmaceutical applications, including the production of amorphous solid dispersions, the encapsulation of medications and essential oils in excipient matrices, and the spray drying of biopharmaceuticals (such as proteins, vaccines, deoxyribonucleic acid (DNA), antibodies, and other substances) [117]. Pharmaceutical powders with particle sizes ranging from the nanometer to micrometer scale are frequently produced by spray drying. Because it enables the modification and control of features, including particle size distribution, shape, density, flowability, moisture content, crystallinity, and dispersibility of the powders, spray drying has been widely employed for the manufacture of inhalation particles [118]. A liquid feed is transformed into a dried particle form in a single phase of manufacturing called spray drying. The process's two main driving forces are the atomization of the liquid input into small droplets and the solvent's evaporation using a hot drying gas. The four steps of the process are preparing the liquid feedstock and turning it into a spray, contact with the

hot drying gas, using evaporative mass to transfer the liquid from the droplet into the drying gas, and finally separating the dried product from the gas [119].

Wet milling

Grinding a substance while it's suspended in a liquid is known as wet milling. According to experimental data on the wet milling of various materials, the breaking rate kinetics (i.e., the median particle size against milling time) seems to follow a first-order exponential. Degradation is brought on by prolonged milling times, which results in finer suspensions. The presence of more cracks and crystal flaws in the larger crystals, which propagate fractures quite easily, can be attributed to the crystals breaking quickly at first. A size decrease continues after the initial quick breakdown stage; however, it slows until a plateau is reached. The slower rate of particle size reduction and eventual development of a plateau (stationary state) indicate that the mechanism of fracture changes as the wet milling progresses. The particle size decreases as the amount of the suspension's shear stress increases with milling time, and attrition takes over as the primary comminution mechanism [120]. To choose the milling period that will produce particles with the desired fineness, it is crucial to understand the breaking kinetics of a particular medication and milling setup. Various mathematical models describing the effects of process variables (such as milling speed, bead concentration, drug loading, etc.) on the breakage kinetics and particle size distribution have been reported. These modelling techniques range from discrete element modelling, population balance models, and microhydrodynamic models to merely descriptive dynamic models. Bilgili et al. contributed a thorough overview of the models that have been created for a better understanding of milling operations [121].

Polymer precipitation from an organic solution is an example of solvent displacement, as is the dispersion of a solvent in an aqueous phase with or without the assistance of a stabilizer. A semipolar aqueous miscible solvent, such as ethanol or acetone, is used to dissolve polymers, polyphenols, and/or hydrophilic surface-active substances. The solution is then poured into a surfactant-containing aqueous solution while being magnetically stirred. Rapid solvent diffusion causes NPs to develop spontaneously. The solvent is then removed from the suspension using low pressure. The pace at which the organic phase is added to the aqueous medium has a significant impact on the size of the produced particles. According to a report, increasing the rate the two phases mix may decrease the particle size and encapsulation effectiveness [122]. Furthermore, adjusting the preparation conditions allowed for efficient tuning of the yield and drug release. Additionally, the generation of smaller particle sizes by the restriction of a particular range of polymer to drug ratios required the proper polymer concentration in the organic phase. The majority of the sparingly soluble chemicals are well suited for NP synthesis via precipitation [123].

5.5.1 Polyphenolic Nanoparticles

5.5.1.1 Quercetin (QT) Nanoparticles

The flavonoid QT is present in a number of fruits and vegetables, including lingonberries, cranberries, and onion bulbs. QT primarily appears as glycosides or aglycones (3-position or/and 4-position) in the leaves, with glucose being the largest sugar group. QT phenolic groups can also be used to bind lactose and rhamnose. QT has a wide range of health benefits, including anticancer, cardiovascular protection, antiviral, allergy, ulcer, anti-inflammatory, anti-infective, and immunomodulatory properties [124, 125]. Additionally, QT has been shown to stimulate angiogenesis and the proliferation of fibroblasts and epithelial cells [126]. Furthermore, if the water solubility and skin penetration qualities can be solved, QT may be a promising agent for wound healing [127]. QT-loaded NPs may be another strategy for efficient and safer topical delivery because they improve topical delivery in vitro and in vivo. QT stimulates the production of anti-inflammatory cytokines and inhibits the synthesis of proinflammatory cytokines [128]. However, the crude extract has reduced bioavailability and is less soluble in water, which limits the capacity of QT to reach a pharmacological threshold concentration after oral administration to demonstrate local anti-inflammatory action and reduces potential medical applications and therapeutic outcomes. In this context, numerous nanoencapsulation techniques have been developed, including albumin nano-assemblies [129], a chitin-glucan-aldehyde-QT conjugate [130], silica NPs [131], halloysite nanotubes [132], and PLGA NPs [133]. Many people are interested in using nanosystems to deliver medications, genes, proteins, and nucleic acids. Highly bioactive chemicals with poor solubility can now be selectively carried and released at the site of the action thanks to drug delivery-based nanocarrier technologies. As a result, the nanocarrier must guard against the encapsulant's loss, interaction with other molecules in the bloodstream, and premature release at an inappropriate location. The qualities of the payload and target site must therefore be met by the nanomatrix as well.

5.5.1.2 Curcumin Nanoparticles

The rhizome of *Curcuma longa* yields curcumin, a hydrophobic yellow plant alkaloid that has long been used in both traditional medicine and the food industry. It is known chemically as 1,7-bis-(4-hydroxy-3-methoxyphenyl)-6-diene-3,5-hepta-1, dione. Recent studies have revealed new health-promoting, anti-inflammatory, antiaging, and antioxidant properties of curcumin [134]. Additionally, it has been shown to have strong anticancer properties against a number of cancer types, including breast cancer [135], gastric cancer [136], and rectal cancer [137, 138]. The phenolic hydroxyl and carbonyl groups that comprise curcumin's highly double-bonded structure is easily reactive, causing the structure to become unstable. Additionally, the structure of curcumin might become unstable in the presence of specific metal ions.

A successful method for overcoming these barriers has emerged: curcumin nanoencapsulation. In particular, protein NPs with high nutritional value, biocompatibility, inherent biodegradability, and nontoxicity have shown tremendous potential as carrier delivery agents. Examples include egg white protein [139], kafirin [140], and wheat gliadin [141]. Scientists have recently looked at the crucial characteristics of nanoformulations that have the potential to deliver drugs and target specific sites. NPs have impressively become a unique delivery technology that provides medicinal chemicals like curcumin with improved water solubility and enhanced bioavailability. Curcumin loading into NPs has been shown to significantly reduce the pH and enzymatic degradation, enhance chemical stability, and extend blood circulation [142]. As a result, numerous nanomaterial formulations have been suggested to enhance the properties of curcumin. Stabilizers, lipid/liposomes, conjugates/polymer conjugates, micelles, NPs, and nano/micro/hydrogels are examples of possible nanoformulation strategies [143].

5.5.1.3 Resveratrol Nanoparticles

Resveratrol is a polyphenol-related natural phytoalexin that is produced by plants as a defense mechanism against externally damaging elements [144]. This polyphenol has been isolated from 72 plant species, including blueberries, grapes, berries, peanuts, and chocolate [145]. It is found in 12 families and 31 genera. The most well-known derivative is pterostilbene, and the three hydroxyl groups attached to the aromatic rings in resveratrol give it the ability to be further regulated. Resveratrol-based nanotechnologies can be produced using a biocompatible, biodegradable, and nontoxic polymer. These nanocarriers are split into two groups based on where they were made: those based on synthetic materials, such as poly (lactic acid), poly (lactic-co-glycolic acid), poly (acrylic acid)-poly (methacrylic acid), and poly (methacrylic acid)-poly (methacrylic acid), and those based on naturally occurring polymers, such as polysaccharides (alginate, chitosan, and Arabic gum). When the payload (resveratrol) is solubilized or disseminated in the polymeric matrix, polymeric nanospheres may result, or nanocapsules may result when the payload is situated in an inner core constrained by the polymeric matrix [146]. Numerous techniques, including coacervation, ionic gelation solvent evaporation, self-assembly emulsion, electrostatic interaction of polymers, and desolvation, can be used to create polymeric NPs [147]. By altering the NP compositions to regulate how they behave in biological mediums in response to various stimulus circumstances, polymers are able to deliver the encapsulant to the target spot. As a result, the design of polymeric NPs is created based on the therapeutic application, delivery method, and target location.

5.5.2 Green Nanoparticles: Drug Delivery Targets for the Management of Glioblastoma Multiforme (GBM)

5.5.2.1 Quercetin (QT)

QT, a unique plant-based phenolic component and a flavonoid, is widely present in a variety of fruits and vegetables and has numerous special qualities [148], including anti-inflammatory, anti-hypertensive, antioxidant, and anticancer activities. Additionally, QT induces cell death and serves as a potent inducer of apoptosis in cancer cells of the liver, breast, and brain [149]. IL-6 is a crucial cytokine that fuels the tumor's inflammatory environment [150]. Additionally, there is a strong correlation between its increased expression and survival in glioblastoma multiforme (GBM) patients. The continual activation of STAT3 and signal transducers by IL-6 in numerous GBM cell lines has also been demonstrated. The inhibition of IL-6 was associated with increased cell proliferation and death [151]. Additionally, as reported by a recent study, QT unquestionably reduced the activation of STAT3 in the U87 and T98G cell lines [152].

5.5.2.2 Resveratrol

Another interesting polyphenolic phytoalexin in GBM is resveratrol. Numerous investigations showed its anti-inflammatory, neuroprotective, antitumorigenic, and antioxidant activities in a range of illnesses [153]. Additionally, it crosses the BBB, working as a treatment for a number of neurodegenerative illnesses. Resveratrol functioned as a chemo-preventive and anti-oncogenic agent in numerous types of cancer [154]. Through NF-dependent signaling, resveratrol had a significant influence on the T98G and U251 cell lines. It also prevented the activity of several anti-apoptotic proteins, including as the X-linked inhibitors of apoptosis protein [155]. Resveratrol frequently results in the apoptosis, autophagy, and senescence of malignant cells. Numerous other investigations revealed that resveratrol increased the oxidative stress in GBM SHG44 cells, accelerating toxicity by activating the AMPK system [156]. Additionally, resveratrol decreased the expression of the Bcl-2 gene in GBM cells after inhibiting the TOR signaling pathway [157]. Further studies revealed that resveratrol was a powerful inhibitor of the development of skin cancer in mice at various stages of carcinogenesis [158]. In addition, resveratrol suppressed DMBA-induced mammary carcinogenesis in a number of animal experiments [159, 160]. In various stages of glioma, it also shows chemoprotective and anticarcinogenic activities [161, 162]. By modifying numerous cell cycle phases (M/G2) in glioma cells, resveratrol also demonstrated its cytostatic and lethal effects that are independent of p53 in medulloblastoma cells [163, 164]. Another investigation demonstrated the impact of resveratrol on the glioma stem cells derived from a number of GBM patients. According to the study's findings, resveratrol demonstrated its positive benefits by influencing the Wnt signaling pathways [165].

5.2.2.3 Curcumin

Curcumin, a different polyphenolic substance, has shown potential growth suppression, prevented angiogenesis, and induced apoptosis in various cancer types [166]. Antioxidant and anti-inflammatory effects are also associated with a decrease in NF-kB and AP-1 activity [167]. Curcumin has been used in a number of clinical trials to treat high-risk malignancies and GBM cells by preventing the growth of new blood vessels and cells [168]. According to reports, curcumin lowers the risk of brain tumor U87 and increases animal survival rates [169]. Curcuminoids (11%), turmerones (45%), and other chemicals found in turmeric force, a derivative of curcumin, had a stronger cytotoxic effect on malignant cells [170]. Additionally, human rhabdomyosarcoma and embryonic cells activity is slowed by curcumin [171]. According to reports, curcumin also modifies AKT signaling, which affects U87 cells' genes and specifically downregulates them [172]. As was previously mentioned, AKT signaling is associated with angiogenesis, cell proliferation, chemotherapeutic tolerance, and suppression of apoptosis [173].

5.6 Concerns about the Fortification and Augmentation of Polyphenols

Consumption could take the place of eating wholesome, healthful things like fruits and vegetables. Additionally, the synergistic effects and health advantages of a diet naturally high in polyphenols may not be present in polyphenol extracts employed in supplementation and fortification [174]. A high-fiber diet, the ingestion of other nutrients and non-nutrients, as well as satiation, are some of these advantages. It is difficult to comprehend the intricate connections behind the practical advantages associated with consuming whole foods high in polyphenols [175]. Consuming isolated polyphenolic chemicals by themselves might not result in the same health advantages as those shown in epidemiological research, or the advantages might be exaggerated by food marketing firms. Concerns have been raised about polyphenol supplementation and fortification. First, any possible anti-obesogenic effects of polyphenols may be negated by the increased energy density of the fortified foods rather than nutrient density, which may result in weight gain [176]. The level at which polyphenols can be added to food for human consumption in a safe and advantageous manner is unknown because the benefits of polyphenols are tested in cellular and animal studies at concentrations much greater than those frequently observed in human diets. Particularly with supplements, there is a risk of consuming harmful amounts of polyphenols. Some producers advocate ingesting quantities more than 100 times greater than those previously associated with a Western diet [177]. Trials of antioxidant supplements have occasionally been linked to negative outcomes, such as higher mortality or stroke rates [178–181]. With the promotion of polyphenol consumption at amounts considerably greater than what occurs naturally, worries about diverse effects in subpopulations and interactions with pharmaceuticals also surface [180]. The fortification of foods with polyphenols cannot be well-informed without a thorough understanding of the safe and advantageous amounts of intake. Prior to conducting polyphenol supplementation experiments in humans, researchers should exercise considerable caution to make sure that the mechanisms and effects in vivo are well characterized.

5.7 Conclusion

The plant kingdom is a rich source of phytoconstituents called polyphenols. Polyphenol classification is a difficult process; however, it can be done based on the chemical makeup of the substance. They are powerful medicinal agents created by plants and some marine species, and they exhibit a beautiful fusion of biological, chemical, and physiological activities. Low permeability, low bioavailability, and insoluble polyphenols are found in nature. They are routinely added to polymeric and lipid NPs to address their primary limitations. However, to fully take advantage of their many applications in human health, their poor bioavailability, unstable solubility, and restricted stability must be addressed. The creation of polyphenol NPs demonstrate a practical way to overcome these restrictions and enhance the stability of the substances. QT, curcumin, and resveratrol are polyphenols that are most frequently encapsulated. No current medication for GBM appears to be curative, and therapeutic techniques are limited because tumor cells may develop resistance to these medicines. The best course of action might be the combination of targeted medicines. Further research must be done to understand the pharmacokinetics and pharmacodynamics of phenolic substances.

References

1 Harborne, J.B. and Williams, C.A. (2000). Advances in flavonoid research since 1992. *Phytochemistry* 55 (6): 481–504.

2 Beckman, C.H. (2000). Phenolic-storing cells: keys to programmed cell death and periderm formation in wilt disease resistance and in general defense responses in plants? *Physiological and Molecular Plant Pathology* 57 (3): 101–110.

3 Graf, B.A., Milbury, P.E., and Blumberg, J.B. (2005). Flavonols, flavones, flavanones, and human health: epidemiological evidence. *Journal of Medicinal Food* 8 (3): 281–290.

4 Arts, I.C.W. and Hollman, P.C.H. (2005). Polyphenols and disease risk in epidemiologic studies. *The American Journal of Clinical Nutrition* 81 (1 Suppl): 317S–325S.

5 van Erk, M.J., Roepman, P., van der Lende, T.R. et al. (2005). Integrated assessment by multiple gene expression analysis of quercetin bioactivity on anticancer-related mechanisms in colon cancer cells in vitro. *European Journal of Nutrition* 44 (3): 143–156.

6 Miklasińska-Majdanik, M., Kępa, M., Wojtyczka, R.D. et al. (2018). Phenolic compounds diminish antibiotic resistance of staphylococcus aureus clinical strains. *International Journal of Environmental Research and Public Health* 15 (10).

7 Debelo, H., Li, M., and Ferruzzi, M.G. (2020). Processing influences on food polyphenol profiles and biological activity. *Food Microbiology • Functional Foods and Nutrition* 32: 90–102.

8 Kondratyuk, T.P. and Pezzuto, J.M. (2004). Natural product polyphenols of relevance to human health. *Pharmaceutical Biology* 42 (supl): 46–63.

9 Spencer, J.P.E., Abd El Mohsen, M.M., Minihane, A.-M., and Mathers, J.C. (2008). Biomarkers of the intake of dietary polyphenols: strengths, limitations and application in nutrition research. *The British Journal of Nutrition* 99 (1): 12–22.

10 Adlercreutz, H. and Mazur, W. (1997). Phyto-oestrogens and Western diseases. *Annals of Medicine* 29 (2): 95–120.

11 de Groot, H. and Rauen, U. (1998). Tissue injury by reactive oxygen species and the protective effects of flavonoids. *Fundamental & Clinical Pharmacology* 12 (3): 249–255.

12 Nursten, H.E. (1996). Food phenolics: sources, chemistry, effects, applications: f. Shahidi & M. Naczk. Technomic Publishing Co. Inc, Lancaster, PA, USA, and basel, 1995. Ix + 331 pp. Price: $125.00. *Food Chemistry* 57: 481–482.

13 Papadopoulou, A. and Frazier, R.A. (2004). Characterization of protein–polyphenol interactions. *NFIF Part 1* 15 (3): 186–190.

14 Williams, R.J., Spencer, J.P.E., and Rice-Evans, C. (2004). Flavonoids: antioxidants or signaling molecules? *Free Radical Biology & Medicine* 36 (7): 838–849.

15 Virgili, F. and Marino, M. (2008). Regulation of cellular signals from nutritional molecules: a specific role for phytochemicals, beyond antioxidant activity. *Free Radical Biology & Medicine* 45: 1205–1216.

16 Serafini, M., Laranjinha, J.A., Almeida, L.M., and Maiani, G. (2000). Inhibition of human LDL lipid peroxidation by phenol-rich beverages and their impact on plasma total antioxidant capacity in humans. *The Journal of Nutritional Biochemistry* 11 (11–12): 585–590.

17 Owen, R.W., Giacosa, A., Hull, W.E. et al. (2000). The antioxidant/anticancer potential of phenolic compounds isolated from olive oil. *European Journal of Cancer (Oxford, England : 1990)* 36 (10): 1235–1247.

18 Kay, C.D. (2010). The future of flavonoid research. *The British Journal of Nutrition* 104 (Suppl 3): S91–95.

19 Gordon, N.C. and Wareham, D.W. (2010). Antimicrobial activity of the green tea polyphenol (-)-epigallocatechin-3-gallate (EGCG) against clinical isolates of Stenotrophomonas maltophilia. *International Journal of Antimicrobial Agents* 36 (2): 129–131.

20 Rudrapal, M. (ed.) (2022). *Dietary Polyphenols in Human Diseases: Advances and Challenges in Drug Discovery*, 1e. CRC Press.

21 Łuczaj, W. and Skrzydlewska, E. (2005). Antioxidative properties of black tea. *Preventive Medicine* 40 (6): 910–918.

22 Milgrom, P., Riedy, C.A., Weinstein, P. et al. (2000). Dental caries and its relationship to bacterial infection, hypoplasia, diet, and oral hygiene in 6- to 36-month-old children. *Community Dentistry and Oral Epidemiology* 28 (4): 295–306.

23 Chen, C., Yu, R., Owuor, E.D., and Kong, A.N. (2000). Activation of antioxidant-response element (ARE), mitogen-activated protein kinases (MAPKs) and caspases by major green tea polyphenol components during cell survival and death. *Archives of Pharmacal Research* 23 (6): 605–612.

24 Shen, S.-Q., Zhang, Y., Xiang, -J.-J., and Xiong, C.-L. (2007). Protective effect of curcumin against liver warm ischemia/reperfusion injury in rat model is associated with regulation of heat shock protein and antioxidant enzymes. *World Journal of Gastroenterology* 13 (13): 1953–1961.

25 Hong, J., Smith, T.J., Ho, C.T. et al. (2001). Effects of purified green and black tea polyphenols on cyclooxygenase- and lipoxygenase-dependent metabolism of arachidonic acid in human colon mucosa and colon tumor tissues. *Biochemical Pharmacology* 62 (9): 1175–1183.

26 Holt, R.R., Lazarus, S.A., Sullards, M.C. et al. (2002). Procyanidin dimer B2 [epicatechin-(4beta-8)-epicatechin] in human plasma after the consumption of a flavanol-rich cocoa. *The American Journal of Clinical Nutrition* 76 (4): 798–804.

27 Du, Y., Guo, H., and Lou, H. (2007). Grape seed polyphenols protect cardiac cells from apoptosis via induction of endogenous antioxidant enzymes. *Journal of Agricultural and Food Chemistry* 55 (5): 1695–1701.

28 Sharma, R.A., Gescher, A.J., and Steward, W.P. (2005). Curcumin: the story so far. *European Journal of Cancer (Oxford, England : 1990)* 41 (13): 1955–1968.

29 Nishinaka, T., Ichijo, Y., Ito, M. et al. (2007). Curcumin activates human glutathione S-transferase P1 expression through antioxidant response element. *Toxicology Letters* 170 (3): 238–247.

30 Becker, L.B. (2004). New concepts in reactive oxygen species and cardiovascular reperfusion physiology. *Cardiovascular Research* 61 (3): 461–470.

31 Gonthier, M.-P., Remesy, C., Scalbert, A. et al. (2006). Microbial metabolism of caffeic acid and its esters chlorogenic and caftaric acids by human faecal microbiota in vitro. *Biomedicine & Pharmacotherapy = Biomedecine & Pharmacotherapie* 60 (9): 536–540.

32 Kerry, N. and Rice-Evans, C. (1999). Inhibition of peroxynitrite-mediated oxidation of dopamine by flavonoid and phenolic antioxidants and their structural relationships. *Journal of Neurochemistry* [illegible]

33 Han, X., Shen, T., and Lou, H. (2007). Dietary polyphenols and their biological significance. *International Journal of Molecular Sciences* 8 (9): 950–988.

34 Ray, P.S., Maulik, G., Cordis, G.A. et al. (1999). The red wine antioxidant resveratrol protects isolated rat hearts from ischemia reperfusion injury. *Free Radical Biology and Medicine* 27 (1–2): 160–169.

35 McAnlis, G.T., McEneny, J., Pearce, J., and Young, I.S. (1999). Absorption and antioxidant effects of quercetin from onions, in man. *European Journal of Clinical Nutrition* 53 (2): 92–96.

36 Tanigawa, S., Fujii, M., and Hou, D.-X. (2007). Action of Nrf2 and Keap1 in ARE-mediated NQO1 expression by quercetin. *Free Radical Biology & Medicine* 42 (11): 1690–1703.

37 Seeram, N.P., Lee, R., and Heber, D. (2004). Bioavailability of ellagic acid in human plasma after consumption of ellagitannins from pomegranate (Punica granatum L.) juice. *Clinica Chimica Acta; International Journal of Clinical Chemistry* 348 (1–2): 63–68.

38 Rangkadilok, N., Sitthimonchai, S., Worasuttayangkurn, L. et al. (2007). Evaluation of free radical scavenging and antityrosinase activities of standardized longan fruit extract. *Food and Chemical Toxicology : An International Journal Published for the British Industrial Biological Research Association* 45 (2): 328–336.

39 Petroni, A., Blasevich, M., Salami, M. et al. (1995). Inhibition of platelet aggregation and eicosanoid production by phenolic components of olive oil. *Thrombosis Research* 78 (2): 151–160.

40 Carluccio, M.A., Siculella, L., Ancora, M.A. et al. (2003). Olive oil and red wine antioxidant polyphenols inhibit endothelial activation: antiatherogenic properties of mediterranean diet phytochemicals. *Arteriosclerosis, Thrombosis, and Vascular Biology* 23 (4): 622–629.

41 Scalbert, A., Manach, C., Morand, C. et al. (2005). Dietary polyphenols and the prevention of diseases. *Critical Reviews in Food Science and Nutrition* 45 (4): 287–306.

42 Vitrac, X., Monti, J.P., Vercauteren, J. et al. (2002). Direct liquid chromatographic analysis of resveratrol derivatives and flavanonols in wines with absorbance and fluorescence detection. *Analytica Chimica Acta* 458: 103–110.

43 Arts, I.C., Jacobs, D.R.J., Harnack, L.J. et al. (2001). Dietary catechins in relation to coronary heart disease death among postmenopausal women. *Epidemiology (Cambridge, Mass)* 12 (6): 668–675.

44 Nakachi, K., Matsuyama, S., Miyake, S. et al. (2000). Preventive effects of drinking green tea on cancer and cardiovascular disease: epidemiological evidence for multiple targeting prevention. *BioFactors (Oxford, England)* 13 (1–4): 49–54.

45 Mink, P.J., Scrafford, C.G., Barraj, L.M. et al. (2007). Flavonoid intake and cardiovascular disease mortality: a prospective study in postmenopausal women. *The American Journal of Clinical Nutrition* 85 (3): 895–909.

46 Rein, D., Lotito, S., Holt, R.R. et al. (2000). Epicatechin in human plasma: in vivo determination and effect of chocolate consumption on plasma oxidation status. *The Journal of Nutrition* 130 (8S Suppl): 2109S–14S.

47 Wan, Y., Vinson, J.A., Etherton, T.D. et al. (2001). Effects of cocoa powder and dark chocolate on LDL oxidative susceptibility and prostaglandin concentrations in humans. *The American Journal of Clinical Nutrition* 74 (5): 596–602.

48 Truelsen, T., Thudium, D., and Grønbaek, M. (2002). Amount and type of alcohol and risk of dementia: the copenhagen city heart study. *Neurology* 59 (9): 1313–1319.

49 Desch, S., Schmidt, J., Kobler, D. et al. (2010). Effect of cocoa products on blood pressure: systematic review and meta-analysis. *American Journal of Hypertension* 23 (1): 97–103.

50 Grassi, D., Necozione, S., Lippi, C. et al. (2005). Cocoa reduces blood pressure and insulin resistance and improves endothelium-dependent vasodilation in hypertensives. *Hypertension (Dallas, Tex. : 1979)* 46 (2): 398–405.

51 Heiss, C., Finis, D., Kleinbongard, P. et al. (2007). Sustained increase in flow-mediated dilation after daily intake of high-flavanol cocoa drink over 1 week. *Journal of Cardiovascular Pharmacology* 49 (2): 74–80.

52 Grassi, D., Mulder, T.P., Draijer, R. et al. (2009). Black tea consumption dose-dependently improves flow-mediated dilation in healthy males. *Journal of Hypertension* 27 (4): 774–781.

53 Erlund, I., Koli, R., Alfthan, G. et al. (2008). Favorable effects of berry consumption on platelet function, blood pressure, and HDL cholesterol. *The American Journal of Clinical Nutrition* 87 (2): 323–331.

54 Papamichael, C., Karatzis, E., Karatzi, K. et al. (2004). Red wine's antioxidants counteract acute endothelial dysfunction caused by cigarette smoking in healthy nonsmokers. *American Heart Journal* 147 (2): E5.

55 Mathur, S., Devaraj, S., Grundy, S.M., and Jialal, I. (2002). Cocoa products decrease low density lipoprotein oxidative susceptibility but do not affect biomarkers of inflammation in humans. *The Journal of Nutrition* 132 (12): 3663–3667.

56 Mao, T.K., van de Water, J., Keen, C.L. et al. (2002). Modulation of TNF-alpha secretion in peripheral blood mononuclear cells by cocoa flavanols and procyanidins. *Developmental Immunology* 9 (3): 135–141.

57 Schramm, D.D., Karim, M., Schrader, H.R. et al. (2003). Food effects on the absorption and pharmacokinetics of cocoa flavanols. *Life Sciences* 73 (7): 857–869.

58 García-Lafuente, A., Guillamón, E., Villares, A. et al. (2009). Flavonoids as anti-inflammatory agents: implications in cancer and cardiovascular disease. *Inflammation Research : Official Journal of the European Histamine Research Society. [Et Al.]* 58 (9): 537–552.

59 Maeda, K., Kuzuya, M., Cheng, X.W. et al. (2003). Green tea catechins inhibit the cultured smooth muscle cell invasion through the basement barrier. *Atherosclerosis* 166 (1): 23–30.

60 Pirola, L. and Fröjdö, S. (2008). Resveratrol: one molecule, many targets. *IUBMB Life* 60 (5): 323–332.

61 Appeldoorn, M.M., Venema, D.P., Peters, T.H.F. et al. (2009). Some phenolic compounds increase the nitric oxide level in endothelial cells in vitro. *Journal of Agricultural and Food Chemistry* 57 (17): 7693–7699.

62 Heiss, C., Dejam, A., Kleinbongard, P. et al. (2003). Vascular effects of cocoa rich in flavan-3-ols. *JAMA* 290 (8): 1030–1031.

63 Woodman, O.L. and Chan, E.C. (2004). Vascular and anti-oxidant actions of flavonols and flavones. *Clinical and Experimental Pharmacology & Physiology* 31 (11): 786–790.

64 Chin-Dusting, J.P., Fisher, L.J., Lewis, T.V. et al. (2001). The vascular activity of some isoflavone metabolites: implications for a cardioprotective role. *British Journal of Pharmacology* 133 (4): 595–605.

65 Stoclet, J.-C., Chataigneau, T., Ndiaye, M. et al. (2004). Vascular protection by dietary polyphenols. *European Journal of Pharmacology* 500 (1–3): 299–313.

66 Yang, C.S., Landau, J.M., Huang, M.T., and Newmark, H.L. (2001). Inhibition of carcinogenesis by dietary polyphenolic compounds. *Annual Review of Nutrition* 21: 381–406.

67 Johnson, I.T., Williamson, G., and Musk, S.R. (1994). Anticarcinogenic factors in plant foods: a new class of nutrients? *Nutrition Research Reviews* 7 (1): 175–204.

68 La Vecchia, C., Chatenoud, L., Franceschi, S. et al. (1999). Vegetables and fruit and human cancer: update of an Italian study. *International Journal of Cancer* 82 (1): 151–152.

69 Corona, G., Deiana, M., Incani, A. et al. (2007). Inhibition of p38/CREB phosphorylation and COX-2 expression by olive oil polyphenols underlies their anti-proliferative effects. *Biochemical and Biophysical Research Communications* 362 (3): 606–611.

70 Corona, G., Deiana, M., Incani, A. et al. (2009). Hydroxytyrosol inhibits the proliferation of human colon adenocarcinoma cells through inhibition of ERK1/2 and cyclin D1. *Molecular Nutrition & Food Research* 53 (7): 897–903.

71 Fini, L., Hotchkiss, E., Fogliano, V. et al. (2008). Chemopreventive properties of pinoresinol-rich olive oil involve a selective activation of the ATM-p53 cascade in colon cancer cell lines. *Carcinogenesis* 29 (1): 139–146.

72 Adams, L.S. and Chen, S. (2009). Phytochemicals for breast cancer prevention by targeting aromatase. *Frontiers in Bioscience (Landmark Edition)* 14 (10): 3846–3863.

73 Khan, W.A., Wang, Z.Y., Athar, M. et al. (1988). Inhibition of the skin tumorigenicity of (±)-7 beta,8 alpha-dihydroxy-9 alpha,10 alpha-epoxy-7,8,9,10-tetrahydrobenzo[a]pyrene by tannic acid, green tea polyphenols and quercetin in Sencar mice. *Cancer Letters* 42 (1–2): 7–12.

74 Zeng, J.-L., Li, Z.-H., Wang, Z.-C., and Zhang, H.-L. (2014). Green tea consumption and risk of pancreatic cancer: a meta-analysis. *Nutrients* 6 (11): 4640–4650.

75 Inoue, M., Tajima, K., Mizutani, M. et al. (2001). Regular consumption of green tea and the risk of breast cancer recurrence: follow-up study from the hospital-based epidemiologic research program at aichi cancer center (HERPACC), Japan. *Cancer Letters* 167 (2): 175–182.

76 Takada, M., Ku, Y., Habara, K. et al. (2002). Inhibitory effect of epigallocatechin-3-gallate on growth and invasion in human biliary tract Carcinoma cells. *World Journal of Surgery* 26 (6): 683–686.

77 Philips, B.J., Coyle, C.H., Morrisroe, S.N. et al. (2009). Induction of apoptosis in human bladder cancer cells by green tea catechins. *Biomedical Research* 30 (4): 207–215.

78 Wu, A.H. and Butler, L.M. (2011). Green tea and breast cancer. *Molecular Nutrition & Food Research* 55 (6): 921–930.

79 Larsen, C.A. and Dashwood, R.H. (2009). Suppression of Met activation in human colon cancer cells treated with (-)-epigallocatechin-3-gallate: minor role of hydrogen peroxide. *Biochemical and Biophysical Research Communications* 389 (3): 527–530.

80 Khan, N., Afaq, F., Saleem, M. et al. (2006). Targeting multiple signaling pathways by green tea polyphenol (-)-epigallocatechin-3-gallate. *Cancer Research* 66 (5): 2500–2505.

81 Athar, M., Back, J.H., Tang, X. et al. (2007). Resveratrol: a review of preclinical studies for human cancer prevention. *Toxicology and Applied Pharmacology* 224 (3): 274–283.

82 Kamaraj, S., Vinodhkumar, R., Anandakumar, P. et al. (2007). The effects of quercetin on antioxidant status and tumor markers in the lung and serum of mice treated with benzo(a)pyrene. *Biological & Pharmaceutical Bulletin* 30 (12): 2268–2273.

83 Rizvi, S.I., Zaid, M.A., Anis, R., and Mishra, N. (2005). Protective role of tea catechins against oxidation-induced damage of type 2 diabetic erythrocytes. *Clinical and Experimental Pharmacology & Physiology* 32 (1–2): 70–75.

84 Matsui, T., Ebuchi, S., Kobayashi, M. et al. (2002). Anti-hyperglycemic effect of diacylated anthocyanin derived from Ipomoea batatas cultivar Ayamurasaki can be achieved through the alpha-glucosidase inhibitory action. *Journal of Agricultural and Food Chemistry* 50 (25): 7244–7248.

85 Matsui, T., Ueda, T., Oki, T. et al. (2001). Alpha-Glucosidase inhibitory action of natural acylated anthocyanins. 1. Survey of natural pigments with potent inhibitory activity. *Journal of Agricultural and Food Chemistry* 49 (4): 1948–1951.

86 Dembinska-Kiec, A., Mykkänen, O., Kiec-Wilk, B., and Mykkänen, H. (2008). Antioxidant phytochemicals against type 2 diabetes. *The British Journal of Nutrition* 99 (Suppl 1): ES109–117.

87 Harikumar, K.B. and Aggarwal, B.B. (2008). Resveratrol: a multitargeted agent for age-associated chronic diseases. *Cell Cycle (Georgetown, Tex.)* 7 (8): 1020–1035.

88 Chen, W.-P., Chi, T.-C., Chuang, L.-M., and Su, M.-J. (2007). Resveratrol enhances insulin secretion by blocking K(ATP) and K(V) channels of beta cells. *European Journal of Pharmacology* 568 (1–3): 269–277.

89 Pandey, K.B. and Rizvi, S.I. (2009). Plant polyphenols as dietary antioxidants in human health and disease. *Oxidative Medicine and Cellular Longevity* 2 (5): 270–278.

90 Hy, L.X. and Keller, D.M. (2000). Prevalence of AD among whites: a summary by levels of severity. *Neurology* 55 (2): 198–204.

91 Nussbaum, R.L. and Ellis, C.E. (2003). Alzheimer's disease and Parkinson's disease. *The New England Journal of Medicine* 348 (14): 1356–1364.

92 de Lau, L.M.L. and Breteler, M.M.B. (2006). Epidemiology of Parkinson's disease. *The Lancet. Neurology* 5 (6): 525–535.

93 Spires, T.L. and Hannan, A.J. (2005). Nature, nurture and neurology: gene–environment interactions in neurodegenerative disease. *The FEBS Journal* 272 (10): 2347–2361.

94 Letenneur, L., Proust-Lima, C., Le Gouge, A. et al. (2007). Flavonoid intake and cognitive decline over a 10-year period. *American Journal of Epidemiology* 165 (12): 1364–1371.

95 Scarmeas, N., Luchsinger, J.A., Mayeux, R., and Stern, Y. (2007). Mediterranean diet and Alzheimer disease mortality. *Neurology* 69 (11): 1084–1093.

96 Commenges, D., Scotet, V., Renaud, S. et al. (2000). Intake of flavonoids and risk of dementia. *European Journal of Epidemiology* 16 (4): 357–363.

97 Dai, Q., Borenstein, A.R., Wu, Y. et al. (2006). Fruit and vegetable juices and Alzheimer's disease: the Kame Project. *The American Journal of Medicine* 119 (9): 751–759.

98 Youdim, K.A. and Joseph, J.A. (2001). A possible emerging role of phytochemicals in improving age-related neurological dysfunctions: a multiplicity of effects. *Free Radical Biology & Medicine* 30 (6): 583–594.

99 Luo, Y., Smith, J.V., Paramasivam, V. et al. (2002). Inhibition of amyloid-beta aggregation and caspase-3 activation by the Ginkgo biloba extract EGb761. *Proceedings of the National Academy of Sciences of the United States of America* 99 (19): 12197–12202.

100 Bastianetto, S., Zheng, W.H., and Quirion, R. (2000). The Ginkgo biloba extract (EGb 761) protects and rescues hippocampal cells against nitric oxide-induced toxicity: involvement of its flavonoid constituents and protein kinase C. *Journal of Neurochemistry* 74 (6): 2268–2277.

101 Tchantchou, F., Xu, Y., Wu, Y. et al. (2007). EGb 761 enhances adult hippocampal neurogenesis and phosphorylation of CREB in transgenic mouse model of Alzheimer's disease. *FASEB Journal : Official Publication of the Federation of American Societies for Experimental Biology* 21 (10): 2400–2408.

102 Datla, K.P., Christidou, M., Widmer, W.W. et al. (2001). Tissue distribution and neuroprotective effects of citrus flavonoid tangeretin in a rat model of Parkinson's disease. *Neuroreport* 12 (17): 3871–3875.

103 Vauzour, D., Corona, G., and Spencer, J.P.E. (2010). Caffeic acid, tyrosol and p-coumaric acid are potent inhibitors of 5-S-cysteinyl-dopamine induced neurotoxicity. *Archives of Biochemistry and. Biophysics* 501 (1): 106–111.

104 Markus, M.A. and Morris, B.J. (2008). Resveratrol in prevention and treatment of common clinical conditions of aging. *Clinical Interventions in Aging* 3 (2): 331–339.

105 Rossi, L., Mazzitelli, S., Arciello, M. et al. (2008). Benefits from dietary polyphenols for brain aging and Alzheimer's disease. *Neurochemical Research* 33 (12): 2390–2400.

106 Aquilano, K., Baldelli, S., Rotilio, G., and Ciriolo, M.R. (2008). Role of nitric oxide synthases in Parkinson's disease: a review on the antioxidant and anti-inflammatory activity of polyphenols. *Neurochemical Research* 33 (12): 2416–2426.

107 Santos, A.C., Pereira, I., Pereira-Silva, M. et al. (2019). Nanotechnology-based formulations for resveratrol delivery: effects on resveratrol in vivo bioavailability and bioactivity. *Colloids and Surfaces. B, Biointerfaces* 180: 127–140.

108 Tsai, Y.M., Chien, C.F., Lin, L.C., and Tsai, T.H. (2011). Curcumin and its nano-formulation: the kinetics of tissue distribution and blood-brain barrier penetration. *International Journal of Pharmaceutics* 416 (1): 331–338.

109 Fang, Z. and Bhandari, B. (2010). Encapsulation of polyphenols – a review. *Trends in Food Science & Technology* 21 (10): 510–523.

110 Choukaife, H., Doolaanea, A.A., and Alfatama, M. (2020). Alginate nanoformulation: influence of process and selected variables. *Pharmaceuticals (Basel, Switzerland)* 13 (11): 335.

111 Deladino, L., Anbinder, P.S., Navarro, A.S., and Martino, M.N. (2008). Encapsulation of natural antioxidants extracted from Ilex paraguariensis. *Carbohydrate Polymers* 71 (1): 126–134.

112 Vieira, R., Souto, S.B., Sánchez-López, E. et al. (2019). Sugar-lowering drugs for type 2 diabetes mellitus and metabolic syndrome-review of classical and new compounds: part-I. *Pharmaceuticals (Basel, Switzerland)* 12 (4): 152.

113 Jose, S., Sowmya, S., Cinu, T.A. et al. (2014). Surface modified PLGA nanoparticles for brain targeting of Bacoside-A. *European Journal of Pharmaceutical Sciences: Official Journal of the European Federation for Pharmaceutical Sciences* 63: 29–35.

114 Bohrey, S., Chourasiya, V., and Pandey, A. (2016). Polymeric nanoparticles containing diazepam: preparation, optimization, characterization, in-vitro drug release and release kinetic study. *Nano Convergence* 3 (1): 3.

115 Sharma, N., Madan, P., and Lin, S. (2016). Effect of process and formulation variables on the preparation of parenteral paclitaxel-loaded biodegradable polymeric nanoparticles: a co-surfactant study. *Asian J Pharm Sci* 11 (3): 404–416.

116 Szczęch, M. and Szczepanowicz, K. (2020). Polymeric core-shell nanoparticles prepared by spontaneous emulsification solvent evaporation and functionalized by the layer-by-layer method. *Nanomaterials (Basel, Switzerland)* 10 (3): 496.

117 Vehring, R., Snyder, H., and Lechuga-Ballesteros, D. (2020). *Spray Drying. In Drying Technologies for Biotechnology and Pharmaceutical Applications*, 1e. (ed. S. Ohtake, K. Izutsu, and D. Lechuga-Ballesteros), 179–216. Weinheim, Germany: Wiley-VCH Verlag GmbH.

118 Vehring, R. (2008). Pharmaceutical particle engineering via spray drying. *Pharmaceutical Research* 25: 999–1022.

119 Masters, K. (1991). *The Spray Drying Handbook*. 329–556. New York, NY, USA: .

120 Stenger, F. and Peukert, W. (2003). The role of particle interactions on suspension rheology – application to submicron grinding in stirred ball mills. *Chemical Engineering & Technology* 26: 177–183.

121 Bilgili, E., Capece, M., and Afolabi, A. (2017). Modeling of milling processes via DEM, PBM, and microhydrodynamics. In: *Predictive modeling of pharmaceutical unit operations*, 159–203. Woodhead Publishing.

122 Fessi, H., Puisieux, F., Devissaguet, J.P. et al. (1989). Nanocapsule formation by interfacial polymer deposition following solvent displacement. *International Journal of Pharmacy [Internet]* 55 (1): R1–4.

123 Pal, S.L., Jana, U., Manna, P.K. et al. (2011). Nanoparticle: an overview of preparation and characterization. *Journal of applied pharmaceutical science*, (Issue), 228–234.

124 Mehrbod, P., Abdalla, M.A., Fotouhi, F. et al. (2018). Immunomodulatory properties of quercetin-3-O-α-L-rhamnopyranoside from Rapanea melanophloeos against influenza a virus. *BMC Complementary and Alternative Medicine* 18 (1): 184.

125 Prakash, O., Singh, R., Singh, N. et al. (2018). Exploring the potentials of quercetin and kaempferol combinations along with regular antibiotics for the effective management of methicillin-resistant staphylococcus aureus (MRSA). *Research and Reviews: A Journal of Microbiology and Virology* 8 (3): 6–9.

126 Gopalakrishnan, A., Ram, M., Kumawat, S. et al. (2016). Quercetin accelerated cutaneous wound healing in rats by increasing levels of VEGF and TGF-β1. *Indian Journal of Experimental Biology* 54 (3): 187–195.

127 Koohpayma, H., Goudarzi, I., Salmani, M.E. et al. (2015). The effect of quercetin on skin wound healing in dexamethasone-treated rats. *Journal of Arak University of Medical Sciences* 18 (9): 66–73.

128 Shegokar, R. and Müller, R.H. (2010). Nanocrystals: industrially feasible multifunctional formulation technology for poorly soluble actives. *International Journal of Pharmaceutics* 399 (1–2): 129–139.

129 Karimi, M., Bahrami, S., Ravari, S.B. et al. (2016). Albumin nanostructures as advanced drug delivery systems. *Expert Opinion on Drug Delivery* 13 (11): 1609–1623.

130 Singh, A., Dutta, P.K., Kumar, H. et al. (2018). Synthesis of chitin-glucan-aldehyde-quercetin conjugate and evaluation of anticancer and antioxidant activities. *Carbohydrate Polymers* 193: 99–107.

131 Aghapour, F., Moghadamnia, A.A., Nicolini, A. et al. (2018). Quercetin conjugated with silica nanoparticles inhibits tumor growth in MCF-7 breast cancer cell lines. *Biochemical and Biophysical Research Communications* 500 (4): 860–865.

132 Yamina, A.M., Fizir, M., Itatahine, A. et al. (2018). Preparation of multifunctional PEG-graft-halloysite nanotubes for controlled drug release, tumor cell targeting, and bio-imaging. *Colloids and Surfaces. B, Biointerfaces* 170: 322–329.

133 Wan, S., Zhang, L., Quan, Y., and Wei, K. (2018). Resveratrol-loaded PLGA nanoparticles: enhanced stability, solubility and bioactivity of resveratrol for non-alcoholic fatty liver disease therapy. *Royal Society Open Science* 5 (11): 181457.

134 Li, X.M., Wu, Z.Z., Zhang, B. et al. (2019). Fabrication of chitosan hydrochloride and carboxymethyl starch complex nanogels as potential delivery vehicles for curcumin. *Food Chemistry* 293: 197–203.

135 Hu, C., Li, M., Guo, T. et al. (2019). Anti-metastasis activity of curcumin against breast cancer via the inhibition of stem cell-like properties and EMT. *Phytomedicine: International Journal of Phytotherapy and Phytopharmacology* 58: 152740.

136 Remedio, L.N., Dos Santos, J.W.S., Maciel, V.B.V. et al. (2019). Characterization of active chitosan films as a vehicle of potassium sorbate or nisin antimicrobial agents. *Food Hydrocoll* 87: 830–838.

137 Kumar, D., Raj, V., Verma, A. et al. (2019). Novel binary grafted chitosan nanocarrier for sustained release of curcumin. *International Journal of Biological Macromolecules* 131: 184–191.

138 Pan, Y., Wu, Z., Zhang, B. et al. (2019). Preparation and characterization of emulsion stabilized by octenyl succinic anhydride-modified dextrin for improving storage stability and curcumin encapsulation. *Food Chemistry* 294: 326–332.

139 Chang, C., Meikle, T.G., Su, Y. et al. (2019). Encapsulation in egg white protein nanoparticles protects anti-oxidant activity of curcumin. *Food Chemistry* 280: 65–72.

140 Li, X., Maldonado, L., Malmr, M. et al. (2019). Development of hollow kafirin-based nanoparticles fabricated through layer-by-the-layer assembly as delivery vehicles for curcumin. *Food Hydrocoll* 96: 93–101.

141 Wu, W., Kong, X., Zhang, C. et al. (2018). Improving the stability of wheat gliadin nanoparticles–effect of gum arabic addition. *Food Hydrocoll* 80: 78–87.

142 Szymusiak, M., Hu, X., Plata, P.A.L. et al. (2016). Bioavailability of curcumin and curcumin glucuronide in the central nervous system of mice after oral delivery of nanocurcumin. *International Journal of Pharmaceutics* 511 (1): 415–423.

143 Ghalandarlaki, N., Alizadeh, A.M., and Ashkani-Esfahani, S. (2014). Nanotechnology-applied curcumin for different diseases therapy. *BioMed Research International* 2014: 394264.

144 Kiselev, K.V. (2011). Perspectives for production and application of resveratrol. *Applied Microbiology and Biotechnology* 90 (2): 417–425.

145 Gambini, J., Inglés, M., Olaso, G. et al. (2015). Properties of resveratrol: in vitro and in vivo studies about metabolism, bioavailability, and biological effects in animal models and humans. *Oxidative Medicine and Cellular Longevity* 2015: 837042.

146 Faridi Esfanjani, A. and Jafari, S.M. (2016). Biopolymer nano-particles and natural nano-carriers for nano-encapsulation of phenolic compounds. *Colloids and Surfaces. B, Biointerfaces* 146: 532–543.

147 Wei, Y., Yang, S., Zhang, L. et al. (0000). Fabrication, characterization and in vitro digestion of food grade complex nanoparticles for co-delivery of resveratrol and coenzyme Q10. *Food Hydrocolloids* 105: 105791.

148 Hong, D.S., Angelo, L.S., and Kurzrock, R. (2007). Interleukin-6 and its receptor in cancer: implications for translational therapeutics. *Cancer* 110 (9): 1911–1928.

149 West, A.J., Tsui, V., Stylli, S.S. et al. (2018). The role of interleukin-6-STAT3 signaling in glioblastoma. *Oncology Letters* 16 (4): 4095–4104.

150 Rahaman, S.O., Harbor, P.C., Chernova, O. et al. (2002). Inhibition of constitutively active Stat3 suppresses proliferation and induces apoptosis in glioblastoma multiforme cells. *Oncogene* 21 (55): 8404–8413.

151 Bischoff, S.C. (2008). Quercetin: potentials in the prevention and therapy of disease. *Current Opinion in Clinical Nutrition and Metabolic Care* 11 (6): 733–740.

152 Perez-Vizcaino, F., Duarte, J., and Andriantsitohaina, R. (2006). Endothelial function and cardiovascular disease: effects of quercetin and wine polyphenols. *Free Radical Research* 40 (10): 1054–1065.

153 El-Readi, M.Z., Eid, S., Abdelghany, A.A. et al. (2019). Resveratrol mediated cancer cell apoptosis, and modulation of multidrug resistance proteins and metabolic enzymes. *Phytomedicine: International Journal of Phytotherapy and Phytopharmacology* 55: 269–281.

154 Vervandier-Fasseur, D. and Latruffe, N. (2019). The potential use of resveratrol for cancer prevention. *Molecules (Basel, Switzerland)* 24 (24): 4506.

155 Jiao, Y., Li, H., Liu, Y. et al. (2015). Resveratrol inhibits the invasion of glioblastoma-initiating cells via down-regulation of the PI3K/Akt/NF-κB signaling pathway. *Nutrients* 7 (6): 4383–4402.

156 Westhoff, M.A., Zhou, S., Nonnenmacher, L. et al. (2013). Inhibition of NF-κB signaling ablates the invasive phenotype of glioblastoma. *Molecular Cancer Research: MCR* 11 (12): 1611–1623.

157 Huang, H., Lin, H., Zhang, X., and Li, J. (2012). Resveratrol reverses temozolomide resistance by downregulation of MGMT in T98G glioblastoma cells by the NF-κB-dependent pathway. *Oncology Reports* 27 (6): 2050–2056.

158 Li, F. and Sethi, G. (2010). Targeting transcription factor NF-kappaB to overcome chemoresistance and radioresistance in cancer therapy. *Biochimica et biophysica acta* 1805 (2): 167–180.

159 Hegi, M.E., Diserens, A.C., Gorlia, T. et al. (2005). MGMT gene silencing and benefit from temozolomide in glioblastoma. *The New England Journal of Medicine* 352 (10): 997–1003.

160 Arepalli, S.K., Choi, M., Jung, J.K., and Lee, H. (2015). Novel NF-κB inhibitors: a patent review (2011–2014). *Expert Opinion on Therapeutic Patents* 25 (3): 319–334.

161 Patel, K.R., Brown, V.A., Jones, D.J. et al. (2010). Clinical pharmacology of resveratrol and its metabolites in colorectal cancer patients. *Cancer Research* 70 (19): 7392–7399.

162 Howells, L.M., Berry, D.P., Elliott, P.J. et al. (2011). Phase I randomized, double-blind pilot study of micronized resveratrol (SRT501) in patients with hepatic metastases–safety, pharmacokinetics, and pharmacodynamics. *Cancer Prevention Research (Philadelphia, Pa.)* 4 (9): 1419–1425.

163 Brown, V.A., Patel, K.R., Viskaduraki, M. et al. (2010). Repeat dose study of the cancer chemopreventive agent resveratrol in healthy volunteers: safety, pharmacokinetics, and effect on the insulin-like growth factor axis. *Cancer Research* 70 (22): 9003–9011.

164 Gangemi, R.M., Griffero, F., Marubbi, D. et al. (2009). SOX2 silencing in glioblastoma tumor-initiating cells causes stop of proliferation and loss of tumorigenicity. *Stem Cells (Dayton, Ohio)* 27 (1): 40–48.

165 Ryu, J., Ku, B.M., Lee, Y.K. et al. (2011). Resveratrol reduces TNF-α-induced U373MG human glioma cell invasion through regulating NF-κB activation and uPA/uPAR expression. *Anticancer Research* 31 (12): 4223–4230.

166 Duvoix, A., Blasius, R., Delhalle, S. et al. (2005). Chemopreventive and therapeutic effects of curcumin. *Cancer Letters* 223 (2): 181–190.

167 Ryskalin, L., Biagioni, F., Busceti, C.L. et al. (2020). The multi-faceted effect of curcumin in glioblastoma from rescuing cell clearance to autophagy-independent effects. *Molecules (Basel, Switzerland)* 25 (20): 4839.

168 Anand, P., Kunnumakkara, A.B., Newman, R.A., and Aggarwal, B.B. (2007). Bioavailability of curcumin: problems and promises. *Molecular Pharmaceutics* 4 (6): 807–818.

169 Suresh, D. and Srinivasan, K. (2010). Tissue distribution & elimination of capsaicin, piperine & curcumin following oral intake in rats. *The Indian Journal of Medical Research* 131: 682–691.

170 Ramachandran, C., Resek, A.P., Escalon, E. et al. (2010). Potentiation of gemcitabine by turmeric force in pancreatic cancer cell lines. *Oncology Reports* 23 (6): 1529–1535.

171 Ramachandran, C., Nair, S.M., Escalon, E., and Melnick, S.J. (2012). Potentiation of etoposide and temozolomide cytotoxicity by curcumin and turmeric force™ in brain tumor cell lines. *Journal of Complementary & Integrative Medicine* 9.

172 Ji, M., Choi, J., Lee, J., and Lee, Y. (2004). Induction of apoptosis by ar-turmerone on various cell lines. *International Journal of Molecular Medicine* 14 (2): 253–256.

173 Ramachandran, C., Quirin, K.W., Escalon, E.A. et al. (2015). Therapeutic effect of supercritical CO2 extracts of curcuma species with cancer drugs in rhabdomyosarcoma cell lines. *Phytotherapy Research: PTR* 29 (8): 1152–1160.

174 Mazzanti, G., Di Sotto, A., and Vitalone, A. (2015). Hepatotoxicity of green tea: an update. *Archives of Toxicology* 89 (8): 1175–1191.

175 Crowe, K.M., Francis, C., and Academy of Nutrition and Dietetics (2013). Position of the academy of nutrition and dietetics: functional foods. *Journal of the Academy of Nutrition and Dietetics* 113 (8): 1096–1103.

176 Williamson, G. and Holst, B. (2008). Dietary reference intake (DRI) value for dietary polyphenols: are we heading in the right direction? *The British Journal of Nutrition* 99 (Suppl 3): S55–S58.

177 García-Cortés, M., Robles-Díaz, M., Ortega-Alonso, A. et al. (2016). Hepatotoxicity by dietary supplements: a tabular listing and clinical characteristics. *International Journal of Molecular Sciences* 17 (4): 537.

178 Whitsett, M., Marzio, D.H., and Rossi, S. (2014). SlimQuick™-associated hepatotoxicity resulting in fulminant liver failure and orthotopic liver transplantation. *ACG Case Reports Journal* 1 (4): 220–222.

179 Martin, K.R. and Appel, C.L. (2010). Polyphenols as dietary supplements: a double-edged sword. *Nutrition and Dietary Supplements* 2: 1–12.

180 Hooper, B. and Frazier, R. (2012). Polyphenols in the diet: friend or foe? *Nutrition Bulletin* 37: 297–308.

181 Bjelakovic, G., Nikolova, D., and Gluud, C. (2014). Antioxidant supplements and mortality. *Current Opinion in Clinical Nutrition and Metabolic Care* 17: 40–44.

6

Polyphenols in Food Products – Nutraceutical Applications

André M. Oliveira[1,] and Mithun Rudrapal[2]*

[1] *Federal Centre of Technological Education of Minas Gerais (CEFET-MG), Alameda das Perdizes, Contagem, MG, Brazil*
[2] *Department of Pharmaceutical Sciences, School of Biotechnology and Pharmaceutical Sciences, Vignan's Foundation for Science, Technology & Research, Guntur, India*
* *Corresponding author*

Polyphenols (PPHs) are compounds derived from the secondary metabolism of plants, and they share a common structure moiety containing a benzene ring bound to multiple hydroxyl groups. The metabolism of shikimic acid and polyacetate is the main biochemical source for the biosynthesis of these compounds [1]. Secondary metabolism is not essential to the plant survival, and it is responsible for the individuality of species [2].

Three principal kinds of secondary metabolites are biosynthesized by plants [3]: phenolic compounds terpenoids/isoprenoids, alkaloids, and glucosinolates. Phenolic compound biosynthesis is promoted by biotic and abiotic stresses such as herbivores, pathogens, temperature effects, pH, saline stress, carbon dioxide, ozone, heavy metal stress, and ultraviolet radiation (Figure 6.1).

The production of phenolic compounds in plants involves a complex biosynthetic system that starts with shikimic acid, leading to the formation of aromatic amino acids, such as L-tryptophan (which is converted into alkaloids) and L-phenylalanine, which after being transformed into L-tyrosine, enables the formation of the other compounds explained here.

6.1 Polyphenols: Concept and Classification

PPHs encompass a large and diverse group of compounds that contain aromatic rings bound to hydroxyl groups. A general classification of them is summarized and illustrated in Table 6.1. Despite the common phenolic substructure that characterizes this family of natural compounds, the differences between the classes are also worth mentioning. Coumarins are lactones (cyclic esters) that can be structurally related to phenolic acids. Flavonoids contain pyran-4-one rings fused with a diphenol ring (or its methyl ethers). Some contain chiral centers such as hesperetin, naringenin, eriodictyol, and epicatechin. Particularly interesting is the class of anthocyanins, which contains an oxygen atom with a formal positive charge, that occurs in plants in the form of salts that influence their solubility and bioavailability. Because of their pH-dependent colors, they are used as colorants, food additives, and acid–base indicators.

Polyphenols: Food, Nutraceutical, and Nanotherapeutic Applications, First Edition. Edited by Mithun Rudrapal.

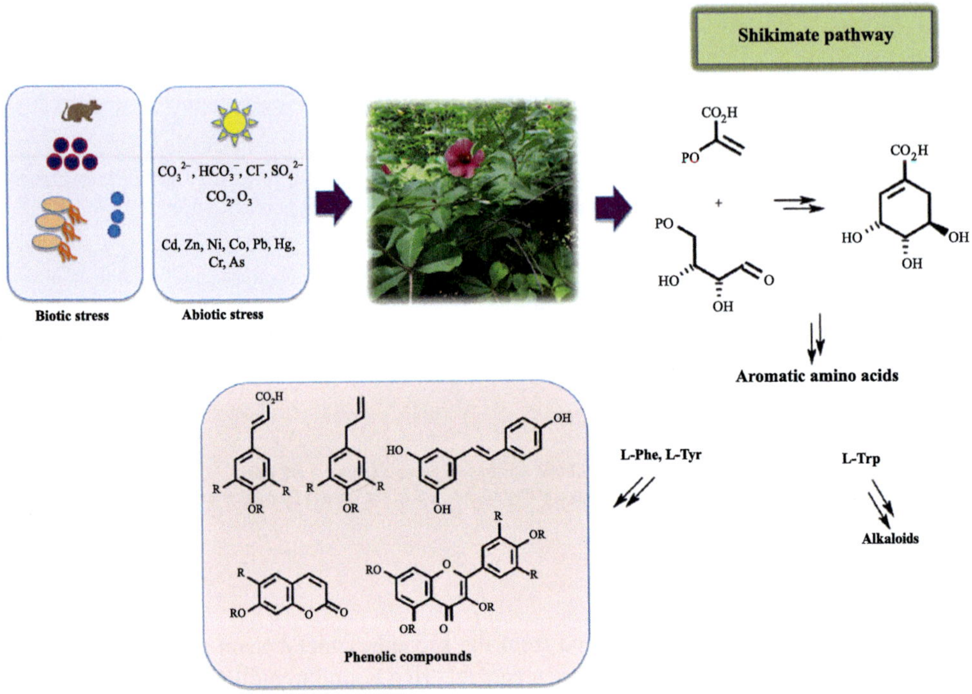

Figure 6.1 Phenolic compound biosynthesis promoted by biotic and abiotic stresses. Reproduced, by permission CC BY 3.0, from Ref [3].

Table 6.1 Main classes of plant polyphenols and typical examples. Adapted from Ref [4, 5].

Polyphenol Class		Examples	
Phenolic acids and coumarins	Hydroxybenzoic acids	Gallic acid	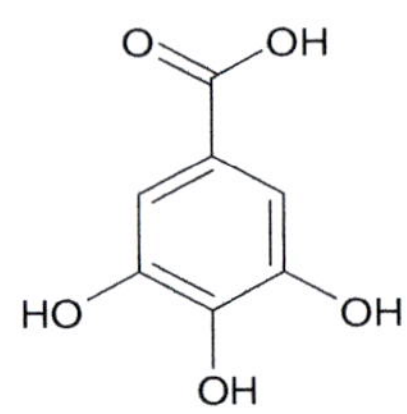
		Vanillic acid	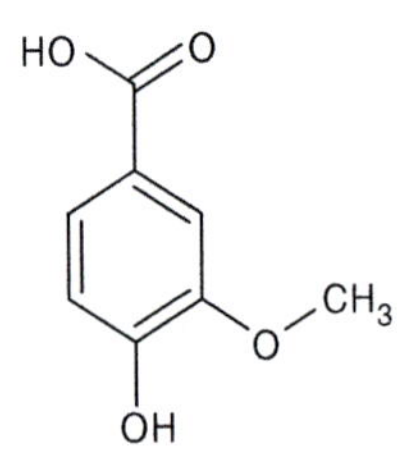
		Protocatechuic acid	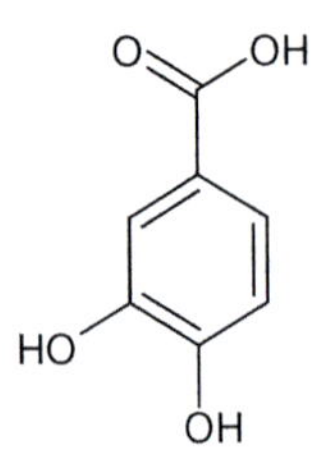

Table 6.1 (Continued)

Polyphenol Class	Examples	
	p-Hydroxybenzoic acid	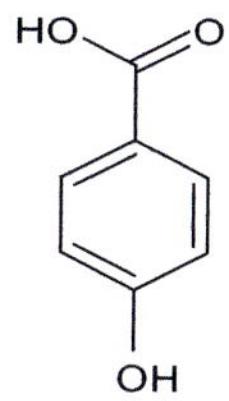
Hydroxycinnamic acids	Caffeic acid	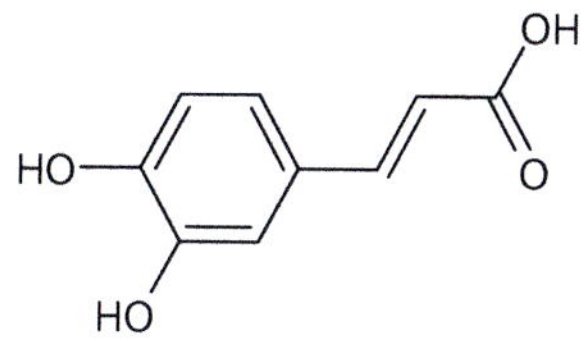
	p-Coumaric acid	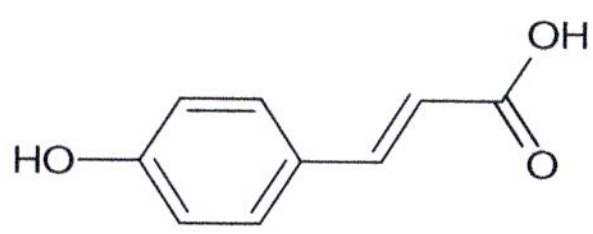
	Sinapic acid	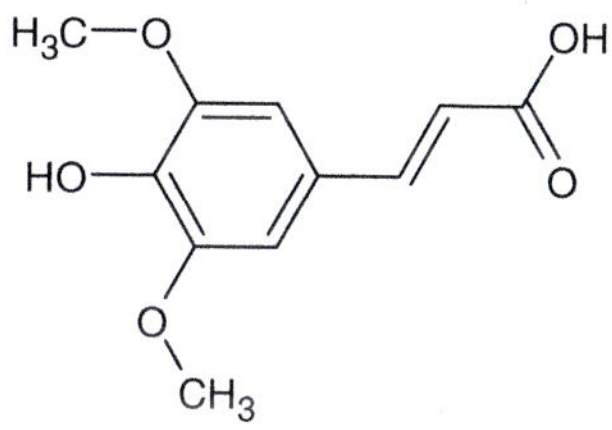
	Ferulic acid	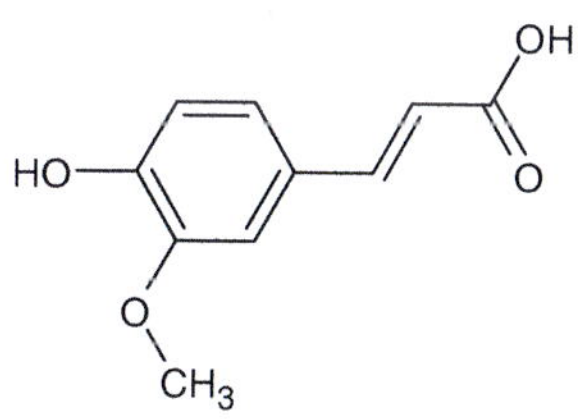
	Umbelliferone	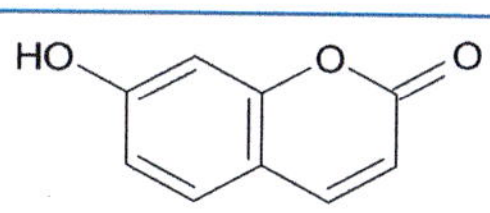

(Continued)

Table 6.1 (Continued)

Polyphenol Class		Examples	
	Coumarins	Aesculetin	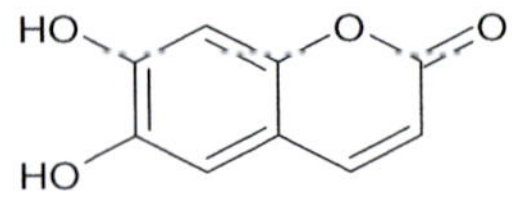
		Scopoletin	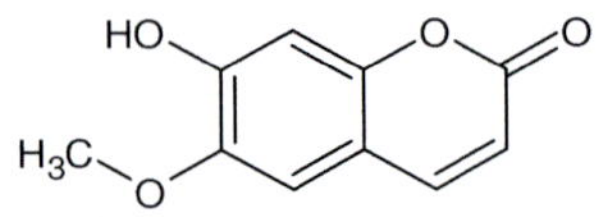
	Stilbenes	Resveratrol	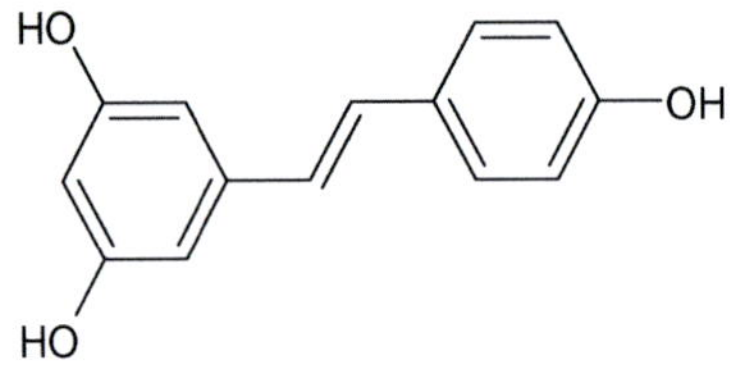
Flavonoids		Myricetin	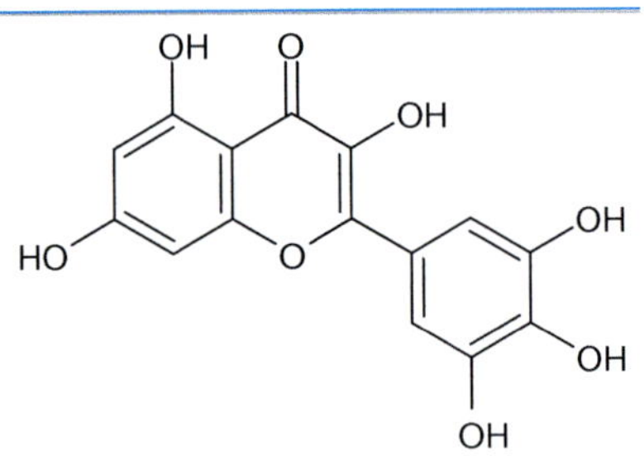
		Quercetin	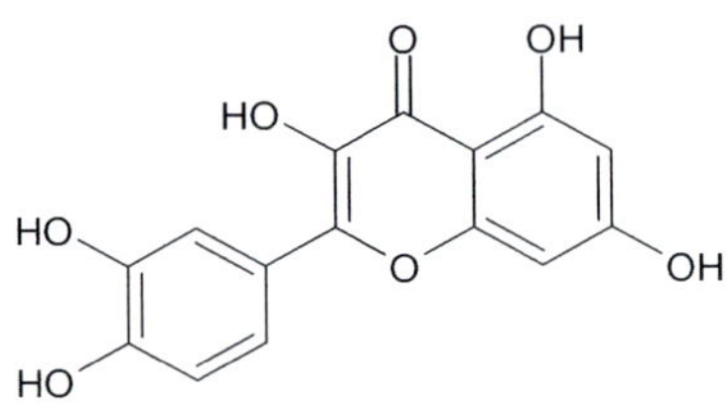
		Kaempferol (and glycosylated forms)	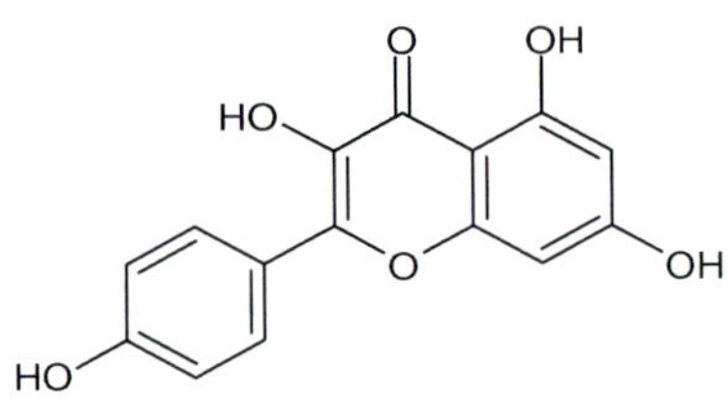

Table 6.1 (Continued)

Polyphenol Class	Examples	
Flavonols	Apigenin	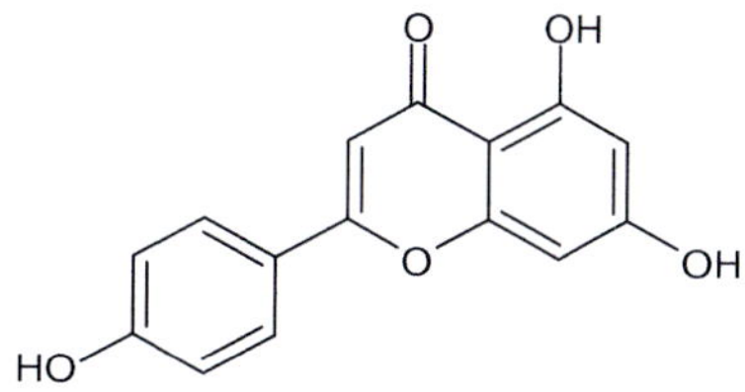
	Luteolin	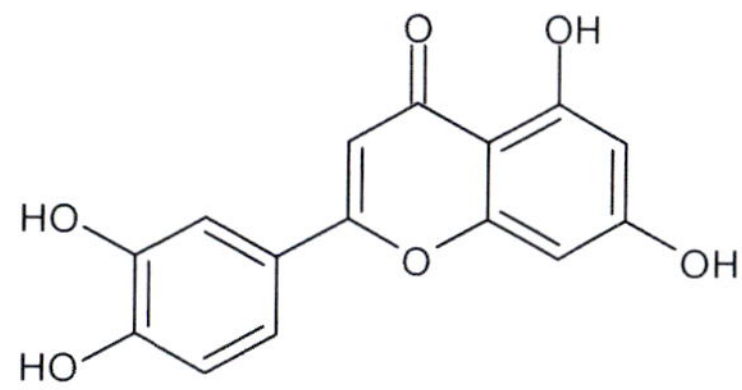
	Tangeretin	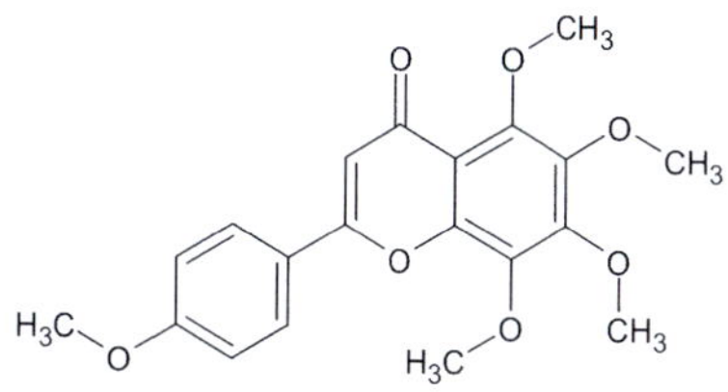
	Nobiletin	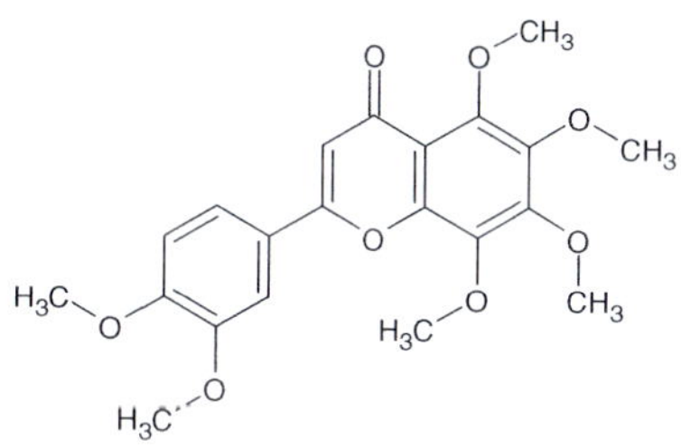
	Sinensetin	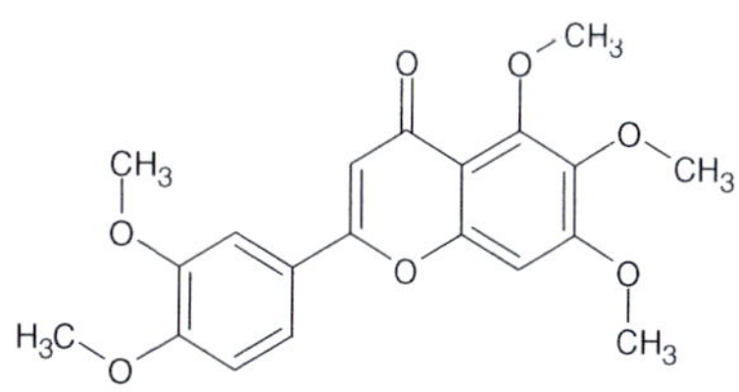
Flavones	Hesperetin	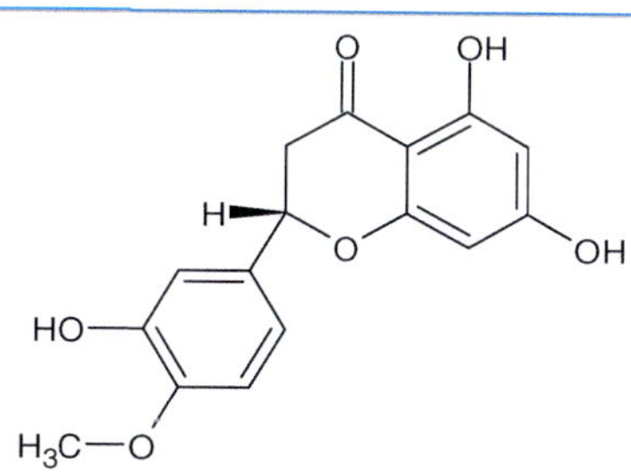

(Continued)

Table 6.1 (Continued)

Polyphenol Class	Examples	
	Naringenin	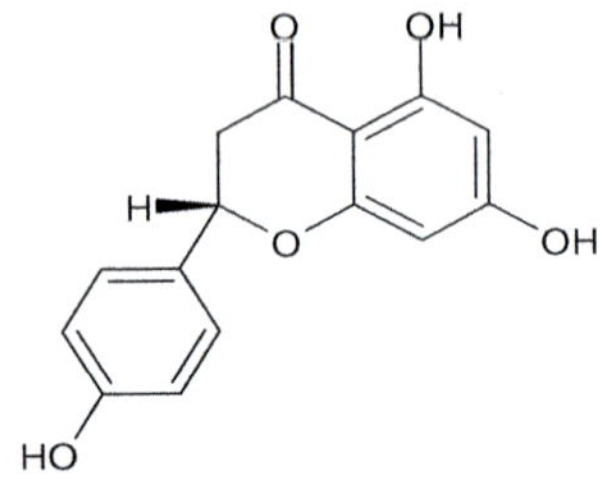
	Eriodictyol	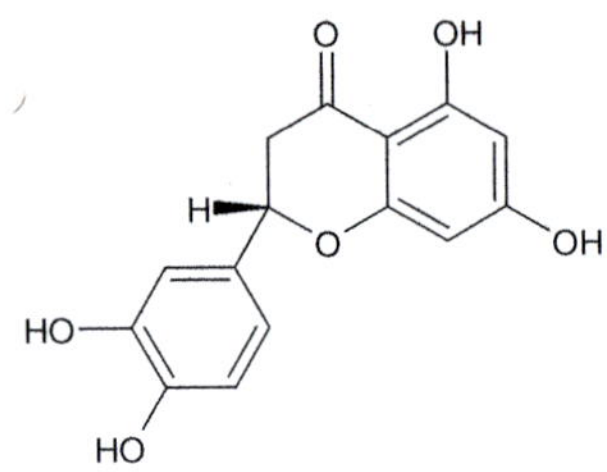
Isoflavones	Genistein	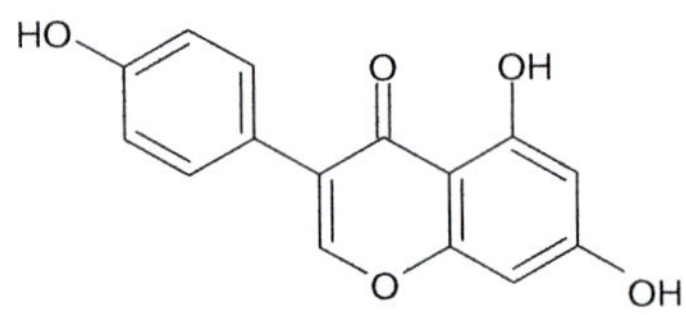
	Daidzein	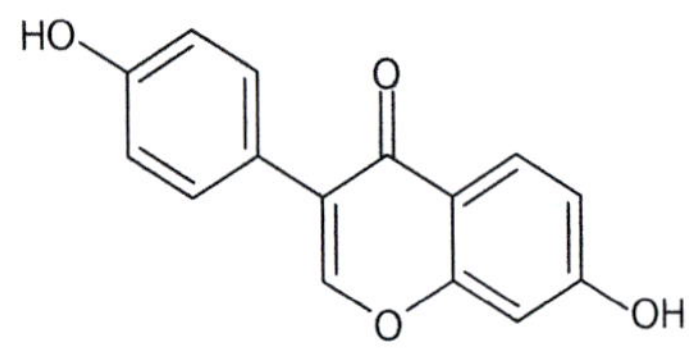
	Glycitein	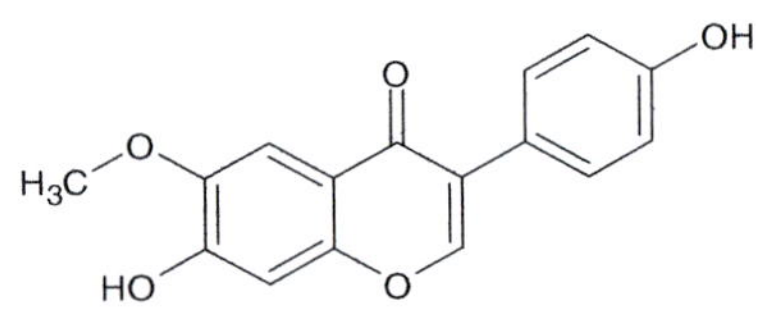
Proanthocyanidins (tannins)	Catechin	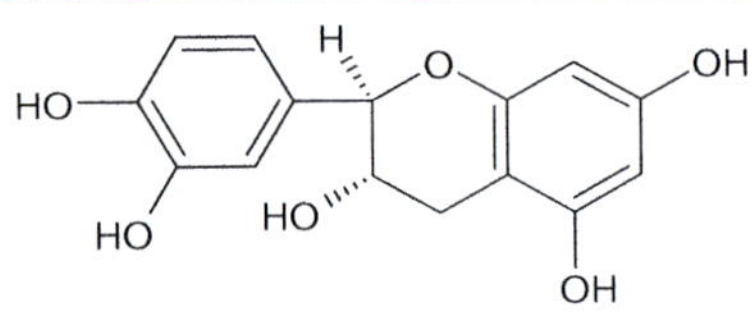
	Epicatechin	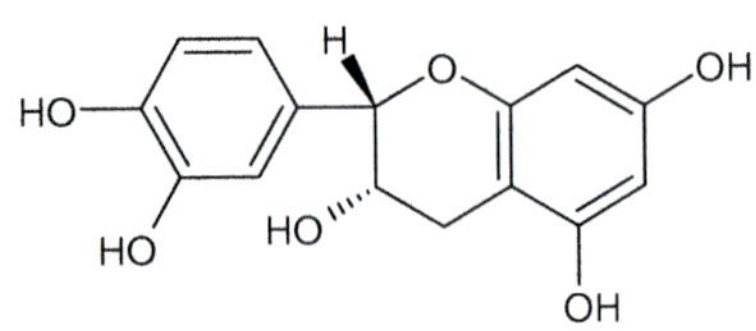

Table 6.1 (Continued)

Polyphenol Class	Examples	
	Castalin	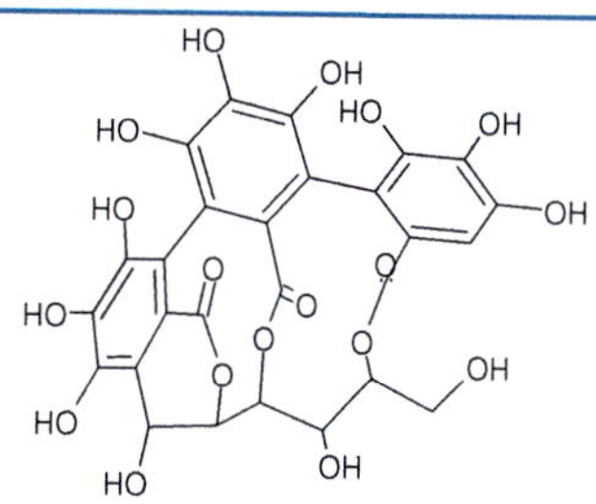
	Vescalin	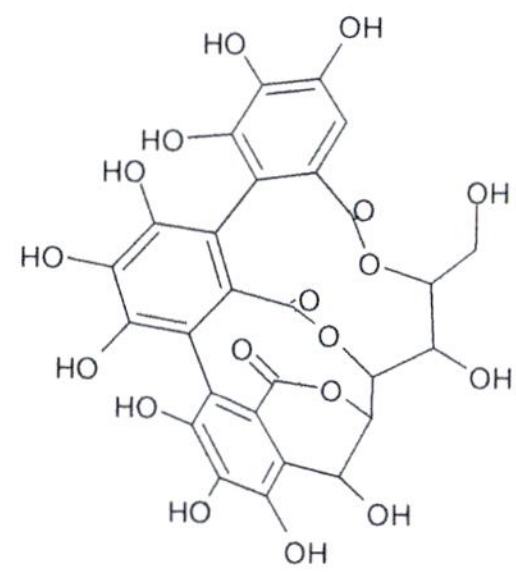
Anthocyanins	Cyanidin	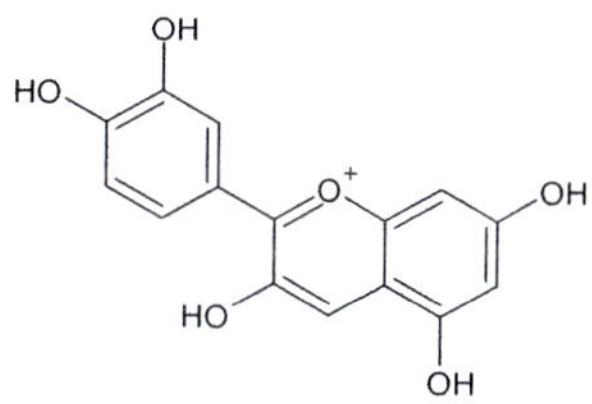
	Pelargonidin	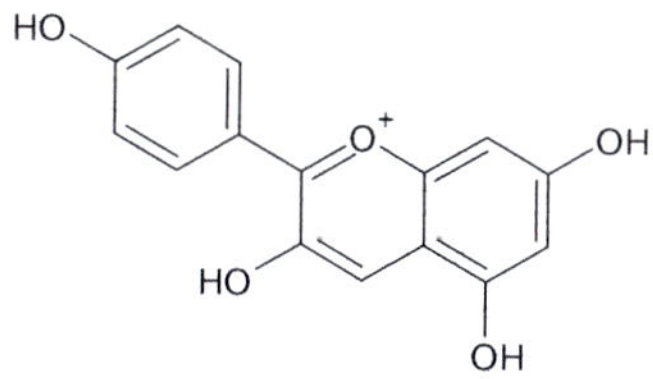
	Delphinidin	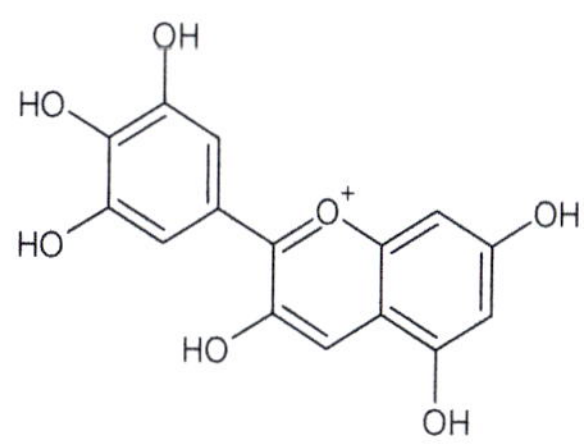
	Petunidin	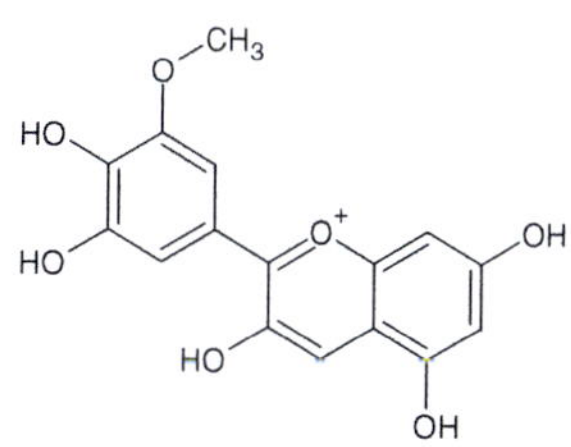

(Continued)

Table 6.1 (Continued)

Polyphenol Class	Examples	
Lignans	Pinoresinol	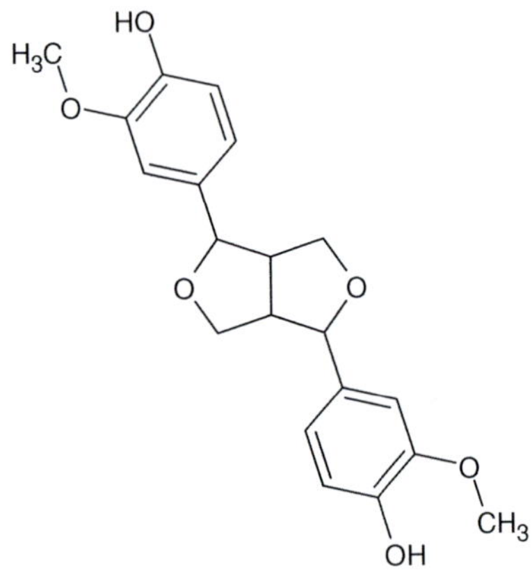
	Podophyllotoxin	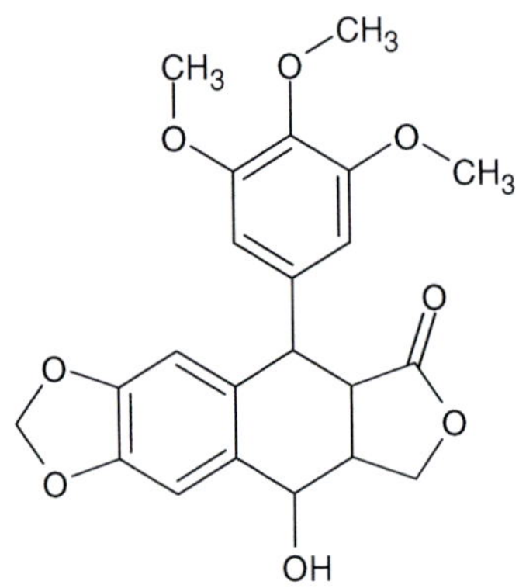
	Steganacin	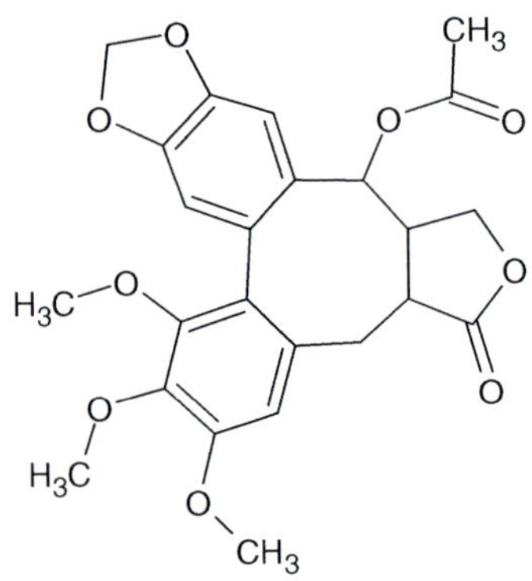

6.2 Natural Sources of Polyphenols

PPHs are obtained from various plant sources [6, 7] such as fruits (oranges, apples, grapes, peaches, grapefruit juice, cherries, blueberries, pomegranate juice, raspberries, cranberries, black elderberries, blackcurrants, plums, blackberries, strawberries, apricots), vegetables (spinach, onions, shallots, potatoes, black and green olives, globe artichoke heads, broccoli, asparagus, carrots), grains (whole grain wheat, rye, oat flours), nuts, seeds and legumes (roasted soybeans, black beans, white beans, chestnuts, hazelnuts, pecans, almonds, walnuts, flaxseed), beverages (coffee, tea, red wine), fats (dark chocolate, virgin olive oil, sesame seed oil), and spices and seasonings (cocoa powder, capers, saffron, dried oregano, dried rosemary, soy sauce, cloves, dried peppermint, star anise, celery seed, dried sage, dried spearmint, dried thyme, dried basil, curry powder, dried ginger, cumin, cinnamon).

Hydroxybenzoic acids can be found in tea and red fruits such as strawberries and raspberries. Hydroxycinnamic acids are obtained from kiwis, blueberries, apples, and cereal grains such as wheat and rice. Hydroxybenzoic and hydroxycinnamic acids have limited therapeutic applications such as antimicrobial activity, fungitoxicity, and anti-inflammatory activity. Anti-inflammatory activity is also observed with coumarins, stilbenes, and flavonoids. Flavonoids can be extracted from diverse sources such as onions, curly kale, leeks, broccoli, blueberries, red wine, tea, celery, citrus fruits (grapefruit, orange, lemon), tomatoes, mint, apricot, cherry, grape, peach, apple, and chocolate.

6.3 Polyphenols in Food

The polyphenol content in vegetables varies largely depending on their origin, harvest, and atmospheric conditions [8, 9].

Table 6.2 summarizes the maximum amounts of these compounds in different types of food according to the portion served. The amount of sinapic acid (from the class of hydroxycinnamic acids) in coffee, anthocyanins in eggplant, blueberries, and black grapes, as well as flavones in parsley, should be highlighted. All are derived from vegetables with strong colors, especially those rich in anthocyanins.

Table 6.2 Polyphenols: food sources. Adapted from Ref [4]. Data refer to the maximum amount for each food.

	SOURCE (serving size)	mg/kg fresh wt (or mg/L)
Hydroxybenzoic acids	Blackberry (100 g)	270
Protocatechuic acid	Raspberry (100 g)	100
Gallic acid	Black currant (100 g)	130
p-Hydroxybenzoic acid	Strawberry (200 g)	90
Hydroxycinnamic acids	Blueberry (100 g)	2,200
Caffeic acid	Kiwi (100 g)	1,000
Chlorogenic acid	Cherry (200 g)	1,150
Coumaric acid	Plum (200 g)	1,150
Ferulic acid	Aubergine (200 g)	660
Sinapic acid	Apple (200 g)	600
	Pear (200 g)	600
	Chicory (200 g)	500
	Artichoke (100 g)	450
	Potato (200 g)	190

(Continued)

Table 6.2 (Continued)

	SOURCE (serving size)	mg/kg fresh wt (or mg/L)
	Corn flour (75 g)	310
	Flour: wheat, rice, oat (75 g)	90
	Cider (200 mL)	500
	Coffee (200 mL)	1,750
Anthocyanins	Aubergine (200 g)	7,500
Cyanidin	Blackberry (100 g)	4,000
Pelargonidin	Black currant (100 g)	4,000
Peonidin	Blueberry (100 g)	5,000
Delphinidin	Black grape (200 g)	7,500
Malvidin	Cherry (200 g)	4,500
	Rhubarb (100 g)	2,000
	Strawberry (200 g)	750
	Red wine (100 mL)	350
	Plum (200 g)	250
Flavonols	Yellow onion (100 g)	1,200
Quercetin	Curly kale (200 g)	600
Kaempferol	Leek (200 g)	225
Myricetin	Cherry tomato (200 g)	200
	Broccoli (200 g)	100
	Blueberry (100 g)	160
	Black currant (100 g)	70
	Apricot (200 g)	50
	Apple (200 g)	40
	Beans, green, or white (200 g)	50
	Black grape (200 g)	40
	Tomato (200 g)	15
	Black tea infusion (200 mL)	45
	Green tea infusion (200 mL)	35
	Red wine (100 mL)	30
Flavones	Parsley (5 g)	1,850
Apigenin	Celery (200 g)	140
Luteolin	Capsicum pepper (100 g)	10
Flavanones	Orange juice (200 mL)	685
Hesperetin	Grapefruit juice (200 mL)	650
Naringenin	Lemon juice (200 mL)	300
Isoflavones	Soy flour (75 g)	1,800
Daidzein	Soy beans, boiled (200 g)	900
Genistein	Miso (100 g)	900
Glycitein	Tofu (100 g)	700

Table 6.2 (Continued)

	SOURCE (serving size)	mg/kg fresh wt (or mg/L)
	Tempeh (100 g)	530
	Soy milk (200 mL)	175
Monomeric flavanols	Chocolate (50 g)	610
Catechin	Beans (200 g)	550
Epicatechin	Apricot (200 g)	250
	Cherry (200 g)	220
	Grape (200 g)	175
	Peach (200 g)	140
	Blackberry (100 g)	130
	Apple (200 g)	120
	Green tea (200 mL)	800
	Black tea (200 mL)	500
	Red wine (100 mL)	300
	Cider (200 mL)	40

The content of these compounds in processed foods differs substantially from that in natural products because of reactions that occur between functional groups and proteins, carbohydrates, and by products from biochemical processes [10], mainly enzymatic oxidation, which act of esterases, glycosidases, and decarboxylases. Enzymatic oxidation is ubiquitous in plant-derived foods, with the reactions in coffee, black tea, and wine the most widely studied [11, 12].

Regarding the properties of P found in foods, they are generally astringent and impart a bitter taste (except for some that are volatile, such as vanillin and eugenol, which make them excellent flavoring agents). Their ability to interact with proteins is also responsible for their precipitation in beverages. Anthocyanins are usually colorful, and their colors vary according to the acidity of the food [13–15].

The interest in these compounds stems from the nutraceutical properties of the plant extracts from which they have been isolated.

6.4 Nutraceutical Benefits of Polyphenols

6.4.1 Against Diabetes

Diabetes is a chronic disease in which the body does not produce insulin or cannot properly use the insulin it produces. Diabetes is characterized by hyperglycemia that is caused by an imbalance between insulin production and demand (either because of a deficiency in pancreatic secretion or resistance mechanisms in its uptake). The International Diabetes Federation estimates that there are currently 285 million diabetic patients worldwide [16].

Diabetes can be classified into main categories: type-1 diabetes (T1D), type-2 diabetes (T2D), gestational diabetes (GD), and specific types because of other causes such as diseases of the exocrine pancreas (cystic fibrosis) or use of immunosuppressive drugs after organ transplantation [17]. T2D is preceded by insulin resistance [18–20], one of the underlying causes of diabetes development. T2D is related to co-morbidities, such as cardiometabolic, as well as severe complications caused by the recent severe SARS-CoV-2 infection because of poor immune health [21, 22].

Concerning the activity of PPHs in the control of diabetes, the effect on postprandial glycemic response is subtle, with a slight effect on the digestion and sugar absorption within the gut. Citrus PPH metabolites also act on the modulation of hepatic glucose metabolism and insulin sensitivity in target tissues [23]. PPH-rich foods, such as decaffeinated green tea, coffee, dark chocolate, blueberry jam, extra-virgin olive oil, and some vegetables, reduce plasma glucose [24].

These are just some of the many mechanisms through which PPHs can interfere with diabetes. Table 6.3 summarizes these biochemical mechanisms. Among the most common benefits of PPHs concerning diabetes are the reduction of glucose uptake, improvement of insulin sensitivity and/or resistance, and increase of insulin secretion [25].

Some of the pharmacological effects of PPHs have synergistic relationships with each other. For example, the action against diabetes, depending on the mechanism, may be related to the antioxidant effect [50].

6.4.2 Against Oxidative Stress

The antioxidant effect of natural polyphenolic compounds is the most documented among known effects. The deleterious effects of molecular oxygen on living organisms involves a reactive oxygen species (ROS). The free radical theory proposed by Harman in 1956 establishes a correlation between aging and dysfunction of the oxygen regulation system [1, 51].

Three mechanisms have been proposed to explain the antioxidant behavior of PPH, as illustrated in Figure 6.2. PPHs can inactivate free radicals via hydrogen atom transfer (HAT) (Figure 6.2a) or single electron transfer (SET) (Figure 6.2b) mechanisms, or can undergo transition metals chelation (TMC) with pro-oxidant proteins by chelation of metal ions such as Fe^{3+}, Al^{3+}, or Cu^{2+} (Figure 6.2c).

Direct antioxidant mechanisms (HAT/SET) of PPHs have been observed with radical species such as superoxide radical ($O_2^{\bullet}$), hydrogen peroxide (H_2O_2), hydroxyl radical ($HO^{\bullet}$), singlet oxygen $^1(O_2)$, or peroxyl radicals ($RO_2^{\bullet}$) [52, 53].

The HAT mechanism depends on the substitution pattern of the phenolic ring, which affects the bond dissociation energy of the OH group. The SET mechanism is more dependent on the ionization potential, although this parameter influences the former mechanism as well. Chelating mechanisms depend on the position of the OH groups (bidentate ligands are expected to be more efficient) and on their nucleophilicity. The high chelating efficiency of PPHs may contribute to their antioxidant activity by preventing redox-active transition metals from catalyzing free radical formation [54].

Concerning the TMC mechanism, the hydroxyl radical, one of the most reactive radical species in a biological environment, can be formed from Fenton reactions with the participation of metal ions (M^{n+}) such as [55]:

$$H_2O_2 + M^{n+} \rightarrow HO^{-} + HO^{\cdot} + M^{(n+1)+}$$

Fenton chemistry takes place in dopaminergic neurons of the nervous tissue, where some hydrogen peroxide is generated from dopamine catabolism [54]. The subsequent formation of radicals has been associated with the occurrence of some neurodegenerative diseases alongside aging (addressed in Section 6.4.3).

Table 6.3 A summary of individual phenolic compounds or extracts exhibiting antidiabetic effects in preclinical and clinical studies. Adapted from Ref [20].

Phenolic Compounds Class	Compounds	Role In Diabetic Complications	References
Stilbenes	Resveratrol	Reduced blood glucose levels, insulin concentration, plasma triglycerides	[26]
		Increased phosphorylation of adenosine mono phosphate kinase, endothelial nitric oxide synthase and protein kinase B, expression of glucose transporter-4 in the myocardium	[27]
		Decreased fasting blood glucose, glycosylated hemoglobin, increased insulin, improved hepatic glycogen content, increased glycogen synthase activity, and reduced glycogen phosphorylase activity	[28]
		Alleviated diabetic nephropathy by reduced renal dysfunction and oxidative stress	[29]
		Ameliorated cognitive decline by inhibition of hippocampal apoptosis via the B-cell lymphoma protein 2-associated X and caspase-3 pathway, improvement of synaptic	[30]
		Enhanced cerebral vasodilator function, improved cognitive performance	[31]
		Reduced fasting blood glucose, increased serum HDL-cholesterol levels, decreased total-/HDL-cholesterol ratio, increased total antioxidant capacity, decreased plasma malondialdehyde (MDA) levels, upregulation of peroxisome proliferator-activated receptor-γ and sirtuin1 in peripheral blood mononuclear cells	[32]
		Reduced foot ulcer size and plasma fibrinogen level	[33]
Phenolic acids	Curcumin (Hydroxycinnamic acid)	Improved diabetes-induced endothelial dysfunction through superoxide reduction and protein kinase C inhibition	[34]
	Ferulic acid (Hydroxycinnamic acid)	Decreased fasting blood glucose levels, reduced level of serum insulin and spleen size. Reduced oxidative stress mediated inflammation and apoptosis	[35]
	Gallic acid (Hydrobenzoic acid)	Reduced fasting serum glucose and lipids, improved hepatic and pancreatic antioxidant capacity, lowered levels of interleukin-6 and tumor necrosis factor-alpha in the liver and pancreas	[36]
Lignans + Phenolic acids	Flaxseed extract + ferulic acid, gallic acid equivalent, p-coumaric acid	Reduced fasting blood glucose, plasma cholesterol, LDL-cholesterol, triglycerides, plasma creatinine, urea and uric acid levels, partially recovers pancreas, liver, and kidney functions	[37]

(Continued)

Table 6.3 (Continued)

Phenolic Compounds Class	Compounds	Role In Diabetic Complications	References
Flavonoids	Luteolin (Flavone)	Improved neuronal injury and cognition by attenuating oxidative stress	[38]
	Fisetin (Flavonol)	Diabetic neuropathy modulation by improved motor nerve conduction velocity and reduced inflammation in sciatic nerves by nuclear factor kappa B inhibition and nuclear erythroid 2-related factor 2-positive modulation	[39]
	Quercetin (Flavonol)	Reduced pancreatic tissue MDA levels, serum nitric oxide concentrations, increased superoxide dismutase, glutathione peroxidase, and catalase enzyme activation in pancreatic homogenates, and preserved pancreatic β-cell integrity	[40]
	Total green tea extract (Flavanol)	Reduced fasting blood glucose level, increased total antioxidant capacity and thiol groups in blood	[41]
	Cocoa (Flavanol)	Decreased fasting plasma glucose, glycosylated hemoglobin, and blood pressure levels	[42]
	Hesperidin (Flavanone)	Attenuated streptozotocin-induced neurochemical alterations, increased norepinephrine, dopamine, and serotonin levels, decreased MDA, increased reduced glutathione, and decreased interleukin-6 in the brain	[43]
	Genistein (Isoflavone)	Reduced hyperglycemia, improved cognition by restoring acetylcholinesterase activity and ameliorated neuroinflammation via decreasing tumor necrosis factor-alpha, interleukin-1β, and nitrites in brain	[44]
	Cyanidin, delphinidin, petunidin, peonidin, malvidin extract (Anthocyanidins)	Reduced fasting plasma glucose and glycosylated hemoglobin levels, elevated serum adiponectin and β-hydroxybutyrate concentrations, improved dyslipidemia, enhanced antioxidant capacity measured in plasma	[45]
	Cyanidin-3-glucoside, delphinidin-3-glucoside, and petunidin-3-glucoside extract (Anthocyanins)	Decreased fed blood glucose, triglycerides levels, enhanced glucose transporter-4 expression and insulin receptor phosphorylation in heart and skeletal muscle, protected pancreatic tissue of apoptosis through regulation of caspase-3, B-cell lymphoma protein 2-associated X, suppressed MDA levels, and restored superoxide dismutase and catalase activities in serum	[46]
	Cyanidin 3-rutinoside, cyanidin 3-glucoside, pelargonidin 3-glucoside and pelargonidin 3-rutinoside extract (Anthocyanins)	Reduced fasting blood glucose, maintains insulin levels and β cell histology	[47]
	Delphinidin and cyanidin extract (Anthocyanidins)	Decreased fasting blood glucose levels and improved glucose tolerance	[48]
	Cyanidin extract (Anthocyanidins)	Inhibited intestinal α-glucosidase activity and decreased post-prandial glycemic response, delayed absorption of carbohydrates	[49].

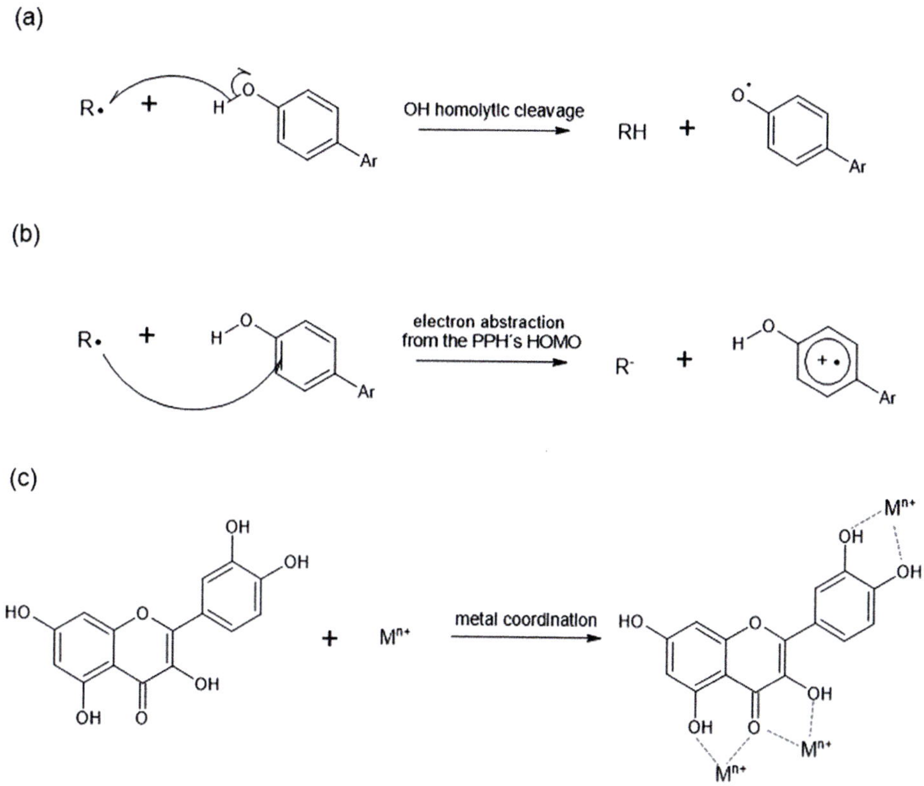

Figure 6.2 Mechanisms for the antioxidant activity: (a) hydrogen atom transfer (HAT), (b) single electron transfer (SET), and (c) transition metal chelation (TMC). Adapted from Ref [54].

Munin and Edwards-Lévy [1] mention the inhibition potential of PPH ROS-generating enzymes, such as xanthine oxydase [56], cyclooxygenase, and lipoxygenase [57], by complexing the protein. This process of PPH complexation is influenced by some protein characteristics such as [58–60]:

- Solubility
- Molecular mass
- Hydrodynamic volume
- Isoelectric point
- Amino-acid composition

Some PPH characteristics also affect their protein complexation ability such as [20, 61–63]:

- Molecular weight
- Chemical structure
- Conformational flexibility
- Water solubility

The physicochemical conditions of the mean properties (pH, solvent, temperature, ionic strength, presence of organic molecules such as polysaccharides) must also be considered [64–67].

6.4.3 Against Degenerative Diseases

Neurodegenerative diseases are caused by the damage or loss of the expected function of neurons, provoking misfunction in the cognitive functions, motor functions, or both. Many of these diseases accomplish aging, despite the fact that the conditions are expected to appear throughout the lifetime. A continuous increase in dementia cases between 2005 and 2030, reaching approximately 50% of the aged population, is predicted [68, 69].

Although there are no treatments beyond the attenuation of symptoms, many neurodegenerative diseases share common mechanisms of neuroinflammation and oxidative stress [70–72] against which PPHs are known to have activity. There is consensus that diet, both in the form of natural or processed foods and in the form of supplements, plays an important role in neuroprotection [73, 74]. Resveratrol (Table 6.1), which attenuate neuron degradation [75, 76], has been investigated for its antioxidative properties and has been prescribed as a supplement against neurodegenerative conditions with low adverse effects [77]. PPHs, such as flavonoids, can reach the brain [78], and epigallocatechin gallate (EGCG) has been proven to be able to significantly cross the human blood–brain barrier and protect cortical neurons from oxidative stress induced apoptosis [79].

The accumulation of free radicals in the tissues of the central nervous system, as well as the presence of high levels of metal ions (mainly iron), have been recognized as the main etiological cause of Parkinson's disease (PD) [55], Alzheimer's diseases (AD), and Huntington's chorea [80–82]. A correlation between basal ganglia ferritin iron content and the occurrence of AD has been described [83, 84]. The content of Fe(III) and hydroxyl radicals in degenerating zones of the brain is probably because of Fenton reactions [85–87].

6.5 Isolation Methods of Polyphenols from Food Products

Isolating PPHs from plants may be difficult; phenolic compounds can be found in plant matrices in two forms, extractable polyphenols (EEPs) and non-extractable polyphenols (NEPs), or bound polyphenols [20]. EEPs do not interact with other plant macromolecules and can be easily extracted after plant tissue disruption, and NEPs can be found crosslinked to cellulose, pectin, hemicellulose, lignin, and rod-shaped structural proteins [88], or entrapped within the food matrix [89]. NEPs cannot usually be extracted with water or mixtures with the most common organic solvents, which makes the use of classic chromatographic techniques impractical. In this scenario, alternative methods, such as high-speed counter current chromatography and microwave-assisted extraction, may be feasible.

6.5.1 Chromatographic Methods

A chromatographic technique with great applicability in the isolation of PPHs is high-speed counter current chromatography (HSCCC) based on continuous liquid–liquid partitioning, which eliminates irreversible adsorption on solid supports [90]. Partitioning of the solute between two non-mixable solvents (stationary and mobile phases) is achieved through hydrodynamic motion in a rotating coiled tube [91].

This technique combines the absorption of two liquid solvents, intensified by constant agitation, with different interactions between the solute and mobile and stationary phases [92]. The solvents are selected such that:

- the settling time should be less than 30 s;
- the analytes should be stable and soluble in the chosen system;

- partition coefficients K should range from 0.5 to 2.0, and the separation factor α ($\alpha = K_2/K_1$, $K_2 > K_1$) for any two components should exceed 1.5;
- the two-phase system should feature similar volumes in both phases.

The combination of two-phase solvent systems for HSCCC usually employs the following classical systems [93]: n-heptane–ethyl acetate–methanol–water (Arizona family), n-heptane–acetonitrile–n-butanol–water (HBAWat family), chloroform–methanol–water (ChMWat family), and n-hexane–ethyl acetate–methanol–water (HEMWat family). The choice will depend on the polarity and distribution constants of the components.

To illustrate the use of chromatographic methods in general (in HSCCC in particular) in the isolation of PPHs, we present two case studies [90]:

Case study 1:

One study was the application of HSCCC to the separation and purification of three phenolic acids (neochlorogenic acid, chlorogenic acid, and 3,5-O-dicaffeoylquinic acid; see Table 6.1) and two *Sorbus pohuashanensis Hedl.* flavonoids (SPF). This isolation was performed under optimal conditions (ethyl acetate–n-butanol–water 3.5:1.5:5) at a ratio of n-hexane–ethyl acetate–methanol–water of 1:3:1:3.5, and the purification yields were > 95%. Further conditions were tested by other researchers, with similar separation performances, such as the isolation of three caffeic acids and two new cinnamic acids using n-hexane–ethyl acetate–methanol–0.5% aqueous acetic acid 1:3:1:4 [94], as well as the isolation of two phenolic acids from white grape skin extract by HSCCC using n-hexane–ethyl acetate–water at a 1:50:50 ratio for > 90% yield [95].

Case study 2:

Another study was the application of HSCCC to the separation and purification of four stilbenes from the roots of Cabernet Sauvignon wine (*Vitis vinifera*). A one-step HSCCC using chloroform–methanol–n-butanol–water 4:3:0.05:2 provided the isolation of trans-resveratrol, δ-viniferin, ε-viniferin, and trans-vitisin B according to their NMR spectra and LC-MS [94]. Trans-vitisin B was obtained with a purity of 78.37%, and the purities of the other species exceeded 90%. Other solvent systems (such as chloroform–methanol–water, cyclohexane ethyl acetate–methanol–water, and ethyl acetate–ethanol–water) have also been used in the separation of resveratrol derivatives and their glucosides [90, 96].

6.5.2 Solvent extraction

The occurrence of NEPs and EEPs in plant-based food has been investigated, and some studies suggest that NEPs are even more abundant than EEPs in many plant foods [6, 97], although conventional solvent extraction methods are most applicable to EEPs [98].

Although the extraction of PPHs with conventional organic solvents (often mixed with some water) is the most common method, it hast some disadvantages [99]:

- It may create hazardous effects that cause health issues.
- Contamination with residues of the solvents may also remain in the final products, requiring additional time-consuming purification steps.
- By using pure organic solvents, very polar phenolic acids (benzoic, cinnamic acids) cannot be extracted completely, requiring aqueous mixtures.

Concerning the polarity of the solvent, methanol has been suggested to be more efficient in the extraction of lower molecular weight PPHs, while aqueous acetone extracts higher molecular weight flavanols more efficiently [100, 101].

Temperature is an important parameter in extraction. Conventional extraction is typically carried out at temperatures ranging from 20°C to 50°C, and temperatures greater than 70°C can cause anthocyanin degradation. Modern extraction methods may be useful to help overcome these drawbacks.

6.5.3 Microwave-assisted Extraction

Classical solvent extraction methods have the disadvantages of being time and solvent consuming. In this scenario, alternative techniques emerge, such as supercritical fluid extraction, microwave-assisted extraction (MAE), pressurized microwave-assisted extraction (PMAE), and Soxhlet solvent extraction [102].

In the comparison of the efficiency of PPH extraction from a *Clinacanthus nutans* Lindau medicinal plant with and without microwave pre-treatment, the pre-treatment resulted in a considerable increase in the extraction yield and a significantly faster extraction [103].

PPH extraction from bamboo shoots with microwave irradiation was approximately eight-fold greater than that obtained with conventional extraction, and the use of microwaves guaranteed a drastic reduction in extraction time [104].

6.6 Nanodelivery Systems for Polyphenol Nutraceutical Applications

6.6.1 Encapsulation

Microencapsulation can be defined as the process of packaging substances in various aggregation states (solids, liquids, gases) within sealed capsules capable of delivering its content at controlled rates under specific conditions [5, 105]. Several formulations can be used to package the substances that will be slowly released in the body in a controlled manner such as sugars, gums, proteins, natural and modified polysaccharides, lipids, and synthetic polymers [106, 107].

The compounds are incorporated in a capsule of approximately 5–300 microns in diameter [108]. The technique for the capsule formation involves the formation of a wall around the material so leakage does not occur and undesired materials are kept out. Figure 6.3 illustrates the various technologies for encapsulation: spray drying, coacervation, liposomes, inclusion, co-crystallization, nanoparticles (more properly discussed in the next topic), freeze drying, yeast encapsulation, and emulsion.

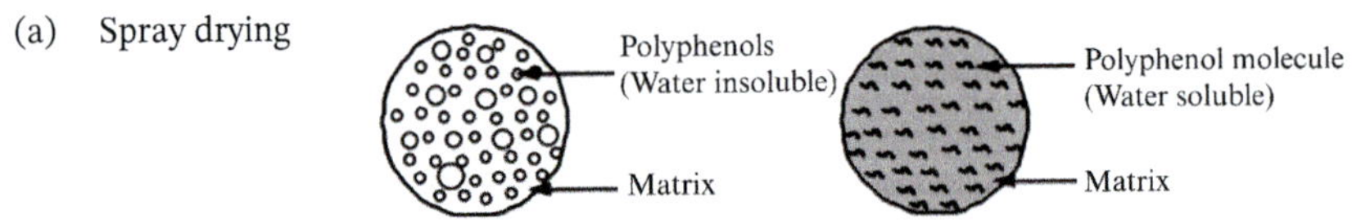

Figure 6.3 Illustration of the characteristics of encapsulated polyphenolic capsules produced by various encapsulation processes. Reproduced, by permission (Elsevier License), from Ref [5].

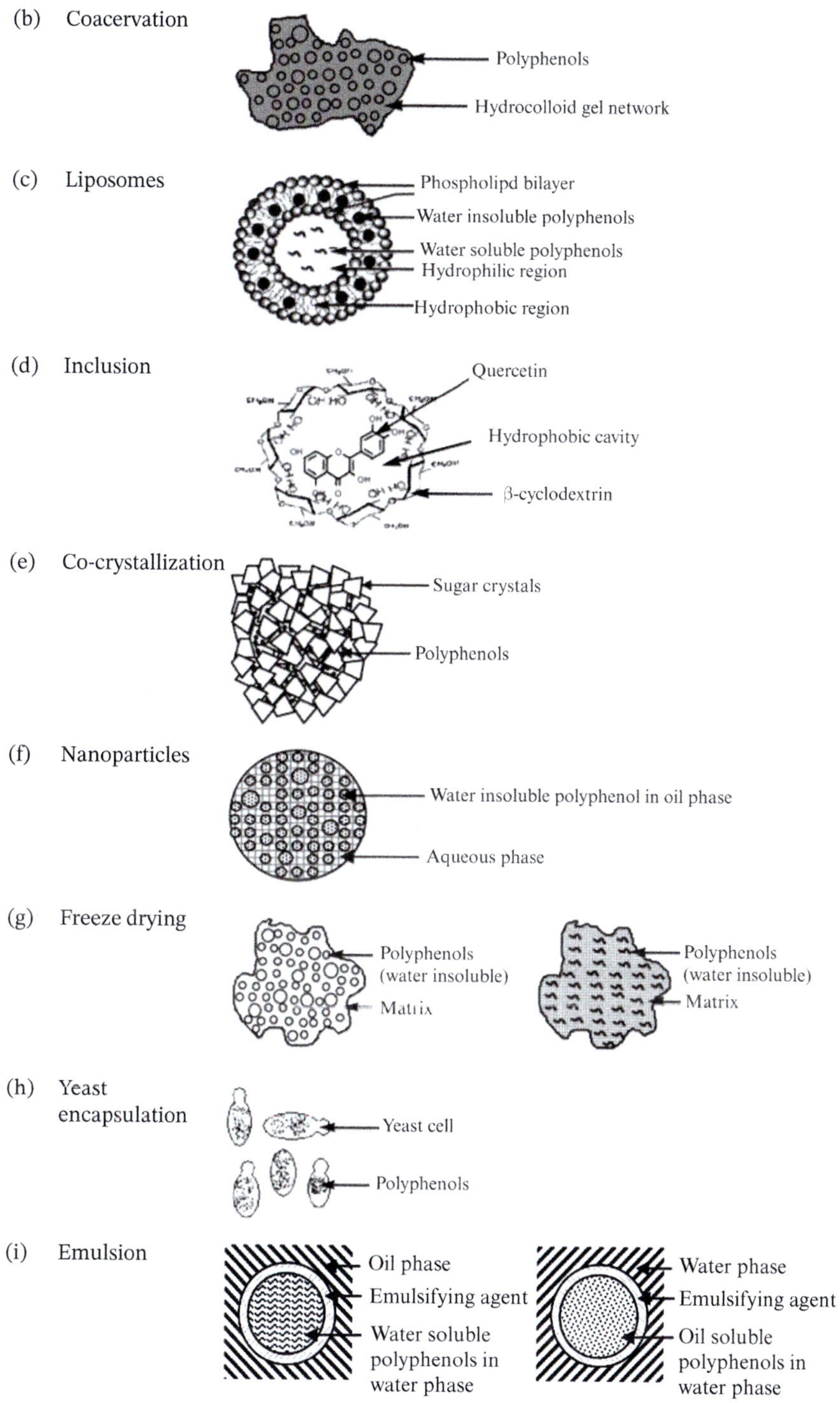

Figure 6.3 **(Cont'd)**

According to Desai and Park [109], spray drying (Figure 6.3a) is a technique comprising soaking a mass of material that will encapsulate the sample with water, and subsequent drying of the agglomerate. Modified starch, maltodextrin, gum, or other materials can be used as the wall materials, which are homogenized with the core materials. The mixture is subject to spray drying and is atomized with a nozzle or spinning wheel, and the water is evaporated by hot air. Coacervation (Figure 6.3b) involves the separation of a liquid phase of the coating material from a polymeric solution followed by the coating of that phase as a uniform layer around the suspended core particles. The overall process involves the following steps:

- Formation of a three-immiscible chemical phase
- Deposition of the coating
- Solidification of the coating

The use of liposomes (Figure 6.3c) as a vehicle for transporting vaccines, hormones, enzymes, and vitamins throughout the body is well known [110]. Liposomes comprise one or more layers of phospholipids capable of encapsulating the exogen compound in a hydrophilic environment. This delivery method has some advantages such that liposomes are non-toxic, stable, and permeable, and their properties can be manipulated by varying the size (from 25 nm to several microns in diameter) and type of lipid chains used [106].

The most common phospholipid is phosphatidyl choline, which is insoluble in water and can be isolated from soy or egg yolk. The liposomes are formed from the mixture of solvents, such as 2:1 chloroform–methanol, and the subsequent decrease in the volume of solvent, followed by the dispersion of the film of lipids/solvent in an aqueous phase. The liposomes are then recovered from the water phase [111]. The liposomes are susceptible to oxidizing, metal chelating, and temperature-driven degradation. Fang et al. [112] described how (+)-catechin and (−)-epicatechin may be entrapped in liposomes with similar encapsulation levels and release rates.

The inclusion technique includes using oligosaccharide cages (Figure 6.3d), the most common example being cyclodextrins (CDs). Inclusion is a low-cost method and is well known in its kinetic and thermodynamic aspects. CDs are naturally occurring cyclic oligosaccharides (with six, seven, or eight glucose residues linked by α (1–4) glycosidic bonds in a cylindrically shaped structure) derived from starch. The most common form that is applied as an inclusion system is β-cyclodextrin. The cylinder structure has non-polar interactions in the inner surface and polar interactions outside [113]. Many works describing the inclusion of PPHs are available such as morin and quercetin in α- and β-CDs [114] and resveratrol in β-CD [115].

Co-crystallization (Figure 6.3e) involves the formation of an irregular porous matrix assembling two or more active ingredients from an original perfect crystal [5, 116]. This encapsulation method improves solubility, wettability, homogeneity, dispersibility, hydration, anticaking, stability, and flowability of the encapsulated moiety [117]. The core materials in liquid form can be converted to a dry powdered form without additional drying [109].

Nanospheres and nanocapsules are a common name for encapsulation by nanoparticles (Figure 6.3f). While nanospheres have a type of structure, such that the active compounds enclosed are either adsorbed on the surface of the sphere or encapsulated in its interior, these compounds are confined in a cavity constituted by an internal liquid core contained by a polymeric membrane [118]. A possible application of this technology to PPHs is the encapsulation of quercetin with lipid nanocapsules by mixing through heating soy bean lecithin, surfactant, NaCl, and distilled water to form a water/oil (W/O) emulsion. The addition of distilled water forms O/W nanocapsules [119].

The freeze drying (Figure 6.3g) process (better known as lyophilization) involves the drastic reduction in pressure and the addition of a minimum amount of heat necessary for the evaporation of the solvent. This procedure prevents the use of severe heating that can cause the sample to decompose. As a result, core materials homogenize in matrix solutions and usually yield amorphous materials [120]. Despite being a safe and efficient method, one of the setbacks is the long drying time (approximately 20 hours).

The use of yeast cells (*Saccharomyces cerevisiae*) as wall material encapsulating the bioactive material has been proven to be low cost and high volume [121]. Yeast encapsulation (Figure 6.3h) allows the exogen substances to be delivery through the cell wall and membrane via passive migration. The cell wall still protects liquid active ingredients inside the matrix against deleterious factors such as evaporation, extrusion, oxidation, and light [122].

An emulsion (Figure 6.3i) as an encapsulation system has at least two immiscible liquids, such as oil (O) and water (W), with one of the liquids being dispersed as small spherical droplets into the other one [123]. There are several combinations of solvents dispersed in a drop-like pattern such as an oil-in-water (O/W) emulsion (oil droplets dispersed in an aqueous phase), water-in-oil (W/O) emulsion (water droplets dispersed in an oil phase), and more complex systems like oil-in-water-in-oil (O/W/O) or water-in-oil-in-water (W/O/W) emulsions [124]. An example of a W/O emulsion applied to PPHs is the mixing of caffeic acid and Fe (III) as an antioxidant system [125].

6.6.2 Protein–Polyphenol Conjugates

The use of protein–polyphenol conjugates (PPCs) as a delivery system is interesting because the conjugation of proteins and non-polar PPHs intensify the hydrophobicity and emulsifying properties of the modified proteins [112]. The proteins and non-polar PPH conjugates exhibit good antioxidant emulsification at the O/W interface [126–128]. The interactions that govern the formation of conjugates between protein and PPHs are generally reversible, weak, and non-covalent in nature [129, 130]. PPHs are hydrogen bond donors, capable of interacting with the C=O groups of proteins, as well as with OH and NH_2 groups [131, 132]. Hydrophobic interactions also occur between non-polar PPHs and hydrophobic amino acids such as leucine, glycine, methionine, alanine, phenylalanine, and tryptophan [133]. Ionic interactions with charged groups are also common [89].

Irreversible covalent interactions between PPHs and proteins are less common because they are subject to factors that allow the formation of free radicals [134]. PPHs can be oxidized under alkaline conditions (pH 9.0) in the presence of oxygen to form semiquinone radicals capable of reacting with nucleophilic protein residues such as methionine, lysine, tryptophan, and cysteine, establishing covalent bonds [38, 135].

The stability of the conjugates depends on some physical–chemical and structural factors. Temperature affects the strength of hydrophobic interactions and intermolecular hydrogen bonds. However, an excessive increase in temperature can cause protein denaturation and have an adverse effect on conjugate stability, as described by Suryaprakash et al. [136], concerning the interaction strength decrease at 10°C –45°C for caffeic acid with lysine, tryptophan, and tyrosine residues of proteins isolated from some sunflower seeds. pH is another relevant factor that must be considered given that PPCs can be formed within a pH of 4.0–10.0 via either covalent or non-covalent interactions, where protein dissociation occurs at pH < 7.0 [137, 138].

Quan et al. [134] also mentions the nature of the PPH and the protein as an important factor the affects the stability of the PPCs. The molecular diversity of the PPHs as depicted in Table 6.1 allows different patterns of binding. Further effects have been described such as the molecular size [139], species from which the protein was isolated [140], and the number of OH groups of the PPH [141].

The surface properties of proteins, related to the amino acid composition, hydrophobicity, and isoelectric point, also affect the binding affinity of proteins with PPHs [137]. Depending on these characteristics, the interaction sites are more or less exposed to bind to the PPHs. Evidence of this behavior is the fact that unfolded proteins present more intense interactions than globular and compact proteins [132].

6.7 Conclusion and Future Perspectives

The study of PPHs introduces perspectives on their nutraceutical applications attributed to the wide spectrum of activities and the availability in the most vegetables. Despite the accumulated knowledge on the nature, occurrence, and applications of these compounds, some challenges remain, such as the improvement of methods for the isolation of non-extractable PPHs, as well as the particular pharmacology of the bound PPHs that may differ substantially from their extractable counterparts.

Research on delivery methods, notably those that employ nanostructures and the formation of conjugates between PPHs and proteins, still promise advances. In this context, studies involving computational chemistry, capable of characterizing the interactions that govern the formation of these systems, are a constant stimulus to academic curiosity and the demand for new products. New approaches include molecular docking, molecular dynamics simulation, and hybrid method calculations of PPHs with β-lactoglobulin, a phenomenon that influences their antioxidant properties [142].

Finally, it is important to pursuit electrochemical studies employing techniques, such as cyclic voltammetry, to accomplish the antioxidant and radical formation feasibility of PPHs related to their influence on other conditions, including diabetes and cancer.

References

1. Munin, A. and Edwards-Lévy, F. (2011). Encapsulation of natural polyphenolic compounds; a review [internet]. *Pharmaceutics* 3 (4): 793–829.
2. Dewick, P.M. (2009). Medicinal natural products: a biosynthetic approach. *Journal of Chemical Information and Modeling* 53: 190–210.
3. Santos-Sánchez, N.F., Salas-Coronado, R., Hernández-Carlos, B., and Villanueva-Cañongo, C. (2019). Shikimic acid pathway in biosynthesis of phenolic compounds. In: *Plant Physiol Asp Phenolic Compd* (ed. M. Soto-Hernández, R. García-Mateos, and M. Palma-Tenango). Rijeka: IntechOpen.
4. Manach, C., Scalbert, A., Morand, C. et al. (2004). Polyphenols: food sources and bioavailability. *The American Journal of Clinical Nutrition* 79 (5): 727–747.
5. Fang, Z. and Bhandari, B. (2010). Encapsulation of polyphenols – a review. *Trends in Food Science and Technology*, Elsevier 21: 510–523.
6. Pérez-Jiménez, J. and Saura-Calixto, F. (2018). Fruit peels as sources of non-extractable polyphenols or macromolecular antioxidants: analysis and nutritional implications. *Food Research International*, Elsevier. 111: 148–152.
7. Olsen, N. (2017). Polyphenols: sources, benefits, and risks [Internet]. *Medical News Today.* Available online: https://www.medicalnewstoday.com/articles/319728 (accessed June 20 2023).
8. Marranzano, M., Rosa, R.L., Malaguarnera, M. et al. (2018). Polyphenols: plant sources and food industry applications. *Current Pharmaceutical Design* 24: 4125–4130.
9. Lorenzo, J.M., Mousavi Khaneghah, A., Gavahian, M. et al. (2019). Understanding the potential benefits of thyme and its derived products for food industry and consumer health: from extraction of value-added compounds to the evaluation of bioaccessibility, bioavailability, anti-inflammatory, and antimicrobial activities. *Critical Reviews in Food Science and Nutrition* 59: 2879–2895.
10. Cheynier, V. (2005). Polyphenols in foods are more complex than often thought. *The American Journal of Clinical Nutrition*, Oxford Academic 81: 223S–229S.

11 Balentine, D.A., Wiseman, S.A., and Bouwens, L.C.M. (1997). The chemistry of tea flavonoids. *Critical Reviews in Food Science and Nutrition* 37: 693–704.

12 Wang, Y. and Ho, C.T. (2009). Polyphenolic chemistry of tea and coffee: a century of progress. *Journal of Agricultural and Food Chemistry*, American Chemical Society. 57: 8109–8114.

13 Wahyuningsih, S., Wulandari, L., Wartono, M.W. et al. (2017). The effect of pH and color stability of anthocyanin on food colorant. *IOP Conference Series: Materials Science and Engineering*, IOP Publishing. 193: 012047.

14 Brouillard, R. and Delaporte, B. (1977). Chemistry of anthocyanin pigments. 2.1 kinetic and thermodynamic study of proton transfer, hydration, and tautomeric reactions of malvidin 3-glucoside. *Journal of the American Chemical Society*, American Chemical Society. 99: 8461–8468.

15 Brouillard, R. and Dubois, J.E. (1977). Mechanism of the structural transformations of anthocyanins in acidic media. *Journal of the American Chemical Society* 99: 1359–1364.

16 The Lancet (2010). Type 2 diabetes-time to change our approach. *Lancet*, Elsevier B.V. 375: 2193.

17 Association, A.D. (2019 1). Improving care and promoting health in populations: standards of medical care in diabetes—2019. *Diabetes Care*, American Diabetes Association. 42: S7–12.

18 O'Rahilly, S., Gray, H., Hattersley, A., and Vaag, A. (1994). Insulin resistance as the major cause of impaired glucose tolerance: a self-fulfilling prophesy? *Lancet*, Elsevier. 344: 585–589.

19 Fritz, J., Bjørge, T., Nagel, G. et al. (2020). The triglyceride-glucose index as a measure of insulin resistance and risk of obesity-related cancers. *International Journal of Epidemiology*, Oxford Academic. 49: 193–204.

20 Fernandes, I., Oliveira, J., Pinho, A., and Carvalho, E. (2022). *The role of nutraceutical containing polyphenols in diabetes prevention*. Metabolites.

21 Azar, W.S., Njeim, R., Fares, A.H. et al. (2020). COVID-19 and diabetes mellitus: how one pandemic worsens the other. *Reviews in Endocrine and Metabolic Disorders 2020 214*, Springer. 21: 451–463.

22 Stefan, N., Birkenfeld, A.L., and Schulze, M.B. (2021). Global pandemics interconnected — obesity, impaired metabolic health and COVID-19. *Nature Reviews. Endocrinologyl 2021 173*, Nature Publishing Group. 17: 135–149.

23 Visvanathan, R. and Williamson, G. (2021). *Citrus polyphenols and risk of type 2 diabetes: evidence from mechanistic studies*. Taylor & Francis. https://DoiOrg/101080/1040839820211971945.

24 Bozzetto, L., Annuzzi, G., Pacini, G. et al. (2015). Polyphenol-rich diets improve glucose metabolism in people at high cardiometabolic risk: a controlled randomised intervention trial. *Diabetologia*, Springer Verlag. 58: 1551–1560.

25 Da Porto, A., Cavarape, A., Colussi, G. et al. (2021). Polyphenols rich diets and risk of type 2 diabetes. *Nutrition*, Multidisciplinary Digital Publishing Institute. 13: 1445.

26 Su, H.C., Hung, L.M., and Chen, J.K. (2006). Resveratrol, a red wine antioxidant, possesses an insulin-like effect in streptozotocin-induced diabetic rats. *American Journal of Physiology - Endocrinology and Metabolism*, American Physiological Society. 290: 1339–1346.

27 Penumathsa, S.V., Thirunavukkarasu, M., Zhan, L. et al. (2008). Resveratrol enhances GLUT-4 translocation to the caveolar lipid raft fractions through AMPK/Akt/eNOS signalling pathway in diabetic myocardium. *Journal of Cellular and Molecular Medicine*, John Wiley & Sons, Ltd. 12: 2350–2361.

28 Palsamy, P. and Subramanian, S. (2009). Modulatory effects of resveratrol on attenuating the key enzymes activities of carbohydrate metabolism in streptozotocin–nicotinamide-induced diabetic rats. *Chemico-Biological Interactions* 179: 356–362.

29 Sharma, S., Anjaneyulu, M., Kulkarni, S.K., and Chopra, K. (2006). Resveratrol, a polyphenolic phytoalexin, attenuates diabetic nephropathy in rats. *Pharmacology, Karger Publishers* 76: 69–75.

30 Tian, Z., Wang, J., Xu, M. et al. (2016). Resveratrol improves cognitive impairment by regulating apoptosis and synaptic plasticity in streptozotocin-induced diabetic rats. *Cellular Physiology and Biochemistry*, Karger Publishers. 40: 1670–1677.

31 Wong, R.H.X., Raederstorff, D., and Howe, P.R.C. (2016). Acute resveratrol consumption improves neurovascular coupling capacity in adults with type 2 diabetes mellitus. *Nutrition*, Multidisciplinary Digital Publishing Institute. 8: 425.

32 Hoseini, A., Namazi, G., Farrokhian, A. et al. (2019). The effects of resveratrol on metabolic status in patients with type 2 diabetes mellitus and coronary heart disease. *Food and Function*, The Royal Society of Chemistry. 10: 6042–6051.

33 Bashmakov, Y.K., Assaad-Khalil, S.H., Abou Seif, M. et al. (2014). Resveratrol promotes foot ulcer size reduction in type 2 diabetes patients. *ISRN Endocrinology*, Hindawi Limited. 2014: 1–8.

34 Rungseesantivanon, S., Thenchaisri, N., Ruangvejvorachai, P., and Patumraj, S. (2010). Curcumin supplementation could improve diabetes-induced endothelial dysfunction associated with decreased vascular superoxide production and PKC inhibition. *BMC Complementary and Alternative Medicine*, BioMed Central. 10: 1–9.

35 Ghosh, S., Chowdhury, S., Sarkar, P., and Sil, P.C. (2018). Ameliorative role of ferulic acid against diabetes associated oxidative stress induced spleen damage. *Food and Chemical Toxicologyl, Pergamon* 118: 272–286.

36 Obafemi, T.O., Jaiyesimi, K.F., Olomola, A.A. et al. (2021). Combined effect of metformin and gallic acid on inflammation, antioxidant status, endoplasmic reticulum (ER) stress and glucose metabolism in fructose-fed streptozotocin-induced diabetic rats. *Toxicol Reports*, Elsevier. 8: 1419–1427.

37 Draganescu, D., Andritoiu, C., Hritcu, D. et al. (2021). Flaxseed lignans and polyphenols enhanced activity in streptozotocin-induced diabetic rats. *Biology*, Multidisciplinary Digital Publishing Institute. 10: 43.

38 Liu, F., Wang, D., Ma, C., and Gao, Y. (2016). Conjugation of polyphenols prevents lactoferrin from thermal aggregation at neutral pH. *Food Hydrocoll*, Elsevier. C: 49–59.

39 Sandireddy, R., Yerra, V.G., Komirishetti, P. et al. (2016). Fisetin imparts neuroprotection in experimental diabetic neuropathy by modulating NRF2 and NF-κB pathways. *Molecular Neurobiology*, Springer New York LLC. 36: 883–892.

40 Coskun, O., Kanter, M., Korkmaz, A., and Oter, S. (2005). Quercetin, a flavonoid antioxidant, prevents and protects streptozotocin-induced oxidative stress and β-cell damage in rat pancreas. *Pharmacological Research*, Academic Press. 51: 117–123.

41 Sharifzadeh, M., Ranjbar, A., Hosseini, A., and Khanavi, M. (2017). The effect of green tea extract on oxidative stress and spatial learning in streptozotocin-diabetic rats. *The Iranian Journal of Pharmaceutical Research*, Brieflands. 16: 201.

42 Rostami, A., Khalili, M., Haghighat, N. et al. (2015). High-cocoa polyphenol-rich chocolate improves blood pressure in patients with diabetes and hypertension. *ARYA Atherosclerosis*, ARYA Atheroscler. 11: 21–29.

43 El-Marasy, S.A., Abdallah, H.M.I., El-Shenawy, S.M. et al. (2014). Anti-depressant effect of hesperidin in diabetic rats. *Canadian Journal of Physiology and Pharmacology* 92 (11): 945–952. https://doiorg/10.1139/cjpp-2014-0281.

44 Rajput, M.S. and Sarkar, P.D. (2017). Modulation of neuro-inflammatory condition, acetylcholinesterase and antioxidant levels by genistein attenuates diabetes associated cognitive decline in mice. *Chemico-Biological Interactions*, Elsevier. 268: 93–102.

45 Li, D., Zhang, Y., Liu, Y. et al. (2015). Purified anthocyanin supplementation reduces dyslipidemia, enhances antioxidant capacity, and prevents insulin resistance in diabetic patients. *The Journal of Nutrition*, Elsevier. 145: 742–748.

46 Nizamutdinova, I.T., Jin, Y.C., Chung, J. et al. (2009). The anti-diabetic effect of anthocyanins in streptozotocin-induced diabetic rats through glucose transporter 4 regulation and prevention of insulin resistance and pancreatic apoptosis. *Molecular Nutrition and Food Research*, John Wiley & Sons, Ltd. 53: 1419–1429.

47 Sarikaphuti, A., Nararatwanchai, T., Hashiguchi, T. et al. (2013). Preventive effects of Morus alba L. anthocyanins on diabetes in Zucker diabetic fatty rats. *Experimental and Therapeutic Medicine*, Spandidos Publications. 6: 689–695.

48 Rojo, L.E., Ribnicky, D., Logendra, S. et al. (2012). In vitro and in vivo anti-diabetic effects of anthocyanins from Maqui Berry (Aristotelia chilensis). *Food Chemistry*, Elsevier. 131: 387–396.

49 Zhang, A.J., Rimando, A.M., Fish, W. et al. (2012). Serviceberry [Amelanchier alnifolia (Nutt.) Nutt. ex. M. Roem (Rosaceae)] leaf extract inhibits mammalian α-glucosidase activity and suppresses postprandial glycemic response in a mouse model of diet-induced obesity and hyperglycemia. *Journal of Ethnopharmacology*, Elsevier. 143: 481–487.

50 Pandey, K.B. and Rizvi, S.I. (2009). Plant polyphenols as dietary antioxidants in human health and disease. *Oxidative Medicine and Cellular Longevity*, Hindawi. 2: 270–278.

51 Harman, D. (1956). Aging: a theory based on free radical and radiation chemistry. *Journal of Gerontology* 11: 298–300.

52 Hanasaki, Y., Ogawa, S., and Fukui, S. (1994). The correlation between active oxygens scavenging and antioxidative effects of flavonoids. *Free Radical Biology and Medicine* 16: 845–850.

53 Dugas, A.J., Castañeda-Acosta, J., Bonin, G.C. et al. (2000). Evaluation of the total peroxyl radical-scavenging capacity of flavonoids: structure-activity relationships. *Journal of Natural Products* 63: 327–331.

54 Leopoldini, M., Russo, N., and Toscano, M. (2011). The molecular basis of working mechanism of natural polyphenolic antioxidants. *Food Chemistry* Elsevier. 125: 288–306.

55 Schulz, J.B., Lindenau, J., Seyfried, J., and Dichgans, J. (2000). Glutathione, oxidative stress and neurodegeneration. *European Journal of Biochemistry* 267: 4904–4911.

56 Anunciato, T.P. and da Rocha Filho, P.A. (2012). Carotenoids and polyphenols in nutricosmetics, nutraceuticals, and cosmeceuticals. *Journal of Cosmetic Dermatology* 11.

57 Kim, H.P., Mani, I., Iversen, L., and Ziboh, V.A. (1998). Effects of naturally-occurring flavonoids and biflavonoids on epidermal cyclooxygenase and lipoxygenase from guinea-pigs. *Prostaglandins Leukotrienes and Essential Fatty Acids*, Churchill Livingstone. 58: 17–24.

58 Benmansour, F., Trist, I., Coutard, B. et al. (2017). Discovery of novel dengue virus NS5 methyltransferase non-nucleoside inhibitors by fragment-based drug design. *European Journal of Medicinal Chemistry* 125: 865–880.

59 Dearden, J.C. (2002). Prediction of environmental toxicity and fate using quantitative structure-activity relationships (QSARs) [Internet]. *Journal of the Brazilian Chemical Society* 13 (6): 754–762.

60 Hagerman, A.E. (1992). Tannin—protein interactions. In: *Phenolic Compounds in Food and Their Effects on Health I* Ch 19, 236–247. ACS Symposium Series. Vol. 506. American Chemical Society.

61 Siebert, K.J., Troukhanova, N.V., and Lynn, P.Y. (1996). Nature of polyphenol–protein interactions. *Journal of Agricultural and Food Chemistry*, American Chemical Society. 44: 80–85.

62 McManus, J.P., Davis, K.G., Beart, J.E. et al. (1985). Polyphenol interactions. Part 1. Introduction; some observations on the reversible complexation of polyphenols with proteins and polysaccharides. *Journal of the Chemical Society, Perkin Transactions 2*, The Royal Society of Chemistry. 9: 1429–1438. https://doi.org/10.1039/P29850001429.

63 Richard, T., Lefeuvre, D., Descendit, A. et al. (2006). Recognition characters in peptide-polyphenol complex formation. *Biochim Biophys Acta*, Biochim Biophys Acta. 1760: 951–958.

64 Naczk, M., Oickle, D., Pink, D., and Shahidi, F. (1996). Protein precipitating capacity of crude Canola Tannins: effect of pH, Tannin, and protein concentrations. *Journal of Agricultural and Food Chemistry*, American Chemical Society. 44: 2144–2148.

65 Debnath, U., Verma, S., Jain, S. et al. (2013). Pyridones as NNRTIs against HIV-1 mutants: 3D-QSAR and protein informatics. *The Journal of Computer-Aided Molecular Design 2013 277*, Springer. 27: 637–654.

66 Bergantini, L.P. (2019). Sinestesia nas artes. *ARS (São Paulo)*, Universidade de Sao Paulo Sistema Integrado de Bibliotecas - SIBiUSP. 17: 225–238.

67 Makkar, H.P.S. and Becker, K. (1996). Nutrional value and antinutritional components of whole and ethanol extracted Moringa oleifera leaves. *Animal Feed Science and Technology*, Elsevier. 63: 211–228.

68 Noble, W. and Burns, M. (2010). Challenges in neurodegeneration research. *Front Psychiatry* 1.

69 Silva, R.F.M. and Pogačnik, L. (2020). Polyphenols from food and natural products: neuroprotection and safety. *Antioxidants*, Multidisciplinary Digital Publishing Institute. 9: 61.

70 Amor, S., Puentes, F., Baker, D., and Van Der Valk, P. (2010). Inflammation in neurodegenerative diseases. *Immunology*, John Wiley & Sons, Ltd. 129: 154–169.

71 Glass, C.K., Saijo, K., Winner, B. et al. (2010). Mechanisms underlying inflammation in neurodegeneration. *Cell*, Cell Press. 140: 918–934.

72 Tarawneh, R. and Galvin, J.E. (2010). Potential future neuroprotective therapies for neurodegenerative disorders and stroke. *Clinics in Geriatric Medicine*, Elsevier. 26: 125–147.

73 Solanki, I., Parihar, P., Mansuri, M.L., and Parihar, M.S. (2015). Flavonoid-based therapies in the early management of neurodegenerative diseases. *Advances in Nutrition*, Elsevier. 6: 64–72.

74 Calabrese, V., Guagliano, E., Sapienza, M. et al. (2006). Redox regulation of cellular stress response in aging and neurodegenerative disorders: role of vitagenes. *Neurochemical Research 2006 324*, Springer. 32: 757–773.

75 Granzotto, A. and Zatta, P. (2014). Resveratrol and Alzheimer's disease: message in a bottle on red wine and cognition. *Frontiers in Aging Neuroscience*, Frontiers Media SA. 6: 95.

76 Sun, A.Y., Simonyi, A., and Sun, G.Y. (2002). The "French paradox" and beyond: neuroprotective effects of polyphenols. *Free Radical Biology and Medicine*, Pergamon. 32: 314–318.

77 Sangeetha, M.K., Eazhisai Vallabi, D., Sali, V.K. et al. (2013). Sub-acute toxicity profile of a modified resveratrol supplement. *Food and Chemical Toxicology* 59: 492–500.

78 Faria, A., Meireles, M., Fernandes, I. et al. (2014). Flavonoid metabolites transport across a human BBB model. *Food Chemistry*, Elsevier. 149: 190–196.

79 Pogačnik, L., Pirc, K., Palmela, I. et al. (2016). Potential for brain accessibility and analysis of stability of selected flavonoids in relation to neuroprotection in vitro. *Brain Research*, Elsevier. 1651: 17–26.

80 Youdim, M.B.H., Ben-Shachar, D., and Riederer, P. (1993). The possible role of iron in the etiopathology of Parkinson's disease. *Journal of Movement Disorders* 8: 1–12.

81 Gerlach, M., Ben-Shachar, D., Riederer, P., and Youdim, M.B.H. (1994). Altered brain metabolism of iron as a cause of neurodegenerative diseases? *Journal of Neurochemistry* 63: 793–807.

82 Hirsch, E.C. (2009). Iron transport in Parkinson's disease. *Parkinsonism and Related Disorders* 15.

83 Bartzokis, G., Sultzer, D., Cummings, J. et al. (2000). In vivo evaluation of brain iron in Alzheimer disease using magnetic resonance imaging. *Archives of General Psychiatry*, Arch Gen Psychiatry. 57: 47–53.

84 Double, K.L., Gerlach, M., Youdim, M.B.H., and Riederer, P. (2000). Impaired iron homeostasis in Parkinson's disease. *Journal of Neural Transmission Supplementum* 60: 37–58.

85 Linert, W., Herlinger, E., Jameson, R.F. et al. (1996). Dopamine, 6-hydroxydopamine, iron, and dioxygen–their mutual interactions and possible implication in the development of Parkinson's disease. *Biochim Biophys Acta* 1316: 160–168.

86 Owen, A.D., Schapira, A.H.V., Jenner, P., and Marsden, C.D. (1997). Indices of oxidative stress in Parkinson's disease, Alzheimer's disease and dementia with Lewy bodies. *Journal of Neural Transmission Supplementum* 51: 167–173.

87 Smythies, J. (2000). Redox aspects of signaling by catecholamines and their metabolites. *Antioxid Redox Signal* 2: 575–583.

88 Shahidi, F., Yeo, J., Cisneros-Zevallos, L., and Jacobo-Velazquez, D. (2016). Insoluble-bound phenolics in food. *Mole*, Multidisciplinary Digital Publishing Institute. 21: 1216.

89 Le Bourvellec, C. and Renard, C.M.G.C. (2012). *Interactions between polyphenols and macromolecules: quantification methods and mechanisms*. http://dxdoiorg/101080/104083982010499808 Taylor & Francis Group. 52: 213–248.

90 Li, L., Zhao, J., Yang, T., and Sun, B. (2022). High-speed countercurrent chromatography as an efficient technique for large separation of plant polyphenols: a review. *Food Research International* 153: 110956.

91 Ito, Y. and Bowman, R.L. (1970). Countercurrent chromatography: liquid-liquid partition chromatography without solid support. *Science (80-)*, American Association for the Advancement of Science. 167: 281–283.

92 Ito, Y., Sandlin, J., and Bowers, W.G. (1982). High-speed preparative counter-current chromatography with a coil planet centrifuge. *Journal of Chromatography A*, Elsevier. 244: 247–258.

93 Ito, Y. (2005). Golden rules and pitfalls in selecting optimum conditions for high-speed counter-current chromatography. *Journal of Chromatography A*, Elsevier. 1065: 145–168.

94 Wei, Y., Li, P., Ma, L., and Li, J. (2014). Separation and purification of four stilbenes from Vitis vinifera L. cv. cabernet sauvignon roots through high-speed counter-current chromatography. *The South African Journal of Enology and Viticulture*, scieloza. 35: 226–233.

95 Luo, L., Cui, Y., Zhang, S. et al. (2016). Preparative separation of grape skin polyphenols by high-speed counter-current chromatography. *Food Chemistry*, Elsevier. 212: 712–721.

96 Chen, L., Han, Y., Yang, F., and Zhang, T. (2001). High-speed counter-current chromatography separation and purification of resveratrol and piceid from Polygonum cuspidatum. *Journal of Chromatography A* 907: 343–346.

97 Arranz, S., Saura-Calixto, F., Shaha, S., and Kroon, P.A. (2009). High contents of nonextractable polyphenols in fruits suggest that polyphenol contents of plant foods have been underestimated. *Journal of Agricultural and Food Chemistry*, American Chemical Society. 57: 7298–7303.

98 Ding, Y., Morozova, K., Scampicchio, M., and Ferrentino, G. (2020). Non-extractable polyphenols from food by-products:current knowledge on recovery, characterisation,and potential applications [internet]. *Processes*.

99 Brglez Mojzer, E., Knez Hrnčič, M., Škerget, M. et al. (2016). Polyphenols: extraction methods, antioxidative action, bioavailability and anticarcinogenic effects. *Mole*, Multidisciplinary Digital Publishing Institute. 21: 901.

100 Labarbe, B., Cheynier, V., Brossaud, F. et al. (1999). Quantitative fractionation of grape proanthocyanidins according to their degree of polymerization. *Journal of Agricultural and Food Chemistry*, American Chemical Society. 47: 2719–2723.

101 Metivier, R.P., Francis, F.J., and Clydesdale, F.M. (1980). Solvent extraction of anthocyanins from wine pomace. *The Journal of Food Science*, John Wiley & Sons, Ltd. 45: 1099–1100.

102 Veggi, P.C., Martinez, J., and Meireles, M.A.A. (2013). Fundamentals of microwave extraction. In: *Microwave-Assisted Extr Bioact Compd Theory Pract* (ed. F. Chemat and G. Cravotto), 15–52. Boston, MA: Springer US.

103 Mustapa, A.N., Martin, A., Gallego, J.R. et al. (2015). Microwave-assisted extraction of polyphenols from Clinacanthus nutans Lindau medicinal plant: energy perspective and kinetics modeling. *Chemical Engineering and Processing: Process Intensification*, Elsevier. 97: 66–74.

104 Milani, G., Curci, F., Cavalluzzi, M.M. et al. (2020). molecules optimization of microwave-assisted extraction of antioxidants from Bamboo shoots of phyllostachys pubescens. *Molecules* 25: 215.

105 Desai, K.G.H. and Park, H.J. (2007). *Recent developments in microencapsulation of food ingredients*. http://dxdoiorg/101081/DRT-200063478, Taylor & Francis Group. 23: 1361–1394.

106 Gibbs, B.F., Kermasha, S., Alli, I., and Mulligan, C.N. (1999). Encapsulation in the food industry: a review. *Journal of Food Sciences and Nutrition* 50: 213–224.

107 Mozafari, M.R. (2006). Bioactive entrapment and targeting using nanocarrier technologies: an introduction. In: *Nanocarrier Technol Front Nanotherapy* (ed. M.R. Mozafari), 1–16. Netherlands, Dordrecht: Springer.

108 Lee, B.H. (2015). *Fundamentals of Food Biotechnology*, 2e. Wiley-Blackwell.

109 Desai, K.G.H. and Jin Park, H. (2005). Recent developments in microencapsulation of food ingredients. *Drying Technology*, Taylor & Francis. 23: 1361–1394.

110 Gregoriadis, G. (2018). *Liposome Technology: Volume I*, 1e. Boca Raton: CRC Press.

111 New, R. (1990). *Liposomes : a practical approach.*

112 Fang, J.Y., Hwang, T.L., Huang, Y.L., and Fang, C.L. (2006). Enhancement of the transdermal delivery of catechins by liposomes incorporating anionic surfactants and ethanol. *International Journal of Pharmaceutics*, Elsevier. 310: 131–138.

113 Loftsson, T., Jarho, P., Másson, M., and Järvinen, T. (2005). Cyclodextrins in drug delivery. *Expert Opinion on Drug Delivery*, Taylor & Francis. 2: 335–351.

114 Calabrò, M.L., Tommasini, S., Donato, P. et al. (2004). Effects of α- and β-cyclodextrin complexation on the physico-chemical properties and antioxidant activity of some 3-hydroxyflavones. *Journal of Pharmaceutical and Biomedical Analysis*, Elsevier. 35: 365–377.

115 Lucas-Abellán, C., Fortea, I., López-Nicolás, J.M., and Núñez-Delicado, E. (2007). Cyclodextrins as resveratrol carrier system. *Food Chemistry* 104 (1): 39–44.

116 Chen, A.C., Veiga, M.F., and Rizzuto, A.B. (1988). Cocrystallization: an encapsulation process. *Food Technology* 42 (11): 87–90.

117 Beristain, C.I., Vázquez, A.L.R., García, H.S., and Vernon-Carter, E.J. (1996). Encapsulation of orange peel oil by co-crystallization. *Food Science and Technology* 29: 645–647.

118 Couvreur, P., Dubernet, C., and Puisieux, F. (1995). Controlled drug delivery with nanoparticles : current possibilities and future trends. *European Journal of Pharmaceutics and Biopharmaceutics* 41: 2–13.

119 Barras, A., Mezzetti, A., Richard, A. et al. (2009). Formulation and characterization of polyphenol-loaded lipid nanocapsules. *International Journal of Pharmaceutics*, Elsevier. 379: 270–277.

120 Berk, Z. (2013). Freeze drying (lyophilization) and freeze concentration. In: *Food Process Engineering Technology* 567–581. Elsevier Inc.

121 Bishop, J.R.P., Nelson, G., and Lamb, J. (1998). Microencapsulation in yeast cells. *Journal of Microencapsulation*, Taylor & Francis. 15: 761–773.

122 Sbehat, M., Mauriello, G., and Altamimi, M. (2022). Microencapsulation of probiotics for food functionalization: an update on literature reviews. *Microorg*, Multidisciplinary Digital Publishing Institute. 10: 1948.

123 Friberg, S.E., Larsson, K., and Sjoblom, J. (2004). *Food emulsions fourth edition, revised and expanded.*

124 Benichou, A., Aserin, A., and Garti, N. (2004). Double emulsions stabilized with hybrids of natural polymers for entrapment and slow release of active matters. *Advances in Colloid and Interface Science*, Elsevier. 108–109: 29–41.

125 Almajano, M.P., Carbó, R., Delgado, M.E., and Gordon, M.H. (2007). Effect of pH on the antimicrobial activity and oxidative stability of oil-in-water emulsions containing caffeic acid. *The Journal of Food Science*, John Wiley & Sons, Ltd. 72: C258–63.

126 You, J., Luo, Y., and Wu, J. (2014). Conjugation of ovotransferrin with catechin shows improved antioxidant activity. *Journal of Agricultural and Food Chemistry* 62: 2581–2587.

127 Yin, C., Yang, L., Zhao, H., and Li, C.P. (2014). Improvement of antioxidant activity of egg white protein by phosphorylation and conjugation of epigallocatechin gallate. *Food Research International* 64: 855–863.

128 Gu, L., Peng, N., Chang, C. et al. (2017). Fabrication of surface-active antioxidant food biopolymers: conjugation of catechin polymers to egg white proteins. *Food Biophysics*, Springer New York LLC. 12: 198–210.

129 Jakobek, L. (2015). Interactions of polyphenols with carbohydrates, lipids and proteins. *Food Chemistry*, Elsevier. 175: 556–567.

130 Czubinski, J. and Dwiecki, K. (2017). A review of methods used for investigation of protein–phenolic compound interactions. *The Journal of Food Science and Technology*, John Wiley & Sons, Ltd. 52: 573–585.

131 Buitimea-Cantúa, N.E., Gutiérrez-Uribe, J.A., and Serna-Saldivar, S.O. (2018). Phenolic-protein interactions: effects on food properties and health benefits. *Journal of Medicinal Food* 21: 188–198.

132 Yildirim-Elikoglu, S. and Erdem, Y.K. (2017). *Interactions between milk proteins and polyphenols: binding mechanisms, related changes, and the future trends in the dairy industry*. https://doi org/101080/8755912920171377225 Taylor & Francis. 34: 665–697.

133 Kanakis, C.D., Hasni, I., Bourassa, P. et al. (2011). Milk β-lactoglobulin complexes with tea polyphenols. *Food Chemistry*, Elsevier. 127: 1046–1055.

134 Quan, T.H., Benjakul, S., Sae-leaw, T. et al. (2019). Protein–polyphenol conjugates: antioxidant property, functionalities and their applications. *Trends in Food Science and Technology*, Elsevier. 91: 507–517.

135 Rawel, H.M., Czajka, D., Rohn, S., and Kroll, J. (2002). Interactions of different phenolic acids and flavonoids with soy proteins. *International Journal of Biological Macromolecules*, Elsevier. 30: 137–150.

136 Suryaprakash, P., Kumar, R.P., and Prakash, V. (2000). Thermodynamics of interaction of caffeic acid and quinic acid with multisubunit proteins. *International Journal of Biological Macromolecules*, Elsevier. 27: 219–228.

137 Ozdal, T., Capanoglu, E., and Altay, F. (2013). A review on protein–phenolic interactions and associated changes. *Food Research International*, Elsevier. 51: 954–970.

138 Wang, X., Zhang, J., Lei, F. et al. (2014). Covalent complexation and functional evaluation of (−)-epigallocatechin gallate and α-lactalbumin. *Food Chemistry*, Elsevier. 150: 341–347.

139 Dubeau, S., Samson, G., and Tajmir-Riahi, H.A. (2010). Dual effect of milk on the antioxidant capacity of green, Darjeeling, and english breakfast teas. *Food Chemistry*, Elsevier. 122: 539–545.

140 Chanphai, P. and Tajmir-Riahi, H.A. (2019). Tea polyphenols bind serum albumins:a potential application for polyphenol delivery. *Food Hydrocoll*, Elsevier. 89: 461–467.

141 Hasni, I., Bourassa, P., Hamdani, S. et al. (2011). Interaction of milk α- and β-caseins with tea polyphenols. *Food Chemistry*, Elsevier. 126: 630–639.

142 Baruah, I., Kashyap, C., Guha, A.K., and Borgohain, G. (2022). Insights into the interaction between polyphenols and β-lactoglobulin through molecular docking, MD simulation, and QM/MM approaches. *ACS Omega*, American Chemical Society. 7: 23083–23095.

7

Functionalization of Food Polyphenols for Nanodeliveries

Koyel Kar

Department of Pharmaceutical Chemistry, BCDA College of Pharmacy & Technology, Hridaypur, West Bengal, India

7.1 Introduction

7.1.1 Nanotechnology

Nanotechnology has broad use in the food industry [1], with application in the safety and quality of food, targeted distribution, increased biological availability, the evolution of newer products, packaging, and improvements in terms of texture and taste [2] (Figure 7.1). Nanotechnology was been used for the preparation of drugs; however, this technology has a recent application in agriculture and the production of food [3]. Nanotechnology greatly helps in the distribution of nutritious food to humans [4]. Green nanotechnology is considered an important outcome of nanotechnology, and it reduces the impact on the environment, which diminishes the cost and environmental hazards. This technique is executed with plant extracts or microorganisms; as a result, it is environmentally friendly and biologically active [5]. This technology has broader implementation in the field of waste management, renewable energy, and environmental cleaning [6]. Green nanotechnology is used in food processes that require less energy and cause damage to food. Nanotechnology is also used in the evolution of functional foods, food nanoparticles, and the packaging of food. Food nanoparticles are a form of delivery of biologically active substances [7]. The nanoencapsulation technique helps in the distribution of biologically active substances to the targeted organ. Different compounds can be used in the nanoencapsulation process; however, polyphenols are especially encapsulated for the food industry.

7.1.2 Role of Nanotechnology in Food and Agriculture

Nanotechnology helps in the evolution of phenol-loaded nanoparticles that are widely used in the processing of food [8]. Nanotechnology enhances a food's physical and chemical characteristics, improves the antioxidant and antimicrobial properties, and also promotes health-related benefits for humans in terms of antitumor, anti-inflammatory, and antiaging activity [9]. Finally, nanotechnology also plays an important role in the packaging of food.

Polyphenols: Food, Nutraceutical, and Nanotherapeutic Applications, First Edition. Edited by Mithun Rudrapal.

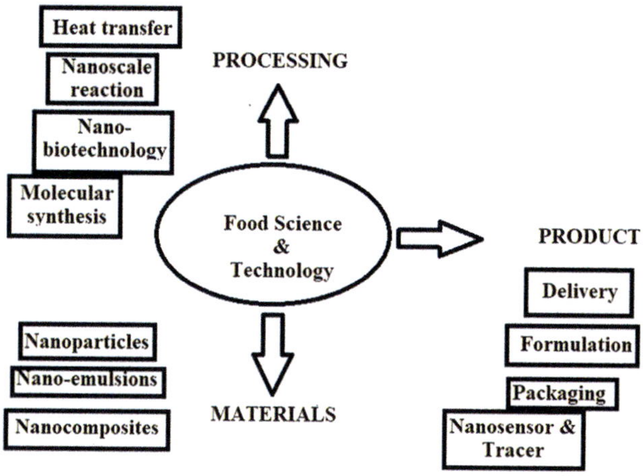

Figure 7.1 Use of nanotechnology in food, science, and technology.

7.1.3 Polyphenols and Nanocarriers

Polyphenols are the result of plant metabolism and they have a great impact on the metabolic process of humans [10]. Polyphenols exhibit different properties such as antioxidation and antimicrobial activities. However, they are quite unstable and tend to degrade. Their dispersibility and biological availability are also less than other components such as glycosides [11]. Thus, phenols added to food may affect their physical and chemical characteristics, stability, dispersibility, and biological availability [12]. Phenolic compounds can be protected from environmental conditions by encapsulation, which also interrupts the activity between phenols and foodstuff. Encapsulation with polyphenols also helps in the evolution of functional food, which has the capability to protect human beings from different diseases. Two types of carriers are used in the nanoencapsulation method: proteins and polysaccharides [13]. Examples of different carriers are cyclodextrins, gelatin, casein, whey, soy bean proteins, chitosan, and zein. Protein and polysaccharide carriers are widely used for the nanoencapsulation of polyphenols. Chitosan is a commonly used carrier as it improves the absorption of phenol compounds in the intestine. However, the main disadvantages exhibited by chitosan are low dispersibility, and inability to distribute particles efficiently. Therefore, the combination of nanoparticles, such as polysaccharide–protein carriers, has proved to be helpful in this genre. The biologically active substance binds to the protein via hydrogen bonding, and the hydrophobicity and carbohydrate prevents the mortification of protein in the gastric environment. Polysaccharide carriers are also used to encapsulate small molecules of polyphenols, and they exhibit biological availability of active substances. Polyphenols, such as catechins, eugenol, curcumin, and quercetin, actively participate in the nanoencapsulation method. Carriers can react with polyphenols. This is supported because catechin reacts with the amino group of chitosan or proline [14].

7.1.4 Nanoencapsulation of Phenolics Important for Food Processing and Therapeutic Applications (Medicine)

Encapsulation with polyphenols also helps in the evolution of functional food that has the capability to protect human beings from different diseases. Encapsulation methods include coacervation, ionic gelation, entrapment of liposome, complexation, and freeze drying. These methods improve the quality of polyphenols, gastric stability, and targeted delivery. These polyphenol-loaded nanoparticles can prove to be useful in the food industry [15]. In consideration of the importance of polyphenols and their encapsulated form, this chapter will focus on the functionalization of food polyphenols for nanodelivery in the food industry.

7.2 Functional Properties of Phenolic Compounds for Nanodelivery

7.2.1 Classification of Dietary Polyphenols

Polyphenols are classified into flavonoids and non-flavonoids. The vast category of phytochemicals are included in both these categories. Flavonoids include flavones, flavan-3-ols, anthocyanidins, flavanones, and isoflavones (Figure 7.2). Examples of non-flavonoids are phenolic acids and stilbenes. Flavonoids are represented by a $C_6eC_3eC_6$ skeleton. Naturally, they exist in the form of glycosides with sugar as a moiety [16]. The sugar and hydroxyl group in flavonoids enhanced their water dispersibility, and the methyl and isopentyl group increases their lipophilicity.

7.2.1.1 Flavonoids

Flavonols are the most common flavonoids found in plants. They exhibit a broad variation in structure and transportation. O-glycosides are the most available flavonols in nature, and they include quercetin, isorhamnetin, kaempferol, and myricetin. Flavones are structurally 2-phenyl-1-benzopyran-4-one (Figure 7.2). This ring also comprises luteolin and apigenin. Different types of substitution reactions occur, including glycosylation, alkylation, methylation, and hydroxylation [17]. Citrus fruits and vegetables comprise nobiletin and tangeretin, and they fall under the class of poly-methoxylated flavones. Flavones are not widely found in plants; however, they are obtained from a few plant sources like parsley and celery. Flavan-3-ols comprise C_3 and they have a complex structure. Catechin and epicatechin are simpler forms of flavan-3-ols and they are capable of undergoing hydroxylation and esterification forming gallocatechins and proanthocyanidins,

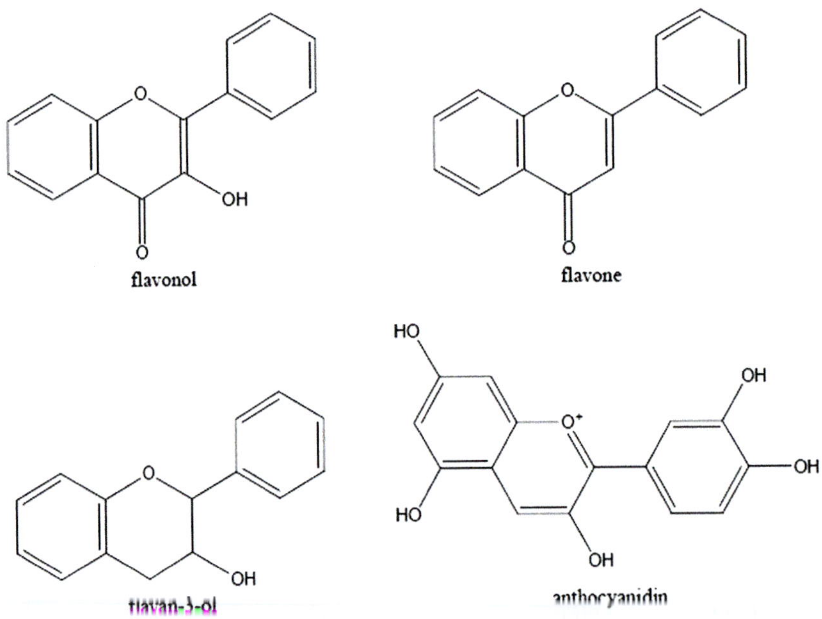

Figure 7.2 Structure of flavonoids.

respectively. The abundant proanthocyanidins are procyanidins and they comprise units of epicatechin. Flavanones are also structurally complex like flavan-3-ols and comprise C_2 elements. They are widely present in citrus fruits and isoflavones; both C_3 and C_2 elements are present and widely found in leguminous plants such as soy beans. The functional activity of isoflavones is very similar to estradiol.

7.2.1.2 Non-flavonoids

Phenolic acids and stilbene are the most commonly available non-flavonoids; examples include gallic acid and resveratrol, respectively. Gallic acid is the precursor of tannins such as hydroxycinammates. Resveratrol (Figure 7.3) is obtained from peanuts, red wine, red cabbage, and berries [18]. It is available both in the cis and trans form. Curcumin (Figure 7.3) falls under the curcuminoid class, which is a type of non-flavonoid phenolic compound that is responsible for the yellow color of turmeric. Chemically, curcumin is 1,7-bis (4-hydroxy-3-methoxy-phenyl)-hepta-1,6-diene-3,5-dione [18].

7.2.2 Functional Properties of Polyphenols

The beneficial effects of phenol compounds were analyzed using in vitro methods with cell lines. The research investigated the effect of flavonoids and phenols on the proliferation of the lymphocytic cell. Phenols showed the highest cytotoxic potentiality in comparison to flavonoids. Another study also investigated the effect of catechin from green tea on lymphocytic cancer cell lines. Catechin significantly diminished the proliferation of cells causing cell death. Recently, research is being conducted with resveratrol. Specifically, the effect of resveratrol on pancreatic cancer cell lines instigated by nicotine was investigated. Resveratrol was successful in eliminating the proliferation of cells by diminishing the generation of malondialdehyde [19]. Sugar cane extract also comprises a high content of polyphenols and these phenols were used to study the anticancer

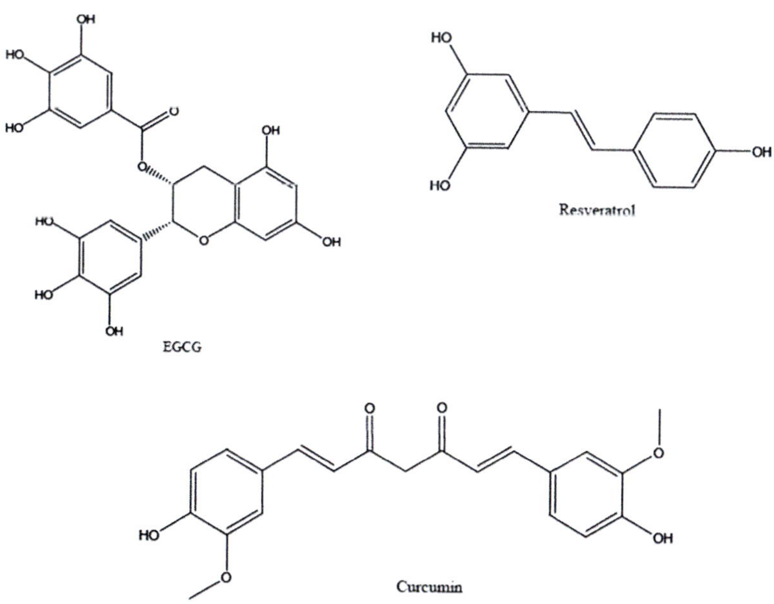

Figure 7.3 Structure of non-flavonoids.

effects on the cell lines of colon cancer. The proliferation seemed to diminish depending on the dose [20]. Phenols obtained from curcumin were also studied on the cell lines of colon cancer, and curcumin seemed to be a favorable method to prevent the methylation of DNA in the colon cancer cell lines and thus may help in its prevention. The polyphenolic effect on the cell lines of leukemia were also studied. Quercetin, emodin, and stilbene diminished the process of cell proliferation leading to cell death.

Furthermore, the anti-inflammatory activity was also explored. An investigation was performed with green tea and acerola, which showed anti-inflammatory responses because of the emanation of cytokines that protect the body from many diseases [21]. Extracts from grapes helped in the concentration of reactive oxygen species (ROS) in a muscle cell line. The anti-inflammatory characteristics of *Limonium algarvense* were studied in comparison with green tea. It exhibited similar antioxidant activity to green tea concerning the protection of cells from the damage of free radicals. However, it exhibited a better iron-chelating property and diminished the generation of nitric oxide.

Phenolics also play an important role in the protection from UV damage. Bilberry extracts were used to evaluate the damage to photoreceptor cells caused by UV. Bilberry protects the cell because of its capability to modulate the level of protein kinase and terminal kinase, as well as inhibit the generation of ROS. The flavonoids from rosemary extract were effective in protecting the keratinocytes from damage by UV. The phenolic component was successful in preventing DNA damage and curtailed the production of ROS [22]. The polyphenols also exhibited antimutagenic activity. Rosmarinic acid from lemon balm extract protected keratinocytes from UV by reducing the generation of ROS.

The results of the in vitro tests were confirmed by in vivo analysis. The primary efficacy and the toxicity level were analyzed by in vivo analysis [23]. Animal testing was used to determine the safety and efficacy of the functionally and biologically active molecules so they can be used safely in the food industry. Furthermore, they were analyzed by human testing to establish the biological activity of the molecules against the diseases. Extracts from broccoli contain phenol compounds. Those phenol compounds exhibit antitumor characteristics with selective activity against the tumor cells without causing toxicity to the non-tumor cells [24].

Phenolics from pomegranate diminished the microgliosis that occurs in Alzheimer's disease in mice. The concentration of the tumor necrosis factor of T-cells was reduced. Thus, it can be concluded that the extract was capable of causing an anti-inflammatory reaction in the brain and reducing the advancement of Alzheimer's disease [24].

7.2.3 Different Nano-driven Strategies for the Encapsulation of Polyphenols

The delivery of biologically active compounds to targeted sites without losing biological activity has been made possible by encapsulation [25]. The phenolic compounds that act as biologically active molecules are entrapped within a membrane. This membrane comprises a combination of chemical substances with stable physicochemical characteristics and it should ensure the safe transportation and distribution of material to the targeted site without getting leaked. Different types of methods are used for encapsulation design (Table 7.1) [26]. In physical techniques, the nanoparticles are coated and then dried. The method is slightly different when it was executed via chemical methods. The chemical method is superior to the physical method because the liquids of polyphenols are entrapped with a core membrane. The factors that are important to be considered for the chemical method are cost, sensitivity of the membrane, size, characteristics of the coating material, the target being used, mechanism of release, the biological activity of the molecule, and the ratio of polymers used [26].

Table 7.1 Processes involved in the nanoencapsulation of phenols.

PHYSICAL	CHEMICAL	PHYSICAL & CHEMICAL
Crystallization	Interfacial condensation	Emulsification
Fluid bed	Normal polymerization	Hot melting
Extrusion	Interfacial polymerization	Entrapment of liposome
Centrifugal	Interfacial crosslinking	Solvent evaporation
Freeze drying		Coacervation
Supercritical fluids		Nanoprecipitation
Spray drying		Electrospinning
Ionic gelation		Ionic gelation

Phenolic compounds need to be encapsulated because of their lower dispersibility in water, rapid metabolism, and lower rate of absorption. All of these factors have a huge impact on the biological availability of the compounds. Encapsulation of phenolic compounds is also required because of some external factors such as microbes of the gut, enzymes present in the stomach, and the metabolism and excretory process of phenols. The coating material is dependent on the following factors: stability, volatility, release characteristics, and environmental responses. Both synthetic and natural polymers may be used. The polymers that are used should be approved by the USA Food and Drug Administration (FDA) or European Food Safety Authority (EFSA) to be considered safe. Starch and its derivatives are used as a core material in the nanoencapsulation process [27]. Carrageenan and alginate are also used. Carbohydrates from animal sources, lipids, proteins, polyvinyl propylene, paraffin, and shellac are also used for the encapsulation method.

Research was also performed on the perpetuation of the health of the gut, short residence time in the gastric environment, diminished dispersibility, and vulnerability in the GI tract. These factors greatly affect the biological availability of polyphenols. The congregation of the protein or carbohydrate molecules and that of ligand molecules is called molecular nanocomplexes, and they are connected by non-covalent bonds. Phenolic compounds are bonded with whey protein by van der Waals forces and hydrogen bonding to form nanocomplexes [27].

7.2.3.1 Biopolymeric Nanoparticles

Biopolymer nanoparticles have an broad application in different fields, such as the food, cosmetic, and pharmaceutical industries, and they are used to distribute biologically active compounds whose size is less than 500 nm [28]. Fucoxanthin nanoparticles comprise a shell that is hydrophilic and a core that is hydrophobic. Different types of hydrophobic proteins, such as whey, casein, and zein, are used to produce the core that is coated with polysaccharides and is hydrophilic in nature. This forms the outer shell [29]. A suitable shell is required for the efficient transportation of the biologically active compound and its stability. Alginate and chitosan are better polymers than pectin for improving the longevity of zein or casein shells. This also improves the water dispersibility and release properties of phenol compounds.

Natural biopolymers are also recommended as the raw materials for the nanoencapsulation of the core material. These polymers help in the enhancement of the oral delivery and biological availability of active molecules. Different nanoencapsulation methods include mixing physically, chemical cross-linking, physicochemical gelation, and enzymatic conjugation. Other methods, such as ionic gelation, solvent evaporation, and emulsification, are also used. Physiochemically, nanoencapsulation occurs by ionic gelation and a coacervation method. Encapsulation is also executed by the electrospinning method, and the main advantage of this method is that it does not involve environmental factors such as temperature or pressure. The particles produced by this method exhibit better release, delivery, improved stability, and more efficient biological availability than other methods, leading to the moderation of functional food [30].

7.2.3.2 Physical Methods

The physical methods for nanoencapsulation are freeze drying, homogenization, emulsification, and sonication. Freeze drying is the most common method for the generation of nanoparticles for the food and pharmaceutical industries with a particle size of less than 200 nm and stability during storage and sterilization. Phenols that are hydrophobic are easily entangled using this method and their biological activity is also preserved [31].

7.2.3.3 Chemical Methods

An antisolvent precipitation method is extensively used in this category. The different chemical methods include solvent evaporation, ion crosslinking, and polymerization. The Maillard reaction is used to generate a polysaccharide–protein complex. Chemical methods also comprise polycondensation, polymerization, and interfacial crosslinking methods [32]. Costly emulsifying instruments are used in the pharmaceutical sector rather than the food sector. In the solvent exchange method, organic solvents are used. Nanoparticles of starch and halloysite were generated by the combination of solvent exchange and a vacuum hybrid method. This process improves the stability of the nanotubes. The encapsulation efficiency and loading capacity were improved [32].

7.2.3.4 Enzymatic Conjugation

This method is used to generate nanoparticles by using enzyme and their mechanisms. The main advantage of this method is its high specificity and the reaction temperature is also low. This process is mainly executed in the presence of trans-glutaminase and oxidase. In this method, protein–polysaccharide complex particles are generated and linked by covalent bonding. Trans-glutaminase helps in transferring the acyl group between the amino group and glutamine. Oxidase helps in catalyzing the oxidation reaction between quinone and polysaccharides [33].

7.2.3.5 Complex Coacervation

This method is widely used for the encapsulation of phenol compounds because it prevents the breakdown of molecules and oxidation of phenols that are unstable in nature [34]. Two biomacromolecules, one positive and another negative, form the coacervate. Protein or carbohydrate molecules can be used for the encapsulation of phenol compounds. The advantage of this method is that the loading capacity is quite large. The electrostatic interaction between the core materials determines the stability of the system. However, the main disadvantage of this method is that it is very costly and it is used for the encapsulation of only a few chemicals.

The ionic gelation method is also very frequently used to generate different types of delivery systems using suitable coating materials. One positive material and one negative material were used to produce a chitosan-sodium tripolyphosphate nanogel, improving the stability and biological availability of phenols [34].

7.3 Food Macromolecule Nanoparticles for the Encapsulation and Delivery of Polyphenols

7.3.1 Food Protein Nanoparticles

Food proteins act as a very good raw material for the generation of nanoparticles in the delivery of drugs and nutraceuticals because of their extraordinary capability to bind different molecules. Food proteins are also considered to be environment friendly, non-antigenic, and have food value. They can also be generated very easily [35].

7.3.1.1 Whey Protein Nanoparticles

Whey protein has gained wide popularity in the generation of food protein nanoparticles of polyphenols. Whey protein is also known as β-lactoglobulin, and it interacts with (−)-epigallocatechin-3-gallate (EGCG) to form nanoparticles with a size of 50 nm. The particles produced are very transparent and they can be used in beverages. Curcumin, when bonded with whey, exhibits improved dispersibility in water and pH stability. The nanoparticles formed are also impenetrable to pepsin; however, they are sensitive to trypsin. The rate of permeation of curcumin was also improved. When whey and resveratrol are bonded, improvised nanocomplexes are produced. The photostability and water dispersibility of resveratrol was increased. Additionally, the nanocomplex also had no effect on the protein structure [36].

7.3.1.2 Casein Nanoparticles

Caseins are open-structured compounds that are rich in proline. They comprise two domains: hydrophobic and hydrophilic. It comprises AS1, AS2, and βcaseins and generates spherical particles having an average size of 150 nm. When curcumin interacts with β-caseins, the particles formed have increased dispersibility. The antioxidant and anticancer characteristics also increased. Casein nanoparticles exhibited higher toxicity to leukemia cell lines in comparison to free curcumin. A solvent-free method was also proposed for the bonding of curcumin and casein, which is economical and consumes less energy. This method uses the properties of dispersibility based on pH [37]. The nanoparticles formed prevented the proliferation of cancer cells in pancreatic and colorectal cancer cell lines. The stability of curcumin was also improvised in combination with casein, and it was also capable of protecting erythrocytes in hemolysis [37].

7.3.1.3 Gelatin Nanoparticles

Gelatin is acquired from collagen by the hydrolysis of acid and alkali, and it is available in denatured form. Gelatin has been approved by the FDA and has been deemed safe. Thus, it is widely used in the food, pharmaceutical, and cosmetics industries. EGCG, when encapsulated with gelatin nanoparticles, exhibits increased biological activity on breast cancer cell lines by blocking the growth of hepatocytes. When resveratrol is bonded with gelatin nanoparticles, it exhibits good loading efficiency, causing cell death by altering various expressions [38].

7.3.1.4 Food Prolamine-based Nanoparticles

Prolamine is categorized as a plant protein with a high percentage of proline, and it is found in the seeds of cereals and grains. Prolamine comprises a high percentage of glutamine and proline and it is dispersible in ethyl alcohol. It is used in the generation of nanoparticles that have a wide application in the food industry. An atomization method is used to generate zein nanoparticles using different concentrations of protein. The particle size is 175–900 nm. Curcumin after encapsulation has good dispersibility in skimmed milk. Resveratrol was bonded with hordein to generate nanoparticles via the liquid–liquid dispersion method, which leads to improved stability compared with free resveratrol. The controlled release mechanism was also improved. The antioxidant activity was also enhanced as depicted by the assay [39]. Resveratrol was also bonded with zein, and the content of phenols in the plasma was increased and protracted. Resveratrol was capable of diminishing endotoxic symptoms in rats.

7.3.2 Food Polysaccharide Nanoparticles

Polysaccharides are considered an important part of our diet, and they contribute to the caloric value of our body. Polysaccharides comprise units of monosaccharides bonded by glycosidic linkages, and they are used extensively as food nanoparticles because of their advantages, including adhesion to the mucosal layer, and are capable of targeting specific cells or organs and protracting the intestinal residence time of the phenol compound. Chitosan is commonly used for the encapsulation process. It is a type of chitin comprising glucosamine. It is advantageous because it is non-toxic, environmentally friendly, and compatible. This polysaccharide has received approval to be used in the food industry in a few countries, including Finland, Japan, and Italy. It is capable of changing the absorption of EGCG in the intestine. Chitosan also reacts with negatively charged functional groups to generate targeted nanoparticles. The alginate–chitosan complex generates nanoparticles when encapsulated with EGCG and it exhibits encapsulation efficiency and free radical scavenging activities [40]. Chitosan bonded with fucose was used for the encapsulation of EGCG. After an in vivo analysis, it was concluded that it diminishes the growth of gastric cancer cells causing cell death. EGCG was complexed with sulfobutyl ether-β-cyclodextrin sodium and then encapsulated with chitosan. This activity increases the antioxidant property and helps extend the storage time. Curcumin–chitosan nanoparticles exhibit an anticancer effect on colon cancer cells. A combination of chitosan and gum of Arabia was used to interact with curcumin for the generation of nanoparticles. The encapsulation efficiency percentage, loading efficiency, and rate of retention were improved with increased antioxidant characteristics and the protracted release in the gastric mucosa [41].

7.3.3 Food Protein–polysaccharide Conjugate (Complex) Nanoparticles

The polysaccharide–protein complex formed by the Maillard reaction causes obstruction or hindrance in the precipitation of protein. Phenolics from tea extract were encapsulated with the combination of gelatine–dextran using the Maillard reaction. When this complex was bonded with EGCG, the distribution and encapsulation efficiency were increased with a controlled loading efficiency. The cytotoxic activity against the cancer cells was also greater than free EGCG [41]. When EGCG was encapsulated with dextran–casein, the stability was improved because EGCG was protected from being degraded in the alkaline pH of the intestine. Protein was also glycosylated with dextran and bovine albumin by the Maillard reaction to transport EGCG [42].

7.3.4 Food Lipid-based Nanoparticles

This type of nanoparticle enhances the dispersibility and biological availability of phenol compounds. The lipid nanoparticles help in the deliberation of curcumin to the cells responsible for the secretion of mucus. The delivery of the curcumin was increased to the junctions of the cell. This type of nanoparticle exhibits antitumor characteristics with improvised biological availability [43].

Resveratrol was complexed with glyceryl behenate to enhance its activity as an antitumor agent. The biological availability of resveratrol in the brain was also enhanced. The solvent evaporation method was engaged for the generation of resveratrol–lipid–poloxamer nanoparticles. The oral distribution was ameliorated compared with free resveratrol.

Lipid nanoparticles were also used for the encapsulation of phenols dispersible in water. EGCG–chitosan coated with lipids generates nanoparticles that can decrease the severity of atherosclerosis [44].

7.3.5 Food Hybrid Nanoparticles

Hybrid nanoparticles were bonded with polysaccharides to facilitate targeted delivery. Chitosan–lipid nanoparticles aid encapsulation of curcumin by prolonging the stability under optimum conditions. They also enhanced the biological availability compared with free curcumin. Curcumin was encapsulated with zein and lipid nanoparticles to improve the oral availability of curcumin. Resveratrol was bonded to casein, leading to the formation of complexes. It was then plated with zein, generating nanoparticles. They also protect resveratrol from polymerization in the presence of UV light. Those nanoparticles plated with casein–dextran were more stable under optimum gastrointestinal (GI) conditions than free casein. The biological availability of resveratrol was also increased under encapsulated conditions [45].

7.4 Role of Polyphenol-loaded Nanoparticles in Food Processing and Therapeutic Applications

7.4.1 Polyphenol-loaded Nanoparticles to Enhance the Physicochemical Properties of Food

Encapsulation is commonly used for conserving different properties of food. Encapsulation focuses on different problems in the food industry, including the refinement of organoleptic characteristics, masking unwanted properties like flavors, and odors, dispersibility, adherence, and emancipation of active substances [46], decomposition, and diminishing of reactions with other substituents and environmental conditions. The most important characteristic that should be addressed is the color because consumers are more worried about the appearance. Polyphenols are important plant metabolites, and they are encapsulated to provide protection. The selection of encapsulation procedures depends on the chemical properties and the size of the particle, dispersibility, and controlled emancipation. Curcumin is a phenol compound obtained from *Curcuma longa*. It provides natural coloring to food and affects the flavor. However, the main disadvantage is that it has a water dispersibility issue [46]. It also has low biological availability and instability in neutral and alkaline pH. These complications are resolved by the nanoencapsulation method. The food developed by this method is novel and is considered functional because they have improvised characteristics. Nanoencapsulation also enhances the antioxidant and antimicrobial activity with improved dispersibility in water. In addition to the physical and chemical properties, the biological property was also enhanced for intestinal absorption. In a recent scenario, casein has gained popularity as a nanocarrier that can easily be altered [47], proving it to be suitable for the distribution of biologically active molecules. Casein can successfully encapsulate carotene and vitamin D, improving the storage stability and biological availability. When polyphenols obtained from tea were added to milk, the basic characteristics of casein, including the gelling and stability, were affected.

Phenol compounds are extracted from the by-products of the food industry. Spray drying is widely used for encapsulation in the food sector. The spray drying method is advantageous because it helps in the conservation of the antioxidant activities of phenols [47]. If we can combine the spray drying method with newly moderated nanocarriers, it may help in the moderation of a new-generation dairy products.

7.4.2 Polyphenol-loaded Nanoparticles to Enhance the Functional Properties of Food

7.4.2.1 Antioxidant Properties

Polyphenol, when encapsulated with nutrients, provides protection and helps in the distribution of active molecules, ensuring biological availability. The polyphenols obtained from tea, such as epigallocatechin and epicatechin, exhibit strong antioxidant and anticancer characteristics; however, their biological availability is less if not encapsulated. Chitosan–poly(glycolic acid (PGA) nanoparticles were generated for improved antioxidant activity. Thus, catechins after encapsulation can be used in the food and dietary sector. In many research studies, EGCG demonstrated significant antioxidant activity; however, its biological availability was low because of its large size and hydrogen bonding. When chitosan was bonded with EGCG, the absorption rate was increased in the intestine. Different nanocarriers, such as dextran, are also used to encapsulate EGCG with high encapsulation efficiency [48]. Pectin–zein–curcumin exhibit improved antioxidant activity and improvised release capacity in the gastric and intestinal fluid. Hydroxycinnamic acids, when instigated in lipid nanoparticles, can be used for the generation of functional foods. The biodistribution of phenols is assessed by in vitro gastric and intestinal digestions. However, it also has disadvantages. When polyphenols are added to dairy products, the biodistribution and function of the phenols are affected [48].

7.4.2.2 Antimicrobial Properties

Diseases related to food are very common and are the main concern of the food sector. That is why consumers are more interested in using natural ingredients for biological safety. If components possessing antimicrobial activity are used in suitable nanocarriers, the pathogen's growth can be prevented [49]. Polyphenols are assumed to exhibit antimicrobial activity because of their absorption and interaction with enzymes. Eugenol is a type of phenolic compound having both antimicrobial and antioxidant activity; however, its water dispersibility is low. Eugenol–zein nanoparticles are generated by the complexation method and have broad application in the food sector because of antimicrobial properties. There is a need for proper emulsifiers. Nanoemulsions have an immense effect against heterotrophic bacteria present in orange juice. Thymol acts on both gram-positive and gram-negative bacteria and it strongly affects the flavor of food; however, the water dispersibility is very poor [50]. Thymol–zein nanoparticles are bonded with chitosan and exhibit antimicrobial activity against bacteria as a potential antimicrobial agent in food [50]. Thymol–chitosan nanoparticles exhibit stronger activity against gram-positive bacteria than gram-negative bacteria. The generation of the complex of chitosan and phenolics against food pathogens has been broadly studied.

7.4.2.3 Promotion of Health-related Properties

Polyphenols are broadly associated with human health, and extensive studies have been performed in recent years. The antioxidant, antitumor, anti-inflammatory, and antiaging activity of resveratrol is improved by nanoencapsulation. However, few in vivo analyses have been performed to ensure the health-promoted activities of resveratrol nanoparticles. As a result, further research is required [51]. When curcumin is bonded with chitosan, it exhibits antidiabetic activities. If curcumin nanoparticles are added to food, then functional activities of the dispersibility stability and decomposition will be maintained, keeping the biological functions intact. Polyphenols are studied for synergistic effects, and a certain combination of phenols exhibits better health-promoting characteristics [52]. Polyphenols obtained from tea were encapsulated with lysozyme, exhibiting antitumor activity and representing a successful carrier as an ingredient of functional food. Polyphenols obtained from green tea exhibit anti-inflammatory, antiproliferative, antihypertensive, antithrombogenic, and lipid-lowering activity. Thus, epicatechin and catechin play an important role in food nanotechnology [52].

7.4.2.4 Anticancer Properties

Over the past 10 to 15 years, there has been regular evaluation and meta analyses of the link between the consumption of polyphenols and cancer risk. In a meta analysis of prospective trials, eating isoflavones was linked to a 19% lower risk of stomach cancer. Recent epidemiological research has shown that consuming soy products reduces the incidence of breast cancer [53].

In cohort and case-control studies, the consumption of isoflavone and flavonol was associated with an estimated 30% decrease in the risk of ovarian and endometrial malignancies. In two Asian population meta analyses, the consumption of soy isoflavones and soy-based foods was linked to a lower risk of colorectal cancer, with one study demonstrating a 23% reduction in risk from 13 case-control studies and four prospective studies. In addition, a case-control Korean study found that consuming a lot of total soy products was associated with a lower risk of colon cancer, especially for distal and rectal site malignancies [54].

Only when seven or more cups of green tea were consumed each day did a meta analysis of green tea polyphenols reveal a 25% reduction in the incidence of prostate cancer. A case-control study in Canada found that consuming a lot of total flavonoids in a diet reduces the risk of developing lung cancer. Polyphenolic compounds are appealing chemicals overall for the treatment of cancer.

7.4.3 Application of Encapsulated Polyphenols in Different Food Products

7.4.3.1 Dairy Products

Tea becomes astringent after the addition of milk because of the interaction between the protein and phenols. This leads to the formation of casein–phenol complexes, which also alters the antioxidant activity of phenols. Phenols also interact with α-lactalbumin, forming phenolic complexes. Both encapsulated and free grape seed extract effect yogurt and improve its water retention activity, antioxidant effect, and phenolic contents [55].

Green tea helps in decreasing rates of obesity and diabetes. The complexes formed help in stabilizing the gastric and intestinal environment while augmenting the antioxidant activity. This also causes changes in the structure of the ingredients. Instigation of capsulated phenols does not affect the potentiality of bacteria. Nine different ingredients of phenols are found in cheese, which increases the strength of the gel, diminishes the moisture, and alters the quality. Other examples include cheddar-type cheese-green tea, hard cheese-catechin, yogurt-black currant extract, cottage cheese-rosemary extract, and cheese-barbera powders [55].

7.4.3.2 Beverages

Fruit juices are popular among customers because they are wholesome, lactose-free, non-alcoholic, and low in fat. By 2023, it is predicted that the global market for functional additives in nutritional beverages will reach $500 billion [56]. In addition to their anti-inflammatory, anticarcinogenic, and antibacterial properties, polyphenols can be employed to boost the antioxidant activity of juices.

In this context, a sensory study showed that the concentration might be 50% greater when encapsulants are added to beverages than when dry green tea extract is used, leading to higher antioxidant samples [57]. According to studies, clear drinks used in high-energy homogenization procedures result in size reduction and uniformity in particles with a diameter of less than 150 nm [57]. Hesperidin's limited solubility in food-grade solvents (water and ethanol), sedimentation, crystallization, and turbidity in liquid food products are some restrictions on the incorporation of polyphenols in various beverages. For instance, the citrus polyphenol naringin crystallizes sediments like aglycones and has limited water dispersibility. The sedimentation, turbidity, loss in bioactivity, and bitterness in liquid products, together with their temperature sensitivity and poorer water solubility, are application issues for polyethoxy flavones. It is crucial to consider the distribution methods that conflict with the molecular properties, polarity, solubility, melting point, and general resilience of polyphenols.

Another study found that cinnamon polyphenol, which was nanoencapsulated in chocolate drinks, improved the antioxidant effects and stabilized the drink. Malvidin-3-O-glucoside and microbial metabolites, such as syringic and gallic acids, have greater biological availability because of nanoencapsulation of wine polyphenols [57].

7.4.3.3 Bakery Products

Additionally, the baking industry changed the ingredients to better suit customer wants and nutritional value. Both in vitro and in vivo studies emphasize the antioxidant and anti-inflammatory characteristics of polyphenols, according to research on cocoa husk polyphenol as a stable bakery component. The country of origin, stage of fermentation, and methods of processing are key determinants of the polyphenol content of cocoa hulls. According to research, the orange juice business also encapsulates stable polyphenols from pomegranate peels, and this by-product is used to produce cookies, increasing their functional value and significantly enhancing their sensory qualities [58].

Similarly, to create 3D-printed printed dough cookies, grape skin derived encapsulated polyphenol extracts were added to the cookie dough. The bioactivity increased by approximately 30%, and the structure's shape was enhanced by approximately 120%. Another study found that a cake made with red onion peel derived maltodextrin had better qualitative properties and a higher polyphenol content than a cake made with soy bean protein isolate that had higher specific volume, moisture absorption, texture, acceptable color, sensory evaluation, and polyphenol concentration.

Functional additives, such as sprouted grains with increased lavender polyphenol content, by-products, tea extracts, and Aronia powders, improve the antioxidant profile of bread while also extending its shelf life. Additionally, this enhances the biological availability of polyphenols (particularly nanoencapsulated polyphenols), which improves the nutritional profile, storage capacity, bioactivity, and marketing appeal of bread [58].

7.4.3.4 Meat Products

Because of a higher likelihood of oxidation, beef has an intricate physical makeup and chemical composition. Additionally, growing consumer demand is driving meat technologists to create healthy substitutes; hence, the use of microencapsulation. Natural antioxidants are thought to increase the product shelf life of meat. Among the different methods available for drying beef, spray drying is thought to be the most efficient process that keeps phenolics stable until they reach their targeted spot, and controlled release is necessary. Additionally, encapsulation allows for the concealment of undesirable meat characteristics, such as aromas, further improving consumer acceptableness in some circumstances. According to research, adding rosemary polyphenol to beef steak boosted the antioxidant profile and extended the shelf life [59]. The fabrication of olive leaf extracts in double emulsions, where they were encapsulated in meat systems as fat replacements, is another application. This method increased oleuropein retention as well as the antioxidant value and oxidative stability. The observation of a 70% retention rate led to the conclusion that encapsulation is a successful method for maintaining a healthy lipid profile and preventing lipid oxidation.

Protein–curcumin nanocomplexes made with various ionic strengths improved the oxidative stability of marinated chicken meat. This extended the shelf life and improved the product's use by preventing the oxidation phenomenon that can occur when marinated chicken meat is stored. Additionally, beef was not quickly degraded because of active nanofibers made of chitosan and pomegranate peel extract, which also enhanced the physicochemical properties of the packaging systems. Pomegranate peel polyphenols also increased the hydrophilicity, which improved the meat's ability to swell [60].

7.4.4 Application of Encapsulated Phenolics in the Active Packaging/Coating of Food Products

In addition to their therapeutic effects in various food compositions, encapsulated polyphenols can be used to increase the shelf life of various foods when they are included in active packaging or coating materials. According to Göksen et al. (2020), the packaging matrix can gradually or deliberately release encapsulated phenolics into the packaged environment, which increases the effectiveness of the compounds as antioxidant and antimicrobial agents [61]. There is a thorough discussion on the use of phenolics in active packaging for fruits, vegetables, dairy, and meat products in Sections 7.4.4.1–7.4.4.3.

7.4.4.1 Fruit and Vegetable Packaging

The primary qualities of fruits and vegetables can be affected by the transpiration, respiration, and ripening processes, which can result in dehydration and microbial degradation. Fruits and vegetables can be shielded against oxidation and microbial growth during storage by bioactive phenolics with antioxidant and antibacterial properties. For instance, more recently, electrospun zein fibers were successfully combined with allyl isothiocyanates to successfully preserve strawberries under 15-day storage conditions [62]. Because the higher concentrations of allyl isothiocyanates were somewhat accompanied by lower levels of ROS and microbial growth, the release rate of the bioactive agent had a significant impact on the antioxidant capacity of the active packaging. Because of their capacity to enhance the physical properties of packaging materials, phenolics can serve as plasticizers in packaging matrices, in addition to their antioxidant and antibacterial activities [62].

7.4.4.2 Meat and Meat Product Packaging

Meat products are typical perishable commodities that, for 24–48 days at room temperature, offer a favorable environment for spoiling and the rapid growth of harmful germs. The *Pseudomonas* species is the main flora found primarily in fresh meat [62]. Under aerobic conditions, these bacteria can also multiply. The minimum temperature for the flora to thrive in chilled meat is roughly 3°C or below, while the minimum temperature for muscle tissue to function without freezing is around 1.5°C. Bacteria that causes spoiling by producing off-odors, changing the color, and changing the flavor. Antimicrobials are frequently used in active packaging and coatings that include phenolics in meats and meat products. Gallic acid encapsulated in zein films can inhibit the growth of *Listeria monocytogenes* and *Campylobacter jejuni*, as well as *Staphylococcus aureus* and *Escherichia coli*. Gallic acid encapsulating materials can be used to package fresh pork with a longer shelf life, as well as broiler carcasses.

Additionally, meat products include saturated fatty acids, which in the presence of oxygen, can lead to lipid oxidation and reduce the quality of meat products. Rancidity in the taste and odor of meat is caused by lipid oxidation [63]. The high concentration and makeup of unsaturated fatty acids have a significant impact on how easily lipids oxidize. The grinding operation or another procedure that destroys the cell tissue and releases ferrous iron and heme components cause lipid oxidation to accelerate. Gallic oil is an example of active packaging that contains phenolics and can function as an oxygen scavenger to stop the oxidation of lipids in meat products. The polypropylene-encapsulated carvacrol and thymol demonstrated a 30% inhibition for the DPPH radical assay in the fatty food simulants. In active packaging that contains olive leaf extract encapsulated by gelatin/tragant gum that release phenolics for two months, oxidative responses in sheep meat burgers were diminished [63].

7.4.4.3 Cheese Packaging

Considering the microbiological, physical, and biochemical changes in cheese during storage, an ideal packaging technique for cheese is needed. The cheese's aroma, flavor, and texture are affected by variations in water activity, fat content, salt content, and microbiological growth. Additional elements that affect the shelf-life include temperature, oxygen, carbon dioxide, concentration, light, and humidity. Oxidative stress is introduced by light, and oxidation reactions are introduced by oxygen. The two contaminants in cheese that are most isolated are *Penicillium* and *Geotrichum candidum* [64]. As a result, most cheese products are wrapped in antimicrobial and antioxidant coatings. The encapsulating materials and the foods are in direct contact when using the wrapping methods. In this instance, the plastic materials contain phenolics that may migrate from the wrapping materials to the food or are released from them. To prevent oxidation or microbial contamination during storage, foods are coated completely. Zein nanofibers infused with essential oils suppress bacteria growth until Day 7, showing a continuous release of bioactivity that results in longer-lasting antibacterial efficacy.

Cheese can be directly consumed with edible film material because the edible coating serves as active packing. The ability of edible films to slowly release phenolics that remain on the surface, maintaining antibacterial action for an extended period, makes them useful among these strategies. In kashar cheese, coatings with grape seed oil that were encased in polyvinyl acid nanofibers prevented the formation of all yeast and mold [65]. Zein nanofibers with clove essential oil are excellent at suppressing *E. coli* and *L. monocytogenes*. The integrity of the bacteria developing in the cheese is disrupted when phenolics are released sustainably from the film material and they interact with the cell membrane of pathogenic microorganisms.

Edible coatings also have an impact on pH, titratable acidity, and moisture content, in addition to microorganisms, which in turn inhibits microbial development and oxidation. We might infer that the cheese's antibacterial and antioxidant properties are gradually lost because of the phenolics-loaded packaging and coating material [65].

7.4.5 Toxicity of Nanomaterial and Polyphenols

It is important to weigh the potential risks of this application against the advantages of nanotechnology. Although there is still no proof that polyphenol-loaded nanoparticles pose a health danger, the risk evaluation of the nanomaterials used to encapsulate them, such as nanoscale silver, zinc oxide, or silicon dioxide, is well known. Alternatively, along with their health advantages, the potential toxicity of polyphenols has been noted in literature [66]

7.4.5.1 Toxicity of Nanomaterials

The term "nanomaterial" primarily considers the nanoparticle's size, ignoring potentially dangerous traits. The study of nanotoxicology focuses on the negative outcomes of exposure to nanoparticles with hazard potential. Particle size, surface area and reactivity, crystal structure, aggregation potential, composition/surface coatings, synthesis and preparation changes, and sample purity are the main physicochemical characteristics of nanoparticles that determine toxicity.

The use of engineered nanoparticles (ENPs) in food packaging materials, such as nanoscale silver, silanated silicon dioxide, titanium dioxide, iron oxide, or zinc oxide, as well as the possibility of their migration into food, are of concern. As recently reviewed by Enescu et al. [66], ambient factors, food type, packing material features, the position of the ENPs in the packaging materials and their interactions, and contact time, all affect migration, in addition to the physicochemical qualities of ENPs. Additionally, the authors provided a summary of the rules that are now in place regarding active food and beverage packaging, as well as the norms, methodologies, and analytical methods used to track the general and targeted migration of ENPs and assess any potential health risks.

These physicochemical characteristics have the potential to induce pro-oxidant conditions in cells, which can result in the production of free radicals, inflammation, or even cell death [66]. Numerous in vivo and in vitro investigations revealed that nanoparticle exposure may impact epigenetic processes, such as methylation of DNA, upgradation of histone, and interference of RNA interference, and it is unquestionably possible for nanoparticles to harm DNA. According to some research, the size of gold nanoparticles has the greatest impact on their cytotoxicity [67]. Because these particles are taken in by cells through pores, another study on silver nanoparticles highlighted the importance of size. Manickam et al. provided a list of some of the most significant silver nanoparticles and genotoxicity reactions. Most of the research demonstrating nanoparticle toxicity is related to their use in pharmacology and medicine; however, the dangers associated with food ingestion have not been sufficiently investigated [68].

It raises questions about whether nanoparticles require unique regulatory frameworks because substances that are normally safe for humans may become harmful when they are made smaller. A summary of the laws governing the use of nanomaterials in the EU and certain non-EU nations has been published [69]. Because of the lack of sufficient scientific data, the biggest uncertainties relate to the toxicity, behavior, and biosorption of nanomaterials.

This must be resolved, and research on risk-and-exposure assessments involving the use of nanomaterials must be conducted [68]. Guidelines for the risk assessment on this subject have been released by the European Food Safety Authority. This advice is the result of a thorough modification of the earlier version regarding nano-specific details. It makes the argument that the current definition of an engineered nanomaterial should not set the parameters for risk assessment, which also considers other types of materials, and it emphasizes the significance of future research to close any gaps and conduct an accurate assessment of the safety of nanomaterials.

7.4.5.5 Toxicity of Polyphenols

Under certain circumstances, polyphenols can exhibit pro-oxidant behavior, resulting in the formation of ROS that can harm DNA, lipids, and other biomolecules. The biological activity of polyphenols is primarily related to their capacity to chelate metals and scavenge free radicals [70]. Numerous variables, including the presence of redox-active substances, the pH of biological tissues, and solubility properties, affect antioxidant and pro-oxidant activity. Drinking large amounts of tea causes an imbalance in the antioxidant and pro-oxidant behavior of tea flavonoids, which has negative effects on human health. Examples include the hepatotoxicity of green tea catechins, the reduction of dietary iron absorption in the intestine, the precipitation of digestive enzymes by black tea tannins, and the reduction of lipase activity by oolong tea polyphenols [70].

Infusions of green tea and other beverages containing green tea catechins are generally safe to consume; however, taking EGCG as a food supplement in doses equal to or greater than 500 mg/day for four months or longer significantly raises serum transaminases in human blood, which is a sign of liver damage.

In addition to pro-oxidant action, polyphenol molecules are considered complex components that can be harmful to human health. Some elements, such as Fe, Mg, and Mn, can be bound by phenolic acids, which can result in a disorder of element-dependent metabolic pathways and decrease the elements' absorption in the gastrointestinal tract and, consequently, their content in blood and tissue [71]. Although nanoparticles have numerous possible uses in the food industry and are not consistent in terms of dangers, the overall conclusion is that they must first undergo rigorous testing before being used. Additionally, consuming polyphenols in large doses over an extended period can harm humans. There is evidence that the coadministration of polyphenols (quercetin) with ENPs (silver nanoparticles) reduces the negative effects of ENPs such as cytotoxicity and oxidative stress. The toxicity of polyphenols encapsulated as nanoparticles have not yet been described. However, more research should be conducted to quantify the possible hazardous consequences of polyphenol-loaded nanoparticles [68–71].

7.5 Conclusion

The functional characteristics, benefits, and encapsulating techniques of phenolics in food products were examined in this chapter. Even the release of phenolics from meals can be managed and sustained to improve the quality of the goods if phenolics are not stable enough after processing. The incorporation of phenolics into capsules holds promise for use in active food packaging, food processing, and food fortification. Food fortification with encapsulated phytochemicals considers several variables, including polyphenol safety, interactions with food components, and sensory impacts.

To extend the shelf life of food items, encapsulated polyphenols have been used in the food processing of dairy, beverages, bakery, and meat goods. Most perishable food goods are packaged using active packaging technologies, including headspace packaging, coating, and wrapping, which contain encapsulated phenolics to prevent lipid oxidation, discoloration, and microbial deterioration. To meet customer demands for fresh and healthy food items and to reduce global food waste, encapsulated phenolics are promising applications in the food business.

References

1 Acevedo-Fani, A., Soliva-Fortuny, R., and Martín-Belloso, O. (2017). Nanostructured emulsions and nanolaminates for delivery of active ingredients: improving food safety and functionality. *Trends in Food Science and Technology* 60: 12–22.

2 Huang, Q., Yu, H., and Ru, Q. (2010). Bioavailability and delivery of nutraceuticals using nanotechnology. *Journal of Food Science* 75: R50–R57.

3 McClements, D.J., Decker, E.A., Park, Y., and Weiss, J. (2009). Structural design principles for delivery of bioactive components in nutraceuticals and functional foods. *Critical Reviews in Food Science and Nutrition* 49: 577–606.

4 Ranjan, S., Dasgupta, N., Chakraborty, A.R. et al. (2014). Nanoscience and nanotechnologies in food industries: opportunities and research trends. *Journal of Nanoparticle Research* 16: 2464.

5 Dasgupta, N. and Ranjan, S. (2018). Nanotechnology in Food Sector. In: *An Introduction to Food Grade Nanoemulsions* (ed. N. Dasgupta and S. Ranjan), 1–18. Singapore: Springer.

6 Ummi, A.S., and Siddiquee, S. (2019). Chapter 15—nanotechnology applications in food: opportunities and challenges in food industry. In: *Applications in Energy, Drug, and Food* (ed. S. Siddique, M.G.J. Hong, and M.M. Rahman), 298–308. Switzerland: Springer International Publishing: Cham.

7 Neo, Y.P., Ray, S., Jin, J. et al. (2013). Encapsulation of food grade antioxidant in natural biopolymer by electrospinning technique: a physicochemical study based on zein–gallic acid system. *Food Chemistry* 136: 1013–1021.

8 Sanguansri, P. and Augustin, M.A. (2006). Nanoscale materials development—A food industry perspective. *Trends Food Sciences. Technology* 17: 547–556.

9 Ahmad, S., Munir, S., Zeb, N. et al. (2019). Green nanotechnology: a review on green synthesis of silver nanoparticles—An eco-friendly approach. *International Journal of Nanomedicine* 14: 5087–5107.

10 Nasrollahzadeh, M., Sajadi, S.M., Sajjadi, M., and Issaabadi, Z. (2019). Chapter 1—an introduction to nanotechnology. In: *An Introduction to Green Nanotechnology*, 28. (ed. M. Nasrollahzadeh, S.M. Sajadi, M. Sajjadi et al.), 1–27. Amsterdam, The Netherlands: Elsevier.

11 Sanjay, S.S. (2019). Chapter 2—Safe nano is green nano. In: *Green Synthesis, Characterization and Applications of Nanoparticles* (ed. A.K. Shukla and S. Iravani), 27–36. Amsterdam, The Netherlands: Elsevier.

12 Ramachandraiah, K., Han, S.G., and Chin, K.B. (2015). Nanotechnology in meat processing and packaging: potential applications—A review. *Asian-Australasian Journal of Animal Sciences* 28: 290–302.

13 Sarkar, P., Choudhary, R., Panigrahi, S. et al. (2017). Nano-inspired systems in food technology and packaging. *Environmental Chemistry Letters* 15: 607–622.

14 Ahmadi, Z., Mohammadinejad, R., and Ashrafizadeh, M. (2019). Drug delivery systems for resveratrol, a non-flavonoid polyphenol: emerging evidence in last decades. *Journal of Drug Delivery Science and Technology* 51: 591–604.

15 Alrgei, H.O.S., Dabic´, D.C.ˇ., Natic´, M.M. et al. (2016). Chemical profile of major taste- and health-related compounds of Oblaˇcinska sour cherry. *Journal of the Science of Food and Agriculture* 96: 1241–1251.

16 Stanisavljevic´, N.S., Ilic´, M.D., Matic´, I.Z. et al. (2016). Identification of phenolic compounds from seed coats of differently colored european varieties of pea (Pisum sativum L.) and characterization of their antioxidant and in vitro anticancer activities. *Nutrition and Cancer* 68: 988–1000.

17 Peši´c, M.B., Milinˇci´c, D.D., Kosti´c, A.Ž. et al. (2019). In vitro digestion of meat- and cereal-based food matrix enriched with grape extracts: how are polyphenol composition, bioaccessibility, and antioxidant activity affected? *Food Chemistry* 284: 28–44.

18 Aguirre, A. and Borneo, R. (2019). Chapter 4—improving bioavailability of polyphenols using nanodelivery systems based on food polymers. In: *Polyphenols in Plants*, 2e. (ed. R.R. Watson), 59–65. Cambridge, MA, USA: Academic Press.

19 Cvetanovi´c, A., Švarc-gaji´c, J., Gaši´c, U. et al. (2017). Isolation of apigenin from subcritical water extracts: optimization of the process. *Journal of Supercritical Fluids* 120: 32–42.

20 Chowdhury, P., Jayroe, J.J., White, B.E., and Fenton, E.R. (2018). Effects of a natural polyphenol on nicotine-induced pancreatic cancer cell proliferation. *Tobacco-Induced Diseases* (16): 1–8. https://doi.org/10.18332/tid/95159.

21 Prakash, M.D., Stojanovska, L., Feehan, J. et al. (2021). Anti-cancer effects of polyphenol-rich sugarcane extract. *PloS One* 16 (3): e0247492. https://doi.org/10.1371/journal.pone.0247492.

22 Souza, N.C., de Oliveira Nascimento, E.N., de Oliveirade Oliveira, I.B. et al. (2020). Anti-inflammatory and antioxidant properties of blend formulated with compounds of *Malpighia emarginata* D.C (acerola) and *Camellia sinensis* L. (green tea) in lipopolysaccharide-stimulated RAW 264.7 macrophages. *Biomedicine&Pharmacotherapy = Biomedecine&Pharmacotherapie* 128: 110277. https://doi.org/10.1016/j.biopha.2020.110277.

23 Mehdizadeh, M., Mushtaq, W., Siddiqui, S. et al. (2021). Assessment of phytotoxins using different technologies. *Current Applied Science and Technology* 22 (4): 1–19. https://doi.org/10.55003/cast.2022.04.22.010.

24 Radunz, M., Hackbart, H.C.D.S., Bona, N.P. et al. (2020). Glucosinolates and phenolic compounds rich broccoli extract: encapsulation by electrospraying and antitumor activity against glial tumor cells. *Colloids and Surfaces. B, Biointerfaces* 192: 111020. https://doi.org/10.1016/j.colsurfb.2020.111020.

25 Siddiqui, S.A., Redha, A.A., Esmaeili, Y., and Mehdizadeh, M. (2022). Novel insights on extraction and encapsulation techniques of elderberry bioactive compounds. *Critical Reviews in Food Science and Nutrition* 1–16. https://doi.org/10.1080/10408398.2022.2026290.

26 Amiri, N., Afsharmanesh, M., Salarmoini, M. et al. (2021). Nanoencapsulation (in vitro and in vivo) as an efficient technology to boost the potential of garlic essential oil as an alternative for antibiotics in broiler nutrition. *Animal* 15 (1): 100022. https://doi.org/10.1016/j.animal.2020.100022.

27 Nawaz, A., Li, E., Khalifa, I. et al. (2021). Effect of structurally different pectin on dough rheology, structure, pasting, and water distribution properties of partially meat-based sugar snap cookies. *Foods* 10: 2692. https://doi.org/10.3390/foods10112692.

28 Chen, S., Li, Q., McClements, D.J. et al. (2020). Co-delivery of curcumin and piperine in zein-carrageenan core-shell nanoparticles: formation, structure, stability and in vitro gastrointestinal digestion. *Food Hydrocolloids* 99: 105334. https://doi.org/10.1016/j.foodhyd.2019.105334.

29 Li, H., Yuan, Y., Zhu, J. et al. (2020). Zein/soluble soybean polysaccharide composite nanoparticles for encapsulation and oral delivery of lutein. *Food Hydrocolloids* 103: 105715. https://doi.org/10.1016/j.foodhyd.2020.105715.

30 Du, Z., Liu, J., Zhang, H. et al. (2020). L-Arginine/L-lysine functionalized chitosan-casein core-shell and pH-responsive nanoparticles: fabrication, characterization and bioavailability enhancement of hydrophobic and hydrophilic bioactive compounds. *Food & Function* 11 (5): 4638–4647. https://doi.org/10.1039/d0fo00005a.

31 Rezvankhah, A., Emam-Djomeh, Z., and Askari, G. (2020). Encapsulation and delivery of bioactive compounds using spray and freeze-drying techniques: a review. *Drying Technology* 38 (1–2): 235–258. https://doi.org/10.1080/07373937.2019.1653906.

32 Sharma, N.S.M.R. and Rafiq, S.I. (2021). Food encapsulation: principles, novel methods, and applications. In: *Handbook of Research on Food Processing and Preservation Technologies: Volume 5: Emerging Techniques for Food Processing, Quality, and Safety Assurance*, 163 (ed. M. Sharma, M.R. Goyal, P. Birwal), Palm Bay, FL: Apple Academic Press.

33 Wei, Z. and Huang, Q. (2019). Assembly of protein-polysaccharide complexes for delivery of bioactive ingredients: a perspective paper. *Journal of Agricultural and Food Chemistry* 67 (5): 1344–1352. https://doi.org/10.1021/acs.jafc.8b06063.

34 Shaddel, R., Hesari, J., Azadmard-Damirchi, S. et al. (2018). Use of gelatin and gum Arabic for encapsulation of black raspberry anthocyanins by complex coacervation. *International Journal of Biological Macromolecules* 107 (Pt B): 1800–1810. https://doi.org/10.1016/j.ijbiomac.2017.10.044.

35 Elzoghby, A.O. (2013). Gelatin-based nanoparticles as drug and gene delivery systems: reviewing three decades of research. *The Journal of Controlled Release* 172: 1075e91.

36 Hemar, Y., Gerbeaud, M., Oliver, C.M., and Augustin, M.A. (2011). Investigation into the interaction between resveratrol and whey proteins using fluorescence spectroscopy. *The Journal of Food Science and Technology* 46: 2137e44.

37 Pan, K., Luo, Y.C., Gan, Y.D. et al. (2014). pH-driven encapsulation of curcumin in self-assembled casein nanoparticles for enhanced dispersibility and bioactivity. *Soft Matter* 10: 6820e30.

38 Karthikeyan, S., Hoti, S.L., and Prasad, N.R. (2015). Resveratrol-loaded gelatin nanoparticles synergistically inhibit cell cycle progression and constitutive NF-kappa B activation and induce apoptosis in non-small cell lung cancer cells. *Biomed Pharmacother* 70: 274e82.

39 Guan, X., Yin, T., and Han, F. (2015). Light stability, controlled-release and antioxidation of resveratrol-hordein composite nanoparticles. *Chemical Journal of Chinese Universities* 36: 1707e12.

40 Park, S.J., Garcia, C.V., Shin, G.H., and Kim, J.T. (2016). Fabrication and optimization of EGCG-loaded nanoparticles by high-pressure homogenization. *Journal of Applied Polymer Science* 133: 43269.

41 Tan, C., Xie, J.H., Zhang, X.M. et al. (2016). Polysaccharide-based nanoparticles by chitosan and gum Arabic polyelectrolyte complexation as carriers for curcumin. *Food Hydrocolloids* 57: 236e45.

42 Xia, S.Q., Li, Y.Q., Xia, Q.Y. et al. (2015). Glycosylation of bovine serum albumin via Maillard reaction prevents epigallocatechin-3-gallate-induced protein aggregation. *Food Hydrocolloids* 43: 228e35.

43 Sun, J.B., Bi, C., Chan, H.M. et al. (2013). Curcumin-loaded solid lipid nanoparticles have prolonged in vitro antitumor activity, cellular uptake, and improved in vivo bioavailability. *Colloids Surf B Biointerfaces* 111: 367e75.

44 Zhang, J., Nie, S.F., and Wang, S. (2013). Nanoencapsulation enhances epigallocatechin-3-gallate stability and its antiatherogenic bioactivities in macrophages. *Journal of Agricultural and Food Chemistry* 61: 9200e9.

45 Ramalingam, P., Yoo, S.W., and Ko, Y.T. (2016). Nanodelivery systems based on mucoadhesive polymer-coated solid lipid nanoparticles to improve the oral intake of food curcumin. *Food Research International* 84: 113e9.

46 Isailovi´c, B., Djordjevi´c, V., Levi´c, S. et al. (2016). Encapsulation of flavors and aromas: controlled release. In: *Edible Films and Coatings: Fundamentals and Applications* (ed. M.P.M. Garcia, M. Carmen Gómez-Guillén, M. Elvira López-Caballero, G.V. Barbosa-Cánovas), 317–344. Boca Raton, FL, USA: CRC Press.

47 Serpa Guerra, A.M., Gómez Hoyos, C., Velásquez-Cock, J.A. et al. (2019). The nanotech potential of turmeric (Curcuma longa L.) in food technology: a review. *Critical Reviews in Food Science and Nutrition* 24: 1–13.

48 Puligundla, P., Mok, C., Ko, S. et al. (2017). Nanotechnological approaches to enhance the bioavailability and therapeutic efficacy of green tea polyphenols. *Journal of Functional Foods* 34: 139–151.

49 Kuswandi, B. (2017). Environmental friendly food nano-packaging. *Environmental Chemistry Letters* 15: 205–221.

50 Tatli Seven, P., Seven, I., Gul Baykalir, B. et al. (2018). Nanotechnology and nano-propolis in animal production and health: an overview. *Italian Journal of Animal Science* 17: 921–930.

51 Santos, A.C., Pereira, I., Pereira-Silva, M. et al. (2019). Nanotechnology-based formulations for resveratrol delivery: effects on resveratrol in vivo bioavailability and bioactivity. *Colloids Surf. B Biointerfaces* 180: 127–140.

52 Di Costanzo, A. and Angelico, R. (2019). Formulation strategies for enhancing the bioavailability of silymarin: the state of the art. *Molecules* 24: 2155.

53 Grosso, G., Godos, J., and Lamuela-Raventos, R. (2017). A comprehensive meta-analysis on dietary flavonoid and lignan intake and cancer risk: level of evidence and limitations. *Molecular Nutrition and Food Research*. https://doi.org/10.1002/mnfr.201600930.

54 Rothwell, J.A., Knaze, V., and Zamora-Ros, R. (2017). Polyphenols: dietary assessment and role in the prevention of cancers. *Current Opinion in Clinical Nutrition and Metabolic Care* 20: 512–521. https://doi.org/10.1097/MCO.0000000000000424.

55 Yadav, K., Bajaj, R.K., Mandal, S. et al. (2018). Evaluation of total phenol content and antioxidant properties of encapsulated grape seed extract in yoghurt. *International Journal of Dairy Technology* 71 (1): 96–104. https://doi.org/10.1111/1471-0307.12464.

56 Caballero, S., Li, Y.O., McClements, D.J., and Davidov-Pardo, G. (2021). Encapsulation and delivery of bioactive citrus pomace polyphenols: a review. *Critical Reviews in Food Science and Nutrition* 1–17. https://doi.org/10.1080/10408398.2021.1922873.

57 Lomovskiy, I.O., Bychkova, E.S., Slavikovskaya, M.S., and Lomovsky, O.I. (2020). Encapsulated material based on polyphenols for manufacturing functional foods. *Materials Today: Proceedings.* 25: 409–411.

58 Papillo, V.A., Locatelli, M., Travaglia, F. et al. (2019). Cocoa hulls polyphenols stabilized by microencapsulation as functional ingredient for bakery applications. *Food Research International (Ottawa, Ont.)* 115: 511–518.

59 RashidaieAbandansarie, S.S., Ariaii, P., and CharmchianLangerodi, M. (2019). Effects of encapsulated rosemary extract on oxidative and microbiological stability of beef meat during refrigerated storage. *Food Science & Nutrition* 7 (12): 3969–3978. https://doi.org/10.1002/fsn3.1258.

60 Robert, P., Zamorano, M., Gonzalez, E. et al. (2019). Double emulsions with olive leaves extract as fat replacers in meat systems with high oxidative stability. *Food Research International (Ottawa, Ont.)* 120: 904–912.

61 Goksen, G., Fabra, M.J., Ekiz, H.I., and Lopez-Rubio, A. (2020). Phytochemical-loaded electrospun nanofibers as novel active edible films: characterization and antibacterial efficiency in cheese slices. *Food Control* 112: 107133. https://doi.org/10.1016/j.foodcont.2020.107133.

62 Colussi, R., da Silva, W.M.F., Biduski, B. et al. (2021). Postharvest quality and antioxidant activity extension of strawberry fruit using allyl isothiocyanate encapsulated by electrospun zein ultrafine fibers. *Lightweight* 143: 111087. https://doi.org/10.1016/j.lwt.2021.111087.

63 Dominguez, R., Pateiro, M., Gagaoua, M. et al. (2019). A comprehensive review on lipid oxidation in meat and meat products. *Antioxidants* 8 (10): 429. MDPI. https://doi.org/10.3390/antiox8100429.

64 Kure, C.F., Skaar, I., and Brendehaug, J. (2004). Mould contamination in production of semi-hard cheese. *International Journal of Food Microbiology* 93 (1): 41–49. https://doi.org/10.1016/j.ijfoodmicro.2003.10.005.

65 Ceylan, Z., Kutlu, N., Meral, R. et al. (2021). Protective effect of grape seed oil-loaded nanofibers: limitation of microbial growth and lipid oxidation in kashar cheese and fish meat samples. *Food Bioscience* 42: 101076. https://doi.org/10.1016/j.fbio.2021.101076.

66 Enescu, D., Cerqueira, M.A., Fucinos, P., and Pastrana, L.M. (2019). Recent advances and challenges on applications of nanotechnology in food packaging. A literature review. *Food and Chemical Toxicology* 134: 110814.

67 Elsebaie, E.M. and Essa, R.Y. (2018). Microencapsulation of red onion peel polyphenols fractions by freeze drying technicality and its application in the cake. *Journal of Food Processing and Preservation* 42 (7): e13654. https://doi.org/10.1111/jfpp.13654.

68 Manickam, V., Velusamy, R.K., Lochana, R. et al. (2017). Applications and genotoxicity of nanomaterials in the food industry. *Environmental Chemistry Letters* 15: 399–412.

69 Hubert, R., Kirsten, R., and Birgit, S. (2017). Regulatory aspects of nanomaterials in the EU. *Chemie Ingenieur Technik* 89 (3): 224–231.

70 Hu, B., Liu, X., Zhang, C., and Zeng, X. (2017). Food macromolecule based nano delivery systems for enhancing the bioavailability of polyphenols. *Journal of Food and Drug Analysis* 25: 3–15.

71 Chang, C., Wang, T., Hu, Q., and Luo, Y. (2017). Caseinate-zein-polysaccharide complex nanoparticles as potential oral delivery vehicles for curcumin: effect of polysaccharide type and chemical cross-linking. *Food Hydrocolloids* 72: 254–262.

8

Nanodeliveries of Food Polyphenols as Nutraceuticals

Santwana Palai[1,*] *and Mithun Rudrapal*[2]

[1] *Department of Veterinary Pharmacology & Toxicology, College of Veterinary Science and Animal Husbandry, Odisha University of Agriculture & Technology, Bhubaneswar, Odisha, India*
[2] *Department of Pharmaceutical Sciences, School of Biotechnology and Pharmaceutical Sciences, Vignan's Foundation for Science, Technology & Research, Guntur, India*
[*] *Corresponding author*

8.1 Introduction

Natural chemicals are still the most important primary sources of bioactives because synthetic medicinal chemistry have not yet generated many alternatives for natural compounds. As a result, the search for new pharmacophores and active chemicals continues [1]. Polyphenols are a diverse group of secondary phytonutrients or metabolites found in all diets. These are antioxidants or dietary bioactives derived from plant-based diets that have antioxidant properties [2]. However, some phytochemicals have low solubility, resulting in poor bioavailability. Encapsulating such bioactive compounds in an appropriate carrier helps to increase their bioavailability to change the pharmacokinetics and biodistribution. Bioactive nanoencapsulation is actively revolutionizing the field of medicine delivery. The medication therapeutic index increases with the use of controlled drug delivery systems, causing localization to specific tissues [3].

New medical breakthroughs have indicated the commercialization potential of nanoparticle-based medicinal therapies. Natural dietary mediators have demonstrated advantages in healthcare because anticancer and antidiabetic properties have lately emerged as an active component of nutraceuticals [4]. Because of the increased interest in traditional medicines among scientists and academics, the World Health Organization has developed a plan to address traditional medicine apprehensions worldwide [5]. Current advances in nutraceutical nanoencapsulation can minimize their limitations while optimizing their health benefits [6]. This chapter focuses on food polyphenols, their benefits, and nutraceutical nanocarriers that can be developed as nanoscale nutraceuticals or functional foods.

8.1.1 Polyphenols in Food

Polyphenol is not a chemical word; instead, it refers to flavonoids, catechins, tannins, and phenolic acids, as well as their chemically modified or polymerized derivatives. Flavanols, flavanones, flavonols, hydroxycinnamic acids, and anthocyanins are important classes of polyphenols in the

Polyphenols: Food, Nutraceutical, and Nanotherapeutic Applications, First Edition. Edited by Mithun Rudrapal.

Table 8.1 Types of polyphenols in foods and drinks.

Sl. No.	Chemical Class	Rich Sources	Examples	References
1	**Ohenolic acids (benzoic acid & cinnamic acid derivatives)**	Red fruits, coffee, chicory, black radish, onions, artichoke, plum, pears	Caffeic, p-coumaric, ferulic, sinapic acids	[9]
2	**Flavonoids**			
A	Flavanols	Tea, cocoa, apple, broad beans	Quercetin, kaempferol	[10]
B	Flavanones	Citrus fruits like orange	Hesperidin, naringenin, taxifolin	[11]
C	Flavones	Dye, vegetables	Luteolin	[12]
D	Flavonols	Tea, apples, onions	Catechins	[13]
E	Anthocyanins	Berry fruits, in fruit and vegetables	Cyanidin colored polyphenols	[14]
F	Isoflavones	Soy or soy-based products	Genistein	[15]
3	**Stilbenes**	Grapes, red wine	Resveratrol	[16]
4	**Lignans**	Linseed, flaxseed	Secoisolariciresinol	[17]

diet [7]. These catechins, flavonoids, anthocyanins, and phenolic acids make up most dietary polyphenols [8] (Table 8.1).

8.1.2 Nutraceuticals

Nutraceuticals are a blend of pharmaceuticals and nutrition. Food or food components provide nutritious and therapeutic aids to the body such as by nutrients, resisting various diseases, and assisting in the treatment of specific conditions [18]. Medical professionals anticipated developing meals that may be used as medicine to prevent and treat ailments. As a result, dietary complements, functional beverages, and functional foods, were introduced as dietary supplements providing protein, mineral, vitamin, herbal, or plant extracts [19].

Probiotics and omega fatty acid foods are examples of functional meals and beverages, while functional brews include sports drinks, fortified juices, and energy drinks. Nutraceutical examples include dietary supplements, multifunctional food, and functional food. Functional foods are basic foods in that they provide nourishment; however, they have been augmented with specific constituents that aid in the maintenance of a healthy body [20].

Recent advancements in food technology have paved the path to functional foods that are made solely for the benefit of human health. The basic operations considered are identification, separation, purification, and characterization of the food properties such as nutritional content and medicinal value.

The basic dietary elements required for the body's optimal function and average energy requirements are carbohydrates, proteins, and lipids [21]. Vitamins have nutritional effects and are not generally changed by the human body. Consequently, they should be ingested for healthy bodily function. Nutraceuticals are small food components that aid the body in battling certain chronic disorders [4].

A nutraceutical product's efficacy is determined by its bioavailability. In nutritional terms, bioavailability implies partially available nutrients in food; however, in pharmacological terms, it is the rate and extent of a drug that reaches its site of action. Bioavailability has increased as the use of nutraceuticals as preventative medicine has become a major concern for health-care regulators and manufacturers [22]. Various parameters, such as inadequate gastric time, lesser solubility, gut permeability, and variability in food, are limiting the health advantage of nutraceuticals when administered orally [23]. Further obstacles, including chemical instability and crystallization, must be overcome before these bioactive compounds can be used in commercial food items, in addition to their limited bioavailability and poor water solubility [24]. Patenting novel delivery technologies, such as nanotechnologies, have grown in popularity as a means of boosting the efficacy of nutraceuticals.

8.1.3 Nanodelivery

Nanotechnology has blossomed in almost every industry, with significant impact to the health and nutrition industries. Encapsulation of dynamic food ingredients in nanoform drug transfer systems is a significant use in nutrition and food. The fundamental concepts of nanotechnology are followed in nutraceutical nanoformulations. Most nanotechnology platforms are employed to develop nutraceutical delivery strategies with low water solubility [25]. By resolving the limitations associated with bioactives, the technology has a high commercialization potential, and thus multifold growth in the next few years is expected. However, before incorporating these newly developed nanoscale delivery techniques into profitable food harvests, they must be safe. A variety of desirable qualities of nanoscale delivery methods must be addressed. When sized to the nanoscale, they will act inversely in the gastrointestinal (GI) tract compared with regularly scaled particulate matter [26].

If the nanoscale delivery method digestion is the same as typical particulate matter, it cannot be more harmful; otherwise, toxicity concerns may develop. It is vital to investigate the toxicity potential of these food-grade nanoscale delivery devices to confirm their safety. It is preferable to build these delivery systems with food-grade materials. These nanoscale devices must be cost effective, sturdy enough to resist storage conditions, and resilient enough to be used in real world applications [27]. Additionally, the delivery method should be physically and chemically resistant to external strains while maintaining useful qualities. The distribution method must be capable of boosting labile bioactive nutrient stomach stability [28] and it maintaining a continuous amount in the systemic blood flow. An extremely lipophilic medication can increase lymphatic mobility and increase the stomach retention time [29].

The nanoparticles are made of lipids, carbohydrates, or proteins, mostly with various bioactive compounds [30]. The incorporation of nanotechnologies for nutraceutical substances improves the oral absorption and bioavailability of phenolic compounds, increasing its nutraceutical effect. Various nutraceutical nanodelivery systems are explored in this article for food polyphenols.

8.2 Food Polyphenols

8.2.1 Classification of Food Polyphenols

Food polyphenols are created in a wide range of foods and have piqued the interest of scientists because of their anti-inflammatory, antioxidant, and anticancer actions. Polyphenols are benzene ring-structured molecules with two or more phenolic hydroxyl groups that are divided as phenolic acids or flavonoids based on their physical properties [31].

Flavonoids are typically present in plant cell vesicles as glycosides. Flavonoids have three cyclic structures as their basic skeleton, which are C6–C3–C6. Based on their chemical assemblies, flavonoids are further classified as flavanones, flavanols, flavones, flavonols, isoflavones, anthocyanins, and chalcones (Table 8.1). Flavonoids are found in most plants and are necessary for plant growth, progress, fruiting, flowering, antioxidant, antibacterial, and disease inhibition. The vast majority of these flavonoids from various natural sources (Table 8.2, Figure 8.1) have antioxidant, antibacterial, anticancer, and anti-inflammatory physiological actions that can benefit the body.

Table 8.2 Chemical structure and pharmacological activities of food polyphenols.

Sl. No.	Flavonoids	Chemical Structure	Pharmacological Activities	References
1	Quercetin	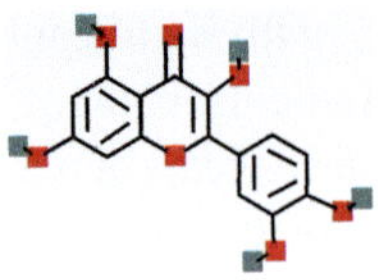	Antioxidant, anticarcinogenic, antidiabetic anti-inflammatory, immunoprotective activities	[33]
2	Rutin	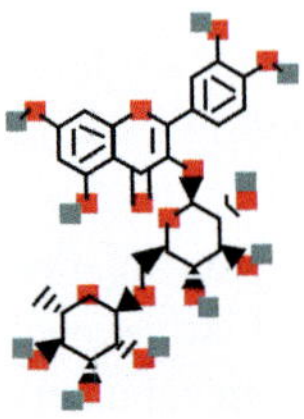	Anti-inflammation, antioxidation, antiallergy, antivirus effects	[34]
3	Resveratrol	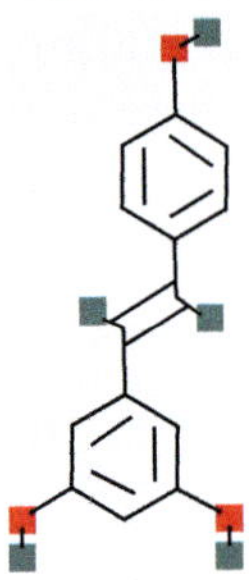	Platelet aggregator, vasoconstrictor in cardiovascular diseases, aids in preeclampsia	[35]
4	Baicalin	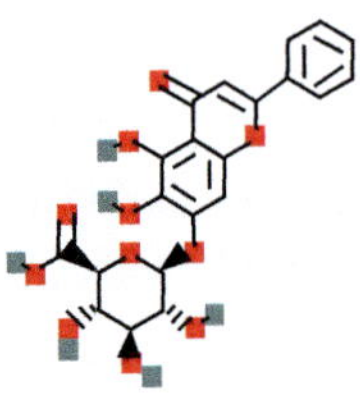	Antibacterial, cholesterol-lowering, diuretic, anti-inflammatory, antiviral antithrombotic, detoxificating, hemostasis, antiallergic effects	[36]
5	Berberine	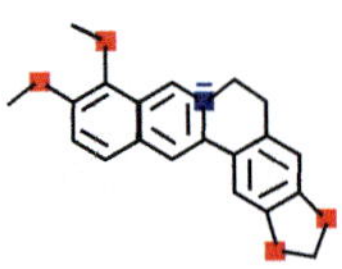	Anticancerous, antioxidant, free radicals eliminator, antidiabetic, anti-inflammatory effect	[37]
6	Myricetin	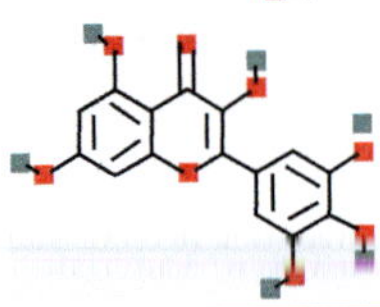	Anti-inflammatory, caries prevention, antitumor, antimutagenic, antioxidant properties	[38]

Table 8.2 (Continued)

Sl. No.	Flavonoids	Chemical Structure	Pharmacological Activities	References
7	Kaempferol	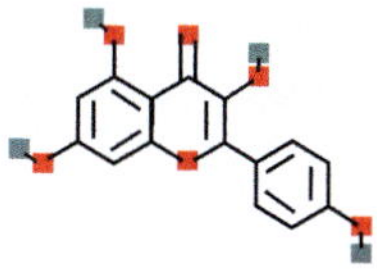	Antifungal, antioxidant, anticancer, anti-inflammatory antiviral, antibacterial effects	[39]
8	Isoliquiritigenin	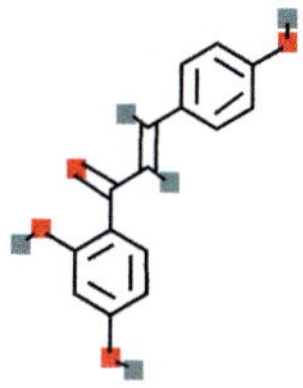	Antibacterial, antispasmodic, anti-ulcer effects, hepatocyte monoamine oxidase inhibitor	[40]
9	Fisetin	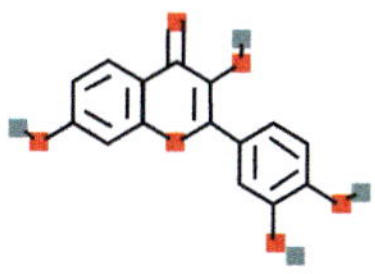	Anti-inflammatory, antithrombotic, antispasmodic, antioxidant, anticoagulant effects, diabetic kidney injury treatment	[41]
10	Apigenin	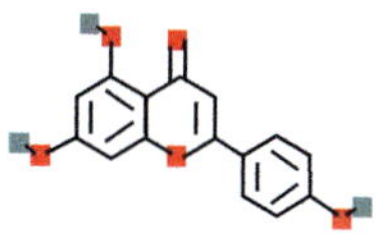	Anticancer, antioxidant, antihypertensive, sedative, tranquilizer, antiviral, anti-inflammatory, drug	[42]
11	Naringenin	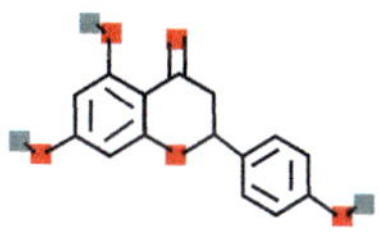	Anti-lipid peroxidation, antioxidant defenses, and modulates immune system and carbohydrate metabolism	[43]
12	Catechin	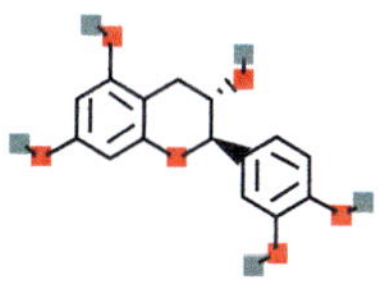	Antidiabetic, improves glucose homeostasis and insulin sensitivity	[44]
13	Anthocyanins	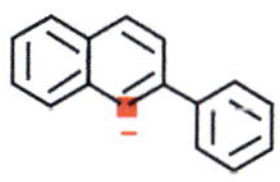	Antioxidant, anti-inflammatory activities, decreases the inflammatory mediators cyclo-oxygenase	[45]
14	Theophylline	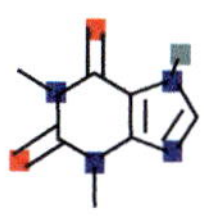	Anti-inflammatory, antioxidant, immunomodulator	[46]
15	Genistein	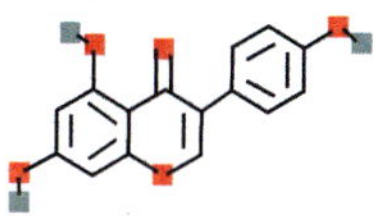	Reduction of cardiovascular disease, osteoporosis, and postmenopausal symptoms	[47]

Phenolic acids are abundant in nuts, vegetables, fruits, and drinks. Low molecular weight phenolic acids are water-soluble throughout processing and human digestion; however, they become water-insoluble when they undertake condensation reactions with quinic acid and glucose. Phenolic acids are classified into three varieties based on their hydroxyl content like

Figure 8.1 Sources of food polyphenols. Credits Andreas Munich/Pixabay, tstock / Adobe Stock, อำนวย หัวคาด/Adobe Stock, Christine Han Photography/Stocksy / Adobe Stock, Terbofast / Wikimedia Commons / CC BY-SA 4.0 "sonatalitravel/Adobe Stock, InfinitePhoto/Adobe Stock, Picture Partners/Adobe Stock, Happy_lark/Adobe Stock Photos, Marylooo/Adobe Stock, Maresol/Adobe Stock, Spline_x/Adobe Stock, Peangdao/Adobe Stock, Gtranquillity/Adobe Stock, volff/Adobe Stock.

dihydroxybenzoic acid, trihydroxybenzoic acid, and monohydroxybenzoic acid. Phosphonic acids are commonly employed in a range of functional foods because of their antiallergic, anti-inflammatory, antioxidant, anticancer cardioprotective actions, and other efficient healthy activities [32].

8.2.2 Properties of Food Polyphenols

Classification can be according to the variation in the chemical structure of polyphenolic compounds. The most common chemical skeleton alterations are oxidation, hydroxylation, methylation, and glycosylation. Polyphenols are separated into groups such as phenolic acids, non-flavonoids with stilbenoids, flavonoids, tannins, lignans, and diaryl heptanoids as subclasses of the latter [48]. Non-flavonoid polyphenol subclasses, such as polyphenolic amides and anthraquinones, are less frequent. Phenolic acids can be classified as hydroxycinnamic or hydroxybenzoic acids in both bound and free forms found in plants, and they account for approximately one-third of dietary polyphenols. Examples include p-hydroxybenzoic acid, vanillic acid, ellagic acid, protocatechuic acid, and gallic acid. Ellagic acid is a benzoic acid derivative, and the cinnamic acid derivatives are ferulic acid, caffeic acid, p-coumaric acid, and sinapic acid [49]. There are numerous effects of each polyphenol class, and their pharmacological benefits are described in Table 8.2.

Flavonoids make up roughly two-thirds of all phenolics found in the human diet. Flavonoids are separated into isoflavonoids, flavonols, anthocyanidins, flavanols, flavones, and flavanones based on their heterocyclic ring variations. The most common phenolics in plant-based foods are flavonols and flavones (Table 8.1). Tannins are a non-flavonoid polyphenol subclass that contains proanthocyanidins, which are non-hydrolysable tannins that have been condensed, and hydrolysable tannins as esters of ellagic acid, ellagitannins, gallotannins, and gallic acid. The lignan non-flavonoid subclass is a result of phenylalanine and they belong to major classes of phytoestrogens, along with isoflavone. Lignans include enterolactone, enterodiol, enterolignans, and secoisolariciresinol. Stilbenoids are phytoalexins that are hydroxylated stilbene derivatives. The most well-known stilbenoid is resveratrol [50].

8.2.3 Advantages of Food Polyphenols

Plant polyphenols have several health-promoting characteristics. The activities, such as antibacterial and antioxidant actions, and their natural accessibility and compatibility in meals capable with unique functional aspects, enhance human health. The likely role of functional foods comprising polyphenolic chemicals in the prevention of several chronic illnesses, such as hypertension, diabetes, and cancer, is extremely important [51].

8.3 Health Benefits of Food Polyphenols

8.3.1 Antioxidant Effect

Plant polyphenolic chemicals possesses high antioxidant action owing to their unique structural properties. Polyphenolic chemicals, which play a significant role in human health, are abundant in fruits, cereals, and vegetables. Polyphenols, as antioxidants, can protect DNA from oxidative damage and thereby prevent a variety of diseases [52]. Curcumin has high antioxidant capacity and they reduce malondialdehyde levels. Glutathione, catechins, and ascorbic acid antioxidant capabilities are related [53]. Catechins are most effective in foraging 2,2-azinobis (3-ethylbenzothiazoline-6-sulfonic acid) radicals with maximum reduction equivalents for trivalent iron ions, and are active in preventing dihydrorhodamine oxidation [54]. Polyphenolic substances, such as catechins, have excellent antioxidant capabilities, making them suitable candidates for antioxidant prophylaxis and therapy [53].

8.3.2 Pro-oxidant Effect

Polyphenolic substances have potent antioxidant properties. However, in large amounts, many polyphenolic chemicals precipitate DNA mutilation causing apoptosis. Resveratrol's pro-oxidant qualities hasten ageing progression, and its pro-oxidant action reduced the sequential life of brewer's yeast [55]. Gallic acid limiting lipid oxidation and protein carbonyl production promotes oxi dative loss of thiol and amine groups, altering protein structural characteristics and biological function [56]. High doses of soy bean added to meat pig feed cause oxidative alterations in pork liver, fat, and plasma. The addition for 64 days enhanced total antioxidant and superoxide dismutase activity, indicating considerable pro-oxidant activity [57]. As a result, the dose of polyphenolic chemicals in various claims is not the same.

8.3.3 Anti-inflammatory Effect

Plant polyphenols suppress and kill many inflammatory cells, also by regulating cytokines and their receptors and modifying their exudation mechanisms. Rutin hydrogels have anti-inflammatory activity like conventional medicines [58]. RAW264.7 cells and a CCl4-made acute liver injury were used to investigate hesperidin's anti-inflammatory efficacy in a liver injury model, and showed that hesperidin effectively lowered interleukin (IL-6), nitric oxide (NO), and tumor necrosis factor (TNF-α) in vivo and in vitro, demonstrating good anti-inflammatory efficacy [59].

8.3.4 Antimicrobial Effect

Polyphenols possess antibacterial properties against numerous microbes. Flavonoids specifically have better antibacterial action than any other polyphenols [60]. According to several studies, polyphenolic substances synergize with antibiotics and have great antibacterial activities. Curcumin coated with chitosan film demonstrated better antibacterial efficacy against *Rhizopus solani* and *Staphylococcus aureus* [53]. Additionally, the antibacterial effects of silymarin, tea polyphenols, and rutin are extensively used [61].

8.3.5 Anticancer Effect

Polyphenols have been shown to protect against certain types of cancer. They can prevent tumor formation as well as have harmful effects on cells, producing apoptosis. Resveratrol, with its repressive action on cell development, apoptotic effect, and antioxidant effects, has the potential to alter the course of cancer and other related illnesses. Quercetin is widely used in the inhibition and treatment of esophageal cancer [62]. Several more research studies have validated the antitumor effects of polyphenolic substances. Silymarin causes apoptosis in liver cancer cells showing better prophylaxis and treatment against liver disorders. Additionally, epigallocatechin gallate (EGCG) and curcumin have better efficacy as anticancerous agents in the treatment of breast cancer [63].

8.3.6 Antidiabetic Effect

A polyphenol-rich diet can reduce the hazard of diabetes. Polyphenols can alter the insulin pathway and increase insulin sensitivity in the peripheral tissues [64]. Several polyphenolic substances block alpha amylase and alpha glucosidase that regulate glucose absorption in the intestine and maintain the blood sugar equilibrium. The nanoforms of polyphenols, such as catechins, have strong antioxidant and antidiabetic properties [53]. The quercetin nanoform has therapeutic promise as an antidiabetic bioactive substance [65]. Bioactive compounds possess their own relevance like resveratrol, naturally found in plant-based foods including peanuts, cranberries, jackfruit, grapes, and mulberries with anti-carcinogenic, hepatoprotection, neuroprotection, antiaging, anti-inflammatory, antiobesity, and antidiabetic effects. Resveratrol's sensitivity to light, heat, and enzymes limits its ability to provide health advantages during consumption. Despite its excellent bioactivity, resveratrol's use as a bioactive agent in the food and pharmaceutical industries is limited for these reasons. The use of resveratrol in its encapsulated form may be able to alleviate these issues helping to increase the bioactivities of resveratrol after fortification in nanoforms [66].

8.3.7 Antihypertensive Effect

Certain polyphenols like cocoa, catechins, and proanthocyanidins have high flavanol concentrations, which can recover endothelial function, lower the oxidative sensitivity of low-density lipoproteins, and boost vasodilation. Anthocyanins, tannins, phenolic acids, flavonoids, flavanols, and other polyphenolic substances like resveratrol can improve vasodilation and regulate blood pressure [67]. The vasodilatory properties of amlodipine, curcumin, and their combination were studied on isolated rat aortic rings, and it was found that hypertensive patients taking amlodipine could take curcumin or turmeric for food or as medicine without impairing the antihypertensive effects of amlodipine. This lays a solid foundation for the use of polyphenols, such as curcumin, as food supplements for hypertension prophylaxis and therapy [68]. Several studies have found that green tea polyphenols like epicatechin and catechin have antihypertensive, antiproliferative, antithrombogenic, anti-inflammatory, and lipid-lowering actions that can be important in developing food through nanotechnology [69].

8.3.8 Antiobesity effect

Polyphenolic compounds can influence obesity by suppressing adipocyte growth, increasing adipocyte death, boosting lipolysis, and fat oxidation. Green tea is a drink with numerous health advantages because it is rich in polyphenolic substances like catechins and gallocatechin gallate and they have a substantial role in obesity management [70]. Soy beans can help people lose weight by reducing the activities of pancreatic protein lipase that inhibits adipocyte development and by triggering hormone-sensitive lipase, which stimulates lipolysis [71]. Quercetin supplementation has a strong inhibitory effect on adipose tissue accumulation in obese rats, suggesting that it may have antiobesity properties [72].

8.4 Common Nanodelivery Systems for Food Polyphenols/ Nanoencapsulation of Nutraceuticals

8.4.1 Lipid-based Nanostructures

Lipid-based assemblies, such as nanoemulsions, liposomes, solid lipid nanoparticles (SLNs), and self-emulsifying systems, can aid in encapsulation and the target delivery of sensitive bioactives nutraceuticals. They can be used as bioreactors to address the issue of aroma and flavor components. They are useful because most of the bioactive compounds can be incorporated in them. Similarly, numerous polyphenolic compounds are targeted to specific areas by active or passive methods [73].

8.4.2 Solid Lipid Nanoparticles

Aqueous surfactant solutions or colloidal dispersions of lipids in water are SLNs. They have numerous advantages, including shielding of the integrated molecule from the exterior atmosphere, improved firmness of the encapsulated component, increased ability for transporting both lipophilic and hydrophilic medicines, controlled release, and simplicity of upscaling. These lipidic arrangements can aid in the enhancement of organoleptic and functional qualities. Additionally, generally regarded as safe (GRAS) chemicals are included in these systems [74]. During storage, disadvantages, such as augmented particle size, aggregation, relatively more water content, and flocculation, may occur [53].

8.4.3 Nanoliposomes

Liposomes are hydrophilic lipid bilayer membrane structures with hydrophobic fatty acid tails that can transport both hydrophobic and hydrophilic bioactives, the former in the aqueous core and the latter in the lipid bilayers. Nanoliposomes are liposomes that are nanoscale. Nanoliposomes could be administered parenterally, orally, topically, or nasally. Its downside is that they are recognized as foreign particles by the circulatory system and thus eliminated quickly by the reticuloendothelial system. Additionally, they can be destroyed by electrostatic, van der Waals, and hydrophobic forces, which can fragment nanoliposomes. As a result, some sort of stability mechanism is necessary such as steric stabilization using inert polymers [75].

8.4.4 Nanoemulsions

Emulsions are biphasic systems made up of an inner phase, or dispersed phase, and an outer phase, or continuous phase. Surfactant molecules make up the interphase. Nanoemulsions are incredibly small emulsions that seem transparent or translucent, with a size range of 50–200 nm, significantly smaller than that of ordinary emulsions. A surfactant molecule is typically 2-nm long; hence, a micelle is typically 5 nm or larger in diameter. However, incorporating an oil phase into a micellar core can result in a significant size increase [76]. It is generally recognized that the most bioactive phytochemicals are either insoluble or lipophilic in nature; hence, nanoemulsions are a possibility for incorporating lesser soluble nutraceuticals into a food matrix. The poor solubility of these active components in water or oil has a significant impact on their systemic bioavailability because their properties, such as lipophilicity, solubility, and partition coefficient, determine their transport, administration route, and target locations. The incorporation of such bioactives into nanoemulsions can be helpful because the small particle size of nanoemulsions increases their surface area, improving digestion rates, faster diffusion across mucus membranes, and amplified epithelial cell permeability [77]. Furthermore, nanoemulsions may shield chemically labile bioactives against oxidation, resulting in an augmented shelf life and condensed GI breakdown. There have been numerous reports on the trapping of bioactives in nanoemulsions, with new trends indicating the usage of food-grade nanoemulsions. Carrier oil, which controls the bioavailability of encapsulated components, is a crucial component in the creation of food-grade nanoemulsions. The carrier oil can form mixed micelles with more capacity for active constituent solubilization, and they are entirely consumable [78].

8.4.5 Polysaccharide-based Nanoparticles

Most natural polysaccharides are inexpensive raw materials suitable for nanoencapsulation of various bioactives or nutraceuticals. Several technologies have been used to create nanoparticles of various shapes and sizes. The method selection and polysaccharides depend on economy, safety, and environmental issues. Depending on the physical and chemical qualities of the nutraceuticals and polysaccharides, different techniques for nanoencapsulation of bioactives can be applied. Polysaccharides can enclose both hydrophilic and hydrophobic substances [4].

Because of their structural plasticity and site-specific digesting capabilities, they are ideal carriers for the targeted and regulated administration of nutraceuticals throughout the human GI tract. Because of their stable structure, reactive site availability for chemical modification, low cost, biocompatibility, non-toxicity, and hydrophilic nature, they are the preferred material. Commonly used polysaccharides are starch, pectin, guar gum, chitosan, chondroitin sulphate, and alginate. They are advantageous because they can be used for the delivery of synergistic combinations; however, when considering food applications, the high molecular weight of polysaccharides sometimes limits their application, for example, for delivery in clear drinks, because transparent systems based on biopolymeric nanoparticles are difficult to prepare [79]. Despite numerous studies on encapsulating a wide range of nutraceuticals and functional meals, their bioactive mechanisms and biological efficacies in vitro or in vivo remain unclear.

8.5 Polyphenol-loaded Nanoparticles for the Enhancement of Functional Properties of Food

8.5.1 Nanopolyphenols Acting as Antidiabetic Agents

Various polyphenols are used in the manufacture of nanopolyphenols with nanocarriers to be used in diabetic animals.

Nano-nutraceuticals can provide antihyperglycemic effects, hypoglycemic, and antidiabetic actions, and address oxidative stress linked with diabetes, as well as prevent diabetic neuropathic pain. They are also effective in diabetic wound healing, affecting other diabetes complications such as diabetes-persuaded learning and memory damage, diabetic cardiomyopathy, diabetic nephropathy, diabetic cataract or retinopathy, and inflammation [80].

Nanocarrier nanoparticles are principally employed, and SLNs, nanoemulsions, nanomicelles, nanospheres, and nanorods are used to a lesser extent. Nanospheres prepared with anti-solvent precipitation and nanorods created in a magnetic field have the smallest reported particle size of approximately 16 nm [81].

The nanoparticles are created using both natural and synthetic polymers, as well as polymeric colloidal stabilizers. Additionally, inorganic species, such as selenium and iron oxide, are used for synergistic actions. The amount of nanomaterial given orally is 1–300 mg/kg/day. In addition, nanopolyphenols were given intravenously as well as topically. The encapsulation efficacy and drug loading measurements of the synthesized nanoparticles ranged from 56% to 97.7% and 4.2% to 53.2%, respectively. A decrease in particle size resulted in a proportional decrease in drug loading capacity. The duration for medicines administered orally were 2–70 days [82].

8.5.2 Nanopolyphenols as Antiobesity Agents

Because of their postulated properties of target specificity and higher efficiency, nanoencapsulation of plant secondary metabolites improves the antiobesity effectiveness of these natural substances. These nanoencapsulated secondary metabolites fight obesity by blocking carbohydrate and lipid metabolizing enzymes, suppressing adipogenesis as well as hunger, and improving energy metabolism. Plants and their secondary metabolites, as well as their nanoencapsulation show antiobesity benefits, with potential modes of action for improved human health [83].

When compared with grape seed powder, the nanoparticles with red grape seed, particularly cellulose nanocrystals, showed a significant positive effect on hyperlipidemia and obesity. The chemical ingredients of the crude leaf extract of *Vitis vinifera* repressed pancreatic lipase, disturbing lipid metabolism and thus obesity [84]. In rat models, quercetin nanoencapsulation with a succinyl chitosan alginate shell and poly(lactic-co-glycolic acid) (PLGA) indicated modulation of the blood lipid profile and glucose level after induction of these nanoencapsulated assemblies [85, 86].

8.5.3 Nanopolyphenols as Anti-atherosclerotic agents

Plant polyphenolic chemicals defend the cardiovascular system by boosting high-density lipoprotein, lowering low-density lipoprotein, and inhibiting low-density lipoprotein oxidation. Nutraceuticals, such as ellagic acid and resveratrol, reduce the risk of atherosclerosis by increasing endothelial barrier function [87]. EGCG has anti-atherosclerotic properties, and its mechanism of action has been established [88]. The betulinic acid-loaded PVA/Lig-g-MA nanoformulation is an excellent nanopolymer that has no harmful effect on normal endothelial cells. Additionally, it reduced lipopolysaccharide-induced inflammation by down regulation of the expression of NFB and MAP/JNK signaling molecules. In vivo studies indicated that the synthesized nanoformulation successfully reduced hyperchlostremia, inflammation, and vasoconstriction that were caused by the overdose of a fat diet. Histopathological examination of cardiac tissues validated the cardioprotective impact of the nanoformulation [89].

8.5.4 Nanopolyphenols as Cancer Prevention Agents

Several nanoformulations, such as nanosuspensions, polymeric nanoparticles, gold nanoparticles, liposomes, and SLNs, have been developed for the delivery of polyphenolic compounds, resulting in improved antineoplastic action, better intracellular concentrations of polyphenols, sustained and slow drug release, and improved proapoptotic activity against tumor cells [90]. The coumarin 4-methyl-7-hydroxycoumarin was synthesized by methylating umbelliferone (7-hydroxycoumarin). In melanoma A375 cell cultures, PLGA nanoparticles containing 4-methyl-7-hydroxy coumarin displayed anti-neoplastic properties by boosting cell death, p53 and caspase-3 (tumor suppressor factors), DNA fragmentation, and reducing cell viability [91].

Dextran micelles containing curcumin have been shown to be a pH-sensitive drug delivery mechanism in C6 glioma cells, displaying higher cellular absorption and decreased cell proliferation [92]. Baicalein nanoparticles, combined with dual-targeted folate and hyaluronic acid ligands, had an anticancer effect on human lung cancer A549 and paclitaxel-resistant lung cancer A549/PTX cells in a xenograft mouse model of A549/PTX by decreasing cell viability and inhibiting tumor growth [93].

8.5.5 Nanopolyphenols as COVID-19 Inhibitors

The nanoformulations of polyphenols (particularly curcumin) can ease clinical manifestations (cough, fever, myalgia, tachypnea, etc.) and aid in recovery because polyphenols can modulate oxidative stress, inflammatory response, and upregulate some proteins tangled in the renin-angiotensin system, which in combination can reduce symptoms caused by COVID-19 [94]. A treatment with nanocurcumin showed a significant reduction in IL-6 and IL-1 expression and secretion in the serum and supernatant. Because nanocurcumin can reduce the elevated inflammatory cytokines (IL-1 and IL-6), mRNA expression, and cytokine production in COVID-19 patients, it can control clinical manifestations and enhance overall recovery [95].

8.6 Use of Polyphenol-loaded Nanoparticles

As implanting agents and defensive barriers for plant polyphenolic compounds, various bio-based nanocarriers, such as nanoemulsions, nanoparticles, nanogels, nanomicelles, and liposomes, can be employed. Numerous nanocarriers have been proven to successfully carry polyphenolic chemicals and improve their bioavailability such as zein, albumin, starch, cellulose, soy protein, and lipids [96]. The chemical and physical character of the carrier are affected differently by different packaging materials. Chitosan and protein are more commonly used as carrier preparation materials, and significant research has shown that they improve intestine absorption and plant polyphenol bioavailability [97].

Polyphenol encapsulation mechanisms using nanocarriers include eutectic, ionic gel, condensation, freeze drying, emulsification, encapsulation, and yeast encapsulation. These encapsulation processes increase targeting, polyphenol dispersion, GI stability, and long-term release. Curcumin and zein exhibit hydrophobicity, hydrogen bonding, and electrostatic interaction, the key factors behind nanoencapsulation [98].

Proteins are often employed for polyphenolic chemical encapsulation and distribution because of their numerous advantageous qualities such as increased nutritional value, lesser toxicity, biodegradability, and biocompatibility (Figure 8.2). They are developed into various forms (hydrogel fibers, thin films, nanoparticles, etc.) and can also be adapted or compounded for surface functioning, resulting in augmented polyphenolic component bioavailability [99].

Studies of quercetin using nanoparticles made of silk proteins, its encapsulation, adsorption, and release, revealed that the encapsulated quercetin had better free radical scavenging activity and achieved a gradual release effect in virtual intestinal fluids [100]. Similarly, whey protein nanoparticles are an effective carrier system for the encapsulation, transport, and gradual GI release of nutraceuticals like curcumin. Quercetin has good stability at pH 7, providing the basis for their use in functional drinks. Supplementary proteins, such as bovine serum albumin and zein, are frequently used to produce nanocarriers to encapsulate nutraceuticals and increase their bioavailability [101].

Polysaccharides are used as raw ingredients to create bio-based nanoparticles with technical approaches that enable the alteration of polysaccharide-based nanomaterials suited for the encapsulation and delivery of bioactive molecules in a variety of fields, particularly for polyphenolic compound shielding [102]. In addition, fluorescence spectroscopy and Fourier infrared transform were used to confirm the successful loading of curcumin into cassava starch nanoparticles. Chitosan is a polysaccharide with high inter-solubility and biodegradability. This natural compound is extensively researched for its potential use in polyphenolic chemical delivery. Chitosan nanoparticles can produce gradual release, regulated release, and enhanced bioavailability of polyphenolic substances such as tea polyphenols [53]. Similar to proteins and polysaccharides, lipids are another type of bio-based nanocarrier often employed to condense

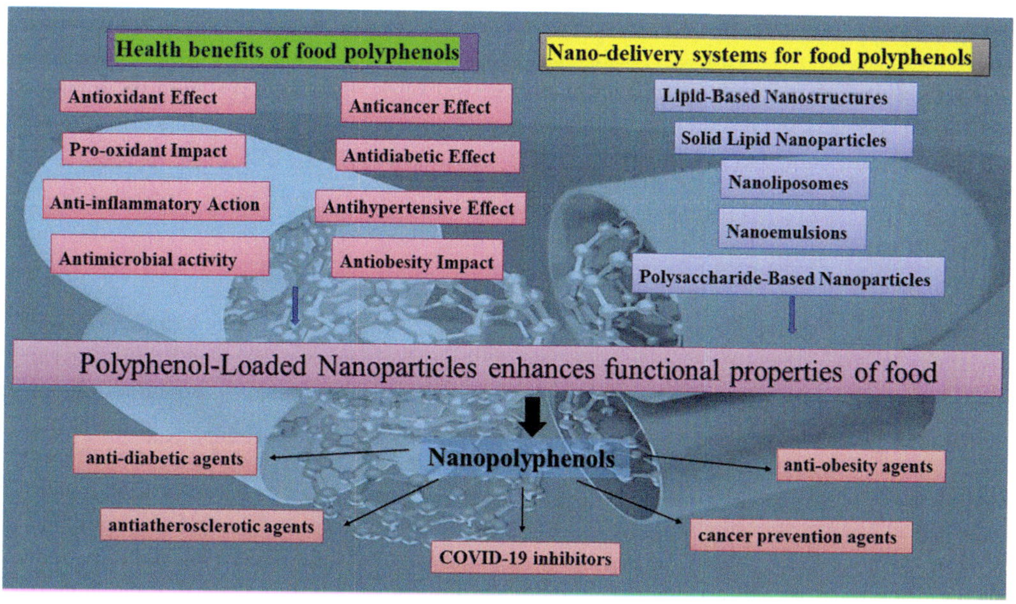

Figure 8.2 Nanodelivery of food polyphenols enhance the properties of food.

polyphenolic chemicals to have biocompatibility and biodegradability. Additionally, bioactive systems based on SLNs are effective in delivering hydrophobic nutraceuticals [103]. For example, resveratrol-loaded SLNs were discovered to effectively enhance antioxidant defense and confer anti-fatigue dimensions after wide-ranging exercise in mice that established the effect of resveratrol delivery systems, which is novel in the area of anti-fatigue nutrition in sports [104]. Correspondingly, the solid lipid nanocarriers increased the bioavailability of other nutraceuticals, such as quercetin, in variable amounts [105]. Furthermore, liposomes and nanoemulsions are good distribution methods of lipophilic polyphenolic chemicals and they deliver outstanding provisions for increasing food polyphenol bioavailability. Nanoemulsions are intensively researched for their ability to encapsulate, preserve, and transport lipophilic functional mechanisms such as pigments, polyphenols, and tastes. Polyphenolic substances given via nanoemulsions can improve the solubility of hydrophobic chemicals, resulting in improved kinetic and biological effects [106]. This provides a critical orientation for the use of nanoemulsions in the food industry for the claim of plant polyphenols as nutritious formulations. Liposomes are also investigated for functional substance delivery through food with respect to their biodegradability, small size, non-toxicity, and exclusive amphiphilic character that initiates good transport. Currently, the use of lipid delivery assemblies to improve the bioavailability of polyphenolic compounds are broadly researched for using polyphenolic compounds in functional foods [107]. Furthermore, many protein-based nanoparticles made up of polysaccharides augments the efficacy of delivery arrangements, like the creation of polysaccharide–protein composite nanoparticles in combination with zein employing pectin, carrageenan, chitosan, and other polysaccharides, which will recover the limitations of single zein nanoparticles tending to congregate, improving the distribution and protective impact of polyphenol compound composite nanoparticles [108].

Zein–rhamnoid composite nanoparticles were created by mixing zein with rhamnolipid. The composite nanoparticles possess strong encapsulation and preservation action on curcumin, offering a substitute for hydrophobic nutraceutical delivery [109]. Finally, biomolecule-based nanoparticles offer numerous advantages, including environmental friendliness, biocompatibility, and availability, making them attractive resources for polyphenol chemical delivery transporter applications in dietary items (Figure 8.2) [110]. Owing to the ongoing development and maturation of nanotechnology, the limits of polyphenolic compound bioavailability have been significantly reduced. The projects and research on food polyphenol-based nanocomposite arrangements would help in developing functional foods [111].

8.7 Future Perspectives

The possible nanotoxicity of polyphenolic nanocomposites is a concern. Commercial polyphenol nanomaterial development must be subjected to rigorous safety testing. The development of consistent safety evaluation procedures and their market entry should be properly supervised.

8.8 Conclusion

Polyphenolic chemicals are gaining popularity because of their widespread dispersion in plants and high biological activity. Although polyphenols have outstanding biological and functional qualities, they have not yet been produced as functional foods. These polyphenolic chemicals with anticancer, antibacterial, antiobesity, anti-atherosclerotic, antioxidant, and antiviral properties can be encapsulated, preserved, and supplied as healthy functional foods via nanocarriers. With the growing interest in food divergence and functionalization, various special dietary foods and functional foods can be established with plant polyphenolic compounds as the primary active constituents.

References

1 Patra, J.K., Das, G., Fraceto, L.F. et al. (2018). Nano based drug delivery systems: recent developments and future prospects. *Journal of Nanobiotechnology* 16 (71): 1–33. https://doi.org/10.1186/s12951-018-0392-8.

2 Williamson, G. (2017). The role of polyphenols in modern nutrition. *Nutrition Bulletin* 42 (3): 226–235. https://doi.org/10.1111/nbu.12278.

3 Chavda, V.P., Patel, A.B., Mistry, K.J. et al. (2022). Nano-drug delivery systems entrapping natural bioactive compounds for cancer: recent progress and future challenges. *Frontiers in Oncology* 12: 1041. https://doi.org/10.3389/fonc.2022.867655.

4 Bernela, M., Kaur, P., Ahuja, M., and Thakur, R. (2018). Nano-based delivery system for nutraceuticals: the potential future. *Advances in Animal Biotechnology and Its Applications* 103–117. https://doi.org/10.1007/978-981-10-4702-2_7.

5 Payyappallimana, U. (2010). Role of traditional medicine in primary health care. *Yokohama Journal of Social Sciences* 14 (6): 57–75.

6 Puttasiddaiah, R., Lakshminarayana, R., Somashekar, N.L. et al. (2022). Advances in nanofabrication technology for nutraceuticals: new insights and future trends. *Bioengineering* 9 (9): 478. https://doi.org/10.3390/bioengineering9090478.

7 Anand, S., Sowbhagya, R., Ansari, M.A. et al. (2022). Polyphenols and their nanoformulations: protective effects against human diseases. *Life* 12 (10): 1639. https://doi.org/10.3390/life12101639.

8 Reyed, R.M. (2023). Focusing on individualized nutrition within the algorithmic diet: an in-depth look at recent advances in nutritional science, microbial diversity studies, and human health. *Food Health* 5 (1): 5. https://doi.org/10.53388/FH2023005.

9 Fleuriet, A. and Macheix, J.J. (2003). Phenolic acids in fruits and vegetables. *Oxidative Stress and Disease* 9: 1–42.

10 Pandey, K.B. and Rizvi, S.I. (2009). Plant polyphenols as dietary antioxidants in human health and disease. *Oxidative Medicine and Cellular Longevity* 2: 270–278. https://doi.org/10.4161/oxim.2.5.9498.

11 Barreca, D., Gattuso, G., Bellocco, E. et al. (2017). Flavanones: citrus phytochemical with health-promoting properties. *BioFactors* 43 (4): 495–506. https://doi.org/10.1002/biof.1363.

12 Yao, L.H., Jiang, Y.M., Shi, J. et al. (2004). Flavonoids in food and their health benefits. *Plant Foods for Human Nutrition* 59: 113–122. https://doi.org/10.1007/s11130-004-0049-7.

13 Guven, H., Arici, A., and Simsek, O. (2019). Flavonoids in our foods: a short review. *Journal of Basic, Clinical and Applied Health Science* 3 (2): 96–106. 30621/jbachs.2019.555?.

14 Paun, N., Botoran, O.R., and Niculescu, V.C. (2022). Total phenolic, anthocyanins HPLC-DAD-MS determination and antioxidant capacity in black grape skins and blackberries: a comparative study. *Applied Sciences* 12 (2): 936. https://doi.org/10.3390/app12020936.

15 Rasheed, S., Rehman, K., Shahid, M. et al. (2022). Therapeutic potentials of genistein: new insights and perspectives. *Journal of Food Biochemistry* 46 (9): 14228. https://doi.org/10.1111/jfbc.14228.

16 Benbouguerra, N., Hornedo-Ortega, R., Garcia, F. et al. (2021). Stilbenes in grape berries and wine and their potential role as anti-obesity agents: a review. *Trends in Food Science and Technology* 112: 362–381.

17 Hajibabaie, F., Abedpoor, N., Safavi, K., and Taghian, F. (2022). Natural remedies medicine derived from flaxseed (secoisolariciresinoldiglucoside, lignans, and α-linolenic acid) improve network targeting efficiency of diabetic heart conditions based on computational chemistry techniques and pharmacophore modeling. *Journal of Food Biochemistry* 14480.

18 Adetuyi, B.O., Odine, G.O., Olajide, P.A. et al. (2022). Nutraceuticals: role in metabolic disease, prevention and treatment. *World News of Natural Sciences* 42: 1–27.

19 Domínguez Díaz, L., Fernández-Ruiz, V., and Cámara, M. (2020). The frontier between nutrition and pharma: the international regulatory framework of functional foods, food supplements and nutraceuticals. *Critical Reviews in Food Science and Nutrition* 60 (10): 1738–1746.

20 Zeng, Z., Centner, C., Gollhofer, A., and König, D. (2021). Effects of dietary strategies on exercise-induced oxidative stress: a narrative review of human studies. *Antioxidants* 10 (4): 542.

21 de Souza Mesquita, L.M., Mennitti, L.V., de Rosso, V.V., and Pisani, L.P. (2021). The role of vitamin A and its pro-vitamin carotenoids in fetal and neonatal programming: gaps in knowledge and metabolic pathways. *Nutrition Research* 79 (1): 76–87.

22 Odoh, U.E., Egbuna, C., Onyegbulam, C.M. et al. (2023). Bioactive compounds in the management of nutritional disorders. In: *Role of Nutrigenomics in Modern-Day Healthcare and Drug Discovery* (ed. G.D. Tupas and C. Egbuna) 343–358.

23 Oyedepo, T.A. and Palai, S. (2021). Herbal remedies, toxicity, and regulations. In: *Preparation of Phytopharmaceuticals for the Management of Disorders* (ed. C. Egbuna, A. Mishra and M. Goyal), 89–127. Academic Press.

24 Palai, S. and Patra, R. (2021). Wound management using phytonanoparticles: an innovative approach. *Journal of Phytonanotechnology and Pharmaceutical Sciences* 1 (3): 1–7.

25 Das, M., Saxena, N., and Dwivedi, P.D. (2009). Emerging trends of nanoparticles application in food technology: safety paradigms. *Nanotoxicology* 3 (1): 10–18.

26 Rudrapal, M., Mishra, A.K., Rani, L. et al. (2022). Nanodelivery of dietary polyphenols for therapeutic applications. *Molecules* 27 (24): 8706.

27 de Souza Simões, L., Madalena, D.A., Pinheiro, A.C. et al. (2017). Micro-and nano bio-based delivery systems for food applications: in vitro behavior. *Advances in Colloid and Interface Science* 243: 23–45.

28 Caballero, S., Li, Y.O., McClements, D.J., and Davidov-Pardo, G. (2022). Encapsulation and delivery of bioactive citrus pomace polyphenols: a review. *Critical Reviews in Food Science and Nutrition* 62 (29): 8028–8044.

29 Tamjidi, F., Shahedi, M., Varshosaz, J., and Nasirpour, A. (2013). Nanostructured lipid carriers (NLC): a potential delivery system for bioactive food molecules. *Innovative Food Science and Emerging Technologies* 19: 29–43.

30 Siraj, A., Naqash, F., Shah, M.A. et al. (2021). Nanoemulsions: formation, stability and an account of dietary polyphenol encapsulation. *International Journal of Food Science and Technology* 56 (9): 4193–4205.

31 Rudrapal, M., Khairnar, S.J., Khan, J. et al. (2022). Dietary polyphenols and their role in oxidative stress-induced human diseases: insights into protective effects, antioxidant potentials and mechanism(s) of action. *Frontiers in Pharmacology* 13: 283.

32 Palai, S. (2022). Nanotherapeutics of phytoantioxidants for inflammatory disorders. In: *Phytoantioxidants and Nanotherapeutics* (ed. M. Rudrapal), 377–404. John Wiley & Son.

33 Kamaraj, S., Vinodhkumar, R., Anandakumar, P. et al. (2007). The effects of quercetin on antioxidant status and tumor markers in the lung and serum of mice treated with benzo (a) pyrene. *Biological and Pharmaceutical Bulletin* 30 (12): 2268–2273. https://doi.org/10.1248/bpb.30.2268.

34 Li, M., Rao, C., Ye, X. et al. (2022). Applications for natural deep eutectic solvents in Chinese herbal medicines. *Frontiers in Pharmacology* 13: 5564. https://doi.org/10.3389/fphar.2022.1104096.

35 Raj, P., Thandapilly, S.J., Wigle, J. et al. (2021). A comprehensive analysis of the efficacy of resveratrol in atherosclerotic cardiovascular disease, myocardial infarction and heart failure. *Molecules* 26 (21): 6600. https://doi.org/10.3390/molecules26216600.

36 Irvin, L., Jackson, C., Hill, A.L. et al. (2019). Skullcaps (Scutellaria spp.): Ethnobotany and current research. In: *Medicinal Plants: From Farm to Pharmacy* (ed. N. Joshee, S. Dhekney, P. Parajuli), 141–168, Cham: Springer.

37 Palai, S. (2022). Berberine. In: *Nutraceuticals and Health Care* (ed. J. Kour and G.A. Nayik), 359–368. Academic Press.

38 Huang, J., He, Z., Cheng, R. et al. (2020). Assessment of binding interaction dihydromyricetin and myricetin with bovine lactoferrin and effects on antioxidant activity. *Spectrochimica Acta, Part A: Molecular and Biomolecular* 243: 118731. https://doi.org/10.1016/j.saa.2020.118731.

39 Yang, L., Gao, Y., Bajpai, V.K. et al. (2021). Advance toward isolation, extraction, metabolism and health benefits of kaempferol, a major dietary flavonoid with future perspectives. *Critical Reviews in Food Science and Nutrition* 1–17. https://doi.org/10.3390/metabo13010063.

40 Akbar, S. and Akbar, S. (2020). Glycyrrhiza glabra L. (Fabaceae/Leguminosae) (Syns.: g. glanduliferaWaldst. & Kit.; G. hirsuta Pall.; G. pallida Boiss. & Noe; G. violaceaBoiss. & Noe). *Handbook of 200 Medicinal Plants: A Comprehensive Review of Their Traditional Medical Uses and Scientific Justifications* 963–980. https://doi.org/10.1007/978-3-030-16807-0_103.

41 Long, L., Han, X., Ma, X. et al. (2020). Protective effects of fisetin against myocardial ischemia/reperfusion injury. *Experimental and Therapeutic Medicine* 19 (5): 3177–3188. https://doi.org/10.3892/etm.2020.8576.

42 Ginwala, R., Bhavsar, R., Chigbu, D.G.I. et al. (2019). Potential role of flavonoids in treating chronic inflammatory diseases with a special focus on the anti-inflammatory activity of apigenin. *Antioxidants* 8 (2): 35.

43 Singh, P., Bansal, S., Kuhad, A. et al. (2020). Naringenin ameliorates diabetic neuropathic pain by modulation of oxidative-nitrosative stress, cytokines and MMP-9 levels. *Food and Function* 11 (5): 4548–4560.

44 Rizvi, S.I., Zaid, M.A., Anis, R., and Mishra, N. (2005). Protective role of tea catechins against oxidation-induced damage of type 2 diabetic erythrocytes. *Clinical and Experimental Pharmacology and Physiology* 32: 70–75.

45 Seeram, N.P., Cichewicz, R.H., Chandra, A., and Nair, M.G. (2003). Cyclooxygenase inhibitory and antioxidant compounds from crabapple fruits. *Journal of Agricultural and Food Chemistry* 51 (7): 1948–1951.

46 Palai, S., Chandra, S., Pandey, N., and Singh, R., 2023. *Theophylline: a bioactive dimethylxanthine alkaloid in the essential guide to alkaloids*. 205–215. ISBN: 979-8-88697-456-0 © 2023 Nova Science Publishers, Inc.

47 Thangavel, P., Puga-Olguín, A., Rodríguez-Landa, J.F., and Zepeda, R.C. (2019). Genistein as potential therapeutic candidate for menopausal symptoms and other related diseases. *Molecules* 24 (21): 3892.

48 Vuolo, M.M., Lima, V.S., and Junior, M.R.M. (2019). Phenolic compounds: structure, classification, and antioxidant power. In: *Bioactive Compounds* (ed. M.M. Vuolo, V.S. Lima, and M.R.M. Junior), 33–50. Woodhead Publishing.

49 Bié, J., Sepodes, B., Fernandes, P.C., and Ribeiro, M.H. (2023). Polyphenols in health and disease: gut microbiota, bioaccessibility, and bioavailability. *Compounds* 3 (1): 40–72. https://doi.org/10.3390/compounds3010005.

50 Durazzo, A., Lucarini, M., Souto, E.B. et al. (2019). Polyphenols: a concise overview on the chemistry, occurrence, and human health. *Phytotherapy Research* 33 (9): 2221–2243.

51 Andrew, M. and Jayaraman, G. (2020). Structural features of microbial exopolysaccharides in relation to their antioxidant activity. *Carbohydrate Research* 487: 107881.

52 Hosseinzadeh, R., Khorsandi, K., and Hemmaty, S. (2013). Study of the effect of surfactants on extraction and determination of polyphenolic compounds and antioxidant capacity of fruits extracts. *PloS one* 8 (3): e57353.

53 Zhang, Z., Li, X., Sang, S. et al. (2022). Polyphenols as plant-based nutraceuticals: health effects, encapsulation, nano-delivery, and application. *Foods* 11 (15): 2189.

54 Dal, H.Ö.G. and Yilmaz, Y. (2022). Use of tea catechins in foods as a functional ingredient. In: *Tea as a Food Ingredient* (ed. H.O. Guler and Y. Yilmaz) 241–257. CRC Press.

55 Yordi, E.G., Pérez, E.M., Matos, M.J., and Villares, E.U. (2012). Antioxidant and pro-oxidant effects of polyphenolic compounds and structure-activity relationship evidence. *Nutrition, Well-being and Health* 2: 23–48.

56 Guo, A. and Xiong, Y.L. (2021). Myoprotein–phytophenol interaction: implications for muscle food structure-forming properties. *Comprehensive Reviews in Food Science and Food Safety* 20 (3): 2801–2824.

57 Zhu, C., Yang, J., Nie, X. et al. (2022). Influences of dietary vitamin E, selenium-enriched yeast, and soy isoflavone supplementation on growth performance, antioxidant capacity, carcass traits, meat quality and gut microbiota in finishing pigs. *Antioxidants* 11 (8): 1510.

58 AbouAitah, K., Higazy, I.M., Swiderska-Sroda, A. et al. (2021). Anti-inflammatory and antioxidant effects of nanoformulations composed of metal-organic frameworks delivering rutin and/or piperine natural agents. *Drug Delivery* 28 (1): 1478–1495.

59 Talhouk, R.S., Karam, C., Fostok, S. et al. (2007). Anti-inflammatory bioactivities in plant extracts. *Journal of Medicinal Food* 10 (1): 1–10.

60 Coppo, E. and Marchese, A. (2014). Antibacterial activity of polyphenols. *Current Pharmaceutical Biotechnology* 15 (4): 380–390.

61 Kalita, B., Das, M.K., and Sharma, A.K. (2013). Novel phytosome formulations in making herbal extracts more effective. *Research Journal of Pharmacy and Technology* 6 (11): 1295–1301.

62 Santos, I.S., Ponte, B.M., Boonme, P. et al. (2013). Nanoencapsulation of polyphenols for protective effect against colon–rectal cancer. *Biotechnology Advances* 31 (5): 514–523. https://doi.org/10.1016/j.biotechadv.2012.08.005.

63 Kidd, P.M. (2009). Bioavailability and activity of phytosome complexes from botanical polyphenols: the silymarin, curcumin, green tea, and grape seed extracts. *Alternative Medicine Review* 14 (3): 226–246.

64 Shahwan, M., Alhumaydhi, F., Ashraf, G.M. et al. (2022). Role of polyphenols in combating type 2 diabetes and insulin resistance. *International Journal of Biological Macromolecules* 206: 567–579. https://doi.org/10.1016/j.ijbiomac.2022.03.004.

65 Wang, S., Du, S., Wang, W., and Zhang, F. (2020). Therapeutic investigation of quercetin nanomedicine in a zebrafish model of diabetic retinopathy. *Biomedicine and Pharmacotherapy* 130: 110573.

66 Ahmad, M. and Gani, A. (2021). Development of novel functional snacks containing nano-encapsulated resveratrol with anti-diabetic, anti-obesity and antioxidant properties. *Food Chemistry* [illegible]. https://doi.org/10.1016/j.foodchem.2021.[illegible].

67 Luna-Vázquez, F.J., Ibarra-Alvarado, C., Rojas-Molina, A. et al. (2013). Nutraceutical value of black cherry *Prunus serotina*Ehrh. fruits: antioxidant and antihypertensive properties. *Molecules* 18 (12): 14597–14612.

68 Milinčić, D.D., Popović, D.A., Lević, S.M. et al. (2019). Application of polyphenol-loaded nanoparticles in food industry. *Nanomaterials* 9 (11): 1629.

69 Puligundla, P., Mok, C., Ko, S. et al. (2017). Nanotechnological approaches to enhance the bioavailability and therapeutic efficacy of green tea polyphenols. *Journal of Functional Foods* 34: 139–151.

70 Saad, B., Ghareeb, B., and Kmail, A. (2021). Metabolic and epigenetics action mechanisms of antiobesity medicinal plants and phytochemicals. *Evidence-Based Complementary and Alternative Medicine* 2021: 1–19. https://doi.org/10.1155/2021/9995903.

71 Rufino, A.T., Costa, V.M., Carvalho, F., and Fernandes, E. (2021). Flavonoids as antiobesity agents: a review. *Medicinal Research Reviews* 41 (1): 556–585.

72 Yun, J.W. (2010). Possible anti-obesity therapeutics from nature–a review. *Phytochemistry* 71 (14–15): 1625–1641.

73 Ngwuluka, N.C., Abu-Thabit, N.Y., Uwaezuoke, O.J. et al. (2021). Natural polymers in micro-and nanoencapsulation for therapeutic and diagnostic applications: part I: lipids and fabrication techniques. In: *Nano.Microencapsul. Tech. Appl.*, 3–54.

74 Borges, A., de Freitas, V., Mateus, N. et al. (2020). Solid lipid nanoparticles as carriers of natural phenolic compounds. *Antioxidants* 9 (10): 998.

75 Hasan, M., Messaoud, G.B., Michaux, F. et al. (2016). Chitosan-coated liposomes encapsulating curcumin: study of lipid–polysaccharide interactions and nanovesicle behavior. *RSC Advances* 6 (51): 45290–45304.

76 Nikam, T.H., Patil, M.P., Patil, S.S. et al. (2018). Nanoemulsion: a brief review on development and application in parenteral drug delivery. *Advance Pharmaceutical Journal* 3 (2): 43–54.

77 McClements, D.J. (2020). Advances in nanoparticle and microparticle delivery systems for increasing the dispersibility, stability, and bioactivity of phytochemicals. *Biotechnology Advances* 38: 107287.

78 Li, Y., Jin, H., Sun, X. et al. (2018). Physicochemical properties and storage stability of food protein-stabilized nanoemulsions. *Nanomaterials* 9 (1): 25. https://doi.org/10.3390/nano9010025.

79 McClements, D.J., Decker, E.A., Park, Y., and Weiss, J. (2009). Structural design principles for delivery of bioactive components in nutraceuticals and functional foods. *Critical Reviews in Food Science and Nutrition* 49 (6): 577–606.

80 Rambaran, T.F. (2020). Nanopolyphenols: a review of their encapsulation and anti-diabetic effects. *SN Applied Sciences* 2 (8): 1335.

81 Das Purkayastha, M. and Manhar, A.K. (2016). Nanotechnological applications in food packaging, sensors and bioactive delivery systems. *Nanoscience in Food and Agriculture* 2: 59–128.

82 Tsivileva, O., Pozdnyakov, A., and Ivanova, A. (2021). Polymer nanocomposites of selenium biofabricated using fungi. *Molecules* 26 (12): 3657.

83 Bhardwaj, M., Yadav, P., Vashishth, D. et al. (2021). A review on obesity management through natural compounds and a green nanomedicine-based approach. *Molecules* 26 (11): 3278.

84 Jaradat, N., Zaid, A.N., Hussein, F. et al. (2017). Anti-lipase potential of the organic and aqueous extracts of ten traditional edible and medicinal plants in Palestine; a comparison study with orlistat. *Medicines* 4 (4): 89. https://doi.org/10.3390/medicines4040089.

85 Mukhopadhyay, P., Maity, S., Mandal, S. et al. (2018). Preparation, characterization and in vivo evaluation of pH sensitive, safe quercetin-succinylated chitosan-alginate core-shell-corona nanoparticle for diabetes treatment. *Carbohydrate Polymers* 182: 42–51. https://doi.org/10.1016/j.carbpol.2017.10.098.

86 Chitkara, D., Nikalaje, S.K., Mittal, A. et al. (2012). Development of quercetin nanoformulation and in vivo evaluation using streptozotocin induced diabetic rat model. *Drug Delivery and Translational Research* 2: 112–123. https://doi.org/10.1007/s13346-012-0063-5.

87 Joiner-Bey, H. (2010). *The Healing Power of Flax: How Nature's Richest Source of Omega-3 Fatty Acids Can Help to Heal, Prevent and Reverse Arthritis.* SCB Distributors.

88 Harishkumar, R., Hans, S., Stanton, J.E. et al. (2022). Targeting the platelet-activating factor receptor (PAF-R): antithrombotic and anti-atherosclerotic nutrients. *Nutrients* 14 (20): 4414.

89 Zhang, J., Almoallim, H.S., Alharbi, S.A., and Yang, B. (2021). Anti-atherosclerotic activity of Betulinic acid loaded polyvinyl alcohol/methylacrylate grafted lignin polymer in high fat diet induced atherosclerosis model rats. *Arabian Journal of Chemistry* 14 (2): 102934.

90 Davatgaran-Taghipour, Y., Masoomzadeh, S., Farzaei, M.H. et al. (2017). Polyphenol nanoformulations for cancer therapy: experimental evidence and clinical perspective. *International Journal of Nanomedicine* 12: 2689.

91 Bhattacharyya, S.S., Paul, S., De, A. et al. (2011). Poly (lactide-co-glycolide) acid nanoencapsulation of a synthetic coumarin: cytotoxicity and bio-distribution in mice, in cancer cell line and interaction with calf thymus DNA as target. *Toxicology and Applied Pharmacology* 253 (3): 270–281.

92 Phan, Q.T., Le, M.H., Le, T.T.H. et al. (2016). Characteristics and cytotoxicity of folate-modified curcumin-loaded PLA-PEG micellar nano systems with various PLA: PEG ratios. *International Journal of Pharmaceutics* 507 (1–2): 32–40.

93 Liu, A., Wang, W., Fang, H. et al. (2015). Baicalein protects against polymicrobial sepsis-induced liver injury via inhibition of inflammation and apoptosis in mice. *European Journal of Pharmacology* 748: 45–53.

94 de Oliveira, G.V., Volino-Souza, M., Dos Santos, K.S. et al. (2021). A perspective on the use of polyphenols nano-formulation as a nutritional strategy to manage the symptoms of the infected patient with COVID-19. *The Research, Society and Development* 10 (13): e400101321471 –e400101321471.

95 Valizadeh, H., Abdolmohammadi-Vahid, S., Danshina, S. et al. (2020). Nano-curcumin therapy, a promising method in modulating inflammatory cytokines in COVID-19 patients. *Int. Immunopharmacol* 89: 107088.

96 Zhang, Z., Qiu, C., Li, X. et al. (2021). Advances in research on interactions between polyphenols and biology-based nano-delivery systems and their applications in improving the bioavailability of polyphenols. *Trends in Food Science and Technology* 116: 492–500.

97 Wang, W., Xue, C., and Mao, X. (2020). Chitosan: structural modification, biological activity and application. *International Journal of Biological Macromolecules* 164: 4532–4546.

98 Abd El-Hack, M.E., El-Saadony, M.T., Swelum, A.A. et al. (2021). Curcumin, the active substance of turmeric: its effects on health and ways to improve its bioavailability. *Journal of the Science of Food and Agriculture* 101 (14): 5747–5762.

99 Esfanjani, A.F. and Jafari, S.M. (2016). Biopolymer nano-particles and natural nano-carriers for nano-encapsulation of phenolic compounds. *Colloids and Surfaces B* 146: 532–543.

100 Lozano-Pérez, A.A., Rivero, H.C., Hernández, M.D.C.P. et al. (2017). Silk fibroin nanoparticles: efficient vehicles for the natural antioxidant quercetin. *International Journal of Pharmaceutics* 518 (1–2): 11–19.

101 Ozkan, G., Kostka, T., Esatbeyoglu, T., and Capanoglu, E. (2020). Effects of lipid-based encapsulation on the bioaccessibility and bioavailability of phenolic compounds. *Molecules* 25 (23): 5545. https://doi.org/10.3390/molecules25235545.

102 Li, Z. and Lin, Z. (2021). Recent advances in polysaccharide-based hydrogels for synthesis and applications. *Aggregate* 2 (2): 21.

103 Nahum, V. and Domb, A.J. (2021). Recent developments in solid lipid microparticles for food ingredients delivery. *Foods* 11. 10 (2): 400.

104 Teskač, K. and Kristl, J. (2010). The evidence for solid lipid nanoparticles mediated cell uptake of resveratrol. *International Journal of Pharmaceutics* 390 (1): 61–69. https://doi.org/10.1016/j.ijpharm.2009.10.011.

105 Li, H., Zhao, X., Ma, Y. et al. (2009). Enhancement of gastrointestinal absorption of quercetin by solid lipid nanoparticles. *The Journal of Controlled Release* 133 (3): 238–244.

106 Yang, S., Liu, L., Han, J., and Tang, Y. (2020). Encapsulating plant ingredients for dermocosmetic application: an updated review of delivery systems and characterization techniques. *International Journal of Cosmetic Science* 42 (1): 16–28.

107 Gonçalves, R.F., Martins, J.T., Duarte, C.M. et al. (2018). Advances in nutraceutical delivery systems: from formulation design for bioavailability enhancement to efficacy and safety evaluation. *Trends in Food Science and Technology* 78: 270–291. https://doi.org/10.3390/molecules26092540.

108 Zuidam, N.J. and Shimoni, E. (2010). Overview of microencapsulates for use in food products or processes and methods to make them. *Encapsulation Technologies for Active Food Ingredients and Food Processing* 410: 3–29.

109 Dai, L., Zhou, H., Wei, Y. et al. (2019). Curcumin encapsulation in zein-rhamnolipid composite nanoparticles using a pH-driven method. *Food Hydrocolloids* 93: 342–350.

110 Dutta, D. and Sit, N. (2022). Application of natural extracts as active ingredient in biopolymer-based packaging systems. *The Journal of Food Science and Technology* 1–15. https://doi.org/10.1007/s13197-022-05474-5.

111 Yin, C., Cheng, L., Zhang, X., and Wu, Z. (2020). Nanotechnology improves delivery efficiency and bioavailability of tea polyphenols. *Journal of Food Biochemistry* 44 (9): e13380.

9

Polyphenol Rich Extracts from Spices and Nanodelivery Systems

Neelesh Kumar Nema[1,], Swapnil Devidas Khamborkar[1], Smitha Sarojam[1], Baby Kumaranthara Chacko[2], and Viju Jacob[1,2]*

[1] *Nutraceuticals Division, Bio Ingredients, Synthite Industries Pvt. Ltd., Synthite Valley, Kadayiruppu, Ernakulam, Kerala, India*
[2] *C.V.J. Creative Centre, Bio Ingredients, Synthite Industries Pvt. Ltd., Synthite Valley, Kadayiruppu-682311, Kerala, India*
[*] *Corresponding author*

9.1 Introduction

9.1.1 Diversity of Plant Polyphenols

Phenolics (flavonoids and non-flavonoids), alkaloids, terpenoids, steroids, stilbenoids, and polysaccharides, among others, are secondary metabolites generated from plants, herbs, and spices that play a vital role in promoting human well-being and desired therapeutic effects [1–3] against various diseases and disorders, i.e., diabetes, heart health, inflammation, Alzheimer's disease, and cancer. Chemically, phenolic acids and polyphenols are the naturally occurring compounds richest with the hydroxyl group in the aromatic ring. They are produced via the shikimic acid/phenylpropanoid and/or polyketide pathways [2,4]. Nearly 10,000 polyphenol compounds have been discovered from different species, with flavonoids being the largest group [5, 6]. According to the frequency of aromatic or phenol rings and the structural components that connect these rings to one another, polyphenols are categorized into several subgroups, i.e., flavonoids, isoflavonoids, neoflavonoids, and tannins.

9.1.2 Polyphenols and Spices

Everyone is familiar with spices. Spices and condiments have been widely used in food preparation and culinary creations for a very long time. Many handling variables, including time of collection, harvesting, and manufacturing, have an impact on the quality of the spices. Moreover, in addition to adding flavor, color, and aroma to food, spices are used all over the world as traditional and folk medicines [7, 2]. The active components, as well as organoleptic and chemical properties, have an impact on the therapeutic effect. Spices are rich in secondary metabolites and are an excellent source of phenolic and polyphenols, which provide nutritional needs and favorable therapeutic benefits by participating in multifunctional enzymatic processes and targets [8]. Spice polyphenols are beneficial in cooking and nutrition; however, they are also used in the beverage, cosmetic, and aromatherapy sectors. Figure 9.1 shows several examples where spice polyphenols are employed.

Polyphenols: Food, Nutraceutical, and Nanotherapeutic Applications, First Edition. Edited by Mithun Rudrapal.

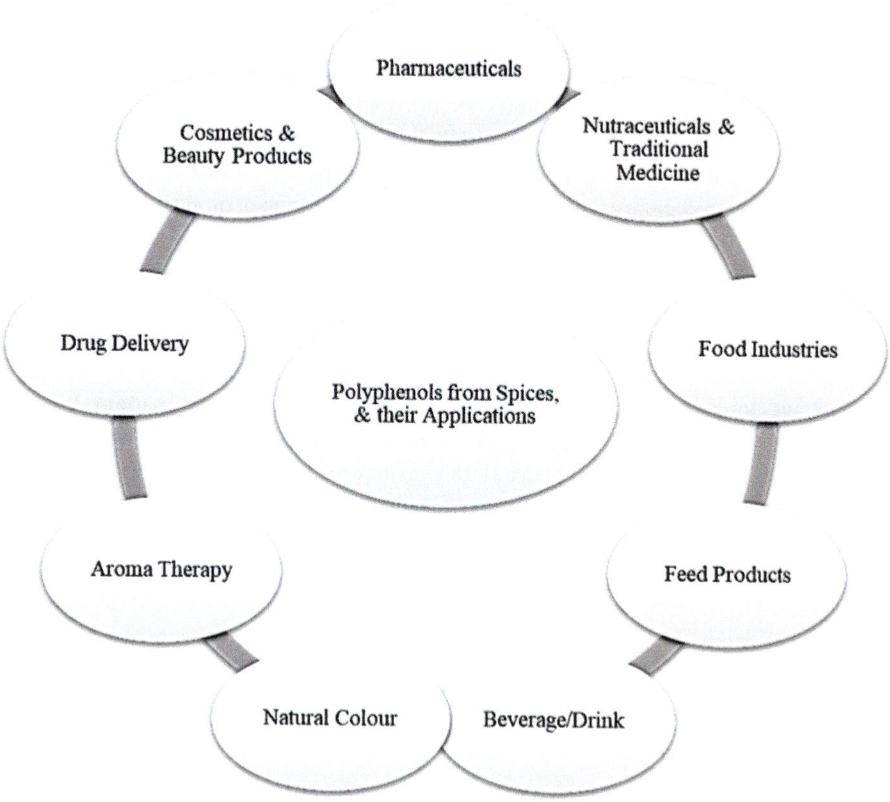

Figure 9.1 Spice polyphenols and their applications.

9.1.3 Therapeutic and Health Benefits

Spice polyphenols are well known for their antioxidant properties, which aid in preventing oxidation caused by free radicals that are formed by a variety of factors in our daily lives. Spices currently have a dual role in the food processing business. On the one hand, they improve the organoleptic characteristics of meals, while on the other hand, they have a strong antioxidant potential because of their overall polyphenol content, particularly flavonoids. Together with their antioxidant qualities, flavonoids demonstrate a variety of therapeutic benefits and effectiveness. During the COVID-19 pandemic, spices and their polyphenols assisted in its control and also showed powerful therapeutic roles in traditional medicine, which eventually helps people [9].

9.2 Nanodelivery Systems

Science-based food research and newer technologies have changed the food industry over the past several decades by solving the difficulties of a balanced diet by "being edible" and "being nutritive", which are the primary needs in the food processing sector [10]. These technologies can enhance a product's attributes, nutritional value, shelf life, food structures and textures, provide control to biochemical, microbiological, and chemical changes, and traceability of the food items, all of which are crucial in the food sector.

Nanodelivery systems are a rapidly emerging interdisciplinary technology for developing molecular food components, food additives, transporters for nutrients/supplements, and food contact materials. Nanodelivery systems generate nanomaterials with sizes ranging from 1–1000 nm and they exhibit physical and chemical properties that differ significantly from those of a single molecule or bulk material. This nanodelivery technology provides a wide range of quality benefits to the client, including innovative qualities and better functional benefits such as improved appearances [11–15].

9.2.1 Diversity of Nanotechnologies

Nanotechnologies have long been used in traditional medicine to enhance patient health. Nanoparticles (NPs), nanoencapsulation, nanoemulsions, nanolipid carriers, nanofibers, nanocomposites, nanolaminates, and similar nanocapsulation structures are among the emerging technologies of modern science that are used in a range of applications. Research is moving towards green nanotechnologies, which are cutting-edge methods in the nanotechnology field that promote environmental sustainability and green methods of synthesis while minimizing costs and environmental dangers. This strategy places a strong emphasis on using environmentally friendly and biocompatible procedures, and frequently uses plant extracts or microorganisms as active ingredients or carriers [16].

9.2.2 Requirements for Modern Delivery Systems

By employing nanomaterials as nanomedicines, nanotechnology can bridge the gap between conventional medicine, drug delivery systems, and current drugs. It also offers a wide range of scientific uses [17].

9.2.3 Nanotechnology and Methods of Preparation

The word "nanomaterial" is defined by the EU Commission Recommendation only in terms of size, ignoring any possible hazards (2011/696/EU) [18]. Some of the commonly used technologies are described here.

Nanoparticles: NPs are one of the most valuable technologies accessible today because of their unique and better surface qualities compared with its bulk counterpart, and they are produced by three primary approaches: physical, chemical, and mechanical methods. NPs are generally made using lipid and polymeric approaches. Lipid NPs, which were developed by Müller and Gasco in the 1961, primarily include solid lipid nanoparticles (SLNs), nanostructured lipid carriers, and liposomes, whereas polymeric NPs are made with a solid colloidal dispersion of biodegradable and biocompatible polymers such as polyalkylcyanoacrylate, polylactide (PLA), polyd,l-lactide-co-glycolic acid (PLGA), and others [19].

Nanoemulsions: There are two type of nanoemulsions. Oil-in-water (O/W) nanoemulsions are generated when a hydrophilic matrix surrounds a lipophilic bioactive agent such as plant sterols, carotenoids (such as β-carotene), α-tocopherol, dietary fats, nutritious oils, or a combination of these agents. Alternatively, water-in-oil (W/O) nanoemulsions, are generated when hydrophilic substances, such as xanthophyll, vitamins, and polyphenols, are disseminated in an oil system.

Nanoencapsulation: This technique involves implanting a liquefy carrier with a secondary material, also known as the matrix, shell, or inert material, to produce "core" or "bioactive" compounds that are tiny particles or miniatures of a solid, liquid, or gas. By using this process, a tiny structure is produced that protects the core components from environmental harm, including oxidation, high temperatures, photosensitivity, light impact, moisture effects, pH fluctuations, and interactions with related substances. Because of their matrix characteristics and universal acceptance, nanoencapsulation technologies are preferred by all commercial industries.

Nanofibers: Polymers, carbon, and semiconductor materials are frequently used to create nanofibers for various purposes. There are several methods for creating nanofibers, including bi-component extrusion, phase separation, template synthesis, drawing, melt blowing, electrospinning, and centrifugal spinning. Electrospun nanofibers are one of the most well-liked and widely-used methods [20–23].

Nanocomposite: In order to achieve certain physical qualities, a nanocomposite is a commercially accessible polymer mixture that is made by connecting or mixing two or more materials with distinct geometries (fibers, flakes, spheres, or particulates) that have a nanometric range (nanopolymers or NPs).

9.2.4 Nanotechnology and Sensory Attributes

Food gives energy to the body, protects it from disease, and regulates body functions for smooth functioning and efficient operation. Everyone enjoys eating wholesome, tasty cuisine. According to the International Food Information Council, taste, price, and healthfulness are the top-ranked buying motives and/or features when selecting any food product [24, 11, 24]. To produce novel food items or introduce new processing technologies, it is crucial to use sensory evaluation methodologies. People are seeking new developing technologies because food and products produced using standard procedures may lose their stability during prolonged storage or may lose sensory qualities during shipping and packing. Newer technology may be used to enhance two processes: choosing quality food and processing to create goods. Products with nanostructures created using nanotechnologies tend to stimulate consumer cognition. With these technologies, the consumer can be persuaded to purchase more wholesome items with predicted organoleptic and/or attractive properties. These technologies may enhance the product's attributes, nutritional value, shelf life, f the food's structures and textures, control the biochemical, microbiological, and chemical changes, and traceability of the food items, all of which are essential criteria in the food sector. Regarding quality, these technologies offer the consumer a wide range of advantages, such as novel qualities and improved sensory attributes, such as enhanced appearance, unique tastes, novel textures, fresh sensations, minimal fat content, improved nutrient absorption, and upgraded packaging, to ensure microbiological safety.

Nanostructured foods, such as NPs, nanoemulsions, nanoencapsulation, and nanofibers, are the industry's answer to producing high-quality food that satisfies consumer demands and offers distinctive sensory experiences without compromising nutritional content. These technologies may offer an alternative to traditional food production methods for creating nutrient-dense foods with a long shelf life without sacrificing the intended sensory experience for customers. These next-generation technologies may also play a role in all aspects of food handling in the future, including ingredient selection, processing, final recipe formulations, decorating, packing, and stability, as well as transportation and storage [10].

9.3 Nanodelivery Platforms with Polyphenols

9.3.1 Polyphenol Extracts from Spices

In addition to herbs and fruits, spices are also a rich source of polyphenols. Solvent and CO_2 extractions are preferred methods for extracting polyphenols from aromatic spices, especially when the polyphenols are in glycoside form. The biosynthetic pathways of shikimic acid and phenylpropanoid metabolism generate polyphenols such as flavones, flavanones, flavonols, flavanols (Flavans), and anthocyanins. Compared with other flavonoids, flavonols like quercetin and kaempferol are more often detected. According to Nutrient Data Laboratory research, dried parsley has the highest concentration of apigenin, followed by celery and oregano (Mexican). The same spice has a lot of isorhamnetin. Celery, thyme, and Mexican oregano have been shown to have the highest luteolin content. Kaempferol and quercetin are the main components of caper plants. Additionally, spices include a significant quantity of flavonoids, which have positive health effects [25]. Table 9.1 lists common spices, the parts that are used, and the associated polyphenols [2].

Table 9.1 Common Indian spices and their polyphenols.

Sl. No	English/ Common Name	Part	Scientific Name (Family)	Polyphenols
1.	Black pepper	Fruit	*Piper nigrum* (Piperaceae)	Apigenin, hydroxytyrosol 4-o-glucoside, kaempferol, rhoifolin, scopoletin
2.	Celery	Fruit, root, leaf	*Apiumgraveolens* (Apiaceae)	Apiin (diglycoside of apigenin), chrysoeriol glycoside (3′-o-methylated flavonoids), luteolin glycosides
3.	Cardamom	Fruit	*Elettaria cardamomum* (Zingiberaceae)	Kaempferol, luteolin, myricetin, quercetin, (+)-gallocatechin, carnosic acid
4.	Coriander	Leaf, seed	*Coriandrum sativum* (Apiaceae)	Apigenin, kaempferol, luteolin, naringin, quercetin dehydrate, quercetin-3-rhamnoside, rutintrihydrate
5.	Dill	Fruit	*Anethum graveolens* (Apiaceae)	Isorhamnetin, kaempferol, myricetin, quercetin
6.	Garlic	Bulb	*Allium sativum* (Lamiaceae)	Hydroxycinnamic acid, kaempferol, myricetin, quercetin
7.	Ginger	Rhizome	*Zingiber officinale* (Zingiberaceae)	6-Gingerol/ shogaol, 8-gingerol/ shogaol, 10-gingerol/ shogaol
8.	Indian mustard	Seed	*Brassica juncea* (Brassicaceae)	Isorhamnetin, kaempferol, p-hydroxybenzoic, quercetin, sinapic acids
9.	Parsley	Seed, leaf, root	*Petroselinum crispum* Syn. *Petroselinum sativum* (Apiaceae)	Apigenin, kaempferol, luteolin, quercetin
10.	Pomegranate seed	Seed dried with flesh	*Punica granatum* (Punicaseae)	Ellagic acid

Table 9.1 (Continued)

Sl. No	English/ Common Name	Part	Scientific Name (Family)	Polyphenols
11.	Rosemary	Terminal shoot, leaf	*Rosmarinus officinalis* (Lamiaceae)	Luteolin, apigenin, naringenin
12.	Saffron	Stigma	*Crocus sativus* (Iridaceae)	Kaempferol, kaempferol sophoroside, quercetin rutinoside
13.	Sweet fennel	Fruit, leaf, twig	*Foeniculum vulgare* (Apiaceae)	Apigenin, quercetin, rutin
14.	Thyme	Leaf	*Thymus vulgaris* (Lamiaceae)	Apigenin, chrysin (5,7-dihydroxyflavone), diosmetin (an o-methylated flavone), luteolin
15.	Turmeric	Rhizome	*Curcuma longa* (Zingiberaceae)	Curcumin

9.3.2 Polyphenol-enriched Extracts and Their Contribution

Natural phenols/polyphenols function in spices as reducing agents and strong antioxidants, delaying enzymatic oxidation processes. Phenolic compounds increase the stability of polyunsaturated fatty acids in fats and oils and reduce oxidation-related changes in food, in addition to maintaining organoleptic qualities [26]. Spices now have two uses in the food processing industry. They enhance the organoleptic qualities of food while also having a strong antioxidant potential because of their total polyphenol concentration. The full potential of compounds derived from spices is now being investigated while preserving the balance between these two essential qualities [27–29].

In India, polyphenol-rich extracts of spices, such as chilies, saffron, and turmeric, are frequently used as natural colorants in cuisine. The Indian Central Food Technological Research Institute has been striving to create innovative processes for producing natural product-based food colors such as kokum (red) and chilies (red) [17].

9.3.3 Polyphenols with Nanodelivery Systems: Beyond the Flavor and Taste

Polyphenols have antioxidant properties, preventing superfluous lipid oxidation that causes rancidity and preserving the flavor and taste of marine foods. High water activity, neutral pH, relatively significant amounts of free amino acids, the presence of autolytic enzymes, and a high percentage of unsaturated fatty acids cause fish products to decay quickly.

Oleoresins and spices are well known for their capacity to protect food taste and act as antioxidants. Fennel is an important medicinal and aromatic spice in everyday cooking because of its estrogenic activities and applications. One study found that encapsulated fennel extract was better than control and pure extract treatments. During storage, encapsulated fennel extract protects silver carp (*Hypophthalmicthys molitrix*) fillets against lipid oxidation and microbiological deterioration [30].

9.4 Polyphenol Spice Extract, Nanodeliveries, and Nutraceuticals in Industry

9.4.1 Application of Polyphenol Spice Extracts in the Food and Pharma Sectors

Natural polyphenol-rich plant extracts and essential oils have also been used to increase the shelf life of fish products because of their natural antibacterial and antioxidant properties. Compared with samples containing the unencapsulated extract, thyme extract that was enclosed in liposomes and mixed with minced silver carp reduced the development of mesophilic and psychrotrophic bacteria during cold storage. *Escherichia coli* were also reduced by liposomes containing 0.5% thyme extract. The overall quality of the minced fish was enhanced by reducing the peroxide value, the total volatile basic nitrogen, and microbial load during storage [31]. This is one example of how the food and food processing sectors benefit from the contribution of spices and their nanoform polyphenol content. Although stored chilled, food products are nonetheless impacted by high temperature cooking. The fish and meat items degrade when they are cooked at temperatures greater than 150°C–160°C because of potentially dangerous chemicals developing such as heterocyclic aromatic amines.

Spice polyphenols and associated metabolite (glycone and aglycone) compounds are highly effective antioxidants. They partially solve the issue; however, because of exposure to light and high temperatures, the powerful antioxidant activity of polyphenols is also diminished and rendered ineffective. One method to improve their stability and promote controlled distribution is liposomal entrapment [19].

Strong antioxidant hesperetin is a flavanone; however, because of its unpleasant sensory properties, it cannot be added to milk or used to manufacture milk drinks. A proposed answer is nanostructured hesperetin encapsulated with lipid carriers, which increases solubility, decreases bitterness, and prevents color change in dairy drinks [19].

Polyphenol spices are also used in various pharmaceutical applications. Ramirez-Nuez conducted research on *Cinnamomun verum* and *Vanilla planifolia* spices, and from both, created Fe_3O_4 magnetic nanoparticles (MNPs) with an average diameter of 10–14 nm. Surface-coated polyphenols provide therapeutic effects using green technology methods [32]. These MNPs have significant biological promise, particularly for drug administration, magnetic resonance imaging, and magnetic hyperthermia.

9.4.2 Nanodelivery Systems of Spice/Spice Extract Polyphenols and Nutraceuticals

In the food processing, healthcare, and pharmaceutical industries, polyphenol-enriched spice extracts are in high demand. Curcumin from *C. longa* has low water solubility, poor bioavailability, and a quick metabolism, all of which are severe detriments to its successful medicinal uses [33]. The nanotechnology-based nanocurcumin formulation dramatically boosts all biological and pharmacological effects of curcumin, which were not previously feasible with conventional processes. To increase curcumin distribution, many curcumin nanoformulations have been developed and some have undergone extensive clinical trials. The keyword "curcumin" produced 20,832 results in the NCBI PubMed database https://pubmed.ncbi.nlm.nih.gov/?term=curcumin, whereas the terms "curcumin nanoformulations" and "nano-curcumin" produced 339 hits and 127 hits, respectively. Many scientists concentrate their research to increase bioavailability for an effective outcome. Some of the curcumin nanoformulations were developed for significant influence on human body applications, including liposomes, polymers, gold NPs, MNPs, SLNs, conjugates, cyclodextrins, solid dispersions, micelles, nanospheres, nanogels, and nanodisks.

Quercetin, which is prevalent in spices, such as dill and chili [2], was encapsulated in PLA NPs to enhance its solubility and stability [34]. Rutin is a glycoside that combines the flavonol quercetin and the disaccharide rutinose, which is also known as rutoside, quercetin-3-O-rutinoside, and sophorin. It is a flavonoid found in many different plants and spices such as star anise. By acidification, nanocomplexes known as nanoPECs were developed using rutin as a model bioactive molecule and three mass ratios of NaCas/pectin [35]. NanoPECs were discovered to have superior encapsulating capabilities. Quercetin has also great potential to prevent atherosclerosis and other cardiovascular diseases [36]. In another study, Giannouli et al. fabricated quercetin-loaded PLGA polymer NPs using a hydrodynamic atomization process for biphasic release to prevent atherosclerosis [37].

Naringenin is a flavorless, colorless flavanone found in a variety of fruits and spices including rosemary leaves. Naringenin binds to a β-casein solution via van der Waals forces, hydrogen bonds, and hydrophobic interactions to generate naringenin β-casein NPs (< 100 nm), which exhibited increased functional properties of naringenins, principally by increasing their solubility [38].

Pulicalagin and ellagic acid are two polyphenols present in the fruit peel of pomegranate, which is commonly used as a spice in some Asian locations. In one study, pomegranate peel polyphenol was added to montmorillonite clay and turned into NPs to improve the bioavailability, half-life, and solubility of the polyphenols and reduce their rate of oxidation. It is conceivable to use these polyphenol-loaded NPs for drug delivery [39]. In a separate work, "mono-dispersed gold nanoparticles (PAuNPs)" with a certain particle size were made using a pomegranate peel extract and gold solution. Additional interactions between PAuNPs and casein, a biocompatible polymer, were used to functionalize folic acid for targeted 5-Fu administration. Using this concept, a high-affinity drug delivery technique against cancer cells was developed [40, 18].

Cumin seeds contain a range of flavonoid glycosides, including apigenin and luteolin derivatives, many free amino acids, and volatile essential oils (5%). Cumin gold NPs (Cu-AuNP) within a size of 10–15 nm were developed using gum Arabic $NaAuCl_4$ and cumin seeds [41].

Curcumin-loaded PLGA NPs, RES-loaded layer-by-layer formulations, quercetin-PLGA NPs, microwave-synthesized apigenin-Pluronic F127 NPs, luteolin-assembled poly (€-caprolactone)-PLGA-nature, and naringenin-loaded chitosan NPs with a particle size of 407 nm are examples developed for diabetes and associated disorders.

Similarly, the scientific community has access to a large number of additional spice/spice extract polyphenols and their nanodelivery systems with regard to the advantages of nutraceuticals and health benefits. These findings should be thoroughly explained and used by the industry for profit.

9.4.3 Market Potential and Commercial Significance

The market for polyphenols is expected to reach 2.9 billion USD in 2030, up from 1.67 billion USD in 2022, a 7.4% CAGR between 2022 and 2030. The functional drinks category, out of all segments, held a 32.7% share of the global polyphenol market in 2021. The second-largest economy in the world, China, is anticipated to grow at a 10.9% CAGR from 2020 to 2027 to reach a projected market size of 362.4 million USD. Japan and Canada are two significant geographic markets that had growth rates of 3.8% and 6.4% in 2020, respectively. Germany is expected to expand at a CAGR of approximately 4.5%. [42]. According to the current Google trend, Asia–Pacific appears to be dominating the polyphenol and antioxidant industries, creating new business opportunities for entrepreneurs. Catechins, quercetin, eugenol, epigallocatechin, epigallocatechin gallate, curcumin, and polyphenols are polyphenols that are most often encapsulated [43]. Apart from fruit and herbs, polyphenols are also obtained from spices and their oleoresins. Because of the availability of a variety of high-quality spices in several states across the nation that are cultivated in various climatic conditions, soil types, and geographic locations, India is known as the "Land of Spices". The Spices Board of India mentioned 52 spices. Polyphenols from spices can be handled by various technologies. In addition to conventional delivery systems, nanoscale materials have been used for many years to deliver spice polyphenols for a variety of applications. These materials include nanocrystals, NPs, liposomes and lipid PEGylated polymeric nanodrugs, protein-based NPs, and metal-based NPs [44]. Some important areas include dietary/health supplements, nutraceuticals, pharmaceuticals, life sciences, bionanotechnology, agriculture, sunscreens and cosmetics, textiles, electronics and communications, chemicals and materials including paints and varnishes, energy technologies, and space exploration [45]. According to "Grand View Research", a market research and consulting database business, revenue generation in 2022 was 10.89 billion USD, with a fast-increasing CAGR of 14.9% from 2022 to 2030 [43].

Spice nanotechnology is crucial for producing high-quality products with the appropriate sensory properties without degrading their nutritional qualities. These technologies may offer an alternative to traditional food production methods for creating nutrient-dense foods with a long shelf life without sacrificing the intended sensory experience for customers. These next-generation technologies may also play a role in all aspects of spice handling in the future, including ingredient selection, processing, final recipe formulations, decorating, packing, and stability, as well as transportation and storage.

9.5 Conclusion and Future Perspectives

Because of their distinctiveness and benefits as a delivery mechanism, nanotechnologies have recently piqued the interest of academics, academicians, industrialists, nutritional scientists, medical professionals, and cosmetologists. The polyphenols in spices have long attracted interest for the creation of NPs, nanoencapsulations, nanoemulsions, nanolipid carriers, nanofibers, nanocomposites, and nanolaminates that are used in a variety of applications, including antimicrobial, wound healing, diabetes prevention, cardiac health, anticancer, as well as defense against susceptible disease and disorders. Although diverse polyphenols from fruits and herbs have been examined along with their nanodelivery mechanism, little is known about the polyphenols in spices and their nanodelivery systems. In addition to the taste, flavor, and scent, the market for spices is quite large; therefore, there are many opportunities to work with newer nanotechnologies and create novel delivery systems for customer advantages. For additional practical advantages, business owners and governmental organizations must continue their study in this field. To explore this topic in front of the scientific community, the food processing industries and pharmaceutical industries require international coordination, and collaborations, only then results will be able to be explored broadly and reach in a mass market.

Acknowledgments

The authors acknowledge the Management of Synthite Industries Pvt. Ltd., India for their support and encouragement

References

1 Leja, K.B. and Czaczyk, K. (2016). The industrial potential of herbs and spices - A mini review. *Acta Scientiarum Polonorum. Technologia Alimentaria* 15: 353–365. https://doi.org/10.17306/J.AFS.2016.4.34.

2 Nema, N.K., Rajan, N., Babu, M. et al. (2022a). Dietary polyphenols from spices and their impact on human health. In: *Dietary Polyphenols in Human Diseases* (ed. M. Rudrapal), 229–254. Boca Raton: CRC Press. https://doi.org/10.1201/9781003251538-12.

3 Viuda-Martos, M., Ruiz-Navajas, Y., Fernández-López, J., and Pérez-Álvarez, J.A. (2010). Spices as Functional Foods. *Critical Reviews in Food Science and Nutrition* 51: 13–28. https://doi.org/10.1080/10408390903044271.

4 Abdel-Shafy, H.I. and Mansour, M.S.M. (2017). Polyphenols: properties, occurrence, content in food and potential effects. In: *Environmnetal Sciences and Engineering: Toxicology* (ed. B.R. Gurjar), 232–261. USA: Studium Press LLC.

5 González-Sarrías, A., Tomás-Barberán, F.A., and García-Villalba, R. (2020). Structural diversity of polyphenols and distribution in foods. In: *Dietary Polyphenols* (ed. F.A. Tomás-Barberán, A. González-Sarrías, and R. García-Villalba), 1–29. Wiley. https://doi.org/10.1002/9781119563754.ch1.

6 Ofosu, F.K., Daliri, E.B., Elahi, F. et al. (2020). New insights on the use of polyphenols as natural preservatives and their emerging safety concerns. *Frontiers in Sustainable Food Systems* 4: 1–14. https://doi.org/10.3389/fsufs.2020.525810.

7 Embuscado, M.E. (2015). Spices and herbs: natural sources of antioxidants - a mini review. *Journal of Functional Foods* 18: 811–819. https://doi.org/10.1016/j.jff.2015.03.005.

8 Sachan, A.K., Kumar, S., Kumari, K., and Singh, D. (2018). Medicinal uses of spices used in our traditional culture: world wide. *Journal of Medicinal Plants Studies* 6: 116–122.

9 Nema, N.K., Mamdapur, G.M.N., Sarojam, S. et al. (2021). Preventive medicinal plants and their phytoconstituents against SARS-CoV-2/COVID-19. *Pharmacognosy Researchs* 13: 173–191. https://doi.org/10.5530/pres.13.4.10.

10 Nema, N.K., Rajan, N., Sabu, S. et al. (2022b). Use of nanotechnology for the improvement of sensory attributes of foods. In: *Application of Nanotechnology in Food Science, Processing and Packaging* (ed. C. Egbuna, J. Jeevanandam, K.C. Patrick-Iwuanyanwu, and E.N. Onyeike), 31–46. Cham: Springer International Publishing. https://doi.org/10.1007/978-3-030-98820-3_3.

11 Chaudhry, Q., Scotter, M., Blackburn, J. et al. (2008). Applications and implications of nanotechnologies for the food sector. *Food Additives and Contaminants* 25: 241–258. https://doi.org/10.1080/02652030701744538.

12 Leena, M.M., Yoha, K.S., Moses, J.A., and Anandharamakrishnan, C. (2021). Nanofibers in Food Applications. In: *Innovative Food Processing Technologies*. 634–650. Elsevier. https://doi.org/10.1016/B978-0-08-100596-5.22952-1.

13 Rashidi, L. and Khosravi-Darani, K. (2011). The applications of nanotechnology in food industry. *Critical Reviews in Food Science and Nutrition* 51: 723–730. https://doi.org/10.1080/10408391003785417.

14 Trotta, F. and Mele, A. (2019). Nanomaterials: classification and properties. *Nanosponges* 1–26. https://doi.org/10.1002/9783527341009.ch1.

15 Yu, H., Park, J.Y., Kwon, C.W. et al. (2018). An overview of nanotechnology in food science: preparative methods, practical applications, and safety. *Journal of Chemistry* 2018: https://doi.org/10.1155/2018/5427978.

16 Ahmad, S., Munir, S., Zeb, N. et al. (2019). Green nanotechnology: a review on green synthesis of silver nanoparticles — an ecofriendly approach. *International Journal of Nanomedicine* 14: 5087–5107. https://doi.org/10.2147/IJN.S200254.

17 Jiwan Chahande, S., Jachak, R., Chahande, R., and Pantawane, P. (2022). Herbal spices and nanotechnology for the benefit of human health. In: *Biogenic Sustainable Nanotechnology* (ed. R.P. Singh, A. Rai, A. Abdala, R. Chaudhary), 107–129. Elsevier. https://doi.org/10.1016/B978-0-323-88535-5.00009-3.

18 Milinčić, D.D., Popović, D.A., Lević, S.M. et al. (2019). Application of polyphenol-loaded nanoparticles in food industry. *Nanomaterials* 9: 1629. https://doi.org/10.3390/nano9111629.

19 Molina, G., Inamuddin, Pelissari, F.M., and Asiri, A.M. (eds.) (2019). *Food Applications of Nanotechnology*, 1e. 33487–2742. Boca Raton, FL: CRC Press. https://doi.org/10.1201/9780429297038.

20 Almetwally, A.A., El-Sakhawy, M., Elshakankery, M.H., and Kasem, M.H. (2017). Technology of nano-fibers : production techniques and properties - critical review. *Journal of the Textile Association* 78: 5–14.

21 Kenry and Lim, C.T. (2017). Nanofiber technology: current status and emerging developments. *Progress in Polymer Science* 70: 1–17. https://doi.org/10.1016/j.progpolymsci.2017.03.002.

22 Okutan, N., Terzi, P., and Altay, F. (2014). Affecting parameters on electrospinning process and characterization of electrospun gelatin nanofibers. *Food Hydrocolloids* 39: 19–26. https://doi.org/10.1016/j.foodhyd.2013.12.022.

23 Tan, E.P.S. and Lim, C.T. (2006). Mechanical characterization of nanofibers – a review. *Composites Science and Technology* 66: 1102–1111. https://doi.org/10.1016/j.compscitech.2005.10.003.

24 International Food Information Council (2021). *Perceptions on Processed*. Consumer Sentiment and Purchasing Habits.

25 Cook, N.C. and Samman, S. (1996). Flavonoids-chemistry, metabolism, cardioprotective effects, and dietary sources. *The Journal of Nutritional Biochemistry* 7: 66–76. https://doi.org/10.1016/S0955-2863(95)00168-9.

26 Kähkönen, M.P., Hopia, A.I., Vuorela, H.J. et al. (1999). Antioxidant activity of plant extracts containing phenolic compounds. *Journal of Agricultural and Food Chemistry* 47: 3954–3962. https://doi.org/10.1021/jf990146l.

27 Srinivasan, K. (2014). Antioxidant potential of spices and their active constituents. *Critical Reviews in Food Science and Nutrition* 54: 352–372. https://doi.org/10.1080/10408398.2011.585525.

28 Wojdyło, A., Oszmiański, J., and Czemerys, R. (2007). Antioxidant activity and phenolic compounds in 32 selected herbs. *Food Chemistry* 105: 940–949. https://doi.org/10.1016/j.foodchem.2007.04.038.

29 Yanishlieva, N.V., Marinova, E., and Pokorný, J. (2006). Natural antioxidants from herbs and spices. *European Journal of Lipid Science and Technology* 108: 776–793. https://doi.org/10.1002/ejlt.200600127.

30 Mazandrani, H.A., Javadian, S.R., and Bahram, S. (2016). The effect of encapsulated fennel extracts on the quality of silver carp fillets during refrigerated storage. *Nutrition and Food Science* 4: 298–304. https://doi.org/10.1002/fsn3.290.

31 Javadian, S.R., Shahosseini, S.R., and Ariaii, P. (2017). The effects of liposomal encapsulated thyme extract on the quality of fish mince and escherichia coli O157:H7 inhibition during refrigerated storage. *Journal of Aquatic Food Product Technology* 26: 115–123. https://doi.org/10.1080/10498850.2015.1101629.

32 Ramirez-Nuñez, A.L., Jimenez-Garcia, L.F., Goya, G.F. et al. (2018). In vitro magnetic hyperthermia using polyphenol coated [illegible] and *Vanilla planifolia*: the concert of green synthesis and therapeutic possibilities. *Nanotechnology* 29: https://doi.org/10.1088/1361-6528/aaa2c1.

33 Karthikeyan, A., Senthil, N., and Min, T. (2020). Nanocurcumin: a promising candidate for therapeutic applications. *Frontiers in Pharmacology* 11: 1–24. https://doi.org/10.3389/fphar.2020.00487.

34 Kumari, A., Yadav, S.K., Pakade, Y.B. et al. (2010). Development of biodegradable nanoparticles for delivery of quercetin. *Colloids Surfaces B Biointerfaces* 80: 184–192. https://doi.org/10.1016/j.colsurfb.2010.06.002.

35 Luo, Y., Pan, K., and Zhong, Q. (2015). Casein/pectin nanocomplexes as potential oral delivery vehicles. *International Journal of Pharmaceutics* 486: 59–68. https://doi.org/10.1016/j.ijpharm.2015.03.043.

36 Cote, B., Carlson, L.J., Rao, D.A., and Alani, A.W.G. (2015). Combinatorial resveratrol and quercetin polymeric micelles mitigate doxorubicin induced cardiotoxicity in vitro and in vivo. *The Journal of Controlled Release* 213: 128–133. https://doi.org/10.1016/j.jconrel.2015.06.040.

37 Giannouli, M., Karagkiozaki, V., Pappa, F. et al. (2018). Fabrication of quercetin-loaded PLGA nanoparticles via electrohydrodynamic atomization for cardiovascular disease. *Materials Today: Proceedings* 5: 15998–16005. https://doi.org/10.1016/j.matpr.2018.05.044.

38 Moeiniafshari, A.A., Zarrabi, A., and Bordbar, A.K. (2015). Exploring the interaction of naringenin with bovine beta-casein nanoparticles using spectroscopy. *Food Hydrocolloids* 51: 1–6. https://doi.org/10.1016/j.foodhyd.2015.04.036.

39 Karimi, M., Sadeghi, R., and Kokini, J. (2017). Pomegranate as a promising opportunity in medicine and nanotechnology. *Trends in Food Science and Technology* 69: 59–73. https://doi.org/10.1016/j.tifs.2017.08.019.

40 Ganeshkumar, M., Sathishkumar, M., Ponrasu, T. et al. (2013). Spontaneous ultra fast synthesis of gold nanoparticles using Punica granatum for cancer targeted drug delivery. *Colloids Surfaces B Biointerfaces* 106: 208–216. https://doi.org/10.1016/j.colsurfb.2013.01.035.

41 Katti, K., Chanda, N., Shukla, R. et al. (2009). Green nanotechnology from cumin phytochemicals: generation of biocompatible gold nanoparticles. *Journal of Biomedical Nanotechnology* 1: 39–52. https://doi.org/10.1080/19430850902931599.

42 Research and Markets (2022). *Polyphenols - Global Strategic Business Report.* https://www.researchandmarkets.com/reports/5030379/polyphenols-global-strategic-business-report (accessed 21 June 2022).

43 GVS (2023). *Nanomaterials Market Size, Share & Trends Analysis Report By Material (Gold, Silver, Iron, Copper), By Application (Aerospace, Automotive, Medical), By Region, And Segment Forecasts, 2023 - 2030.* https://www.grandviewresearch.com/industry-analysis/nanotechnology-and-nanomaterials-market (accessed 20 February 2023a).

44 Islam, F., Khadija, J.F., Islam, M.R. et al. (2022). Investigating polyphenol nanoformulations for therapeutic targets against diabetes mellitus. Evidence-based complement. *Alternative Medicine* https://doi.org/10.1155/2022/5649156.

45 Rambaran, T. and Schirhagl, R. (2022). Nanotechnology from lab to industry - a look at current trends. *Nanoscale Advances* 4: 3664–3675. https://doi.org/10.1039/d2na00439a.

10

Nanodelivery of Polyphenols as Nutraceuticals in Anticancer Interventions

Kamoru A. Adedokun[1,#,], Sikiru O. Imodoye[2,#,*], Zwanden S. Yahaya[3,#], Ifeoluwa T. Oyeyemi[4,5], Ibrahim O. Bello[6], Mujidat T. Adeyemo-Imodoye[7,¥], Malik A. Sanusi[8,¥], and Ramat T. Kamorudeen[9]*

[1] *Roswell Park Comprehensive Cancer Center, University at Buffalo, Buffalo, NY, USA*
[2] *Huntsman Cancer Institute, University of Utah, Salt Lake City, UT, USA*
[3] *Department of Pharmaceutics and Industrial Pharmacy, Faculty of Pharmaceutical Sciences, Kaduna State University, Kaduna, Nigeria*
[4] *Department of Biosciences and Biotechnology, University of Medical Sciences, Ondo, Nigeria*
[5] *Department of Chemistry and Biochemistry, Old Dominion University, Hampton Boulevard, Norfolk, VA, USA*
[6] *Department of Biological Sciences, Southern Illinois University Edwardsville, Edwardsville, IL, USA*
[7] *Department of Medical Laboratory Science, Somolu General Hospital, Lagos Health Service Commission, Lagos State, Nigeria*
[8] *Department of Biochemistry and Molecular Biology, Obafemi Awolowo University, Ile-Ife, Osun State, Nigeria*
[9] *Department of Public Health, University of South Wales, Pontypridd, United Kingdom*
[*] *Corresponding authors*
[#, ¥] *These authors share the same position*

10.1 Introduction

Plants have remained an integral component of the human diet for a long time, serving as rich reservoirs of important minerals, vitamins, or dietary fiber that regulate metabolism. As their secondary metabolites, plants produce a variety of biologically active compounds called phytochemicals that play crucial roles in protecting against ultraviolet radiation and other foreign attacks. Interestingly, plant metabolites have a long therapeutic history in human existence [1]. The consumption of plants and their derived analogs has been proposed to prevent and cure diseases, thus providing a long and healthy life [2, 3]. Phytochemicals are grouped into polyphenols, alkaloids, terpenoids, and thiols based on their chemical features [1]. Polyphenols are the most diverse bioactive phytochemicals, comprising more than 10,000 chemical species that occur naturally in fruits, vegetables, berries, coffee, and tea. Owing to their health benefits, in addition to providing important nutritive values, polyphenols are often referred to as nutraceuticals. Structurally, polyphenols possess aromatic rings with at least one hydroxyl functional group. Based on their chemical structures, they can be further classified as flavonoids and non-flavonoids. Flavonoids have 15 carbon atoms at their core and are divided into flavanols, flavonols, anthocyanidins, flavones, flavanones, and chalcones. Non-flavonoids contain an aromatic ring with one or more hydroxyl groups. Examples are stilbene, phenolic acids, saponin, and other polyphenols such as curcumin and tannins.

Conventional cancer therapies largely rely on the use of cytotoxic agents to destroy rapidly dividing cells, targeting one of the hallmarks of cancers. However, this approach is rapidly becoming less valuable as these agents destroy all other rapidly proliferating cells, such as in the hair follicles, bone marrow, and intestinal epithelium, leaving patients with side effects.

Polyphenols: Food, Nutraceutical, and Nanotherapeutic Applications, First Edition. Edited by Mithun Rudrapal.

To overcome this downside, researchers are exploring alternatives to selectively destroy cancer cells with minimal or no adverse effects on healthy cells. Several preclinical, clinical, and in silico studies show that polyphenols possess antioxidant potential, preventing oxidative stress-related chronic diseases such as chronic inflammation, atherosclerosis, and cancers [4–6]. In particular, the aromatic rings of polyphenols act as electron acceptors and donors and facilitate their interaction with peptides to modify protein activity through which they alter cell behaviors. Additionally, the presence of hydroxyl groups in polyphenols has been associated with their antioxidant properties [7]. They do this by blocking the formation of reactive oxygen species (ROS), quenching reactive radicals, and chelating metal ions. Through the free radical scavenging effect, polyphenols mitigate oxidative damage to biomolecules such as protein, DNA, and membrane lipids [8], thereby preventing DNA replication errors, promoting genomic stability and preventing cancer.

It is strongly believed that the consumption of polyphenolic-rich diets can reduce the risk of developing cancer and other chronic diseases [9]. The World Health Organization (WHO) reported that approximately 35% of cancer mortalities are due to the five leading behavioral and dietary risks, which are high body mass index, low fruit and vegetable intake, lack of physical activity, tobacco use, and alcohol use [10]. Collectively, these observations reinforce the critical role of plant-derived compounds, such as polyphenols, in chemoprevention.

Polyphenolic compounds exhibit antiproliferative effects on many cancer types. However, despite these elegant chemopreventive and therapeutic activities, various pharmacokinetic data revealed that polyphenols have several significant shortcomings, such as non-specific tumor targeting, low aqueous solubility, and poor bioavailability, which has restricted their application in therapeutics. Importantly, the poor bioavailability of polyphenols is due in part to their total reliance on passive diffusion across the epithelial cells of the small intestine because they lack specific receptors for active transport. A promising approach to improve the pharmacokinetics of polyphenolic compounds is via the application of nanoparticulate drug delivery systems. Nanodelivery of phenolic compounds represents an effective tool that permits their effective transport across the gastrointestinal (GI) tract and in tumor tissue [11–13]. These systems can be customized to deliver phenolics to specific tissues, provide controlled release delivery, decrease off-target effects, and increase patient compliance with less frequent dosing. Because of their small size and large surface area, drug nanoparticles show increased solubility and thus enhanced bioavailability, ability to cross the blood–brain barrier (BBB), enter the pulmonary system, and be absorbed through the tight junctions of endothelial cells [14]. Nanoparticles can reversibly open up tight junctions of the epithelial cells via a receptor-mediated process to enhance cargo delivery. Mechanistically, nanoparticles enhance intestinal permeability by binding to integrins and activating myosin light chain kinase [15]. Additionally, through parallel mechanisms, such as endocytosis or macropinocytosis, nanoparticles could be internalized via transcellular pathways [16]. Macropinocytosis-associated transport is aided by actin cytoskeletal networks, while endocytosis can be clathrin or caveolae-mediated to release nanoparticles directly inside the cell. After internalization, the nanoparticles may be translocated to an endolysosome, where they are degraded to release the bioactive compound to fulfill their pharmacological actions. Sometimes, the compound may undergo further metabolic transformation within the cell before they become pharmacologically active. The focus of this chapter is to highlight important mechanisms through which polyphenols control tumor growth and discuss various carrier platforms that can safeguard them from GI degradation, thereby enhancing their bioavailability and tissue specificity via nanoparticulate-based delivery systems.

10.2 Nutraceuticals and Cancer

10.2.1 Role of Polyphenols in Chemoprevention

People from both developing and industrialized countries continue to suffer from cancer. Cancer is the second leading cause of mortality globally [17]. At the cellular and molecular levels, cancer is very diverse and heterogeneous, leading to broad metabolic and behavioral alterations in cancer cells. As a result, polyphenols, which are structurally diversified, can interact with and inactivate oncogenic peptides to prevent tumorigenesis. Consumption of polyphenol-rich foods has been linked to a lower risk of cancer [18]. The use of specific natural (dietary), synthetic, or biological agents to prevent, delay, or slow the carcinogenic process is referred to as chemoprevention [19]. Cancer prevention using various approaches has been recommended as a cost effective approach against cancer [20]. Chemoprotective agents can disrupt one or more stages of cancer [21]. The fact that cancer goes through different stages and develops over time before getting established has made chemoprevention a promising strategy against cancer.

Several polyphenols have demonstrated chemopreventive potential. Clinical studies have demonstrated the benefits of some of them in preventing primary cancer, secondary cancer, progression, and the recurrence of cancer [22]. Polyphenols, among other phytochemicals, have been shown to exert chemopreventive effects via various mechanisms including mitigation of oxidative DNA damage, inhibition of inflammation, inhibition of cell cycle, angiogenesis, and metastasis, as well as induction of apoptosis [1]. They have the ability to alter the expression of many protein targets and carcinogenesis-related pathways. Over the past 20 years, a lot of research has been done on the connection between dietary polyphenol consumption and the risk of developing cancer. In a meta-analysis study, intake of isoflavone reduced the risk of gastric cancer by 19% compared with the cohort that did not consume the compound [23]. Similarly, an epidemiologic study showed that consuming soy products reduced the risk of developing breast cancer [24]. Additionally, a case control study found that consuming isoflavone and flavonol led to a 30% decrease in the risk of ovarian and endometrial cancer [25]. Together, these findings offer compelling evidence that dietary polyphenol consumption may lower the risk of later-life cancer development.

10.2.2 Phenolic Compounds and their Anticancer Activities

In the fight against cancers, increased attention is being paid to the development of anticancer agents from polyphenolic compounds because of their low toxicities and widespread abundance. Several polyphenolic agents with anticancer properties have been reported (Table 10.1).

The following section focuses on anticancer mechanisms of polyphenols through various cancer hallmarks.

10.2.2.1 Pro-apoptotic Effects

Apoptosis, also known as programmed cell death, is essential for normal cell physiology or homeostasis; however, it is frequently dysregulated in cancer [56]. Apoptosis in cancer cells can be induced by either intrinsic or extrinsic pathways, both of which cause cell death by activating the caspase cascade [57].

Bcl-2 family proteins mediate cell death and survival. They are made up of anti-apoptotic (Bcl-2 and Bcl-xl) and pro-apoptotic members (Bax, Bad, and Bid). These proteins regulate programmed cell death by regulating the permeabilization of the outer mitochondrial membrane. Anti-apoptotic

Table 10.1 Various anticancer mechanisms of polyphenolic compounds.

S/N	Polyphenol	Cancer Type	Experimental Model	Mechanism	References
1	Quercetin	Colon cancer	In vivo and in vitro	Inhibits hypoxia-induced adenosine monophosphate-activated protein kinase (AMPK)	[26]
2	Resveratrol	Melanoma	In vivo and in vitro	Induces apoptosis and autophagy, inhibits the P13K/AKT/mTOR pathway	[27]
3	Curcumin	Ovarian cancer	In vivo and in vitro	Induces apoptosis and autophagy, inhibits the AKT/mTOR/p7056K pathway	[28]
4	Kaempferol	Osteosarcoma	In vivo and in vitro	Inhibits metastasis via suppression of the MAPK pathway	[29]
5	Anthocyanin	Leukemia, colon cancer	In vitro	Stimulation of apoptosis through regulation of cytochrome c release, caspases, expression of FAS and FASL	[30]
6	Apigenin	Cervical cancer	In vivo and in vitro	Inducing cell cycle arrest at the G2/M phase by modulating cyclin B1/CDK1 AND $p21^{clp1}$	[31]
7	Chrysin	Colon cancer	In vivo and in vitro	Induces apoptosis via upregulation of caspase-3 and caspase-9 activities and inhibition of sall4 expressions	[32]
8	Rutin	Breast cancer	In vitro	Inhibits growth through the induction of apoptosis and G2/M phase cell cycle arrest	[33]
9	Acacetin	Prostate cancer	In vivo	Blocking the activities of STAT3 through phosphorylation	[34]
10	Baicalein	Lung cancer	In vivo	Upregulation of apoptotic genes (ITGB3 and TNFRSF25) and downregulation of angiogenic genes (VEGF)	[35]
11	Luteolin	Lung cancer	In vivo	Inhibition of P13/AKT/mTOR	[36]
12	Tangeretin	Gastric cancer	In vivo and in vitro	Stimulation of apoptosis through activation of *RARB*, caspase-3, -9, and PARP1	[37]
13	Wogonin	Ovarian cancer, colorectal cancer	In vivo	Suppression of tumor growth by inhibiting glucose metabolism in a p53-dependent manner	[38]
14	Dieckol	Fibrosarcoma	In vitro	Downregulation of FAK signaling through scavenging intracellular ROS. Inhibition of the complex formation of FAK-SRc-p130Cas and expression of MMP-2, 9, and 13	[39]
15	Gallic acid	Gastric carcinoma	In vitro	Suppression of the MMP-2/9 expression through the inhibition of NF-jB activity	[40]

(Continued)

Table 10.1 (Continued)

S/N	Polyphenol	Cancer Type	Experimental Model	Mechanism	References
16	3,4-Dihydroxy-benzalacetone	Chronic myeloid leukemia, kidney cancer, multiple myeloma	In vitro	Inhibition of NF-jB activity, Suppression of TNF-induced IKK activation, modulation of NF-jB-dependent gene products involved in tumor cell invasion (e.g., MMP-9, intracellular adhesion molecule- 1 and VEGF)	[41]
17	Resveratrol	Breast cancer and hepatocellular carcinoma	In vitro In vitro	Inhibition of IGF-1-mediated cell migration in part through the suppression of activation of the PI3K/Akt signaling pathway Inhibition of TNF-a-mediated MMP-9 expression in part through the downregulation of the NF-jB signaling pathway. Inhibition of HRG-b1-mediated MMP-9 expression in part through the downregulation of MAPK/ERK signaling pathway	[42] [43] [42]
18	Xanthohumol	Breast cancer	In vitro	Inactivation of S100A4, MLC2, selectin E, and paxillin. Inhibition of ICAM-1 and NF-jB. Reduction of CYP1A1 activity	[44]
19	baicalein	Breast cancer	In vitro and in vivo	Suppression of cell proliferation, induction of autophagy and apoptosis through downregulation of mTOR, p-AKT, p-IκB and NF-κB	[45]
20	Allicin	Lung cancer	In vitro	Inhibition of cell adhesion, migration and invasion Suppression of the PI3K/AKT signaling pathway by reducing phosphorylation of AKT	[46]
21	Daidzein	Breast cancer	In vitro and In vivo	Induction of apoptosis via suppression of Bcl-xl and upregulation of Fas, FADD, Bax, cleaved caspases 3 and 9	[37]
22	Myricetin	Prostate cancer	In vitro	Pro-apoptotic and antimetastatic effects via inhibition of PIM1 and disrupting the PIM1/CXCR4 interaction. Knockdown of the interaction between P1M1/CXCR4	[47]
23	Gossypol	Prostate cancer	In vitro and in vivo	Proapoptotic effect: induces caspase-9- and caspase-3 and inhibits Bcl-2	[48]
24	Narenginin	Breast cancer	In vitro	Antiangiogenic activities through suppression of VEGF expression	[49]

Table 10.1 (Continued)

S/N	Polyphenol	Cancer Type	Experimental Model	Mechanism	References
25	Epigallocatechin-gallate	Glioblastoma	In vitro	Increased ROS production	[50]
26	Baicalin	Colon cancer	In vitro and in vivo	Induction of cell cycle arrest via induction of apoptosis and senescence	[51]
27	Carnosol	Breast cancer	In vitro	Cell cycle blockade through increased p21/WAF1 expression and decreased expression of p27	[52]
28	Genistein	Human promyelocytic leukaemic	In vitro and in vivo	Stimulation of apoptosis by inducing G2/M phase arrest: increases Bax, PARP-cleavage, caspase-9, and -3 and decreases Bcl-2 and Bid	[53]
29	gamma-tocotrienol	Prostate cancer		Antiangiogenic and stemness effects by downregulating angiopoietin-1	[54]
30	Ginkgetin	Osteosarcoma	In vitro	Inhibition of apoptosis through suppression of STAT3 and upregulation of caspase-3/9	[55]

Bcl-2 proteins interact directly with pro-apoptotic proteins, inhibiting their activity and apoptosis [58]. Pro-apoptotic proteins, such as Bax, directly permeabilize the mitochondrial outer membrane, resulting in the release of cytochrome c during apoptosis [59]. Bax is a p53 target transactivated in a variety of systems during p53-mediated apoptosis [60]. Polyphenols directly target the anti-apoptotic pathway by inhibiting Bcl-2 family members [57].

Polyphenols can also target cancer cells by generating ROS within the cell [61]. Excessive ROS activates the permeability transition pore, resulting in mitochondrial membrane potential loss and cytochrome c release. After its release, cytochrome c activates caspase 9, which then activates downstream caspase-3 [62].

Polyphenols also induce apoptosis by modulating a variety of apoptosis-related genes and pathways. Polyphenols trigger apoptosis by increasing the expression of the PI3K/AKT proteins. The PI3K/AKT signaling pathway is involved in a variety of cellular activities, including apoptosis. The PI3K/AKT pathway's main downstream target, AKT, directly phosphorylates caspase-9 and BAD, inhibiting their activity and thus promoting cell survival [63]. It also interacts with the BCl-2 family, which typically sends survival signals to cells [64]. These anti-apoptotic means, which rescue tumor cells from programmed cell death, are regulated by polyphenolic compounds. In this way, polyphenols inhibit PI3K/AKT signaling, thereby committing malignant cells to death [65, 66].

Polyphenols can also modulate the mitogen-activated protein kinase (MAPK) family involved in the regulation of cell growth, differentiation, and apoptosis [67, 68]. p38, a member of the MAPK superfamily, targets phospho-p53 and phospho-Jun [69], and the FasL gene is a target of c-jun activation, which leads to FasL binding to the Fas receptor (CD95) [70]. An increase in c-jun phosphorylation results in the activation of FasL, an AP-1 target gene. FasL activation induces apoptosis via the Fas death receptor, which then activates caspase-8 and Bid [68]. Taking advantage of this pathway, polyphenols have been shown to induce apoptosis by activating MAPKs [68, 71–73].

The Hippo/YAP signaling pathway is another conserved signaling pathway whose activation promotes survival, proliferation, and migration, and inhibits apoptosis [74, 75]. YAP, a critical member of the pathway, is implicated in mediating these biological functions and its high expression has been linked to tumorigenesis and a poor cancer prognosis [76]. Various studies have shown that polyphenols inactivate this pathway by promoting apoptosis [76–78].

p53, encoded by the TP53 gene, induces apoptosis through caspase activation, mediated by an increase in the expression of pro-apoptotic Bcl-2 family proteins, Fas/Fas L, and reducing the expression of the pro-survival Bcl-2 [79]. Inactivation of p53 is a common feature of different cancer types and restoration of p53 has been proposed as a therapeutic strategy against cancer [80, 81]. Polyphenols have exerted their chemopreventive effect by activating the p53 gene [82–84]. Polyphenols also regulate the expression of other apoptosis-related genes such as the adenosine triphosphate binding cassette subfamily G member 2 (ABCG2) [85].

Polyphenols also exhibit their chemopreventive effect by modulating miRNA, which regulates a wide variety of biological processes, including apoptosis and cell cycle. Alteration of the expression of these miRNAs has been linked to the progression of almost all diseases, including cancer [86, 87]. The expression of miR-195-5p/miR-497-5p reduces the expression of Bcl-2 while miR-1247-3p reduces the expression of caspase-2 [88]. Polyphenols have been reported to modulate the expression of these miRNAs and thereby restore the activities of Bcl-2 [88, 89].

10.2.2.2 Anti-angiogenic Effects

Angiogenesis, the development of new vasculature from pre-existing blood vessels, is an important process occurring in both healthy and disease states. It is critical to normal growth and development, and plays an important role in tumor growth and development [90, 91]. Several phytochemicals from natural products, such as plants, have been investigated for their ability to suppress angiogenesis. Polyphenols have been studied extensively on their potential benefits to human health. More importantly, anticancer activities and specifically, anti-angiogenic activities of polyphenols, have been investigated. Polyphenols, such as flavonoids present in several fruits and vegetables, have been reported to inhibit angiogenesis and tumor metastasis. Flavonoids modulate regulation of angiogenesis and metastasis via multiple signaling pathways. Flavonoids regulate the expression of VEGF, matrix metalloproteinases (MMPs), EGFR, and inhibit NFB, PI3-K/Akt, and ERK1/2 signaling pathways, thereby causing strong antiangiogenic effects [92]. Hypoxia and hypoxia-induced factor (HIF) signaling pathways are important in the initiation of angiogenesis in tumor cells. Transcription of pro-angiogenic genes, such as Ang1, Ang2, angiopoietin TIE-2, and VEGF, are controlled by HIF-1α and HIF-2α [93]. Furthermore, expression of VEGF is induced in endothelium by HIF-1α, which leads to activation of the autocrine signaling pathway associated with VEGF. Hence, endothelial cells (ECs) proliferate and survive with the involvement of this signaling cascade. In contrast, VEGF upregulates the permeability of blood vessels and regulates the secretion of degrading enzymes that break down the extracellular matrix, which in turn results in the expansion of the vascular system [94]. In addition, biological effects of VEGF are exerted via binding to VEGFR1 AND VEGFR2 receptors on ECs to eventually activate them. When activated, ECs produce MMPs that facilitate the migration of ECs by breaking down the basement membrane. After extracellular matrix degradation and rearrangement, tube formation and angiogenesis are stimulated by TIE-2 [95].

There are reported limitations of the therapeutic effects of polyphenolic compounds because of their poor solubility in aqueous solution. The solubility and bioavailability of natural or synthetic bioactive compounds can be increased using nanotechnologies, and this would provide potential benefits, such as selective targeting, in cancer therapy [12]. As an anticancer agent, the stability and cellular absorption of polyphenols can be increased by encapsulation. Dube et al. [96] reported that encapsulation significantly increased intestinal absorption. Balakrishnan et al. [97] used gold nanoparticle-based drug delivery systems that were formulated to improve the therapeutic effects of polyphenols as anticancer agents, and these nanoconjugates were tested against cancer cells in vitro and ex vivo. In the in vitro experiment, the nanoconjugates inhibited angiogenesis by suppressing the capillary-like tube formation in cancer cells, while it inhibited neovascularization compared with free polyphenolic compounds ex vivo. Generally, nanoconjugates of polyphenols inhibit expression of VEGF-2 protein and the downstream signaling pathway in cancer cell lines [79]. Furthermore, reductions in the MMPs have been reported among molecular mechanisms [98] suggesting inhibition of invasion, migration, and angiogenesis [99]. The mechanisms of cellular uptake include diffusion, endocytosis, and specific molecular target nanoparticles. Cancer cells have lower pH and higher ROS content, and ROS-dependent [100] and pH-dependent [101] drug delivery and/or cell death mechanisms have been suggested. There have been suggestions on ROS-dependent and pH-dependent drug delivery mechanisms as a result of the increased ROS level and decreased pH of cancer cells.

10.2.2.3 Antiproliferative Effects

Foods containing polyphenols are used in a regular diet to achieve good health. The metabolic processes, transport, and distribution of phenolics to their target organs might alter their structure and bioactivities. Bio-accessibility and bioavailability of polyphenols in the GI are essential for their uptake [102]. Furthermore, the interaction of polyphenols with other dietary components influences their bioavailability, which is defined as the ability to be absorbed and distributed throughout the body [103]. Poor availability of polyphenols is caused by several features that restrict their metabolism such as the level of polymerization, complexity of their chemical structure, solubility, and their interaction with other compounds.

There is improved effectiveness of polyphenol nanoformulations as bioactive compounds for the prevention and treatment of cancer. Currently, there are various nanoformulations designed for the delivery of polyphenols, including nanosuspensions, solid lipid nanoparticles, liposomes, gold nanoparticles, and polymeric nanoparticles, which have yielded improved antitumor activity, increased intracellular concentration of polyphenols, slow and steady release of drugs, and enhanced antiproliferative activity against cancer cells [104]. One of the various mechanisms by which polyphenols accomplish an anticancer effect is through induction of arrest during phases of the cell cycle in tumor cells [105]; specifically, the induction of cell cycle arrest in the G2/M phase. Regulation of the cell cycle, often referred to as the "accelerating and braking" process of tissue growth, is a major molecular event that orchestrates proliferation of tumor cells. Importantly, the tremendous degree of conservation of this system across eukaryotes underscores its early origins and significance for life. A typical cell cycle is controlled by cyclin-dependent kinases (CDKs) and their cyclin associates, and is divided into four phases: G1, S, G2, and M. According to a detailed mechanism by Duronio and Xiong [106], the G1 phase of the cell cycle, which is controlled by the retinoblastoma tumor suppressor (RB) pathway, is fundamentally connected to many cellular signaling processes. The RB-pathway involves RB protein, CDK inhibitors (CKIs) and activators, and the E2F family of transcription factors. The modulation of cyclins and CKIs is primarily responsible for signaling to the RB pathway and thus G1 control by various cellular events. Even though the regulation of both G1 cyclins and CKIs is evolutionarily conserved, the RB pathway plays a critical role in the mammalian system. During the cell cycle, the mitogenic signals initiate the synthesis of D-type cyclins, which progressively activates cyclin-D-dependent CDK4 and CDK6, and subsequently induces the synthesis of E-type cyclins, which activates CDK2. Additionally, cyclin-D-CDK4/6 and cyclin-E-CDK2 cooperate to phosphorylate RB-family proteins, allowing transcription of E2F-target genes and thereby accelerating the G1/S transition. As a "braking" mechanism, cell cycle regulation is well coordinated in normal tissue growth, where CDK4 and CDK6 are particularly inhibited by INK4 proteins, while the p21 (CIP/KIP) family of CKIs inhibits various CDKs.

In a study by Cilibrasi and colleagues [107], a polyphenolic compound (resveratrol) had a significant antiproliferative and antimigratory effect in glioma stem cells, usually in a dose- and time-dependent manner. It was able to modulate the expression of Wnt signaling pathway-related genes, decreasing nuclear β catenin levels, and inducing a transcriptional upregulation of *MYC*. Rosarin et al. [108] reported that silver nanoparticles of polyphenols from a *Phyllanthus embla* extract reduced the cytotoxicity and proliferation of tumor cells in a Hep2 cell line in vitro.

10.2.2.4 Epigenetic Regulation

Epigenetics refers to the regulation of gene expression and subsequent alteration in biological functions without modifying the nucleotide sequence in the DNA. The main epigenetic programs that alter gene expression include ATP-dependent chromatin remodeling, DNA methylation, and modification of histone proteins such as acetylation, methylation, and phosphorylation. Generally, DNA methylation, catalyzed by DNA methyltransferases (DNMTs), such as DNMT1, DNMT3a and DNMT3b, results in the repression of genes by blocking the transcription start site and restricting access by RNA polymerases. Hypermethylation of tumor suppressor genes and hypomethylation of oncogenes is a common feature of cancer. Histone modification is another epigenetic driver of cancer. The main histone modifications include methylation and acetylation of histone tails. Histone acetylation is catalyzed by histone acetylases (HAT) that add acetyl groups to the lysine or arginine residues of the histone tail resulting in gene expression. Conversely, histone deacetylases (HDACs) remove acetyl groups from histone tails to repress the target gene. Histone methylation is much more complex, it can inhibit or enhance gene expression depending on the target amino acid site, the number of bound methyl groups, and the specific enzymes present in the lysine residue being methylated with how many methyl groups. Overall, epigenetic modifications involving the overexpression of oncogenes and suppression of tumor suppressor genes are the main mechanisms by which epigenetics drive tumorigenesis. For instance, hypoxia-induced downregulation of mixed-lineage leukemia 1 (MLL1) by histone methyltransferase (HMT) is associated with glioblastomas [109].

Scientists have proposed that nutritional supplementation containing folic acid and methionine may alter the epigenome in favor of reducing the disease burden. All flavonoids have a unique flavan nucleus in their structure, which is responsible for their inhibitory activities on DNA methylation enzymes. The major donor of the methyl group is s-adenosylmethionine (SAM) formed from methionine and folic acid; hence supplementation of these compounds in a diet may regulate DNA methylation. For example, kaempferol (Kae) is a natural flavonoid that is found abundantly in many fruits and green leafy vegetables. It suppresses DNA methylation in a murine model of bladder cancer [110]. Kae reduces de novo DNA methylation by selectively inhibiting the levels of DNMT3B protein through the ubiquitin-proteasome pathway [110]. In another study, Kae was found to reverse DNA methylation at the promoter region of dapper homolog 2 (DACT2) to inhibit nuclear β-catenin expression and suppress colorectal proliferation and migration in Wnt/β-catenin dependent manner [111]. Quercetin is involved in the inhibition of important epigenetic writers such as DNMT, HDAC, and HMT in human cervical cancer cells [112], resulting in enhanced apoptosis by inhibiting DNA demethylation, histone HDAC inhibition, and H3Ac and H4Ac enrichment in promoter regions of apoptotic genes. In acute myeloid leukemia cell culture experiments, quercetin eliminated the methylation effects of DNMT1 and DNMT3a [113], which increases apoptosis by DNA demethylation, HDAC inhibition, and H3Ac and H4Ac enrichment in promoter regions of apoptotic genes. Furthermore, gallic acid has been shown to suppress cell viability, proliferation, invasion, and angiogenesis in cancer. Gallic acid is found in all parts of plants such as leaves, fruits, roots, back, and seeds. In a lung tumor model experiment, gallic acid demethylates cyclins, such as CCNE2 and CCNB1 [114], to induce growth arrest. Additionally, gallic acid inhibits DNA methylating enzymes, including DNMT1, DNMT3A, and DNMT3B to repress the expression of tumor suppressors [115].

Additionally, epigenetics plays a significant role in the development, regulation, and maintenance of cancer stem cells (CSCs) [116] that are involved in chemoresistance, tumor dormancy, and tumor relapse. The signaling pathways that control these features are diverse and they include Wnt/β-catenin, Hedgehog, Notch, WNT/β-catenin, and TGF-β, as well as stemness factors such as Oct4, Sox2, Klf4, and Nanog [117]. The self-renewal and heterogeneous differentiation capacity of CSCs make them a key target in cancer treatment. Interestingly, unlike mutations, epigenetic alterations are generally reversible and have been regarded as potential therapeutic targets for cancer treatment. As a result, we can reverse CSC formation by erasing epigenetic marks that induce and maintain CSCs through several signaling pathways. Recent reports showed that curcumin inhibits the growth of CSCs in breast [118], pancreatic [119], brain [120], and liver [121] cancers. Further mechanistic studies showed that the anti-CSC effects of curcumin are mediated by increasing apoptotic proteins, such as Bcl-2 and Bcl-w, and suppress anti-apoptotic protein levels (Bax, Bak, Bad, Bik, and Bim) [122]. This suggests that co-administration of curcumin with conventional chemotherapy can reduce chemoresistance by eliminating a subset of tumor cells that otherwise confer immortality on tumors.

10.2.2.5 Antimetastatic Effects

Metastasis is one of the hallmarks of cancer that allows neoplastic cells to spread to secondary sites where they establish a novel metastatic niche. It is the most dreaded feature of cancer behavior, accounting for more than 90% of cancer-related deaths [123]. This is in part due to our limited understanding of the processes, lack of tools to predict the ability of cancer to spread to other sites, and limited therapeutic options. Several studies at different levels have investigated metastasis in terms of gene expression [124], changes in proteomics [125], extracellular matrix breakdown [126], angiogenesis [127], and immune evasion that encourages cell migration, invasion, and metastasis.

The complexity of cancer metastasis involves the interaction between numerous genes and signaling pathways. The initial step is characterized by the loss of epithelial traits and the acquisition of a mesenchymal phenotype during which they become motile in a process called epithelial–mesenchymal transition (EMT). Following their detachment from the basement membrane and degradation of the extracellular matrix by MMPs, such as MMP-2 and MMP-9, invasive cancer cells enter the blood vessels (intravasation) and on reaching the target tissue, extravasate to form a pre-metastatic niche, where they may either establish novel cancer immediately or remain dormant for a long time depending on the growth condition at the new site. To survive the circulatory stress, circulating tumor cells often express thrombin, cathepsin B, and procoagulant chemicals, suggesting that designing a specific target to these proteins may sensitize circulating tumor cells to T-cell cytotoxicity. Consequently, based on the available information, researchers are determined to develop effective and less toxic plant-based therapeutic agents halt tumor spread.

From the initial EMT program to the final colonization step, several signaling pathways influence, maintain, and ensure the successful seeding of metastatic cells at target sites. These pathways are discussed below, including some notable polyphenols that can inhibit limiting molecular candidates to prevent tumor invasion. The Ras/Raf/MAPK/ERK pathway is an adaptor for extracellular signals, such as growth factors, hormones, or oncogenic factors, allowing growth stimuli to be transmitted intracellularly. This pathway includes cell proliferation, differentiation, motility, and survival. The key molecular events involve receptor-ligand interaction, and activation of the rat sarcoma virus oncogene (RAS) via phosphorylation, which in turn recruits and activates the rapidly accelerated fibrosarcoma (RAF). Activated RAF ignites the MAPK and finally, activated MAPK phosphorylates extracellular signal-regulated kinases (ERK 1 and 2) become activated. This pathway is highly involved in metastasis by stimulating cell migration and inhibiting apoptosis [128]. Curcumin can disrupt this pathway by inhibiting EGFR, a receptor tyrosine kinase that initiates the Ras/Raf/MAPK/ERK pathway. This inhibition is conveyed in cancer cell proliferation, adhesion, migration, and differentiation [129].

The phosphoinositide 3-kinase (PI3K)/Akt/mTOR pathway is another key pathway that regulates cell division, migration, and survival. PI3K is activated via its recruitment to the growth factor receptors, which then generates phosphatidylinositol 3,4,5 triphosphate (PIP3) by phosphorylating PIP2. PIP3 then recruits the protein kinase, AKT, to the plasma membrane where it is activated by 3-phosphoinositide-dependent kinase 1 (PDK1) and the second mTOR complex, mTORC2. The effector molecule, AKT, regulates several genes involved in cell survival, proliferation, and motility while mTORC2 regulates actin cytoskeleton dynamics via protein kinase C alpha (PKC), GTPases, and focal adhesion proteins, promoting cellular motility and invasion. Alternatively, protein kinase B (Akt) maintains cellular viability by inhibiting pro-apoptotic proteins such as Bcl-2 and procaspase-9 [130]. Resveratrol induces autophagy and inhibits cell migration by suppression of the Akt/mTOR pathway in B16 melanoma cells [131, 132], as well as other malignant cell lines [133]. Phosphatase and tensin homolog (PTEN) are a PIP3-specific phosphatase and suppressor of the PI3K/Akt/mTOR pathway. Resveratrol suppresses prostate cancer by increasing PTEN levels [134]. In another study, fisetin, a flavonoid, inhibits triple-negative breast cancer cell line metastasis by suppressing the PI3K-Akt-GSK-3β signaling pathway, which leads to EMT reversion. Fisetin reduces lung metastasis and affects the expression of EMT molecules and PTEN/Akt/GSK-3β in a metastatic breast cancer xenograft model similar to in an in vitro model [135]. Moreover, in a cell-based experiment with colorectal cancer cell lines, luteolin suppresses cell migration and invasion by downregulating the expression of ECM degrading enzymes (MMP-2, MMP-3, MMP-9, and MMP-16) because of interference with the PI3K/AKT signaling pathway, as indicated by luteolin-induced suppression of phosphorylated Akt1 and PI3K [136]. Taken together, the pleiotropic actions of polyphenols on Ras/Raf/MAPK/ERK and PI3K/Akt/mTOR pathways suggest that when used in combination with other phenolic compounds or with a standard drug, it may increase the sensitivity of cancer cells to chemotherapy while also reducing toxicity.

The vascular endothelial growth factor (VEGF) pathway controls the formation of new blood vessels in response to hypoxic signals to nourish metastatic cancer. The binding of VEGF to VEGF receptor tyrosine kinase leads to receptor autophosphorylation and signaling cascade activation. In addition to angiogenesis, VEGFR-1 stimulates tumor cell proliferation, migration, and invasion by activating ERK-1/-2 and c-Jun NH2-terminal kinase (JNK). Additionally, activation of VEGFR-3 enhances cell survival, migration, and proliferation by inducing p42/p44, MAPK, and Akt signaling [137]. As a potential class for cancer-related angiogenesis, polyphenols offer a novel way to counteract VEGF-related cell signaling in cancer cells. Kae inhibits cancer neovascularization in human cancer cell lines by inhibiting VEGF release. In ovarian cancer cells, Luo and colleagues found that Kae dampens the expression of cMyc at both mRNA and protein levels through which VEGF is regulated [138]. Anthocyanin-isolated *Vitis coignetiae* Pulliat was found to suppress the NF-κB pathway, and its associated proteins such as COX-2, C-myc, MMP-2, MMP9, ICAM-1 and VEGF, thereby inhibiting invasion and angiogenesis in a breast cancer cell line [139]. This compound also blocks EMT by increasing the expression of E-cadherin.

The Wnt/−catenin pathway controls the expression of E-cadherin, which is a key regulator of EMT and by extension, metastasis. Wnt ligands bind to the Frizzled receptor, prevent proteasomal degradation of catenin, and regulate the expression of Wnt-related genes including E-cadherin and S100A4, an important member of the S100 family of proteins. Activated S100A4 promotes the Ras/Raf/MAPK/ERK pathway and triggers tumor metastasis, suggesting that S100A4 could be a promising target in cancer therapy [140]. This provides a mechanistic understanding of an earlier report that found that oral administration of green tea downregulated S100A4 expression and boosted E-cadherin expression in the prostate of TRAMP, resulting in a significant decrease in the ratio of S100A4 to E-cadherin expression in mouse prostate tissues [141]. In a parallel study, the prenylated flavonoid xanthohumol attenuated tumor cell-mediated breaching of the lymph endothelial barrier and inhibited tumor cell intravasation and metastasis by inhibiting the expression of ICAM-1, paxillin, S100A4, and selectin E, and decreasing the activity of MLC2, NF-jB, and CYP1A1 [44].

10.2.2.6 Immunomodulatory and Immunometabolic Effects

Cancer is characterized by chronic inflammation, which is also one of its hallmarks [142]. Although it is not fully clear how inflammation drives cancer development or promotes tumor growth, in tumor progression, it is common knowledge that the entire immune cell repertoire within the tumor microenvironment (TME) is profoundly altered over time [143]. Harnessing immune cells in cancer treatment is one of the leading cutting-edge methods currently available in the toolkit to fight cancer. Despite the opportunity provided by many immunomodulatory strategies, phytochemicals have received little attention.

It is becoming increasingly clear that polyphenols play a significant role in many of the desirable properties of phytochemical compounds. The most prevalent antioxidants in human diets are polyphenols, and the flavonoids, which include several compounds, are the largest and best researched class. A large body of evidence indicates that dietary polyphenols can affect several immune systems in the body. In tumor growth, polyphenols impact immune cells and the inflammatory state in the TME. Through the modulation of various cytokines, polyphenols control innate and adaptive immunity and have significant anti-inflammatory effects [144]. Additionally, polyphenols enhance dendritic cell (DC) activities, modulate macrophage function, promote B and T cell proliferation, and regulate Type 1 T helper (Th1), Th2, Th17, and Th9 cells [145]. These effects show that polyphenols play a direct role on the number and differentiation of some immune cells.

When male C3H/HeN mice are administered oral doses of polyphenols extracted from the fruit of dates, the lymphoid tissues show an increase in Th1, natural killer (NK), DCs, and macrophages [146]. Macrophages are one of the most prevalent leukocyte populations and can be broadly divided into M1 and M2 types. Tumor-associated macrophages (TAMs) possess tumor-promoting properties and express an M2-like phenotype in the tumor [147]. TAMs are important agents in cancer development, metastasis, and therapeutic resistance because of their anti-inflammatory phenotype [148]. Various phenolic substances have been found to modulate the immune system by inhibiting the signaling pathways for NF-kB and MAPK. The anti-inflammatory effects of several flavonoids have been related to the suppression of the NF-kB signal transduction pathway. Fisetin (7,3′, 4′-flavon-3-ol), a plant flavonol from the flavonoid group of polyphenols, exhibits anti-inflammatory activity in lipopolysaccharide (LPS)-treated macrophages by repressing NF-kB and MKP-1-dependent signaling pathways [149]. A more recent study also demonstrated that chrysin exerted potent properties by acting as an antagonist of NF-kB, which downregulated the production of iNOS and COX-2 [150].

Similarly, NF-kB has been associated with monocyte chemoattractant protein-1/chemokine CC motif receptor-2 (MCP-1/CCR2) expression in tumor-associated bone pain [151]. MCP-1 production controls the vicious cycle between tumor cells and TAMs (macrophages) that promotes the growth of tumors [152], in particular, the primary function of MCP-1 involves recruitment of monocytes to tumors and inflammatory sites [153]. MCP-1 is currently a molecular target for cancer treatment because its production in tumors promotes the accumulation of TAMs, which are tumor-promoting and immunosuppressive. Interestingly, another study also established that under LPS stimulation, a cocoa flavonoid-enriched extract and the monomers epicatechin and isoquercitrin reduce the expression of inflammatory molecules such as MCP-1 and TNF [154].

In addition, by activating the intestinal Treg cells, suppressing TNF-α, and triggering apoptosis, polyphenols reduce inflammation by reducing the pro-inflammatory cytokines in inflammatory bowel disease, which ultimately reduces DNA damage [145]. For instance, the proinflammatory cytokine IL-23, which is primarily produced by dendritic cells, macrophages, and neutrophils during intestinal inflammation, is a key player in the pathogenesis of both inflammatory bowel disease and colitis-associated cancer [155]. Moreover, inflammatory cytokines and chemokines aid in the survival of cancer cells by promoting tumor cell growth and interfering with their differentiation [156]. Taken together, this suggests that the TME, which harbors immune cells to tumor growth advantage, could be a primary target both for polyphenol preventive and therapeutic measures, particularly given that tumor cells develop a dependence on the inflammatory stroma. In this way, proinflammatory cytokines (like TNF, IL-6, and IL-1) and their transcription factors, including NF-kB and STATs, could be targets to develop anticancer therapies from polyphenolic compounds.

Additionally, some tumor cells can evade detection and progression to cancer by suppressing or evading immune surveillance. Cancer cells express antiphagocytic factors, such as CD47, MHCI, and CD24 [157], to avoid being engulfed by macrophages. In other words, cancer cells that express these signals are more susceptible to phagocytosis by macrophages when the signals are blocked. While there is little data regarding the effect of polyphenols on immune evasion, an important hallmark of cancer, literature shows that TME may contribute to immune evasion by modifying cancer metabolism (via the Warburg effect) that may interfere with the immune response. In this way, cancer cells drive an increased glycolytic rate and consequently increase the extracellular acidification rate [158], which ultimately reduces the pH. Cancer cells benefit from an acidic extracellular microenvironment in terms of proliferation, survival, metastasis, and signal transduction [159]. According to a report, macrophages play a novel role in identifying cancer cells and engulfing them through a process known as "Programmed Cell Removal" (PrCR) [160], which avoids induction of cell death and are therefore, are directly phagocytosed by macrophages. Yet, cancer cells frequently evade PrCR in established tumors and metastases by developing self-protective mechanisms, the best known of which is the upregulation of antiphagocytic signals to specifically inhibit PrCR [160].

In tumorigenic cells, a study shows that polyphenols more specifically target glycolysis, pro-oxidant pathways, oxidative phosphorylation, mitochondrial membrane potential regulation, and antioxidant (adaptive) stress responses [158]. Furthermore, polyphenols have been demonstrated to interrupt cancer cells during metabolic reprogramming [161]. When combined with chemotherapy, such metabolic reprogramming may have synergistic effects by targeting both the anabolic phenotype and mitotic division of proliferating tumor cells [158]. Meanwhile, interfering with glucose metabolism in cancer cells by downregulating glycolysis may result in a cascade of events on adaptive immune cells, modulating the immune response. Tumor cells are highly glucose-dependent because of the switch from aerobic respiration to glycolysis as the primary ATP-producing pathway. Because it produces enough NAD+ equivalents to maintain glycolysis, the conversion of pyruvate into lactate, by lactate dehydrogenase (LDH), is crucial in tumor cells [162]. LDH is inhibited by polyphenols [163, 164]. The effect of polyphenolic compounds on metabolic alteration in tumor cells has been demonstrated in many in vitro and in vivo studies. Galloflavin and urolithin M6, an ellagitannin metabolite produced by gut microbes, were shown by Rupiani et al. [164] to inhibit the human LDH-A cell line.

Lowering glucose consumption, as shown in Figure 10.1, typically lowers the pH level in the TME, increases IFN-ɣ, downregulates HIF-1α, and decreases the number and activity of myeloid-derived suppressor cells (MDSCs), increasing T-cytotoxic cells. While the effect of polyphenolic compounds on tumor immune evasion is so far underexplored, evidence from immunometabolic alterations suggests that these compounds are potential immunotherapeutic agents that could benefit cancer patients. More importantly, TME promotes tumor heterogeneity. The heterogeneous nature of various cancer cell phenotypes, which causes tumor relapse, is a major contributor to cancer treatment resistance. It has repeatedly been demonstrated that polyphenols have a multifaceted strategy for addressing cancer hallmarks, with a strong emphasis on TME. Examining the therapeutic potential of phytochemicals, especially with regard to polyphenols, may open up a new avenue for novel cancer treatment approaches.

10.2.2.7 Microbiome Interactions

Various microorganism species, including bacteria, yeast, and viruses, make up the gut microbiota. However, *Firmicutes*, *Bacteroidetes*, *Actinobacteria*, *Proteobacteria*, *Fusobacteria*, and *Verrucomicrobia* are the predominant gut microbial phyla, and the main phyla, *Firmicutes* and *Bacteroidetes*, account for 90% of the gut microbial community [165]. The gut microbiota

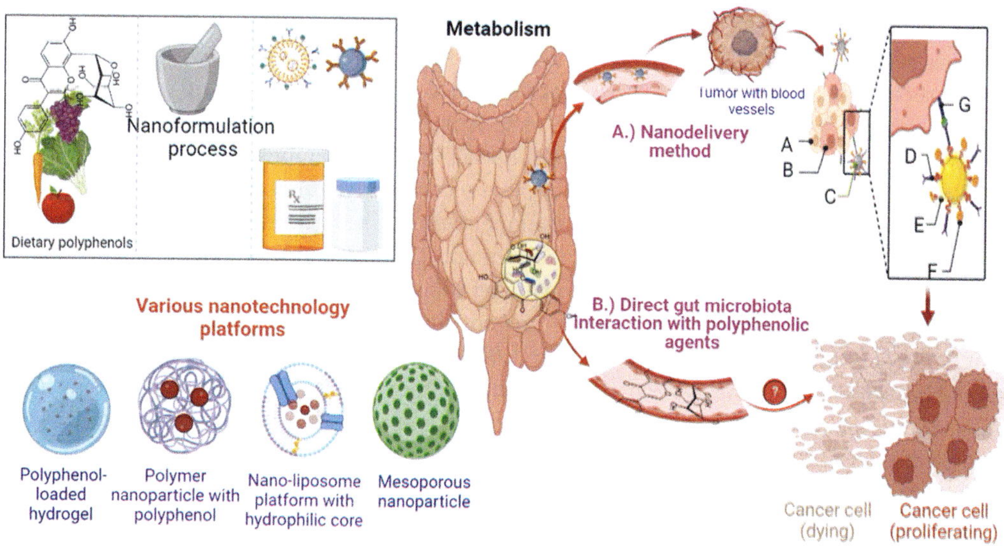

Figure 10.1 Active targeting of nanoformulated polyphenolic agents to cancer cells. A: Differentiated tumor cell; B: Cancer stem cell; C: Nanoformulated polyphenolic agents; D: Targeting moiety; E: Linker chain; F: Chemotherapeutic arm; G: CSC marker: Question mark (?): Potential reduction of the active agents, and possible therapeutic inefficacy (debatable among several factors).

functional composition, variation, and diversity (across a range of species) at any particular time are crucial for human health and the development of many diseases including cancer, obesity, infection, and neurological diseases. The interaction between polyphenols and the gut microbiota can have an effect on the microbial ecology in the gut. Additionally, polyphenol compounds may affect the gut morphology and integrity, which may preclude the growth of gut cancers such as gastric and colorectal cancers [166]. Recent reports have shown various roles that polyphenols play in the anticancer arsenal and the impact these have on the microbiome [167]. Polyphenols have been linked to a lower risk in many cancers, including lung, prostate, and colon cancers, through their activities on many cancer hallmarks [1]. Unfortunately, their roles in cancer prevention through microbiome modulation remain elusive. Investigation of various biological and molecular mechanisms by which polyphenols act as anticancer preventive and/or therapeutic agents may lead to the development of novel anticancer strategies. As a result, it is critical to explore various mechanisms of polyphenol-preventive anticancer activity through the gut microbiome.

Helicobacter pylori is neither suppressed or eradicated by antibiotic treatment; instead, antibiotic resistance can develop [68]. *H. pylori* is linked to severe and chronic stomach and duodenum diseases, such as peptic ulcer, non-cardiac adenocarcinoma, and gastric lymphoma, making it a dreadful cancer-causing bacterial pathogen [168]. The spiral-shaped bacterium coats the inside of the gastric mucus layer to increase the risk of stomach cancer. Despite the fact that immune cells that typically recognize and attack invasive bacteria amass close to the sites of *H. pylori* infection, they are unable to get to the protective niche of *H. pylori* inside the stomach. Moreover, evidence shows that *H. pylori* persists in its biome despite a strong immune response by the gastric mucosa, encouraging conditions like chronic gastritis and ultimately, adenocarcinoma [169]. Worse still, the organism secretes a toxigenic protein, vacuolating cytotoxin A (VacA). Interestingly, research shows that the polyphenols in red wine and green tea effectively block the VacA, which is the primary toxin produced by *H. pylori* [170]. VacA is a key virulence factor of the common stomach bacteria *H. pylori*. By forming anion-selective, urea-permeable channels in the plasma and endosomal membranes, it causes cell vacuolation and tissue damage. The ability of VacA toxigenicity to induce cell vacuolation and damage stomach tissue architecture by causing programmed cell death has been reported [171].

Many dietary polyphenols inhibit urease activity, affect bacterial growth and cell membrane integrity, making bacteria more susceptible to xenobiotics like antibiotics, and cause the collapse of the proton motive force through loss of H+ ATPase and membrane-associated activities in their inhibition of *H. pylori* [68]. According to a recent study, curcumin oxidative products, cyclobutylcyclopentadione and dihydroxy cyclopentadione, have been shown to have a high affinity for the active site 5 of the p55 protein domain (a fragment of the p88 toxigenic VacA protein) [172], which is essential for VacA-mediated vacuole formation. This raises the possibility that some polyphenolic compounds can prevent the growth of cancer. While many treatment strategies fall short in their ability to completely limit or eradicate *H. pylori* from resurfacing, polyphenols can synergistically create the most successful treatment strategy and so halt the development of the non-cardia adenocarcinoma, gastric lymphoma and other gut associated cancers.

Additionally, the mutualistic relationship between gut bacteria and polyphenols promotes the biotransformation of the compounds into their absorbable metabolites, which increases the bioavailability of the compounds [173, 174]. The bioavailability and bioactivity of metabolites may vary depending on the composition of the gut microbiota between individuals. These variations are linked to various metabotypes, which are determined by a person's ability to generate particular metabolites [175]. Flavonols (such as Kae, quercetin, and myricetin), isoflavones, flavanones (naringenin, hesperidin), flavan-3-ols, and proanthocyanidins are a few examples of polyphenols that have been reported to be metabolized by various gut microbiota [176]. Due to the cleavage of the C ring, gut bacteria extensively hydrolyzes flavonols into their metabolite-derivative products at the A and B rings. Phloroglucinol, 3-(3,4-dihydroxyphenyl) propionic acid, and 3-(3-hydroxyphenyl) propionic acid are made from the A ring of phloroglucinol, whereas 2-(3,4-dihydroxyphenyl) acetic acid, 2-(3-hydroxyphenyl) acetic acid, and 3,4-dihydroxybenzoic acid are produced by quercetin from the B ring [173].

Intriguingly, molecular studies of mechanisms of anticancer products have shown that many of these metabolites are helpful as anticancer chemopreventive and therapeutic agents. Of particular interest, an in vivo experiment on HeLa cells shows that 3,4-dihydroxyphenyl acetic acid inhibits mitogenic signaling by suppressing EGF-induced phosphorylation of EGFR and its downstream signal molecules [177], suggesting its roles in the treatment of cervical cancer. Gut bacteria can modify the metabolic pathways of polyphenols, possibly leading to reduced levels of cancer-causing chemicals in the gut. There is still a lot to understand about how gut microbiota affects polyphenols and their anticancer properties. However, studies examining the relationships between gut microbiota, polyphenols, and cancer are still in progress and are anticipated to make a significant contribution to our understanding of this crucial subject.

Additionally, while polyphenols have been regarded as a promising anticancer therapeutic agent because of their chemopreventive effects, several issues, including adequacy of their bioavailability to attain effective cytotoxic concentrations, have made their practical application difficult. One of the best ways to increase efficacy and reduce the hormetic effect is through novel formulation strategies such as nanoparticle-based delivery systems.

10.3 Strategies for the Nanodelivery of Polyphenols as Nutraceuticals for Cancer

Although polyphenolic compounds have been shown to have beneficial effects, there are few human applications for them in therapeutics. Their effects in vivo are limited to the effects of natural polyphenols in vitro. This disparity is often caused by inconsistent and lower amounts of polyphenols in vivo compared with the effective concentrations derived from the in vitro findings or their poor oral bioavailability [178, 179]. Notably, polyphenols bind with salivary proteins that are high in proline to produce insoluble complexes. Polyphenols are subjected to extensive first-pass phase II metabolism, primarily glucuronidation, sulfation, and methylation in the small intestine, and are then are metabolized in the liver. These conditions, along with exposure to stomach acid and the alkaline environment of the small intestine, are detrimental to their stability [180]. Significant modifications to the polyphenol structure and biological activity result from these metabolic processes. As a result, the forms that can enter the blood and tissues differ greatly from those found in consumed food [178, 180]. Other factors influencing polyphenol bioavailability are their poor water solubility and cell-membrane permeability [178].

To improve the solubility, chemical stability, and permeability rate of polyphenolic compounds, a variety of techniques have been employed. Among these methods are the suppression of phase I and phase II metabolism enzymes to stop biotransformations, the addition of agents that can promote solubility, the use of chemical additives to maintain the structure of the compound, and the addition of complementary molecules (lipids or proteins) [180]. Recently, the use of nanoparticle-based delivery systems has emerged as a cutting-edge and effective method to increase polyphenol bioavailability [7, 178, 179]. The focus of this subsection is to provide relevant knowledge regarding the application of nanodelivery techniques to improve the delivery of polyphenols as nutraceuticals for cancer therapy. It focuses on nanoparticle-based delivery systems.

At Tokyo Science University, Norio Taniguchi coined the term "nanotechnology" for the first time in 1974 [181]. In pharmaceuticals or formulations, the prefix "nano" designates the drug's particles that range in size from 10 nm to 1000 nm. However, the lymphatic system is likely to be stimulated, causing particles 200 nm or larger to be expelled through blood circulation [181]. Consequently, nanoparticles with diameters between 10 nm and 200 nm have the desired properties for effective drug delivery [182, 183]. In contrast to bigger particles, nanoparticles can move more freely within the human body because of their distinct structural, electrical, chemical, mechanical, biological, and magnetic properties. Therefore, nanoformulations are widely utilized to transport bioactive components to target tissues in a more precise and controlled manner [184, 185]. The use of nanotechnology to improve the efficacy of plant extracts has been extensively discussed because nanostructured systems could improve selectivity and efficiency, protect against chemical, thermal, and photodegradation, promote the sustained release of active ingredients, lower dose requirements, and reduce side effects [186, 187]. Depending on the features of the active ingredient and the targeted organ, nanoparticles are created in a variety of sizes, compositions, shapes, and functionalities and are physiochemically modified to produce desired characteristics.

Polyphenols can be encapsulated by nanocarriers via a variety of mechanisms, including ionic condensation, eutectic, ionic gel, emulsification, encapsulation, yeast encapsulation, and freeze drying. These encapsulating techniques can enhance polyphenol targeting, GI environmental stability, water dispersion, and sustained release properties [188]. Biodegradable and biocompatible nanostructures (as shown in Figure 10.1), including liposomes, solid lipid nanoparticles, phytosomes, polymeric nanoparticles, micelles, protein-based nanocarriers, and dendrimers, are the most often used nanoparticle delivery vehicles for polyphenol delivery [7, 179, 181, 186, 188, 189]. They are briefly detailed here. These nanoparticles can be administered in a variety of ways, including orally, topically, intravenously, and intraperitoneally.

A: Differentiated tumor cell; B: Cancer stem cell; C: Nanoformulated polyphenolic agents; D: Targeting moiety; E: Linker chain; F: Chemotherapeutic arm; G: CSC marker: Question mark (?): Potential reduction of the active agents, and possible therapeutic inefficacy (debatable among several factors).

10.3.1 Liposomes

Liposomes are sphere-shaped, biodegradable phospholipid bilayer vesicles with an aqueous core. They comprise one or many concentric lipid bilayers that are separated by an aqueous phase. Liposomes exhibit many of the characteristics of naturally membrane-bound structures because the lipids used are primarily glycerophospholipids and/or sphingophospholipids [190]. As a result of their well-organized structure, hydrophilic drug molecules can be loaded into the aqueous section of liposomes, whereas lipophilic molecules can be incorporated into the membrane [178]. Lamella count, size, and surface charge are used to categorize liposomes. Based on their surface charge, liposomes are classified as anionic, cationic, or neutral. Liposomes can be categorized as oligo, uni-, or multilamellar, as well as small, large, or giant depending on their form, size, and number of lamellae. Small unilamellar liposomes with diameters of roughly 25–100 nm, large unilamellar liposomes with diameters of 100 nm–1 μm, and very large unilamellar liposomes with diameters greater than 1 μm are different sizes of unilamellar liposomes that contain a single bilayer. Multilamellar liposomes are made up of more than five lamellae and frequently have an onion-like structure, whereas oligolamellar liposomes are vesicles made up of 2–5 concentric lamellae [7, 186, 187, 191].

Several techniques have been thoroughly examined for the manufacture of liposomes, some of which include the thin film hydration method, reverse phase evaporation, solvent injection, and detergent removal [192–195]. The lipid composition, the sensitivity of the material to be encapsulated, and the desired morphology are usually considered when choosing a technique. Additionally, the same kind of liposomes made using different methodologies may exhibit different properties in vitro and in vivo, including bilayer permeability, encapsulation efficacy, and vesicular stability. The use of liposomes as drug delivery vehicles has a significant impact on the pharmacokinetics and biodistribution of the compound. This has the potential to enhance treatment outcomes while reducing undesirable drug-related effects in several situations. Liposomes can be made into lyophilized powder for reconstitutions, suspensions, creams, lotions, or aerosols, making them essentially suitable for all routes of administration into the body. However, a significant number of liposomal medicinal applications are parenteral [191].

Numerous studies and reviews on polyphenol and other phytoconstituent-based liposomal formulations have reported advances in their aqueous solubility, stability, bioavailability, efficacy, and capability to offer targeted and/or sustained release [7, 179, 181, 186, 190]. Additionally, polyphenols are known to impart an undesirable taste, such as astringency, when they are added to nutraceuticals. They may also interact with food ingredients, such as proteins, leading to significant aggregation and precipitation as well as a loss in the polyphenol quantity and/or functionality. Encapsulation via liposome formulation may address these issues and increase the content of antioxidants in food products to enhance their antioxidant effects and shelf life [190].

Niosomes and polymersomes are two structures that are similar to liposomes. Niosomes (also known as nonionic liposomes) are made in the same ways as liposomes from mixtures of diacyl or monoacyl polyglycerol or polyoxyethylene-based amphiphiles and cholesterol. Polymersomes are polymer shells that self-assemble from block copolymer amphiphiles [191].

10.3.2 Solid lipid Nanoparticles

Solid lipid nanoparticles (SLNs) are colloidal nanocarriers that are sub-micron (50–1000 nm) in size and comprise physiologically acceptable lipids distributed in an aqueous solution containing surfactants. They are made from solid lipids, or combinations of them, and are stabilized by surfactants. The capacity to target pharmaceuticals, preserve the integrated drug from degradation, increase bioavailability, load both lipophilic and hydrophilic compounds, and their relative ease for large-scale manufacturing are just a few benefits that SLNs can provide for drug administration. Furthermore, SLNs normally entrap the lipophilic compound in a stable form without the need for hazardous chemical solvents, resulting in low toxic effects and acceptable tolerance. Alternatively, they are frequently characterized by a low drug loading capacity and potential stability issues during long-term storage (aggregation, component degradation, tendency to form a gel) [7, 191]. In SLNs, drug payload is delivered in two ways. First, the drug can be incorporated into the polymeric core, and then it can be attached to the surface of the polymeric component. SLNs are made using a variety of methods such as solvent displacement, phase inversion, emulsification-solvent evaporation, emulsification-solvent diffusion, and ultrasonication [187]. The following methods can increase the oral bioavailability of the medicinal compounds [191]:

- Drug protection against gastric fluid and enzymatic breakdown
- Transportation into the lymphatic system avoiding the first-pass metabolism
- Lipids can increase the absorption of insufficiently aqueous-soluble compounds
- Increased intestinal tract residence time because of the adhesion of nanoparticles to the intestinal wall
- Effects of surfactants that may increase the permeability of the intestinal membrane or promote the binding between lipid particles and the membrane

Many studies have shown that SLNs can successfully be used to enhance the physicochemical characteristics of various polyphenolic substances; for example, a study by Jourghanian et al. [196] showed prolonged release and increased antibacterial characteristics of curcumin when formulated in SLNs. Additionally, the in vitro cytotoxicity of emodin (formulated as SLNs) against human breast cancer was greatly enhanced [197], while the SLNs of resveratrol showed a significant reduction in mitochondrial ROS generation, lipid peroxidation, and protein carbonyls [198]. In a review by Teja et al. [181], 26 SLNs of different phytoconstituents, some of which were polyphenols, were reported to have shown improvement in different physicochemical characteristics and enhanced release and/or bioavailability profiles.

10.3.3 Self-Emulsifying Drug Delivery Systems (SEDDS) and Nanoemulsions

Self-emulsifying drug delivery systems (SEDDS) are micro/nanoemulsion preconcentrates made up of a drug, surfactant, oil, and co-surfactant. The pre-concentrate spontaneously emulsifies to yield a fine oil-in-water emulsion with smaller globule sizes (< 100 nm) when placed in an aqueous medium with gentle agitation. The agitation required for self-emulsification in vivo is produced by the digestive motility of the stomach and intestine [199, 200]. Throughout its passage through the digestive system, SEDDS can present and keep the drug in a dissolved state. SEDDS are an effective method for delivering hydrophobic medications orally because of this characteristic [201]. In addition to having poor aqueous solubility, stomach acid is known to degrade polyphenolic substances [178]. SEDDS can emulsify into nanodroplets, which will offer gastroprotection to the entrapped drug solution and thus prevent contact between the polyphenols and stomach acid. Co-administration of medicinal substances with lipids increase the formation of triglyceride-rich lipoproteins in the enterocyte, which are then transported from the intestine via the intestinal lymph, avoiding the first-pass metabolism. Furthermore, lipids and the other SEDDS ingredients (co-solvents and surfactants) increase intestinal permeability by inhibiting efflux transporters, loosening tight junctions, and enhancing trans-cellular permeability and membrane solubilization [201]. All of these characteristics will limit the physicochemical and metabolic activities that negatively impact polyphenol bioavailability, thereby enhancing effective delivery. As an innovative, commercially viable lipid-based delivery system, SEDDS have attracted a lot of attention because they have the potential to increase the oral bioavailability and therapeutic efficacy of a variety of medicinal compounds [201].

Numerous studies have demonstrated how well SEDDS or nanoemulsions can be used to improve the physicochemical characteristics of different polyphenolic compounds. For example, polyphenolic compounds, such as resveratrol, silymarin, carvacrol, curcumin, quercetin, and zedoary essential oil, have been developed as SEDDS formulations to improve their solubility, stability, permeability, bioavailability, and respective pharmacological effects [101, 105, 202–204].

The general procedure for producing SEDDS is as follows: first, the drug's solubility in various oils, surfactants, and co-solvents is assessed; second, the results are used to select the appropriate oils, surfactants, and co-solvents. A phase titration study will be conducted by Tripathi et al. [205] and the phase diagram plotted. The quantity of oil, surfactant, and co-surfactant (determined from the phase diagram) is then combined by first mixing the oil and surfactants (Smix) at 50°C–60°C, then dissolving the drug in the Smix retained at 50°C until a clear solution is achieved. Finally, the co-surfactant is added. As previously stated, the micro/nanoemulsion is formed in vivo upon dilution in the GI fluids, followed by mild agitation provided by the digestive motility of the stomach and intestine. Alternatively, methods such as microfluidization, high-pressure homogenization, and solvent evaporation could be used to form nanoemulsions in vitro [181]. The majority of SEDDS are liquids or semisolids at ambient temperature. However, lipid-based formulations can be converted into solid forms using a range of processes, including adsorption onto solid carriers, spray drying, melt granulation, melt extrusion, freeze drying, and solvent evaporation [201].

10.3.4 Phytosomes

Phytosomes (phyto–phospholipid complexes) are a complex of phospholipids and natural active phytochemicals, bound in their structures, obtained by the reaction between phosphatidylcholine (or any hydrophilic polar head group) and plant extracts in aprotic solvent. Polyphenols are ideal candidates for a phytosomal drug delivery system because only compounds with an active hydrogen atom (–COOH, –OH, –NH2, –NH, etc.) can be incorporated into a phytosome structure. To form a phyto–phospholipid complex, the hydroxyl groups of polyphenols would bind with the nitrate and phosphate groups of phospholipids [206]. The stability, pharmacokinetics, and pharmacological properties of components are significantly enhanced after the formation of the phospholipid complex [207]. The antisolvent precipitation technique, rotary evaporator method, salting-out technique, and freeze drying co-solvency technique, are just a few of the methods for making phytosome [206]. In many studies and reviews, the improvements introduced by various polyphenols and other phytochemical-based phytosome formulations have been well reported [181, 207].

10.3.5 Polymeric Nanoparticles

Polymeric nanoparticles are nanosized biocompatible and biodegradable polymeric matrices that function as drug delivery systems by directing drug release to certain sites. Vesicular systems (nanocapsules) and matrix systems (nanospheres) make up a large class of polymeric nanoparticles [208, 209]. Nanospheres are systems in which the drug is dispersed throughout the polymer matrix, whereas nanocapsules are systems in which the drug is restricted to a cavity and enclosed by a specific polymeric membrane [210].

In addition to being stable, non-toxic, non-thrombogenic, non-immunogenic, and non-inflammatory, polymeric nanoparticles also avoid the reticuloendothelial system (RES) and do not activate neutrophils. They can improve a drug's bioavailability and absorption of the active ingredients, boost component solubility, provide the possibility for controlled release, shield compounds with biological activity from the environment, and lower the therapeutic dose. A high concentration of pharmacological substances may also be delivered to the area of interest via polymeric nanoparticles that have been designed for specificity. Large quantities of polymeric nanoparticles may be produced easily and affordably using a variety of techniques [186, 209, 210]. The most commonly used polymers include poly-D, L-lactic acid (PLA), poly-D, L-lactic-co-glycolic acid (PLGA), poly (ε-caprolactone) (PCL), and some naturally occurring polymers such as chitosan, albumin, and alginates. Several techniques have been used to create polymeric nanoparticles depending on their intended use and the physicochemical properties of the drug. The most frequently used techniques for preparing nanospheres are solvent evaporation, emulsification/solvent diffusion, nanoprecipitation, and emulsification/reverse salting-out, while nanoprecipitation is the commonly used method for preparing nanocapsules. Xie et al. [211] and Zielińska et al. [209] reported 640-fold increases in aqueous solubility and 5.6-fold increases in oral bioavailability by curcumin PLGA nanoparticles, respectively. Studies by Manikkam and Pitchai [212] and Singh et al. [213] have demonstrated increased stability of catechins when formulated as polymeric nanoparticles.

10.3.6 Micelles

Micelles are self-assembling, 5–100 nm-diameter colloidal nanoparticles. They comprise molecules called amphiphiles, which spontaneously assemble in an aqueous medium at a specific concentration and temperature levels. When the surfactant concentration exceeds the critical micelle concentration, micelles form [178]. The threshold surfactant concentration necessary for the self-aggregation process is the critical micelle concentration. A reduction in free energy is, according to theory, what causes micelles to develop. The removal of hydrophobic fragments from the aqueous environment re-establishes the hydrogen bond network in water, which decreases the free energy of the system and results in the formation of micelles [214]. Drugs can be loaded into micelles using a dialysis method, oil-in-water emulsion solvent evaporation method, or solid dispersion method. Other methods are direct dissolution, complexation, chemical conjugation, and various solvent evaporation procedures [214]. The micelles drug delivery system has several advantages, including enhanced drug solubility, circulation, tissue permeability, targeted drug delivery, reduced toxicity, and a simple preparation method [181]. There are different studies that have shown the successful deployment of micellar drug delivery systems to improve the stability, bioavailability, and cell permeability of phenolic compounds such as curcumin [215], silymarin [216], and *Arctium lappa* [197]. Teja et al. [181] showed that many studies have utilized polymeric nanoparticulate drug delivery systems to improve the physicochemical properties of many phenolic compounds and other phytoconstituents.

10.3.7 Protein-based Nanocarriers

Proteins are a suitable carrier for the encapsulation and delivery of medicinal compounds because of their distinctive molecular structure, which offers a wealth of functional groups, high affinity hydrophobic binding sites, binding ligands, and bioactive chemicals. They have a high nutritional value, low toxicity, biocompatibility, and biodegradability, and they are non-antigenic [188]. They have a high clearance rate and low rates of opsonization by the RES. Protein-based nanocarriers are able to establish a variety of interactions within the therapeutic molecules, leading to the development of 3D networks, which, in turn, reverses the binding of active molecules, ensuring their protection and targeted action. This is made possible by the presence of numerous functional groups in the structural sequences of polypeptides [188]. To boost the bioavailability of polyphenolic compounds, they can be prepared in a variety of ways (nanoparticles, thin films, hydrogel fibers, etc.) and are modified or compounded to enhance their surface functioning [179]. Proteins, including human serum albumin, collagen, gliadin, gelatin, and silk fibroin, can be converted into protein nanoparticles via chemical processes (like emulsion), physical processes (like electrospray), self-assembly processes (like desolvation), and other processes [188, 217]. A variety of data are available on the use of protein-based nanocarriers for the delivery of polyphenols [160, 179, 218–221]

10.3.8 Dendrimers

Dendrimers are radially symmetric, nano-sized structures with a well-defined, homogenous, and monodisperse structure with tree-like arms or branches [222]. Dendrimers are fabricated from monomers using either convergent or divergent step growth polymerization [181, 210]. With dendrimers, a level of control in preparation is possible that is not possible with most linear polymers, resulting in macromolecules that are almost monodispersed, spherical, and have a lot of peripheral groups [222]. The ability to control the number of branches in these polymer-based nanoparticles allows them to be manufactured in very small sizes (1–5 nm). A spherical polymerization process that creates cavities inside the dendrimer molecule can be used to construct the dendrimers. Additionally, dendrimers have a free end group that can be easily modified to enhance the delivery of medicinal compounds to specific targets [223]. With the possibility to adapt the carrier to the unique requirements of the active material and its therapeutic applications, the bioactive substances may be encapsulated inside the interior of the dendrimers or physically adsorbed/chemically attached to the dendrimer surface [224]. By forming covalent bonds and engaging in host–guest interactions, dendrimers have the rare capacity to entrap high molecular weight hydrophobic and/or hydrophilic phytoconstituents. They also have a large surface area, which is advantageous for drug release and trapping. These characteristics make it the perfect carrier for the targeted delivery of herbal extracts [181]. Formulations of curcumin [225] and quercetin [226] dendrimers have been reported to enhance the anticancer activity of the hydrophobic curcumin molecule and improve the anti-inflammatory activity of quercetin, respectively, compared with pure compounds.

10.3.9 Formulation Challenges, Chemical Functionalization, and Characterization

The chemical classes, solubility, molecular weight, and medicinal potential of components originating from plants, such as polyphenols, vary widely. They could have partial aqueous or lipid solubility with partial polarity, median polarity, or non-polarity. These differences make the design and characterization of the nanoformulation strategy challenging. These extracts can be fractionated using bioassays to produce fractions with a similar solubility profile and/or chemical class, which offers a solution to these issues [181].

Nanoparticles have many benefits; however, there are some disadvantages as well. After systemic delivery, nanoparticles, for example, may interact with proteins, substrates, and other compounds moving through the bloodstream, all of which may reduce their therapeutic efficacy [178]. These proteins significantly alter the characteristics of nanoparticles, leading them to be rapidly cleared from the bloodstream by RES macrophages, which are mostly found in the spleen and liver. Functionalizing the nanoparticle surface with various hydrophilic surfactants, such as polyethylene glycol (PEG), polysorbates 80, or different types of ligands (such as aptamers, proteins, or antibodies against specific endothelial receptors), can address these limitations. The most common surfactant, PEG, for instance, forms a steric barrier on a nanoparticle surface that slows the opsonization process and, as a result, immune system clearance. A protective polymer can be physically adsorbed on the surface of a polymeric nanoparticle or chemically bonded to a particle to accomplish surface functionalization. The targeted delivery of the needed molecule to brain cells is made possible by nanoparticle functionalization with ligands (such as antibodies) specific to the BBB cell surface characteristics. Chemical conjugation of the ligand to premade nanoparticles is the current approach for adding ligands to the nanoparticle surface [178, 223].

The produced nanocarriers have a high tendency to be morphologically polydisperse despite following the correct methodological flow, which could possibly influence their intended or desired activities. As a result, characterization techniques are crucial for studies involving nanoparticles because they can give us the knowledge to comprehend and ascertain the physicochemical characteristics of manufactured products. The most often employed techniques are surface area analysis, droplet/particle size analysis, polydispersity index, drug loading capacity, release rate, cell permeability, the percentage composition of the polymer, surfactant, cryoprotectant, zeta potential, nuclear magnetic resonance analysis for complex formation, and transform infrared spectroscopy. These are all crucial process parameter and optimization approaches [181, 217].

10.4 Summary and Perspectives

Evidence suggests that polyphenol-enriched diets, which are found in fruits and vegetables, can reduce the risk of many malignancies. Additionally, all phenolic compounds can function as supportive therapy to slow the growth of several human cancer types. As a result, the use of polyphenolic compounds for anticancer drug discovery has received special attention because of their ubiquity and unique interactions with disease causing proteins. The anticancer mechanisms of polyphenols cover a wide range of tumor hallmarks, including anti-apoptosis, antimetastasis, immunomodulation, and anti-angiogenesis. However, recapitulating these impressive in vitro activities of polyphenols in animal model experiments remains a challenge because of their GI instability.

As the application of nanotechnology in the delivery of bioactive ingredients is becoming increasingly recognized, coupling of polyphenols with nanocarriers for effective drug design and development has been optimized. Nanodelivery systems safeguard polyphenols from premature degradation, promote tissue specific delivery for prolonged pharmacological actions and lessen off-target effects. Because polyphenols display a tight peptide binding affinity and nanoparticles can safely deliver them intracellularly, they are considered promising anticancer agents that may inhibit intracellular oncogenic peptides such as the aryl hydrocarbon receptor, PI3K, MAPK, JAK/STAT and other oncogenic scaffolds. In addition to the existing polyphenols that have shown potent anticancer effects in vitro, the structure–activity relationships may be employed to elucidate other phenolics with potent anticancer activities that may be useful for advanced tumors. The use of nanoparticle-based agents for clinical therapy faces significant barriers. The uncertainty in the interactions of nanomaterials with biological systems is of great concern and is impeding the massive commercialization of these technologies. There is a need for further clinical trials of nanoparticulate drug delivery to examine the desired biological effectiveness and evaluate biocompatibility to conform with regulatory guidelines.

References

1 Adedokun, K.A., Imodoye, S.O., Bello, I.O. et al. (2023). Therapeutic potentials of medicinal plants for the treatment of cancer and significance of computational tools in anticancer drug discovery. In: *Phytochemistry, Computational Tools, and Databases in Drug Discovery*, 1e (ed. C. Egbuna, M. Rudrapal, and H. Tijjani), 393–455. Amsterdam: Elsevier. ISBN: 9780323905930.

2 Teiten, M.H., Gaascht, F., Dicato, M., and Diederich, M. (2013). Anticancer bioactivity of compounds from medicinal plants used in European medieval traditions. *Biochemical Pharmacology* 86 (9): 1239–1247.

3 Adedosu, O.T., Adeleke, G.E., Imodoye, S.O. et al. (2016). Aqueous extract of Hibiscus sabdariffa calyx showed antioxidative and ameliorate acetaminophen (paracetamol)-induced hepatotoxicity and nephrotoxicity in Rats. *Nature Science* 14 (8): 28–37.

4 Olejnik, A., Kowalska, K., Olkowicz, M. et al. (2015). Anti-inflammatory effecs of gastrointestinal digested Sambucus nigra L. fruit extract analysed in co-cultured intestinal epithelial cells and lipopolysaccharide-stimulated macrophages. *Journal of Functional Foods* 19: 649–660.

5 Egbuna, C., Patrick-Iwuanyanwu, K.C., Onyeike, E.N. et al. (2023). Wnt/β-catenin signaling pathway inhibitors, glycyrrhizic acid, solanine, polyphyllin I, crocin, hypericin, tubeimoside-1, diosmin, and rutin in medicinal plants have better binding affinities and anticancer properties: Molecular docking and ADMET study. *Food Science and Nutrition* 11 (7): 4155–4169. https://doi.org/10.1002/fsn3.3405.

6 Egbuna, C., Patrick-Iwuanyanwu, K.C., Onyeike, E.N. et al. (2023). Phytochemicals and bioactive compounds effective against acute myeloid leukemia: a systematic review. *Food Science and Nutrition* 11 (7): 4191–4210. https://doi.org/10.1002/fsn3.3420.

7 Rudrapal, M., Khairnar, S.J., and Khan, J. (2022). Dietary polyphenols and their role in oxidative stress-induced human diseases: insights into protective effects, antioxidant potentials and mechanism (s) of action. *Frontiers in Pharmacology* 13: 283.

8 Zhang, H. and Tsao, R. (2016). Dietary polyphenols, oxidative stress and antioxidant and anti-inflammatory effects. *Current Opinion Food Science* 8: 33–42.

9 Anand, S., Sowbhagya, R., Ansari, M.A. et al. (2022). Polyphenols and their nanoformulations: protective effects against human diseases. *Life* 12 (10): 1639.

10 World Health Organization (2022). *Cancer Geneva: World Health Organization*. https://www.who.int/news-room/fact-sheets/detail/cancer (accessed 27 June 2022).

11 Strugała, P., Loi, S., Bazanów, B. et al. (2018). A Comprehensive study on the biological activity of elderberry extract and cyanidin 3-O-Glucoside and their interactions with membranes and human serum albumin. *Molecules* 23: 2566.

12 Aiello, P., Consalvi, S., Poce, G. et al. (2021). Dietary flavonoids: nano delivery and nanoparticles for cancer therapy. *Semin Cancer Biology* 69: 150–165.

13 Sadhukhan, P., Kundu, M., Chatterjee, S. et al. (2019). Targeted delivery of quercetin via pH-responsive zinc oxide nanoparticles for breast cancer therapy. *Materials Science and Engineering C* 100: 129–140.

14 Kohane, D.S. (2006). Microparticles and nanoparticles for drug delivery. *Biotechnology and Bioengineering* 96 (2): 203–209. https://doi.org/10.1002/bit.21301.

15 Lamson, N.G., Berger, A., Fein, K.C., and Whitehead, K.A. (2020). Anionic nanoparticles enable the oral delivery of proteins by enhancing intestinal permeability. *Nature Biomedical Engineering* 4 (1): 84–96.

16 Zhang, L., Xu, Y., Cao, W. et al. (2018). Understanding the translocation mechanism of PLGA nanoparticles across round window membrane into the inner ear: a guideline for inner ear drug delivery based on nanomedicine. *International Journal of Nanomedicine* 13: 479.

17 Sung, H., Ferlay, J., Siegel, R.L. et al. (2021). Global cancerstatistics 2020: GLOBOCAN estimates of incidence and mortality worldwide for 36 cancersin 185 countries. *CA Cancer Journal for Clinicians* 71 (3): 209–249.

18 Touvier, M., Druesne-Pecollo, N., Kesse-Guyot, E. et al. (2012). Dual association between polyphenol intake and breast cancer risk according to alcohol consumption level: a prospective cohort study. *Breast Cancer Research Treatment* 137: 225–236. https://doi.org/10.1007/S10549-012-2323-Y.

19 Cimino, S., Sortino, G., Favilla, V. et al. (2012). Polyphenols: key issues involved in chemoprevention of prostate cancer. *Oxidative Medicine and Cellular Longevity*. https://doi.org/10.1155/2012/632959.

20 Tran, K.B., Lang, J.J., Compton, K. et al. (2022). The global burden of cancer attributable to risk factors, 2010–19: a systematic analysis for the global burden of disease study 2019. *Lancet* 400: 563–591. https://doi.org/10.1016/S0140-6736(22)01438-6.

21 Piccolella, S. and Pacifico, S. (2015). Plant-derived polyphenols: a chemopreventive and chemoprotectant worth-exploring resource in toxicology. *Advances in Molecular Toxicology* 9: 161–214. https://doi.org/10.1016/B978-0-12-802229-0.00005-0.

22 Zhou, Y., Zheng, J., Li, Y. et al. (2016). Natural polyphenols for prevention and treatment of cancer. *Nutrients* 8 (8): 515. https://doi.org/10.3390/nu8080515.

23 Grosso, G., Godos, J., Lamuela-Raventos, R. et al. (2017). A comprehensive meta-analysis on dietary flavonoid and lignan intake and cancer risk: level of evidence and limitations. *Molecular Nutrition and Food Research*. https://doi.org/10.1002/mnfr.201600930.

24 Messina, M. (2016). Impact of soy foods on the development of breast cancer and the prognosis of breast cancer patients. *ForschKomplementmed* 23: 75–80.

25 Rothwell, J.A., Knaze, V., and Zamora-Ros, R. (2017). Polyphenols: dietary assessment and role in the prevention of cancers. *Current Opinion in Clinical Nutrition and Metabolic Care* 20: 512–521.

26 Kim, H.S., Wannatung, T., Lee, S. et al. (2012). Quercetin enhances hypoxia-mediated apoptosis via direct inhibition of AMPK activity in HCT116 colon cancer. *Apoptosis* 17: 938–949.

27 Meng, X., Zhou, J., Zhao, C.N. et al. (2020). Health benefits and molecular mechanisms of resveratrol: a narrative review. *Foods* 14 (3): 9. 340.

28 Akbik, D., Ghadiri, M., Chrzanowski, W., and Rohanizadeh, R. (2014). Curcumin as a wound healing agent. *Life Science* 22 (1): 116. 1–7.

29 Kim, T.W., Lee, S.Y., Kim, M. et al. (2018). Kaempferol induces autophagic cell death via IRE1-JNK-CHOP pathway and inhibition of G9a in gastric cancer cells. *Cell Death Disease* 9 (9): 875.

30 Wang, L.S. and Stoner, G.D. (2008). Anthocyanins and their role in cancer prevention. *Cancer Letters* 269 (2): 281–290. https://doi.org/10.1016/j.canlet.2008.05.020.

31 Chen, Y.H., Wu, J.X., and Yang, S.F. (2022). Anticancer effects and molecular mechanisms of apigenin in cervical cancer cells. *Cancers* 14 (7): 1824.

32 Bahadori, M., Baharara, J., and Amini, E. (2016). Anticancer properties of chrysin on colon cancer cells, in vitro and in vivo with modulation of Caspase-3,-9, Bax and Sall4. *Iranian Journal of Biotechnology* 14 (3): 177.

33 Kowalczyk, T., Sitarek, P., and Skała, E. (2019). Induction of apoptosis by in vitro and in vivo plant extracts derived from Menyanthes trifoliata L. in human cancer cells. *Cytotechnology* 71: 165–180.

34 Yun, S., Lee, Y.J., Choi, J. et al. (2021). Acacetin inhibits the growth of STAT3-activated DU145 prostate cancer cells by directly binding to signal transducer and activator of transcription 3 (STAT3). *Molecules* 26 (20): 6204.

35 Cathcart, M.C., Useckaite, Z., and Drakeford, C. (2016). Anti-cancer effects of baicalein in non-small cell lung cancer in-vitro and in-vivo. *BMC Cancer* 16 (1): 1–3.

36 Hong, Z., Cao, X., Li, N. et al. (2014). Luteolin is effective in the non-small cell lung cancer model with L 858 R/T 790 M EGF receptor mutation and erlotinib resistance. *British Journal of Pharmacology* 171 (11): 2842–2853.

37 Wang, B., Xu, H., Hu, X. et al. (2020). Synergetic inhibition of daidzein and regular exercise on breast cancer in bearing-4T1 mice by regulating NK cells and apoptosis pathway. *Life Science* 245: 117387.

38 Zhao, Y., Zhang, L., Wu, Y. et al. (2018). Selective anti-tumor activity of wogonin targeting the Warburg effect through stablizing p53. *Pharmacology Research* 135: 49–59.

39 Park, S.J. and Jeon, Y.J. (2012). Dieckol from Ecklonia cava suppresses the migration and invasion of HT1080 cells by inhibiting the focal adhesion kinase pathway downstream of Rac1-ROS signaling. *Molecular Cells* 33: 141–149.

40 Ho, H.H., Chang, C.S., Ho, W.C. et al. (2010). Anti- metastasis effects of gallic acid on gastric cancer cells involves inhibition of NF-kappa B activity and downregulation of PI3K/AKT/ small GTPase signals. *Food Chemical Toxicology* 48: 2508–2516.

41 Sun, H.R., Wang, S., Yan, S.C. et al. (2019). Therapeutic strategies targeting cancer stem cells and their microenvironment. *Frontiers in Oncology* 9: 1104.

42 Tang, F.Y., Su, Y.C., Chen, N.C. et al. (2008). Resveratrol inhibits migration and invasion of human breast-cancer cells. *Molecular Nutrition and Food Research* 52: 683–691.

43 Yu, H., Pan, C., Zhao, S. et al. (2008). Resveratrol inhibits tumor necrosis factor-alpha-mediated matrix metalloproteinase-9 expression and invasion of human hepatocellular carcinoma cells. *Biomedicine and Pharmacotherapy* 62: 366–372.

44 Viola, K., Kopf, S., Rarova, L. et al. (2013). Xanthohumol attenuates tumour cell-mediated breaching of the lymphendothelial barrier and prevents intravasation and metastasis. *Archives of Toxicology* 87: 1301–1312.

45 Yan, W., Ma, X., Zhao, X., and Zhang, S. (2018). Baicalein induces apoptosis and autophagy of breast cancer cells via inhibiting PI3K/AKT pathway in vivo and vitro. *Drug Design, Development and Therapy* 12: 3961–3972.

46 Huang, L., Song, Y., Lian, J., and Wang, Z. (2017). Allicin inhibits the invasion of lung adenocarcinoma cells by altering tissue inhibitor of metalloproteinase/matrix metalloproteinase balance via reducing the activity of phosphoinositide 3-kinase/AKT signaling. *Oncology Letters* 14 (1): 468–474. https://doi.org/10.3892/ol.2017.6129.

47 Ye, C., Zhang, C., Huang, H. et al. (2018). The natural compound myricetin effectively represses the malignant progression of prostate cancer by inhibiting PIM1 and disrupting the PIM1/CXCR4 interaction. *Cell. Physiology and Biochemistry* 48 (3): 1230–1244.

48 Loberg, R.D., McGregor, N., Ying, C. et al. (2007). In vivo evaluation of AT-101 (R-(—)-gossypol acetic acid) in androgen-independent growth of VCaP prostate cancer cells in combination with surgical castration. *Neoplasia* 9 (12): 1030–1037.

49 Li, Q., Wang, Y., Zhang, L. et al. (2016). Naringenin exerts anti-angiogenic effects in human endothelial cells: involvement of ERRα/VEGF/KDR signaling pathway. *Fitoterapia* 111: 78–86.

50 Grube, S., Ewald, C., Kögler, C. et al. (2018). Achievable central nervous system concentrations of the green tea catechin EGCG induce stress in glioblastoma cells in vitro. *Nutrition Cancer* 70 (7): 1145–1158.

51 Dou, J., Wang, Z., Ma, L. et al. (2018). Baicalein and baicalin inhibit colon cancer using two distinct fashions of apoptosis and senescence. *Oncotarget* 9 (28): 20089–20102. https://doi.org/10.18632/oncotarget.24015.

52 Al Dhaheri, Y., Attoub, S., Ramadan, G. et al. (2014). Carnosol induces ROS-mediated beclin1-independent autophagy and apoptosis in triple negative breast cancer. *PloS One* 9 (10): e109630.

53 Hsiao, Y.C., Peng, S.F., Lai, K.C. et al. (2019). Genistein induces apoptosis in vitro and has antitumor activity against human leukemia HL-60 cancer cell xenograft growth in vivo. *Environmental Toxicology* 34 (4): 443–456.

54 Tang, K.D., Liu, J., Russell, P.J. et al. (2019). Gamma-tocotrienol induces apoptosis in prostate cancer cells by targeting the Ang-1/Tie-2 signaling pathway. *International Journal of Molecular Sciences* 20 (5): 1164.

55 Xiong, M., Wang, L., Yu, H.L. et al. (2016). Ginkgetin exerts growth inhibitory and apoptotic effects on osteosarcoma cells through inhibition of STAT3 and activation of caspase-3/9. *Oncology Reports* 35 (2): 1034–1040.

56 Reed, J.C. (2000). Mechanisms of apoptosis. *The American Journal of Pathology* 157: 1415. https://doi.org/10.1016/S0002-9440(10)64779-7.

57 Carneiro, B.A. and El-Deiry, W.S. (2020). Targeting apoptosis in cancer therapy. *Nature Reviews Clinical Oncology* 17 (7): 395–417. https://doi.org/10.1038/s41571-020-0341-y.

58 Hafezi, K., Hemmati, A.A., Abbaszadeh, H. et al. (2020). Anticancer activity and molecular mechanisms of α-conidendrin, a polyphenolic compound present in Taxus yunnanensis, on human breast cancer cell lines. *Phytotherapy Research* 34: 1397–1408. https://doi.org/10.1002/PTR.6613.

59 Czabotar, P.E., Lessene, G., Strasser, A., and Adams, J.M. (2014). Control of apoptosis by the BCL-2 protein family: implications for physiology and therapy. *Nature Reviews. Molecular Cell Biology* 15 (1): 49–63. https://doi.org/10.1038/nrm3722.

60 Pietsch, E.C., Sykes, S.M., McMahon, S.B., and Murphy, M.E. (2008). The p53 family and programmed cell death. *Oncogene* 27; 27 (50): 6507–6521. https://doi.org/10.1038/onc.2008.315.

61 Liang, S., Kisseleva, T., and Brenner, D.A. (2016). The role of NADPH oxidases (NOXs) in liver fibrosis and the activation of myofibroblasts. *Frontiers in Physiology* 7: 17. https://doi.org/10.3389/fphys.2016.00017.

62 Huang, J., Lv, C., Hu, M., and Zhong, G. (2013). The mitochondria-mediate apoptosis of lepidopteran cells induced by Azadirachtin. *PLoS One* 8: e58499. https://doi.org/10.1371/JOURNAL.PONE.0058499.

63 Kauffmann-Zeh, A., Rodriguez-Viciana, P., Ulrich, E. et al. (1997). Suppression of c-Myc-induced apoptosis by Ras signaling through PI(3)K and PKB. *Nature* 385: 544–548. https://doi.org/10.1038/385544A0.

64 Song, G., Ouyang, G., and Bao, S. (2005). The activation of Akt/PKB signaling pathway and cell survival. *Journal of Cellular and Molecular Medicine* 9: 59–71. https://doi.org/10.1111/J.1582-4934.2005.TB00337.X.

65 Chen, X.-M., Yang, W.-Q., Wang, X. et al. (2022). Effects of natural dihydrochalcones in sweet tea (Lithocarpuspolystachyus) on diabetes: a systematical review and meta-analysis of animal studies. *Food and Function*. https://doi.org/10.1039/D2FO00245K.

66 Parra-Perez, A.M., Pérez-Jiménez, A., Gris-Cárdenas, I. et al. (2022). Involvement of the PI3K/AKT intracellular signaling pathway in the anticancer activity of hydroxytyrosol, a polyphenol from olea europaea, in hematological cells and implication of HSP60 levels in its anti-inflammatory activity. *International Journal of Molecular Sciences* 23: 7053. https://doi.org/10.3390/IJMS23137053.

67 Cobb, M.H. and Goldsmith, E.J. (1995). How MAP kinases are regulated. *The Journal of Biological Chemistry* 270: 14843–14846. https://doi.org/10.1074/JBC.270.25.14843.

68 Lin, Y.T., Kwon, Y.I., Labbe, R.G., and Shetty, K. (2005). Inhibition of helicobacter pylori and associated urease by oregano and cranberry phytochemical synergies. *Applied and Environmental Microbiology* 71 (12): 8558–8564. https://doi.org/10.1128/AEM.71.12.8558-8564.2005.

69 Herlaar, E. and Brown, Z. (1999). p38 MAPK signaling cascades in inflammatory disease. *Molecular Medicine Today* 5: 439–447. https://doi.org/10.1016/S1357-4310(99)01544-0.

70 Mansouri, A., Ridgway, L.D., Korapati, A.L. et al. (2003). Sustained activation of JNK/p38 MAPK pathways in response to cisplatin leads to Fas ligand induction and cell death in ovarian carcinoma cells. *The Journal of Biological Chemistry* 278: 19245–19256. https://doi.org/10.1074/JBC.M208134200.

71 Bhattacharya, U., Halder, B., Mukhopadhyay, S., and Giri, A.K. (2009). Role of oxidation-triggered activation of JNK and p38 MAPK in black tea polyphenols induced apoptotic death of A375 cells. *Cancer Science* 100: 1971–1978. https://doi.org/10.1111/J.1349-7006.2009.01251.X.

72 Nani, A., Belarbi, M., Murtaza, B. et al. (2019). Polyphenols from Pennisetum glaucum grains induce MAP kinase phosphorylation and cell cycle arrest in human osteosarcoma cells. *Journal of Functional Foods* 54: 422–432. https://doi.org/10.1016/J.JFF.2019.01.042.

73 Yuan, Z., Yang, Z., Li, W. et al. (2022). Triphlorethol-A attenuates U251 human glioma cancer cell proliferation and ameliorates apoptosis through JAK2/STAT3 and p38 MAPK/ERK signaling pathways. *The Journal of Biochemical and Molecular Toxicology* 36: e23138. https://doi.org/10.1002/JBT.23138.

74 Geng, H., Liu, G., Hu, J. et al. (2021). HOXB13 suppresses proliferation, migration and invasion, and promotes apoptosis of gastric cancer cells through transcriptional activation of VGLL4 to inhibit the involvement of TEAD4 in the Hippo signaling pathway. *Molecular Medicine Reports* 24: 1–12. https://doi.org/10.3892/MMR.2021.12361/HTML.

75 Harvey, K.F., Zhang, X., and Thomas, D.M. (2013). The Hippo pathway and human cancer. *Nature Reviews Cancer* 13 (4): 246–257. https://doi.org/10.1038/nrc3458.

76 Qin, X., Luo, H., Deng, Y. et al. (2022). Resveratrol inhibits proliferation and induces apoptosis via the Hippo/YAP pathway in human colon cancer cells. *Biochemical and Biophysical Research Communications* 636: 197–204. https://doi.org/10.1016/J.BBRC.2022.10.077.

77 Fu, C.Y., Chen, M.C., Tseng, Y.S. et al. (2019). Fisetin activates Hippo pathway and JNK/ERK/AP-1 signaling to inhibit proliferation and induce apoptosis of human osteosarcoma cells via ZAK overexpression. *Environmental Toxicology* 34: 902–911. https://doi.org/10.1002/TOX.22761.

78 Li, A., Gu, K., Wang, Q. et al. (2018). Epigallocatechin-3-gallate affects the proliferation, apoptosis, migration and invasion of tongue squamous cell carcinoma through the hippo-TAZ signaling pathway. *International Journal of Molecular Medicine* 42: 2615. https://doi.org/10.3892/IJMM.2018.3818.

79 Li, F., Bai, Y., Zhao, M. et al. (2015). Quercetin inhibits vascular endothelial growth factor-induced choroidal and retinal angiogenesis in vitro. *Ophthalmic Research* 53 (3): 109–116.

80 Junttila, M.R., Karnezis, A.N., Garcia, D. et al. (2010). Selective activation of p53-mediated tumour suppression in high-grade tumours. *Nature* 468: 567–571. https://doi.org/10.1038/NATURE09526.

81 Jung, E.J., Lee, W.S., Paramanantham, A. et al. (2020). p53 enhances artemisia annua L. polyphenols-induced cell death through upregulation of p53-dependent targets and cleavage of PARP1 and Lamin A/C in HCT116 colorectal cancer cells. *International Journal of Molecular Sciences* 21: 1–17. https://doi.org/10.3390/IJMS21239315.

82 Khan, H., Reale, M., Ullah, H. et al. (2020). Anti-cancer effects of polyphenols via targeting p53 signaling pathway: updates and future directions. *Biotechnology Advances* 38: 107385. https://doi.org/10.1016/J.BIOTECHADV.2019.04.007.

83 Gupta, K., Thakur, V.S., Bhaskaran, N. et al. (2012). Green tea polyphenols induce p53-dependent and p53-independent apoptosis in prostate cancer cells through two distinct mechanisms. *PLoS One* 7: e52572. https://doi.org/10.1371/JOURNAL.PONE.0052572.

84 Misir, S., Aliyazicioglu, Y., Demir, S. et al. (2019). Effect of Turkish propolis on miRNA expression, cell cycle, and apoptosis in human breast cancer (MCF-7). *Cells* 72: 133–145. https://doi.org/10.1080/01635581.2019.1616100.

85 Liu, L., Ju, Y., Wang, J., and Zhou, R. (2017). Epigallocatechin-3-gallate promotes apoptosis and reversal of multidrug resistance in esophageal cancer cells. *Pathology Research and Practice* 213: 1242–1250. https://doi.org/10.1016/J.PRP.2017.09.006.

86 de Oliveira, J.C., Oliveira, L.C., Mathias, C. et al. (2019). Long non-coding RNAs in cancer: another layer of complexity. *The Journal of Genetic Medicine* 21. https://doi.org/10.1002/JGM.3065.

87 Yoshioka, Y., Ohishi, T., Nakamura, Y. et al. (2022). Anti-cancer effects of dietary polyphenols via ROS-mediated pathway with their modulation of microRNAs. *Molecules* 27. https://doi.org/10.3390/MOLECULES27123816.

88 Wu, H., Li, C., Cui, M. et al. (2021). Polyphenols from Hippophaerhamnoides suppressed colon cancer growth by regulating miRNA-mediated cell cycle arrest and apoptosis in vitro and in vivo. *Journal of Functional Foods* 87: 104780. https://doi.org/10.1016/J.JFF.2021.104780.

89 Tsang, W.P. and Kwok, T.T. (2010). Epigallocatechin gallate up-regulation of miR-16 and induction of apoptosis in human cancer cells. *The Journal of Nutritional Biochemistry* 21: 140–146. https://doi.org/10.1016/J.JNUTBIO.2008.12.003.

90 Hillen, F. and Griffioen, A.W. (2007). Tumour vascularization: sprouting angiogenesis and beyond. *Cancer Metastasis Reviews* 26 (3): 489–502.

91 Carmeliet, P. and Jain, R.K. (2011). Molecular mechanisms and clinical applications of angiogenesis. *Nature* 473 (7347): 298–307.

92 Subbaraj, G.K., Kumar, Y.S., and Kulanthaivel, L. (2021). Antiangiogenic role of natural flavonoids and their molecular mechanism: an update. *The Egyptian Journal of Internal Medicine* 33 (1): 1–10.

93 Choi, J.Y., Jang, Y.S., Min, S.Y., and Song, J.Y. (2011). Overexpression of MMP-9 and HIF-1α in breast cancer cells under hypoxic conditions. *Journal of Breast Cancer* 14 (2): 88–95.

94 Lewis, C.E., De Palma, M., and Naldini, L. (2007). Tie2-expressing monocytes and tumor angiogenesis: regulation by hypoxia and angiopoietin-2. *Cancer Research* 67 (18): 8429–8432.

95 Askari, V., Shamlou, S., Mostafaie, A., and Khaleqi, S. (2019). Ethyl acetate fraction of Teucrium polium extract abolishes human umbilical vein endothelial cells (HUVEC) tubulogenesis in collagen bed through suppression of cell proliferation/VEGF secretion. *The Iranian Journal of Allergy, Asthma and Immunology* 18: 281–288.

96 Dube, A., Nicolazzo, J.A., and Larson, I. (2010). Chitosan nanoparticles enhance the intestinal absorption of the green tea catechins (+)-catechin and (−)-epigallocatechin gallate. *European Journal of Pharmaceutical Sciences* 41 (2): 219–225.

97 Balakrishnan, S., Bhat, F.A., Raja Singh, P. et al. (2016). Gold nanoparticle–conjugated quercetin inhibits epithelial–mesenchymal transition, angiogenesis, and invasiveness via EGFR/VEGFR-2-mediated pathway in breast cancer. *Cell Proliferation* 49 (6): 678–697.

98 Mandal, A.K., Ghosh, D., Sarkar, S. et al. (2014). Nanocapsulated quercetin downregulates rat hepatic MMP-13 and controls diethylnitrosamine-induced carcinoma. *Nanomedicine* 9 (15): 2323–2337.

99 Merchant, N., Nagaraju, G.P., Rajitha, B. et al. (2017). Matrix metalloproteinases: their functional role in lung cancer. *Carcinogenesis* 38 (8): 766–780.

100 Storjohann, A. (2005). The shifted number system for fast linear algebra on integer matrices. *Journal of Complex* 21 (4): 609–650.

101 Xing, L., Lyu, J.Y., Yang, Y. et al. (2017). pH-Responsive de-PEGylated nanoparticles based on triphenylphosphine–quercetin self-assemblies for mitochondria-targeted cancer therapy. *Chemical Communications* 53 (62): 8790–8793.

102 Ahmad, N., Qamar, M., Yuan, Y. et al. (2022). Dietary polyphenols: extraction, identification, bioavailability, and role for prevention and treatment of colorectal and prostate cancers. *Molecules* 27 (9): 2831.

103 Grgić, J., Šelo, G., Planinić, M. et al. (2020). Role of the encapsulation in bioavailability of phenolic compounds. *Antioxidants* 9 (10): 923.

104 Andújar, I., Recio, M.C., Giner, R.M., and Ríos, J.L. (2012). Cocoa polyphenols and their potential benefits for human health. *Oxidative Medicine and Cellular Longevity* 2012: 906252. https://doi.org/10.1155/2012/906252

105 Davatgaran-Taghipour, Y., Masoomzadeh, S., Farzaei, M.H. et al. (2017). Polyphenol nanoformulations for cancer therapy: experimental evidence and clinical perspective. *International Journal of Nanomedicine* 12: 2689.

106 Duronio, R.J. and Xiong, Y. (2013). Signaling pathways that control cell proliferation. *Cold Spring Harbor Perspectives in Biology* 5 (3): a008904.

107 Cilibrasi, C., Riva, G., Romano, G. et al. (2017). Resveratrol impairs glioma stem cells proliferation and motility by modulating the wnt signaling pathway. *PLoS One* 12 (1): e0169854.

108 Rosarin, F.S., Arulmozhi, V., Nagarajan, S., and Mirunalini, S. (2013). Antiproliferative effect of silver nanoparticles synthesized using amla on Hep2 cell line. *Asian Pacific Journal of Tropical Medicine* 6 (1): 1–10.

109 Muñoz, P., Ilioum, M.S., and Esteller, M. (2012). Epigenetic alterations involved in cancer stem cell reprogramming. *Molecular Oncology* 6 (6): 620–636.

110 Qiu, W., Lin, J., Zhu, Y. et al. (2017). Kaempferol modulates DNA methylation and downregulates DNMT3B in bladder cancer. *Cellular Physiology and Biochemistry* 41 (4): 1325–1335.

111 Lu, L., Wang, Y., Ou, R. et al. (2018). DACT2 epigenetic stimulator exerts dual efficacy for colorectal cancer prevention and treatment. *Pharmacological Research* 129: 318–328.

112 Alvarez, M.C., Maso, V., Torello, C.O. et al. (2018). The polyphenol quercetin induces cell death in leukemia by targeting epigenetic regulators of pro-apoptotic genes. *Clinical Epigenetics* 10 (1): 1-1.

113 Priyadarsini, R.V., Vinothini, G., Murugan, R.S. et al. (2011). The flavonoid quercetin modulates the hallmark capabilities of hamster buccal pouch tumors. *Nutrition and Cancer* 63 (2): 218–226.

114 Lee, W., Lee, S.Y., Son, Y.J. et al. (2015). Gallic acid decreases inflammatory cytokine secretion through histone acetyltransferase/histone deacetylase regulation in high glucose-induced human monocytes. *Journal of Medicinal Food* 18 (7): 793–801.

115 Soflaei, S.S., Momtazi-Borojeni, A.A., Majeed, M. et al. (2018). Curcumin: a natural pan-HDAC inhibitor in cancer. *Current Pharmaceutical Design* 24 (2): 123–129.

116 Liu, Z., Ren, Y., Meng, L. et al. (2021). Epigenetic signaling of cancer stem cells during inflammation. *Frontiers in Cell and Developmental Biology* 9: 772211.

117 Yang, L., Shi, P., Zhao, G. et al. (2020). Targeting cancer stem cell pathways for cancer therapy. *Signal Transduction and Targeted Therapy* 5 (1): 8.

118 Hu, C., Li, M., Guo, T. et al. (2019). Anti-metastasis activity of curcumin against breast cancer via the inhibition of stem cell-like properties and EMT. *Phytomedicine* 58: 152740.

119 Yoshida, K., Toden, S., Ravindranathan, P. et al. (2017). Curcumin sensitizes pancreatic cancer cells to gemcitabine by attenuating PRC2 subunit EZH2, and the lncRNA PVT1 expression. *Carcinogenesis* 38 (10): 1036–1046.

120 Kuo, Y.C., Wang, L.J., and Rajesh, R. (2019). Targeting human brain cancer stem cells by curcumin-loaded nanoparticles grafted with anti-aldehyde dehydrogenase and sialic acid: colocalization of ALDH and CD44. *Materials Science & Engineering C-Materials for Biological Applications* 102: 362–372.

121 Marquardt, J.U., Gomez-Quiroz, L., Arreguin Camacho, L.O. et al. (2015). Curcumin effectively inhibits oncogenic NF-κB signaling and restrains stemness features in liver cancer. *Journal of Hepatology* 63 (3): 661–669.

122 Zhou, Q.M., Sun, Y., Lu, Y.Y. et al. (2017). Curcumin reduces mitomycin C resistance in breast cancer stem cells by regulating Bcl-2 family-mediated apoptosis. *Cancer Cell International* 17: 84.

123 Imodoye, S.O., Adedokun, K.A., Muhammed, A.O. et al. (2021). Understanding the complex milieu of epithelial-mesenchymal transition in cancer metastasis: new insight into the roles of transcription factors. *Frontiers in Oncology* 11: 762817.

124 Park, J.W. and Han, J.W. (2019). Targeting epigenetics for cancer therapy. *Archives of Pharmacal Research* 42: 159–170.

125 Nikolaou, S. and Machesky, L.M. (2020). The stressful tumour environment drives plasticity of cell migration programmes, contributing to metastasis. *The Journal of Pathology* 250 (5): 612–623.

126 Hamidi, H. and Ivaska, J. (2018). Every step of the way: integrins in cancer progression and metastasis. *Nature Reviews Cancer* 18 (9): 533–548.

127 Haibe, Y., Kreidieh, M., El Hajj, H. et al. (2020). Resistance mechanisms to anti-angiogenic therapies in cancer. *Frontiers in Oncology* 10: 221.

128 McCubrey, J.A., Steelman, L.S., Chappell, W.H. et al. (2007). Roles of the Raf/MEK/ERK pathway in cell growth, malignant transformation and drug resistance. *Biochimica et Biophysica Acta (BBA) - Molecular Cell Research* 1773 (8): 1263–1284.

129 Starok, M., Preira, P., Vayssade, M. et al. (2015). EGFR inhibition by curcumin in cancer cells: a dual mode of action. *Biomacromolecules* 16 (5): 1634–1642.

130 Zhou, H. and Huang, S. (2011). Role of mTOR signaling in tumor cell motility, invasion and metastasis. *Current Protein and Peptide Science* 12 (1): 30–42.

131 Wang, M., Yu, T., Zhu, C. et al. (2014). Resveratrol triggers protective autophagy through the ceramide/Akt/mTOR pathway in melanoma B16 cells. *Nutrition and Cancer* 66 (3): 435–440.

132 Bhattacharya, S., Darjatmoko, S.R., and Polans, A.S. (2011). Resveratrol modulates the malignant properties of cutaneous melanoma via changes in the activation and attenuation of the anti-apoptotic proto-oncogenic protein Akt/PKB. *Melanoma Research* 21 (3): 180.

133 Leo, M.S. and Sivamani, R.K. (2014). Phytochemical modulation of the Akt/mTOR pathway and its potential use in cutaneous disease. *Archives of Dermatological Research* 306: 861–871.

134 Dhar, S., Kumar, A., Rimando, A.M. et al. (2015). Resveratrol and pterostilbene epigenetically restore PTEN expression by targeting oncomiRs of the miR-17 family in prostate cancer. *Oncotarget* 6 (29): 27214.

135 Li, J., Gong, X., Jiang, R. et al. (2018). Fisetin inhibited growth and metastasis of triple-negative breast cancer by reversing epithelial-to-mesenchymal transition via PTEN/Akt/GSK3β signal pathway. *Frontiers in Pharmacology* 9: 772.

136 Yao, X., Jiang, W., Yu, D. et al. (2019). Luteolin inhibits proliferation and induces apoptosis of human melanoma cells in vivo and in vitro by suppressing MMP-2 and MMP-9 through the PI3K/AKT pathway. *Food and Function* 10 (2): 703–712.

137 Achen, M.G. and Stacker, S.A. (2008). Molecular control of lymphatic metastasis. *Annals of the New York Academy of Sciences* 1131 (1): 225–234.

138 Luo, H., Rankin, G.O., Juliano, N. et al. (2012). Kaempferol inhibits VEGF expression and in vitro angiogenesis through a novel ERK-NFκB-cMyc-p21 pathway. *Food Chemistry* 130 (2): 321–328.

139 Paramanantham, A., Kim, M.J., Jung, E.J. et al. (2020). Pretreatment of anthocyanin from the fruit of vitiscoignetiaepulliat acts as a potent inhibitor of TNF-α effect by inhibiting NF-κB-regulated genes in human breast cancer cells. *Molecules* 25 (10): 2396.

140 Boye, K. and Mælandsmo, G.M. (2010). S100A4 and metastasis: a small actor playing many roles. *The American Journal of Pathology* 176 (2): 528–535.

141 Saleem, M., Adhami, V.M., Ahmad, N. et al. (2005). Prognostic significance of metastasis-associated protein S100A4 (Mts1) in prostate cancer progression and chemoprevention regimens in an autochthonous mouse model. *Clinical Cancer Research* 11 (1): 147–153.

142 Coussens, L.M. and Werb, Z. (2002). Inflammation and Cancer. *Nature* 420 (6917): 860–867. https://doi.org/10.1038/nature01322. PMID: 12490959; PMCID: PMC2803035.

143 Hiam-Galvez, K.J., Allen, B.M., and Spitzer, M.H. (2021). Systemic immunity in cancer. *Nature Reviews Cancer* 21 (6): 345–359. https://doi.org/10.1038/s41568-021-00347-z. Epub 2021 Apr 9. PMID: 33837297; PMCID: PMC8034277.

144 Mileo, A.M., Nisticò, P., and Miccadei, S. (2019). Polyphenols: immunomodulatory and therapeutic implication in colorectal cancer. *Frontiers in Immunology* 10: 729. https://doi.org/10.3389/fimmu.2019.00729. PMID: 31031748; PMCID: PMC6470258.

145 Shakoor, H., Feehan, J., Apostolopoulos, V. et al. (2021). Immunomodulatory effects of dietary polyphenols. *Nutrients* 13 (3): 728. https://doi.org/10.3390/nu13030728.

146 Karasawa, K., Uzuhashi, Y., Hirota, M. et al. (2011). A matured fruit extract of date palm tree (Phoenix dactylifera L.) stimulates the cellular immune system in mice. *Journal of Agricultural and Food Chemistry* 59 (20): 11287–11293. https://doi.org/10.1021/jf2029225. Epub 2011 Oct 3. PMID: 21936496.

147 Malfitano, A.M., Pisanti, S., Napolitano, F. et al. (2020). Tumor-associated macrophage status in cancer treatment. *Cancers (Basel)* 12 (7): 1987. https://doi.org/10.3390/cancers12071987. PMID: 32708142; PMCID: PMC7409350.

148 Nishida-Aoki, N. and Gujral, T.S. (2022). Fisetin inhibits lipopolysaccharide-induced macrophage activation and dendritic cell maturation. *Cancer Research* 82 (3): 433–446. 10.1158/0008-5472.CAN-21-1428. Epub 2021 Dec 13. PMID: 34903600; PMCID: PMC8847322.

149 Liu, S.H., Lin, C.H., Hung, S.K. et al. (2010). Fisetin inhibits lipopolysaccharide-induced macrophage activation and dendritic cell maturation. *Journal of Agricultural and Food Chemistry* 58 (20): 10831–10839. https://doi.org/10.1021/jf1017093. Epub 2010 Oct 5. PMID: 20923145.

150 Tekin, İ.Ö. and Marotta, F. (2018). Polyphenols and immune system. *Polyphenols: Prevention and Treatment of Human Disease (Second Edition)* 263–276. https://doi.org/10.1016/b978-0-12-813008-7.00022-9.

151 Wang, Y., Ni, H., Li, H. et al. (2018). Nuclear factor kappa B regulated monocyte chemoattractant protein-1/chemokine CC motif receptor-2 expressing in spinal cord contributes to the maintenance of cancer-induced bone pain in rats. *Molecular Pain* 14: 1744806918788681. https://doi.org/10.1177/1744806918788681. Epub 2018 Jun 29. PMID: 29956585; PMCID: PMC6055241.

152 Yoshimura, T., Howard, O.M., Ito, T. et al. (2013). Monocyte chemoattractant protein-1/CCL2 produced by stromal cells promotes lung metastasis of 4T1 murine breast cancer cells. *PLoS One* 8 (3): e58791. https://doi.org/10.1371/journal.pone.0058791. Epub 2013 Mar 18. PMID: 23527025; PMCID: PMC3601078.

153 Yoshimura, T., Galligan, C., Takahashi, M. et al. (2014). Non-myeloid cells are major contributors to innate immune responses via production of monocyte chemoattractant protein-1/CCL2. *Frontiers in Immunology* 4: 482. https://doi.org/10.3389/fimmu.2013.00482. PMID: 24432017; PMCID: PMC3882876.

154 Andujar, I., Recio, M.C., Giner, R.M. et al. (2012). Cocoa polyphenols and their potential benefits for human health. *Oxidative Medicine and Cellular Longevity* 2012: Article ID 906252. 23.

155 Neurath, M.F. (2019). IL-23 in inflammatory bowel diseases and colon cancer. *Cytokine and Growth Factor Reviews* 45: 1–8. https://doi.org/10.1016/j.cytogfr.2018.12.002. Epub 2018 Dec 12. PMID: 30563755.

156 Klampfer, L. (2011). Cytokines, inflammation, and colon cancer. *Current Cancer Drug Targets* 11 (4): 451–464. https://doi.org/10.2174/156800911795538066. PMID: 21247378; PMCID: PMC3540985.

157 Moradinasab, S., Pourbagheri-Sigaroodi, A., Ghaffari, S.H. et al. (2022). Targeting macrophage-mediated tumor cell phagocytosis: an overview of phagocytosis checkpoints blockade, nanomedicine intervention, and engineered CAR-macrophage therapy. *International Immunopharmacology* 103: 108499. https://doi.org/10.1016/j.intimp.2021.108499. Epub 2021 Dec 28. PMID: 34972068.

158 Stevens, J.F., Revel, J.S., and Maier, C.S. (2018). Mitochondria-centric review of polyphenol bioactivity in cancer models. *Antioxid Redox Signal* 29 (16): 1589–1611. https://doi.org/10.1089/ars.2017.7404. Epub 2017 Dec 11. PMID: 29084444; PMCID: PMC6207154.

159 Boedtkjer, E. and Pedersen, S.F. (2020). The acidic tumor microenvironment as a driver of cancer. *Annual Review of Physiology* 82: 103–126. https://doi.org/10.1146/annurev-physiol-021119-034627. Epub 2019 Nov 15. PMID: 31730395.

160 Chen, W., Ju, X., Aluko, R.E. et al. (2020). Rice bran protein-based nanoemulsion carrier for improving stability and bioavailability of quercetin. *Food Hydrocoll* 108: 106042. https://doi.org/10.1016/j.foodhyd.2020.106042.

161 del Mar Blanquer-Rossello, M., Herna´ndez-Lo´pez, R., Roca, P. et al. (2017). Resveratrol induces mitochondrial respiration and apoptosis in SW620 colon cancer cells. *BiochimBiophys Acta* 1861: 431–440.

162 Fantin, V.R., St-Pierre, J., and Leder, P. (2006). Attenuation of LDH-A expression uncovers a link between glycolysis, mitochondrial physiology, and tumor maintenance. *Cancer Cell* 9 (6): 425–434. https://doi.org/10.1016/j.ccr.2006.04.023. Erratum in: Cancer Cell. 2006 Aug;10(2):172. PMID: 16766262.

163 Mo, H., Chen, Y., Huang, L. et al. (2013). Neuroprotective effect of tea polyphenols on oxyhemoglobin induced subarachnoid hemorrhage in mice. *Oxidative Medicine and Cellular Longevity* 2013: 743938. https://doi.org/10.1155/2013/743938. Epub 2013 Jun 3. PMID: 23840920; PMCID: PMC3686094.

164 Rupiani, S., Guidotti, L., Manerba, M. et al. (2016). Synthesis of natural urolithin M6, a galloflavin mimetic, as a potential inhibitor of lactate dehydrogenase A. *Organic & Biomolecular Chemistry* 14: 10981–10987.

165 Arumugam, M., Raes, J., Pelletier, E. et al. (2011). Enterotypes of the human gut microbiome. *Nature* 473 (7346): 174–180. https://doi.org/10.1038/nature09944. Epub 2011 Apr 20. Erratum in: *Nature*. **474**(7353):666. Erratum in: Nature. 2014 Feb 27;506(7489):516. PMID: 21508958; PMCID: PMC3728647.

166 Wang, K., Sun, G., Conlon, M.A. et al. (2021). Editorial: dietary polyphenols for improving gut health: volume 1. *Frontiers in Nutrition* 8: 760917. https://doi.org/10.3389/fnut.2021.760917. PMID: 34671635; PMCID: PMC8520993.

167 Zhao, Y. and Jiang, Q. (2021). Roles of the polyphenol-gut microbiota interaction in alleviating colitis and preventing colitis-associated colorectal cancer. *Advances in Nutrition* 12 (2): 546–565. https://doi.org/10.1093/advances/nmaa104. PMID: 32905583; PMCID: PMC8009754.

168 Junaid, M., Linn, A.K., Javadi, M.B. et al. (2016). Vacuolating cytotoxin A (VacA) - A multi-talented pore-forming toxin from Helicobacter pylori. *Toxicon* 118: 27–35. https://doi.org/10.1016/j.toxicon.2016.04.037. Epub 2016 Apr 20. PMID: 27105670.

169 Gobert, A.P. and Wilson, K.T. (2016). The immune battle against helicobacter pylori infection: no offense. *Trends in Microbiology* 24 (5): 366–376. https://doi.org/10.1016/j.tim.2016.02.005. Epub. PMID: 26916789; PMCID: PMC4841705.

170 Tombola, F., Campello, S., De Luca, L. et al. (2003). Plant polyphenols inhibit VacA, a toxin secreted by the gastric pathogen Helicobacter pylori. *FEBS Letters* 543 (1–3): 184–189. https://doi.org/10.1016/s0014-5793(03)00443-5. PMID: 12753930.

171 Ansari, S. and Yamaoka, Y. (2020). Role of vacuolating cytotoxin A in Helicobacter pylori infection and its impact on gastric pathogenesis. *Expert Review of Anti-infective Therapy* 18 (10): 987–996. https://doi.org/10.1080/14787210.2020.1782739. Epub 2020 Jun 27. PMID: 32536287.

172 Chaturvedi, M., Mishra, M., Pandey, A. et al. (2022). Oxidative products of curcumin rather than curcumin bind to helicobacter pylori virulence factor vaca and are required to inhibit its vacuolation activity. *Molecules* 27 (19): 6727. https://doi.org/10.3390/molecules27196727. PMID: 36235264; PMCID: PMC9572645.
173 Ozdal, T., Sela, D.A., Xiao, J. et al. (2016). The reciprocal interactions between polyphenols and gut microbiota and effects on bioaccessibility. *Nutrients* 8 (2): 78. https://doi.org/10.3390/nu8020078. PMID: 26861391; PMCID: PMC4772042.
174 Makarewicz, M., Drożdż, I., Tarko, T., and Duda-Chodak, A. (2021). The interactions between polyphenols and microorganisms, especially gut microbiota. *Antioxidants (Basel)* 10 (2): 188. https://doi.org/10.3390/antiox10020188. PMID: 33525629; PMCID: PMC7911950.
175 González-Sarrías, A., García-Villalba, R., Romo-Vaquero, M. et al. (2017). Clustering according to urolithin metabotype explains the interindividual variability in the improvement of cardiovascular risk biomarkers in overweight-obese individuals consuming pomegranate: a randomized clinical trial. *Molecular Nutrition and Food Research* 61 (5): https:doi.org/10.1002/mnfr.201600830. https:doi.org/10.1002/mnfr.201600830.
176 Corrêa, T.A.F., Rogero, M.M., Hassimotto, N.M.A., and Lajolo, F.M. (2019). The two-way polyphenols-microbiota interaction and their effects on obesity and related metabolic diseases. *Frontiers in Nutrition* 6: 188. https://doi.org/10.3389/fnut.2019.00118.
177 Jo, M.J., Bae, S.J., Son, B.W. et al. (2013). 3,4-dihydroxyphenyl acetic acid and (+)-epoxydon isolated from marine algae-derived microorganisms induce down regulation of epidermal growth factor activated mitogenic signaling cascade in Hela cells. *Cancer Cell International* 13 (1): 49. https://doi.org/10.1186/1475-2867-13-49. PMID: 23706036; PMCID: PMC3673896.
178 Squillaro, T., Cimini, A., Peluso, G. et al. (2018). Nano-delivery systems for encapsulation of dietary polyphenols: an experimental approach for neurodegenerative diseases and brain tumors. *Biochemical Pharmacology* 154: 303–317. https://doi.org/10.1016/j.bcp.2018.05.016.
179 Zhang, Z., Li, X., Sang, S. et al. (2022). Polyphenols as plant-based nutraceuticals: health effects, encapsulation, nano-delivery, and application. *Foods* 11 (15): 2189. https://doi.org/10.3390/foods11152189.
180 Conte, R. (2016). Polyphenols nanoencapsulation for therapeutic applications. *Journal of Biological Researchand Therapy* 5 (2): https://doi.org/10.4172/2167-7956.1000139.
181 Teja, P.K., Mithiya, J., Kate, A.S. et al. (2022). Herbal nanomedicines: recent advancements, challenges, opportunities and regulatory overview. *Phytomedicine* 96: 153890. https://doi.org/10.1016/j.phymed.2021.153890.
182 Hoshyar, N., Gray, S., Han, H., and Bao, G. (2016). The effect of nanoparticle size on in vivo pharmacokinetics and cellular interaction. *Nanomedicine* 11 (6): 673–692. https://doi.org/10.2217/nnm.16.5.
183 Rizvi, S.A.A. and Saleh, A.M. (2018). Applications of nanoparticle systems in drug delivery technology. *The Saudi Pharmaceutical Journal* 26 (1): 64–70. https://doi.org/10.1016/j.jsps.2017.10.012.
184 de Jong, W.H. (2008). Drug delivery and nanoparticles: applications and hazards. *International Journal of Nanomedicine* 133. https://doi.org/10.2147/IJN.S596.
185 Khan, I., Bahuguna, A., Kumar, P. et al. (2018). In vitro and in vivo antitumor potential of carvacrol nanoemulsion against human lung adenocarcinoma A549 cells via mitochondrial mediated apoptosis. *Scientific Reports* 8 (1): 144. https://doi.org/10.1038/s41598-017-18644-9.
186 Silva, P., Bonifácio, B., Ramos, M. et al. (2013). Nanotechnology-based drug delivery systems and herbal medicines: a review. *International Journal of Nanomedicine* 1. https://doi.org/10.2147/IJN.S52634.

187 Ahmed, H.M., Nabavi, S., and Behzad, S. (2021). Herbal drugs and natural products in the light of nanotechnology and nanomedicine for developing drug formulations. *Mini-Reviews in Medicinal Chemistry* 21 (3): 302–313. https://doi.org/10.2174/1389557520666200916143240.

188 Zhang, Z., Qiu, C., Li, X. et al. (2021). Advances in research on interactions between polyphenols and biology-based nano-delivery systems and their applications in improving the bioavailability of polyphenols. *Trends in Food Science and Technology* 116: 492–500. https://doi.org/10.1016/j.tifs.2021.08.009.

189 Annaji, M., Poudel, I., Boddu, S.H.S. et al. (2021). Resveratrol-loaded nanomedicines for cancer applications. *Cancer Reports* 4 (3). https://doi.org/10.1002/cnr2.1353.

190 Emami, S., Azadmard-Damirchi, S., Peighambardoust, S.H. et al. (2016). Liposomes as carrier vehicles for functional compounds in food sector. *Journal of Experimental Nanoscience* 11 (9): 737–759. https://doi.org/10.1080/17458080.2016.1148273.

191 Bhushan, B. (2016). *Encyclopedia of Nanotechnology*, 2e. Springer.

192 Akbarzadeh, A., Rezaei-Sadabady, R., Davaran, S. et al. (2013). Liposome: classification, preparation, and applications. *NRL* 8 (1): 102. https://doi.org/10.1186/1556-276X-8-102.

193 Huang, Z., Li, X., Zhang, T. et al. (2014). Progress involving new techniques for liposome preparation. *AJPS* 9 (4): 176–182. https://doi.org/10.1016/j.ajps.2014.06.001.

194 Patil, Y.P. and Jadhav, S. (2014). Novel methods for liposome preparation. *Chemistry and Physics of Lipids* 177: 8–18. https://doi.org/10.1016/j.chemphyslip.2013.10.011.

195 Maja, L., Željko, K., and Mateja, P. (2020). Sustainable technologies for liposome preparation. *Journal of Supercritical Fluids* 165: 104984. https://doi.org/10.1016/j.supflu.2020.104984.

196 Jourghanian, P., Ghaffari, S., Ardjmand, M. et al. (2016). Sustained release curcumin loaded solid lipid nanoparticles. *APB* 6 (1): 17–21. https://doi.org/10.15171/apb.2016.04.

197 Wang, S., Chen, T., Chen, R. et al. (2012). Emodin loaded solid lipid nanoparticles: preparation, characterization and antitumor activity studies. *International Journal of Pharmaceutics* 430 (1–2): 238–246. https://doi.org/10.1016/j.ijpharm.2012.03.027.

198 Yadav, A., Sunkaria, A., Singhal, N., and Sandhir, R. (2018). Resveratrol-loaded solid lipid nanoparticles attenuate mitochondrial oxidative stress in vascular dementia by activating Nrf2/HO-1 pathway. *Neurochemistry International* 112: 239–254. https://doi.org/10.1016/j.neuint.2017.08.001.

199 Atef, E. and Belmonte, A.A. (2008). Formulation and in vitro and in vivo characterization of a phenytoin self-emulsifying drug delivery system (SEDDS). *European Journal of Pharmaceutical Sciences* 35 (4): 257–263. https://doi.org/10.1016/j.ejps.2008.07.004.

200 Yahaya, Z.S., Oyi, A.R., Allagh, T.S. et al. (2020). Solid self-emulsifying drug delivery system for artemether with improved physicochemical properties: design and characterization. *West African Pharmacist* 31 (1): 164–175.

201 Feeney, O.M., Crum, M.F., McEvoy, C.L. et al. (2016). 50 years of oral lipid-based formulations: provenance, progress and future perspectives. *Advanced Drug Delivery Reviews* 101: 167–194. https://doi.org/10.1016/j.addr.2016.04.007.

202 Zhao, Y., Wang, C., Chow, A.H.L. et al. (2010). Self-nanoemulsifying drug delivery system (SNEDDS) for oral delivery of Zedoary essential oil: formulation and bioavailability studies. *International Journal of Pharmaceutics* 383 (1–2): 170–177. https://doi.org/10.1016/j.ijpharm.2009.08.035.

203 Shukla, M., Jaiswal, S., Sharma, A. et al. (2017). A combination of complexation and self-nanoemulsifying drug delivery system for enhancing oral bioavailability and anticancer efficacy of curcumin. *Drug Development and Industrial Pharmacy* 43 (5): 847–861. https://doi.org/10.1080/03639045.2016.1239732.

204 Li, G., Zhang, Z., Liu, H., and Hu, L. (2021). Nanoemulsion-based delivery approaches for nutraceuticals: fabrication, application, characterization, biological fate, potential toxicity and future trends. *Food and Function* 12: 1933–1953. https://doi.org/10.1039/d0fo02686g. rsc.li/food-function.

205 Tripathi, C.B., Beg, S., Kaur, R. et al. (2016). Systematic development of optimized SNEDDS of artemether with improved biopharmaceutical and antimalarial potential. *Drug Delivery* 23 (9): 3209–3223. https://doi.org/10.3109/10717544.2016.1162876.

206 Barani, M., Sangiovanni, E., Angarano, M. et al. (2021). Phytosomes as innovative delivery systems for phytochemicals: a comprehensive review of literature. *International Journal of Nanomedicine* 16: 6983–7022. https://doi.org/10.2147/IJN.S318416.

207 Lu, M., Qiu, Q., Luo, X. et al. (2019). Phyto-phospholipid complexes (phytosomes): a novel strategy to improve the bioavailability of active constituents. *Asian Journal of Pharmaceutical Sciences* 14 (3): 265–274. https://doi.org/10.1016/j.ajps.2018.05.011.

208 Mallakpour, S. and Behranvand, V. (2016). Polymeric nanoparticles: recent development in synthesis and application. *EXPRESS Polymer Letters* 10 (11): 895–913. https://doi.org/10.3144/expresspolymlett.2016.84.

209 Zielińska, A., Carreiró, F., Oliveira, A.M. et al. (2020). Polymeric nanoparticles: production, characterization, toxicology, and ecotoxicology. *Molecules* 25 (16): 3731. https://doi.org/10.3390/molecules25163731.

210 Velavan, P., Karuppusamy, C., and Venkatesan, P. (2016). Nanoparticles as drug delivery systems. *Journal of Pharmaceutical Sciences* 7 (12): 1118–1122.

211 Xie, X., Tao, Q., Zou, Y. et al. (2011). PLGA nanoparticles improve the oral bioavailability of curcumin in rats: characterizations and mechanisms. *Journal of Agricultural and Food Chemistry* 59 (17): 9280–9289. https://doi.org/10.1021/jf202135j.

212 Manikkam, R. and Pitchai, D. (2014). Catechin loaded chitosan nanoparticles as a novel drug delivery system for cancer – synthesis and in vitro and in vivo characterization. *World Journal of Pharmacy and Pharmaceutical Sciences* 3: 1553–1577.

213 Singh, N.A., Mandal, A.K.A., and Khan, Z.A. (2017). Fabrication of PLA-PEG nanoparticles as delivery systems for improved stability and controlled release of catechin. *Journal of Nanomaterials*. https://doi.org/10.1155/2017/6907149.

214 Xu, W., Ling, P., and Zhang, T. (2013). Polymeric micelles, a promising drug delivery system to enhance bioavailability of poorly water-soluble drugs. *Journal of Drug Delivery* 1–15. https://doi.org/10.1155/2013/340315.

215 Naseri, S., Darroudi, M., Aryan, E. et al. (2017). The antiviral effects of curcumin nanomicelles on the attachment and entry of Hepatitis C virus. *Iranian Journal of Virology* 11: 29–35.

216 El-Far, M., Salah, N., Essam, A. et al. (2018). Silymarin nanoformulation as potential anticancer agent in experimental Ehrlich ascites carcinoma-bearing animals. *Nanomedicine* 13 (15): 1865–1858. https://doi.org/10.2217/nnm-2017-0394.

217 Samrot, A.V., Sean, T.C., Kudaiyappan, T. et al. (2020). Production, characterization and application of nanocarriers made of polysaccharides, proteins, bio-polyesters and other biopolymers: a review. *International Journal of Biological Macromolecules* 165: 3088–3105. https://doi.org/10.1016/j.ijbiomac.2020.10.104.

218 Lozano-Pérez, A.A., Rivero, H.C., Pérez Hernández, M.D.C. et al. (2017). Silk fibroin nanoparticles: efficient vehicles for the natural antioxidant quercetin. *International Journal of Pharmaceutics* 518 (1–2): 11–19. https://doi.org/10.1016/j.ijpharm.2016.12.046.

219 Solghi, S., Emam-Djomeh, Z., Fathi, M., and Farahani, F. (2020). The encapsulation of curcumin by whey protein: assessment of the stability and bioactivity. *Journal of Food Process Engineering* 43 (6): e13403. https://doi.org/10.1111/jfpe.13403.

220 Xiong, J., Chan, Y.H., Rathinasabapathy, T. et al. (2020). Enhanced stability of berry pomace polyphenols delivered in protein-polyphenol aggregate particles to an in vitro gastrointestinal digestion model. *Food Chemistry* 331: 127279. https://doi.org/10.1016/j.foodchem.2020.127279.

221 Lei, C., Liu, X.-R., Chen, Q.-B. et al. (2021). Hyaluronic acid and albumin-based nanoparticles for drug delivery. *The Journal of Controlled Release* 331: 416–433. https://doi.org/10.1016/j.jconrel.2021.01.033.

222 Abbasi, E., Aval, S.F., Akbarzadeh, A. et al. (2014). Dendrimers: synthesis, applications, and properties. *Nanoscale Research Letters* 9 (1): 247. https://doi.org/10.1186/1556-276X-9-247.

223 Yetisgin, A.A., Cetinel, S., Zuvin, M. et al. (2020). Therapeutic nanoparticles and their targeted delivery applications. *Molecules* 25 (9): 2193. https://doi.org/10.3390/molecules25092193.

224 Svenson, S. and Tomalia, D. (2005). Dendrimers in biomedical applications—reflections on the field. *Advanced Drug Delivery Reviews* 57 (15): 2106–2129. https://doi.org/10.1016/j.addr.2005.09.018.

225 Mollazade, M., Nejati-Koshki, K., Akbarzadeh, A. et al. (2013). PAMAM dendrimers augment inhibitory effects of curcumin on cancer cell proliferation: possible inhibition of telomerase. *APJCP* 14 (11): 6925–6928. https://doi.org/10.7314/APJCP.2013.14.11.6925.

226 Madaan, K., Kumar, S., Poonia, N. et al. (2014). Dendrimers in drug delivery and targeting: drug-dendrimer interactions and toxicity issues. *Journal of Pharmacy and Bioallied Sciences* 6 (3): 139. https://doi.org/10.4103/0975-7406.130965.

11

Nanodelivery of Polyphenols as Nutraceuticals for CNS Disorders

*Puneet K. Samaiya, Rakesh Sagar, Sharad P. Pandey, Gourav Jain, and Abhishek K. Sah**

Department of Pharmacy, Shri Govindram Seksariya Institute of Technology & Science (SGSITS), Indore, Madhya Pradesh, India
** Corresponding author*

The prevalence of central nervous system (CNS) disorders is on the rise, presenting significant challenges to human survival and resulting in a surge in healthcare costs. Neurological disorders are the leading cause of global disability-adjusted life-years (DALYs), accounting for 45%–11.6% of global DALYs and 16.5% of deaths in 2019. Serious threats to human health include major CNS disorders such as brain tumors, neurodegenerative diseases, and cerebrovascular diseases (Figure 11.1). The effective treatment of CNS disorders is impeded by the blood–brain barrier (BBB), which comprises various cell types such as endothelial cells, astrocytes, microglial cells, and pericytes.

11.1 Gliomas

A glioma is a type of tumor that arises from the supportive cells of the CNS, known as glial cells. These tumors are the most common type of primary CNS tumors, accounting for approximately 24% of all cases [1]. There are various types of gliomas, including ependymomas, astrocytomas (including glioblastoma [GBM]), oligodendrogliomas, mixed gliomas, as well as optic nerve and brain stem gliomas [2]. In the recent classification of gliomas, in addition to histological findings supported by ancillary tissue-based tests (e.g., immunohistochemical, ultrastructural), molecular biomarkers have gained importance in providing both ancillary and defining diagnostic information. In the recent 5th edition of the World Health Organization (WHO) classification of CNS tumors, classification of gliomas has been based on key diagnostic genes, molecules, pathways, and/or their combinations in gliomas [3]. The median survival time of patients suffering from gliomas is approximately 10 months. The United States Food and Drug Administration (US FDA) granted accelerated approval of bevacizumab, a humanized monoclonal antibody against vascular endothelial growth factor, for the treatment of recurrent glioblastomas [4]. However, in recent years, oncolytic virus therapy, stem cell therapy, immunotherapy, and electric field therapy have been available [5]. These emerging treatments are expected to improve the prospect of treating recurrent high-grade gliomas.

Polyphenols: Food, Nutraceutical, and Nanotherapeutic Applications, First Edition. Edited by Mithun Rudrapal.

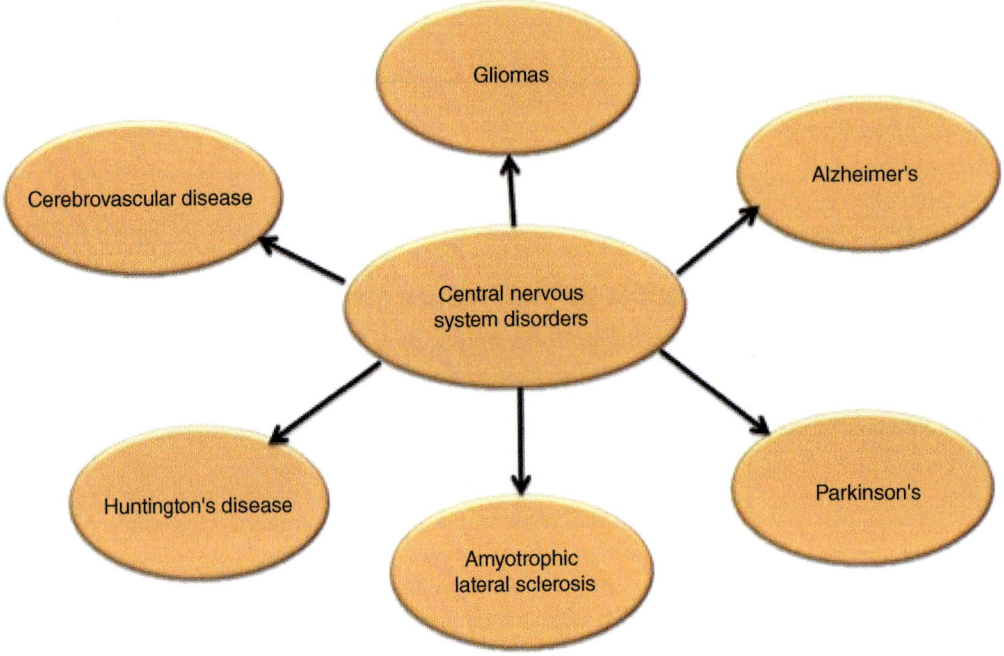

Figure 11.1 Types of central nervous system (CNS) disorders.

11.2 Neurodegenerative Disorders

11.2.1 Alzheimer's Disease

Alzheimer's disease (AD) is the most common cause of dementia in adults. AD is a progressive disorder with the pathological hallmarks of extracellular amyloid-β (Aβ) plaques and neurofibrillary tangles of intraneuronal hyperphosphorylated tau protein. According to The World Alzheimer Report 2021 [6], AD-related dementia is the 7th leading cause of mortality globally. Currently, approximately 55 million people have dementia worldwide, with Alzheimer's accounting for nearly 70% of diagnoses. According to the World Health Organization, 10 million cases are reported each year, with projections of 78 million people affected by 2030 and 139 million by 2050[7]. The progression from being an average person to an AD patient typically occurs in the following manner: first, there is no cognitive impairment (NCI), then there is mild cognitive impairment (MCI), and finally, there is AD (dementia) [8, 9]. The earliest stages of AD are asymptomatic, which are no different from normal aging. The mild cognitive impairment (MCI) stage is characterized by a slight impairment of cognition (orientation, language, attention, and executive functions might be affected), usually in memory, though not demented [10]. The last stage is AD dementia, characterized by a significant loss of memory and cognitive functions [11]. Two forms of AD have been reported: early onset AD occurring before 65 years of age, manifesting in 5%–10% of the population, and late onset of AD manifesting after 65 years of age [12]. The clinical trials have failed so far as it is very difficult to detect the pathology in the early state in most high-risk individuals [13]. This can also be why therapeutic interventions can only offer symptomatic relief to AD patients and do not prevent AD from originating in the individual or curing it. Researchers have proposed various hypotheses for the progression and development of AD; however, the cause and actual mechanism of AD remains known. A fast response to the first symptoms of AD is crucial in diagnosing the early stages of the disease so that treatment can be initiated and the quality of life extended [14]. In the conventional pharmacological treatment of AD, cholinesterase inhibitors (ChEIs), such as donepezil, tacrine, berberine, and galantamine, are employed for symptomatic treatment of AD. Many other therapeutic strategies have been explored for several decades in clinical trials; however, the currently available treatments are primarily treatments of symptoms rather than actual curative therapies [15]. Because of this, attention has turned towards prevention or reducing AD risk.

11.2.2 Parkinson's Disease

Parkinson's disease (PD) is the second most prevalent neurodegenerative disorder, impacting 1%–2% of the population aged 65 and older [16]. The characteristic features of PD include the specific degeneration of dopaminergic neurons in the substantia nigra, as well as the accumulation of misfolded and aggregated alpha-synuclein in the brainstem, resulting in challenges with the motor function. Some other molecular pathogenic mechanisms include mitochondrial dysfunction, impairment of protein clearance (associated with deficient ubiquitin-proteasome and autophagy-lysosomal systems), neuroinflammation, and oxidative stress [17]. Enhancing the level of dopamine in the brain is the most common treatment strategy to improve the symptoms in PD patients. However, this treatment does not alter the progression of the disease or restore the affected dopaminergic neurons. L-dopa, amantadine, metatyrosine, melatonin, thyrotrophin-releasing hormone, lithium, baclofen, electroconvulsive shock therapy, vitamin E, and marijuana are some regimes to provide symptomatic relief in PD [18]. One of the biggest challenges in the development of potential neuroprotective therapies has been the lack of reliable and sensitive biomarkers of disease progression. Immunotherapies, such as the use of vaccination or monoclonal antibodies directed against aggregated, toxic α-synuclein as well as anti-aggregation or protein clearance strategies, are currently being investigated in clinical trials [17, 19].

11.2.3 Amyotrophic Lateral Sclerosis

Amyotrophic lateral sclerosis (ALS) is a degenerative neurological condition that gradually causes motor neurons in the CNS to deteriorate, ultimately resulting in paralysis [20]. It has been reported that ALS affects approximately 16,000 individuals, with a prognosis for survival of 2–5 years [21]. There are two types of ALS diseases differentiated by genetics: familial and sporadic (idiopathic) [22]. Diagnosis is determined by excluding other conditions and utilizing clinical examinations, laboratory tests, and nerve conduction/electromyography studies [20]. The pathogenesis of ALS comprises of failure of the upper motor neurons leading to abrupt reflexes and reduced coordination of the limbs with stiffness of the muscles. Alternatively, the failure of lower motor neurons causes progressive atrophies when the synapses connecting the muscles are lost, which tends to begin in the limbs and progresses to the eye and sphincter muscle neurons in the late stages. Overall, the patients has difficulty with speaking, chewing, or swallowing [23].

11.2.4 Huntington's Disease

Huntington's disease (HD) is an inherited neurodegenerative disease characterized by neuropsychiatric symptoms, a movement disorder (most commonly choreiform) and progressive cognitive impairment [24]. HD is triggered by mutation of the Huntingtin gene, leading to anomalies in the protein functioning, deleterious effects, and ultimately the demise of specific neuronal cells. The disease is inherited in an autosomal dominant manner and leads to a battery of neuropathological conditions and neuronal cell death, mainly in the striatal and cortical regions of the brain [25]. Structural imaging has demonstrated significant volume loss across multiple white and gray matter regions in HD, particularly within subcortical structures [26]. However, there have been no significant improvements in treatments of patients suffering from this ailment and most of the treatment is still symptomatic. HD warrants more attention towards a better understanding and treatment as more advancements in molecular diagnostics and therapeutic interventions become available.

11.3 Cerebrovascular Disease

Cerebrovascular disease is a group of brain disorders related to diseases of the blood vessels supplying the brain. Cerebrovascular disease is a leading cause of physical disabilities, the second most common cause of mortality and the primary reason for the admission of a large proportion of hospital patients. The most common cerebrovascular disease is stroke. As the third leading cause of death in the industrialized world, stroke causes 15 million injuries and 5 million deaths each year. Strokes can be categorized as ischemic or hemorrhagic. Ischemic stroke accounts for 80% of all stroke insults that are precipitated by hypoperfusion, thrombosis, or embolism. The remaining 20% of strokes are hemorrhagic in etiology and can be attributed to an underlying vascular lesion or hypertension.

11.3.1 Ischemic Stroke

Ischemic strokes are classified according to the Trial Org 10172 in Acute Stroke Treatment (TOAST) system, which subdivides them into categories such as cardioemboli, small-vessel occlusions, large-artery atherosclerosis, and strokes of undetermined etiology. These types of strokes are caused by a decrease in blood flow to the brain resulting from either a thrombotic or embolic event [27].

11.3.2 Hemorrhagic Stroke

Hemorrhagic stroke occurs when a blood vessel ruptures and causes bleeding in the brain. This type of stroke can be classified as either intracerebral hemorrhage (ICH) or subarachnoid hemorrhage (SAH). The primary cause of hemorrhagic stroke is often hypertension. The presence of a hematoma can disrupt the functioning of neurons and glial cells, leading to oligaemia, release of neurotransmitters, mitochondrial dysfunction, and cellular swelling. In addition, the activation of microglia by thrombin can result in inflammation and edema [28]. The goal of therapy in acute ischemic stroke is to preserve tissue in areas where perfusion is decreased, but sufficient to avoid infarction. The AHA/ASA recommends intravenous alteplase for patients who satisfy the inclusion criteria and have symptom onset or a last known baseline within three hours. Use of a mechanical thrombectomy should be considered for all patients, even in those who received fibrinolytic therapy.

11.4 Polyphenols

Polyphenols are a large category or class of phenolic compounds, largely distributed as phytoconstituents with diverse properties and bioactivities. Since ancient times, phenolic compounds are widely used in the Indian system of medicines for the treatment of various disorders because of their health benefits. The basic monomer in polyphenols is a phenolic ring and generally these are classified as phenolic acids and phenolic alcohols. Depending on the strength of the phenolic ring, polyphenols can be classified into many classes (Figure 11.2); however, the main classes in the polyphenols are phenolic compounds that are subdivided into five major groups, i.e., phenolic acid, flavonoids, stilbins, tannins, lignans, and coumarins. Phenolic compounds are naturally synthesized by two metabolic pathways, i.e., shikimic acid and acetate pathways. Phenyl propanoids are formed by the shikimic acid pathway, whereas simple phenols are formed by the acetic acid pathway. The junction of these pathways leads to the formation of flavonoids, and after the condensation process, non-hydrolysable tannins are formed. Phenolic compounds are present in bound form with sugars and proteins that are inside the cell vacuoles. In free form, they are toxic in nature [29]. Phenols are very sensitive to enzymatic oxidation, thus extraction with the boiling alcohol may limit enzymatic oxidation [30]. These compounds are embedded in the human diet and originate from plants such as fruits, vegetables, cereals, and coffee. Polyphenols are also known as being preventive for degenerative diseases. Investigations on polyphenols are delayed because of their characteristics and structural complexity. The most frequent antioxidants in our diet are polyphenols. These hinder the oxidative change in low-density lipoprotein and this is the basic mechanism of endothelial lesions taking place in atherosclerosis [31–33]. Studies demonstrated the role of polyphenols in the remedy of cardiovascular disease, osteoporosis, neurogenerative disease, cancer, and diabetes mellitus [31, 32].

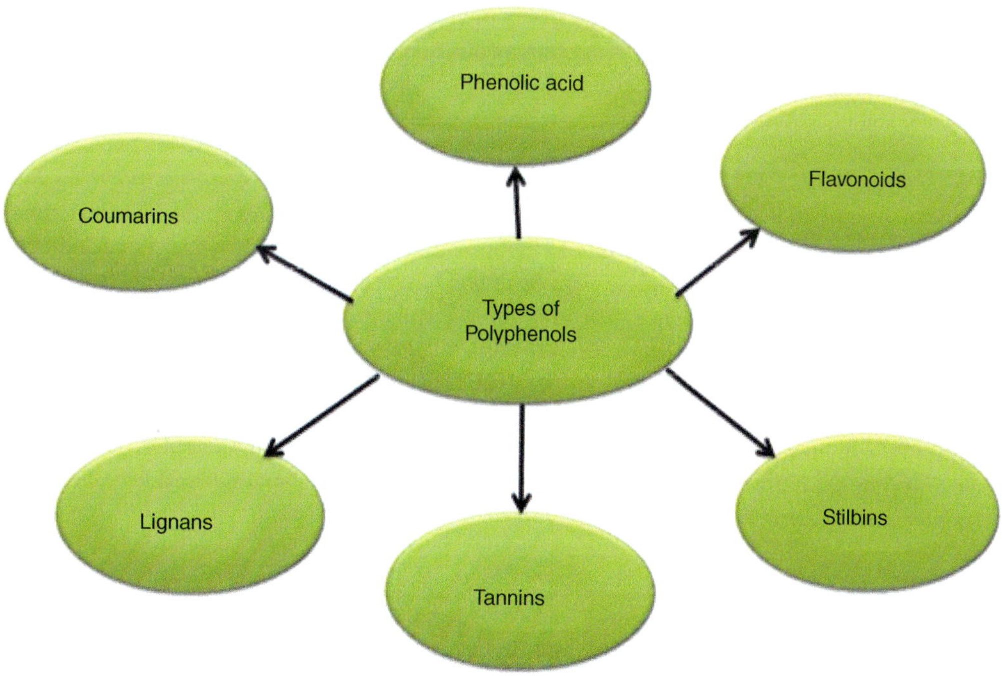

Figure 11.2 Different types of polyphenols.

11.4.1 Polyphenol Sources and Classes

11.4.1.1 Phenolic Acids

Phenolic acids comprise aromatic rings with a carboxylic acid group (–COOH). These phenols represent the main class of plant-based phenolic compounds. Because phenolic acid has antioxidant properties, it protects against cardiovascular diseases [34–36]. The aromatic ring can lose its electron and form free hydrogen radicals, which acts as a reducing agent and quenches free radicals; this mechanism protects against cardiovascular diseases. Plant foods, such as seeds, fruit peels, and leafy vegetables, are the main source of phenolic acids. These phenolic acids offer a wide range of cosmetic, food, and therapeutic applications [37]. Phenolic acid is divided into hydroxybenzoic acid and hydroxycinnamic acid. Hydroxybenzoic acid comprises C6–C1 derived from benzoic acid ($C_7H_6O_2$). Salicylic acid, vanillic acid, protocatechuic acid, gallic acid, benzoic acid, and ellagic acid are classified in the subcategory of hydroxybenzoic acid [38], which is a monomer of structurally more complex compounds, such as hydrolyzed tannins, and is usually available in bound form. Some hydroxybenzoic acids are also present in olive products, which have antioxidant, anti-inflammatory, and cardioprotective effects [36, 39, 40]. Hydroxycinnamic acid represents the class of aromatic acids (C6–C3) derived from cinnamic acid [41]. Caffeic acid, ferulic acid, coumaric acid, sinapic acid, and cinnamic acid are common examples of hydroxycinnamic acid [40]. This cinnamic acid is the structural and biological component that makes up the cell organelle. Tea and grape seeds are a rich source of hydroxybenzoic acids, and coffee, berries, apples, cereals, and kiwi are a sufficient source of hydroxycinnamic acids [42, 43]. Tea and grape seeds (gallic acid), coffee (caffeine and chlorogenic acid), berries, apples (caffeic acid), cereals (ferulic acid) contain the respective acids, while most citrus fruits contain cinnamic acid. This phenolic acid is an easily digestible, excellent source of phytochemicals that provides numerous health benefits, including anti-inflammatory properties, protection against cellular damage, reactive oxygen species (ROS), oxidative stress, cardiovascular effects, anticancer, antidiabetic, neuroprotective effects, and neuropreservative [41].

11.4.1.2 Flavonoids

Flavonoids are a major group of plant metabolites of polyphenolic compounds. The structural composition of these flavonoids comprises 15 carbon atoms and 2 aromatic rings connected by a chain of 3 carbon atoms [44]. Flavonoids can be divided into several categories depending on the C-ring to which the B-ring is attached, its configuration, and the oxidation of the C-ring. The main classes of flavonoids, including flavones (chrysin, apigenin, baicalein), flavonols (quercetin, kaempferol), isoflavones (daidzein, glycitein), flavan-3-ols (gallocatechin, catechin, epicatechin), flavanones (hesperetin, naringenin), and anthocyanidins (delphinidin, peonidin, cyanidin, pelargonidin) [35, 40]. Flavonoids are mainly found in berries, onions, tea, grapes, apples, berries, and cocoa, and they exhibit numerous health properties, including cell signaling, antithrombogenic, and neuroprotective properties [45].

11.4.1.3 Flavonols

Flavonols belong to the flavonoid family and comprise C2–C3 double bonds and C4 carbonyl. Flavonols include quercetin, kaempferol, and myricetin, which are ubiquitous plant flavonoids found in numerous fruits and vegetables such as kale, onions, lettuce, and tomatoes [46]. Recent evidence from clinical studies show that these flavonols have the potential for pretreatment and prevention of cardiovascular disease, cell regeneration, prevention of blood plaque, human gum, and other heart related diseases [34, 40]

11.4.1.4 Isoflavones

Isoflavones are a type of phytoestrogen that are non-steroidal compounds derived from plants [47]. Isoflavones are derived from the *Fabaceae* family. Isoflavones have 2 phenyl-4H-1benzopyr-4-ones as their structure instead of a phenyl group [40]. Isoflavones are obtained via the phenylpropanoid pathway, which contributes to the formation of flavone groups in higher plants. Isoflavones are extracted from soy beans and soy products. The isoflavone content in soy beans is high because they contain two key components, daidzein and genistein [48].

11.4.1.5 Flavanones

Citrus fruits, some aromatic plants, and tomatoes contain flavanones, which are a small part of flavonoids. Apart from their taste properties, they are considered an important component of human health. Eriodyctiol from lemons, hesperidin from oranges, and naringenin from grapefruit are flavanones [49].

11.4.1.6 Anthocyanidins

Anthocyanidins are the pigments mainly responsible for the color (red, pink, purple) of fruits and vegetables [50]. The different colors of fruits, vegetables, and flowers are because of the presence of anthocyanidin in their epidermis. Various fruits and vegetables, such as radish, beet, berries, strawberries, and cherries, are good sources of anthocyanidins [51]. Among these, aronia berries are well known for their increased antioxidant properties [52]. They are obtained from prickly pears and have various components such as flavanols, kaempferol, cyanidin glycosides, and other bioactive compounds [40]. Honey, olive oil, and berries have proven their immense benefits for human health over time.

11.4.1.7 Flavones

Flavones are commonly found in cereals, celery, parsley, and broccoli. They are also present in large amounts in the outer layers of citrus fruits. Flavones have been found to have an inverse effect on Coronary artery disease [53].

11.4.1.8 Other Polyphenols

Other polyphenols, including stilbenes (resveratrol, piccatannol), lignans (sesamol, pinoresinol, sinol, enterodiol), tannins (hydrolyzable, non-hydrolyzable, and condensed tannins), and lignins, have a wide range of therapeutic and industrial applications depending on their mode of action [35, 40, 54].

11.4.1.9 Lignin

Lignins are heterogeneous polymers derived from a few signal precursors that are crosslinked in various forms [40]. There are three types of crosslinks derived from phenylpropane, and these crosslinks are coniferyl alcohol, sinapyl alcohol, and paracoumaryl alcohol [55]. Some foods rich in lignin are flaxseed, tomatoes, peaches, apples, and some berries. Lignin silymarin is a type of flavonolignan that has antioxidant properties. This type of lignin is obtained from milk thistle seeds, which are obtained from certain varieties of daisies and other herbaceous plants [56].

11.4.1.10 Stilbenes

A category of metabolites derived from phenols are the stilbenes ($C_{14}H_{12}$). Their biological activity and health benefits are of great interest to researchers and have been the focus of several studies [57]. Stilbenes are organic compounds that have a compact structure with a central ethylene moiety and a phenyl group. The phenyl group is located at the ends of the carbon double bonds [58]. Stilbenes are found in grapes, berries, and few other plants. The most commonly known type of stilbene is resveratrol. Resveratrol is more extensively known because of its properties in disease prevention, and it is derived from stilbene [59].

11.4.2 The Role of Polyphenols in the Treatment of CNS Disorders

Curcumin is a diaryl hepatanoid polyphenol isolated from the rhizomes of *Curcuma longa*. It is a potent antioxidant because of its capacity to scavenge free radicals because of its unique structure that can donate H atoms or transfer electrons from the phenolic sites. It also has anti-inflammatory activity as it inhibits lipopolysaccharide-induced morphological changes of microglia and decreases the production of pro-inflammatory factors. Curcumin restores glutathione levels, which protects neurons against protein oxidation and preserves mitochondrial complex-I activity [60]. Eicosapentaenoic acid (EPA) is a type of omega-3 fatty acid that is commonly found in fatty fish such as salmon and mackerel. EPA has anti-inflammatory properties, reduces the production of pro-inflammatory cytokines, decreases beta-amyloid secretion, and has a positive impact on dopamine-producing neurons in the basal ganglia. Sulforaphane is a naturally occurring compound found in cruciferous vegetables such as broccoli and cabbage. It has various health benefits, including its ability to improve cognitive function by increasing acetylcholine levels and decreasing acetylcholinesterase activity in the brain. Additionally, it has antioxidant and anti-inflammatory effects by reducing levels of ROS and inhibiting the pro-inflammatory signaling pathway through NF-kβ. Anthocyanin is a type of polyphenol found in various fruits and vegetables, including blueberries and raspberries, and it has anti-inflammatory properties. Apigenin, a flavonoid commonly found in chamomile, celery, parsley, and peppermint, has potent antioxidant properties and protects neuronal cells from damage. Coenzyme Q 10 is also known as ubiquonone. It is a coenzyme that is ubiquitous in animals and most bacteria. Coenzyme Q 10 acts as an antioxidant by scavenging free radicals and regenerating other antioxidants such as vitamin C and vitamin E. Additionally, coenzyme Q 10 improves insulin sensitivity and glucose uptake by cells and enhances energy metabolism by assisting in the function of mitochondrial enzymes. Coenzyme Q 10 has potential therapeutic effects in AD by increasing acetylcholine production, inhibiting the formation of harmful free radicals, and scavenging toxic byproducts of lipid peroxidation [61].

11.4.2.1 Polyphenols in Alzheimer's Disease and Dementia

Polyphenols found in green tea and grape extracts have been shown in various studies in animal models to have neuroprotective properties. These polyphenols have been shown to inhibit acetylcholinesterase, protect primary rat cortical neurons against Aβ-induced cytotoxicity, improve cognitive functions in AD mouse models, improve synaptic transmission, and inhibit the oligomerization of Aβ peptides and abnormal folding of tau proteins. These findings indicate the potential of polyphenols in treating age-related disorders such as AD and dementia. Resveratrol, a polyphenol abundant in grapes and red wines, has been shown to inhibit Aβ 42 fibril formation [62] and protect against Aβ neurotoxicity by inhibiting inducible nitric oxide synthase [63]. Resveratrol, which may have high bioavailability in lipid-core nanocapsules, has demonstrated therapeutic potential in AD [64]. Flavonoids, such as fisetin and its analogues, has been shown to inhibit the formation of Aβ fibrils and they are emerging as new drug candidates for the treatment of AD [65]. Morin (2,3,4,5,7-pentahydroxyflavone) has also been shown to prevent neuronal cell death by protecting neurons against tau hyperphosphorylation induced by Aβ [66]. A flavonoid, 7,8 dihydroxyflavone, has been shown to improve cognitive abilities in the 5XFAD transgenic mouse model of AD by activating the tyrosine receptor kinase B, resulting in a reduction in β-secretase enzyme levels and synthesis of amyloid beta (Aβ) [67]. Rutin has been found to control oxidative stress, reduce malondialdehyde levels, and prevent the formation of glutathione disulfide in SH-SY5Y neuroblastoma cells. Additionally, rutin has mitigated the inflammatory cascade by decreasing levels of cytokines such as TNF-α and IL-1β [68].

11.4.2.2 Polyphenols in Multiple Sclerosis

Quercetin was found to control the immune response through modulation of IL-1β and TNF-α and to reduce the proliferation of peripheral blood mononuclear cells isolated from multiple sclerosis patients [69]. The epigallocatechin-3-gallate (EGCG) exhibited neuroprotective effects by modulating neuroinflammation and reducing neural damage [70]. Quercetin [71], apple polyphenols [72], myricetin, and piceatannol [73] have also activated SIRT1, showing potential in the treatment of multiple sclerosis. Preclinical data have shown that polyphenols have the potential to block neural inflammation and damage by activating the SIRT1 pathway and modulating inflammatory cytokines. The potential of polyphenols to limit demyelination makes them promising therapeutic agents for age-related multiple sclerosis and ALS.

11.4.2.3 Polyphenols in Ischemic Stroke

Green tea polyphenol, EGCG, exhibited neuroprotective action through downregulation of matrix metalloproteinases (MMP) in a mice model of cerebral ischemia [74]. Green tea polyphenols have also been found to protect neurons against hypoxia-induced ischemic injury by controlling the inflammation cascade and attenuating the decline in transmembrane potential [75]. Rutin has been reported to mitigate neural damage and further necrosis in middle cerebral artery occlusion rat model through down regulation of the p53 gene [76]. The flavonoid fisetin has been shown to have neuroprotective action during cerebral ischemia as it stops infiltration of macrophages and dendritic cells into the ischemic hemisphere, thus controlling neural inflammation and damage [77]. Another flavonoid, baicalin, has been shown to reduce ischemic stroke damage by targeting multiple therapeutic targets such as MMP-9 [78], caspase-3, oxidative stress [79], and p38 mitogen-activated protein kinase (MAPK) [80], as well as by downregulating the toll-like receptor (TLR2/4) pathway [81].

11.4.2.4 Polyphenols in Parkinson's Disease

Resveratrol has been shown to inhibit the loss of dopaminergic neurons in a rat model of Parkinson's disease [82]. It has also been shown to reduce neural inflammation in Parkinson's disease by lowering mRNA levels of cyclooxygenase-2 (COX-2) and TNF-α in the substantia nigra [83]. Additionally, resveratrol has been demonstrated to reduce oxidative stress, lipid peroxidation, and protein carbonyl in a rat model of Parkinson's disease [84]. Other polyphenols, such as baicalein [85], kaempferol [86], caffeic acid [87], and EGCG [82], have been shown to provide neuroprotection in PD studies. Similarly, polyphenolic extracts from various plants have established a pharmacological role in PD studies. For example, polyphenol-rich mulberry fruit extracts have displayed antioxidant and anti-apoptotic effects in SH-SY5Y cells by modulating caspase-3, B-cell lymphoma (Bcl-2), and BCL2-associated X protein (Bax) [88].

11.4.2.5 Polyphenols in Huntington's Disease

Resveratrol has been found to exhibit positive effects in a transgenic mouse model of HD through activation of SIRT1, peroxisome proliferative activated receptor-gamma, and coactivator 1-alpha (PGC-1α) signaling pathways. Studies have further demonstrated that the neuroprotective potential observed in HD models was because of the activation of Ras-extracellular signal-regulated kinase by resveratrol and fisetin [89]. Additionally, hesperidin and naringenin, which are abundant in citrus fruits, induce neuroprotection in rats, possibly through the inhibition of nitric oxide synthase [90]. Curcumin has been shown to control Huntington aggregates and improve various transgene-dependent parameters, promising therapeutic action in HD [91]. Polyphenols from grapes and green tea have also shown potential in treating or preventing HD disease pathogenesis [92].

11.4.2.6 Polyphenols in Psychiatric Diseases

Psychiatric disorders, including major depression, attention deficit hyperactivity disorder (ADHD), and schizophrenia, contribute largely to mental problems in children, adolescents, and adults. It is believed that oxidative stress also plays an important role in the pathology of psychiatric disorders [93]. Curcumin, a non-flavonoid phenolic compound found in *C. longa* and widely used in Indian traditional medicine, has been shown to significantly decrease depression-like behavior in rats through the improvement of BDNF levels. When combined with the alkaloid piperine, curcumin also inhibits monoamino-oxidase (MAO) activity and increases serotonin and dopamine levels in mice. Clinical evidence from a randomized, placebo-controlled trial of 60 patients with major depression showed that curcumin (at a dose of 1000 mg/day for 6 weeks) may be an effective and safe treatment option for depression without concurrent suicidal ideation [94]. The flavonoid derived from catechin, EGCG, which is present in green tea, has been used in traditional Chinese medicine for at least 4,000 years. EGCG is currently recognized for its potent antioxidant properties and for its ability to mitigate stress and depression. In an experimental study on mice, an increase in BDNF levels was observed after long-term administration of green tea polyphenols [95], or reduced serum corticosterone and adrenocorticotropic hormone levels were found after a forced swimming test [96]. In vitro experiments with cultured hippocampal neurons confirmed that the GABA-A receptor benzodiazepine site can be specifically modulated by the application of EGCG [97]. In a double-blind, randomized, and placebo-controlled human study involving 74 subjects, long-term supplementation with green tea extract was found to increase reward learning and prevent depressive symptoms [98]. The polyphenols from *Ginkgo biloba*, comprising flavonols quercetin and kaemferol, were shown to have antidepressant-like effects in mice, likely through increasing BDNF levels, promoting neuronal survival and plasticity, and inhibiting MAO towards serotonin [99]. In vitro studies have also shown that MAO is inhibited by anthocyanins from berries, flavone apigenin from celery, and stilbene trans-resveratrol from red wine [100]. Furthermore, flavonoids from cocoa have demonstrated antidepressant-like effects in rats using the forced swimming test [101] and reduced symptoms of chronic fatigue in ten subjects enrolled in a double-blind, randomized, clinical pilot crossover study [102]. Attention deficit hyperactivity disorder (ADHD) is a polygenic condition [103] that is a complex and heterogeneous disorder influenced by multiple etiological factors. Oroxylin A, a chemical compound found in the medicinal plant *Scutellaria baicalensis* and the *Oroxylum indicum* tree, is an O-methylated flavone. It has been shown to inhibit dopamine reuptake; however, not noradrenaline reuptake. Its analogue, 5,7-dihydroxy-6-methoxy-4-phenoxyflavone, demonstrated remarkable inhibition of dopamine reuptake, comparable to methylphenidate; however, it did not show modulation of the GABA pathway in a spontaneously hypertensive rat model of ADHD [103, 104]. Preclinical studies suggest that green tea extract, which contains the main polyphenol EGCG, may benefit patients with schizophrenia, and they investigated the efficacy of EGCG doses of 600 mg/day as an adjunctive treatment with antipsychotic medication in 34 patients in a double-blind, placebo-controlled study [105]. The flavonoid epicatechin was found to inhibit lipid peroxidation in human plasma caused by haloperidol in an ex vivo experiment [106]. Similarly, polyphenols from berries isolated from *Aronia melanocarpa* were shown to inhibit plasma lipid peroxidation induced by the atypical antipsychotic drug ziprasidone in ex vivo experiments [107]. The extract from *Ginkgo biloba* (EGb-761), whose components are mostly lipophilic, crosses the blood–brain barrier and protects the brain against the damaging effects of oxidative stress. This improvement may be because of the well-known antioxidant properties of the extract [108]. Genistein, a polyphenol belonging to phytoestrogens, along with the amino acid leucine, can better potentiate haloperidol-induced catalepsy in rats than a haloperidol-treated group, and it reduces the number of fights and increases the latency to fights in foot shock-induced aggression [109]. Resveratrol has been found to extend protection against ischemic injury by improving brain energy metabolism and controlling oxidative stress during ischemic injury in animal model studies [110]. It also modulates the release of multiple therapeutic neurotransmitters and neuromodulators during ischemic injury [111].

11.4.3 Polyphenols in Nanodeliveries

Polyphenols are also used in nanodelivery applications. In recent years, various nanocarrier-based formulations were investigated for the treatment of CNS disorders, including polymeric nanoparticles, solid lipid nanoparticles (SLN), liposomal formulations, phytosomes™, and aquaosomes.

11.4.3.1 Polymeric Micelles

Polymeric micelles are tiny structures formed from amphiphilic polymers, which comprise both hydrophobic and hydrophilic segments. These structures can be used to encapsulate a variety of molecules, including drugs, dyes, and nutrients. Polyphenols are a type of nutrient that are found in many plant-based foods and they have been shown to have a variety of health benefits, including antioxidant, anti-inflammatory, and anticancer properties. However, polyphenols can be difficult to deliver to the body effectively because of their poor solubility and low bioavailability. Polymeric micelles offer a potential solution to this problem. When polyphenols are encapsulated within polymeric micelles, they become more stable and soluble, this can increase their bioavailability and improve their delivery to target tissues [112]. The hydrophobic core of micelles provides a protective environment for the polyphenols, shielding them from degradation and increasing their stability. The hydrophilic outer shell of the micelles makes them more water-soluble and enables them to be transported through the bloodstream to reach their intended target tissues [113]. One major advantage of using polymeric micelles for the delivery of polyphenols is that they can improve the absorption of polyphenols in the body because their small size allows them to pass through the intestinal barrier and enter the bloodstream more easily. In addition, the protective environment provided by the micelles can help to prevent the breakdown of polyphenols by digestive enzymes, increasing their bioavailability and efficacy [114]. Another advantage of using polymeric micelles for the delivery of polyphenols is that they can improve the stability of the polyphenols. Polyphenols are prone to oxidation and degradation, which can reduce their effectiveness. Encapsulating them within polymeric micelles can protect them from these processes and help to maintain their activity over time. In the same context, Mohanty et al. (2010) developed polymeric micelles made of methoxy poly(ethylene glycol) (MePEG)/ poly-e-caprolactone (PCL) de block to encapsulate curcumin, and they reported improved solubility and stability of curcumin [115]. They also mentioned that it may improve the bioavailability. In a similar work, Khonkarn et al. (2011) also reported the development of polymeric micelles made of (poly(ethylene glycol)-b-oligo(e-caprolactone)) used to encapsulate quercetin [116]. They mentioned an improved solubility and stability of quercetin with its increased arrest of the cell cycle and can be used for the management of tumors. Polymeric micelles of other polyphenols, such as resveratrol and EGCG, are also reported. Polyphenol-loaded polymeric micelles have also been studied for their potential use for topical application for skin health; however, they were contradictory. Smejklova et al. (2017) noted a decrease in size and an increase in the skin penetration [117]; in contrast, Lapteva et al. (2014) reported the impermeability of MPEG-dihex PLA poly lactic copolymer micelles through the stratum corneum [118]. However, at the same time, several reported polymeric micelles being used for the delivery of various polyphenols, such as silibinin [119] and EGCG [120], through topical or transdermal routes. Finally, polymeric micelles are a safe and biocompatible delivery system for polyphenols. The polymers used to form the micelles are typically non-toxic and non-immunogenic, which means they do not provoke an immune response in the body. This makes them a suitable option for the delivery of polyphenols as nutraceuticals, which are intended to promote health and prevent disease. Polymeric micelles offer a promising delivery system for polyphenols as nutraceuticals. By encapsulating polyphenols within micelles, their solubility, stability, and bioavailability can be improved, making them more effective as dietary supplements. The use of polymeric micelles for the delivery of polyphenols has the potential to enhance their therapeutic properties and contribute to the development of new and effective nutraceutical products.

11.4.3.2 Metallic Nanocarriers

Metallic nanocarriers are a promising approach for delivering the medicaments, including polyphenols, because of their unique physicochemical properties such as high surface area, high reactivity, and ability to be easily functionalized. Polyphenols are a class of natural compounds that are found in many plant-based foods and have been shown to have numerous health benefits, including antioxidant, anti-inflammatory, and anticancer properties. However, the poor solubility and low bioavailability of polyphenols have limited their effectiveness as dietary supplements. Metallic nanocarriers can potentially overcome these challenges and improve the delivery of polyphenols to target tissues. Metal–polyphenol network (MPN) based nanoparticles are one of the most important examples that can also be used for drug loading and theragnostic purposes. In this context, Jiang et al. (2023) reported the MPN-coated magnetic hydroxyapatite loaded with docetaxel and mentioned the use of MPN in improving the biocompatibility and therapeutic efficacy [121]. Metallic nanocarriers containing polyphenols can be prepared using a variety of metals such as gold, silver, iron, and zinc. These nanocarriers are typically synthesized using chemical or physical methods such as reduction, precipitation, electrospinning, and microwave irradiation. The surface of the nanocarriers can be functionalized with various coatings or ligands to enhance their stability, solubility, and targeting specificity [122]. One major advantage of metallic nanocarriers for polyphenols is their ability to improve the bioavailability of the polyphenols. The high surface area of the nanocarriers allows for a large number of polyphenols to be loaded onto their surface, increasing their concentration and availability for absorption. In addition, the protective environment provided by the nanocarriers can help to prevent the breakdown of polyphenols by digestive enzymes, enhancing their bioavailability and efficacy [114]. There are several examples of metallic nanocarriers being used to deliver polyphenols. In this context, gold and silver nanocarriers have been used for the delivery of resveratrol with improved antibacterial activity [123]. Resveratrol-loaded cyclodextrin metal–organic framework/chitosan (CD-MOF/CS) nanocapsules have also been reported with improved water dispersibility, photostability, and antioxidant activity [124]. EGCG-conjugated gold NPs is another very good example that can be used for the management of different types of tumors [122]. In conclusion, metallic nanocarriers offer a promising approach for delivering polyphenols. By conjugating the polyphenols with metallic nanocarriers, their stability, solubility and bioavailability can be improved, making them more effective. Different types of novel nanocarriers for the effective delivery of polyphenols in the treatment of neurological disorders are listed in the Table 11.1.

11.4.3.3 Phytosomes™

Phytosomes™ are circular vesicles made up of phytoconstituents encapsulated with phospholipids (Figure 11.3). Various phytosomes™ formulations were prepared for the improvement of bioavailability of polyphenols for the use of nutraceuticals and immunomodulators, which includes panax ginseng [125]. Phytosome-loaded *Ginko biloba* was reported for the treatment of neurological disorders [126].

Table 11.1 Various nanocarrier-loaded polyphenols for the treatment of neurological disorders.

Drug / Polyphenols	Nanocarrier	Diseases	Outcomes	References
Resveratrol	Nanoparticle	Parkinson's Disease	Drug loaded nanoparticles improved the bioavailability and therapeutic effect in rotenone-induced oxidative stress, motor deficits, mitochondrial dysfunction in experimental animal as compared with free drug	[132]
Resveratrol	Lipid core nanocapsule	Alzheimer's Disease	Enhanced distribution of the drug in the liver, brain, kidney; better memory and learning capacity, better neuroprotective efficacy against reactive oxygen species development and cell death	[133, 134]
Resveratrol	Polymeric micelles	Alzheimer's Disease	Improved prevention of Aβ-induced oxidative stress in PC-12 cells	[135]
Resveratrol	Solid lipid nanoparticle (SLN)	Alzheimer's Disease	Drug-loaded SLNs showed better anti-aggregation properties	[136]
Resveratrol	Liposomes	Parkinson's Disease	Improved drug efficacy compared with free drug in the nigral cell	[137]
Resveratrol	Polymeric Nanoparticle	Parkinson's Disease	Improved neuroprotective effects in C57BL/6 mice compared with free drug	[138]

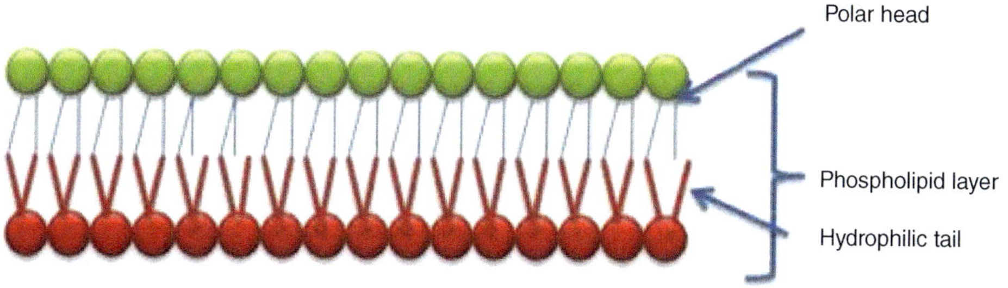

Figure 11.3 The bilayer of a phospholipid complex.

11.4.3.4 Ethosomes

Ethosomes are phospholipid vesicles comprising a drug molecule in the inner core and ethanol at a higher concentration. Because of these features, ethanol can solubilize with a variety of bioactives and improve the penetration profile through the skin [127–129]. Kayal et al. (2019) and Faisal et al. (2016) prepared EGCG-loaded ethosomes for effective delivery and better therapeutic antioxidant potential and photostability [130, 131].

11.4.3.5 Nutraceuticals as a Source for Polyphenols

Nutraceuticals are an organic compound comprising various polyphenols obtained from plants and they have wide application in the treatment of various human disorders (Figure 11.4). These are basically used for health benefits, prevention, and treatment of various medical conditions with no side effects. These nutraceutical products are currently available in the global market with different brand names. Some of these patents are listed in Table 11.2. Marketed products are listed in Table 11.3. Overall, these nutraceuticals are a multibillion-dollar industry [141].

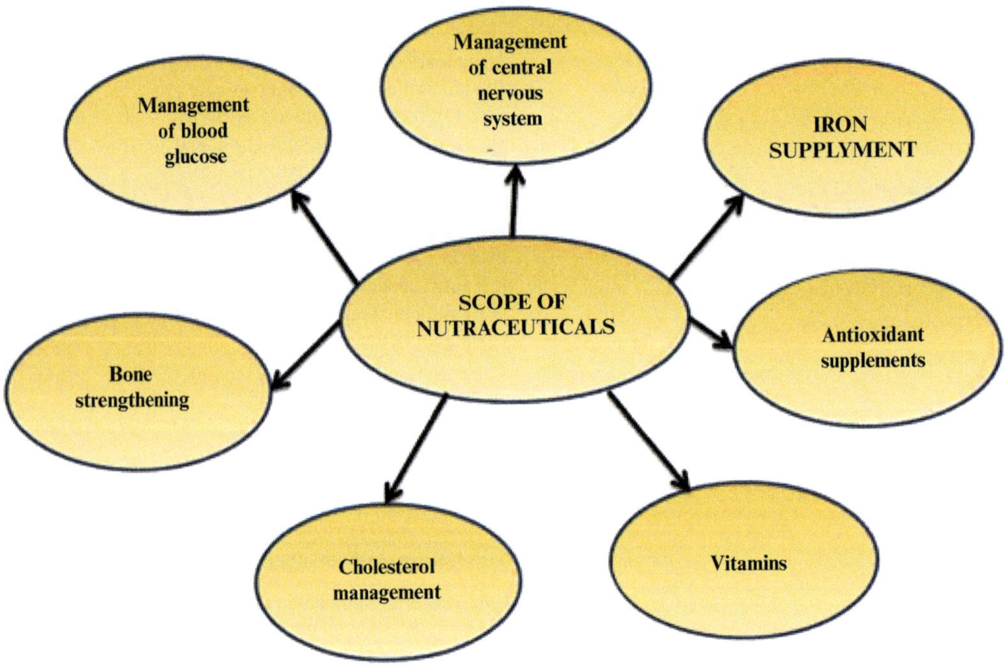

Figure 11.4 Scope of nutraceuticals.

Table 11.2 Patents of polyphenolic drug delivery for central nervous system disorders.

Patent No.	Polyphenol Used	Formulation	Diseases	References
WO2008/005577	Epigallocatechin-3-gallate, epicatechin (Green tea polyphenols)	–	Dementia	[139]
CN111803632A	Flavone polyphenol	Phospholipid complex	–	[140]

Table 11.3 Marketed polyphenols used as nutraceuticals.

Brand Name	Manufacturer	Available Dose	Benefits
Magnum Big C™	Magnum® Nutraceuticals	Capsule	Muscle gainer, water balance
Iron Folic	Nutrilite®	Tablet	Iron supplements, anemia
Acai berry pure	Natures Craft®	Capsule	Vitamins, minerals, antioxidant supplements
Nutrela™ Omega	Patanjali® Ayurved	Capsule	Improve heart, skin, eye, brain function, and cholesterol management
Nutrela™ Bone health	Patanjali® Ayurved	Capsule	Improve bone, fracture healing, and bone strengthening
Nutrela™ Diabetic Care	Patanjali® Ayurved	Powder	Management of blood glucose and weight
BeneFlora® S Probiotic	Shreya Life science	Capsule	Pro- and pre-biotic supplement

11.5 Conclusion and Future Perspectives

Polyphenols are nutraceuticals that have a wide scope in the treatment of various CNS disorders, including Alzheimer', Parkinson's, migraines, and other medical conditions. The main challenge of polyphenols is their poor aqueous solubility, which causes low bioavailability and stability. Therefore, in recent years, various nanocarriers have been investigated for the delivery of polyphenols, including nanoparticles, SLNs, nanostructured lipid carriers, metallic nanoparticles, vesicular carriers, liposomes, niosomes, and micelles. These nanoengineered carriers introduce new avenues over the conventional formulations for improving stability and bioavailability. These agents are promising for the nanodelivery of polyphenols because of their unique profiles. In addition, formulation development and commercialization of these nanocarriers for polyphenols as nutraceuticals must be performed.

Acknowledgment

The authors are thankful to the Librarian of the Central Library and the Director of Shri Govindram Seksaria Institute of Technology and Science (SGSITS), Indore for providing necessary e-resources, Inflibnet and lab facilities.

References

1 Davis, M.E. (2018). Epidemiology and overview of gliomas. *Seminars in Oncology Nursing* 34 (5): 420–429.

2 Ostrom, Q.T., Gittleman, H., Liao, P. et al. (2021). CBTRUS Statistical report: primary brain and other central nervous system tumors diagnosed in the United States in 2010–2014. *Neuro Oncology* 19 (suppl_5): v1–v88.

3 Louis, D.N., Perry, A., Wesseling, P. et al. (2021). The 2021 WHO classification of tumors of the central nervous system: a summary. *Neuro-oncology* 23 (8): 1231–1251.

4 Wen, P.Y. and Brandes, A.A. (2009). Treatment of recurrent high-grade gliomas. *Current Opinion in Neurology* 22 (6): 657–664.

5 Xiong, L., Wang, F., and Qi, X.X. (2019). Advanced treatment in high-grade gliomas. *Journal of BUON: Official Journal of the Balkan Union of Oncology* 24 (2): 424–430.

6 Gauthier, S., Rosa-Neto, P., Morais, J.A., and Webster, C. (2021). *World Alzheimer Report 2021: Journey Through the Diagnosis of Dementia*. London, England: Alzheimer's Disease International.

7 Cahill, S. (2020 Feb). WHO's global action plan on the public health response to dementia: some challenges and opportunities. *Aging and Mental Health* 24 (2): 197–199. https://doi.org/10.1080/13607863.2018.1544213. Epub 2019 Jan 2. PMID: 30600688.

8 Kuang, J., Zhang, P., Cai, T. et al. (2021). Prediction of transition from mild cognitive impairment to Alzheimer's disease based on a logistic regression–artificial neural network–decision tree model. *Geriatrics & Gerontology International* 21 (1): 43–47.

9 Pagani, M., Giuliani, A., Öberg, J. et al. (2016). Predicting the transition from normal aging to Alzheimer's disease: a statistical mechanistic evaluation of FDG-PET data. *Neuroimage* 141: 282–290.

10 Thawkar, B.S. and Kaur, G. (2019). Inhibitors of NF-κB and P2×7/NLRP3/Caspase 1 pathway in microglia: novel therapeutic opportunities in neuroinflammation induced early-stage Alzheimer's disease. *Journal of Neuroimmunology* 326: 62–74.

11 Abdivalievna, A.N. (2022). Features of cognitive disorders. *Innovative Society: Problems, Analysis and Development Prospects* 101–105.

12 Pandey, P. and Sharma, P. (2011). Analysis of early onset of Alzheimer's disease genes: disease causing and risk factors. *European Journal of Biological Research* 11 (2): 251–259.

13 Schneider, L.S., Mangialasche, F., Andreasen, N. et al. (2014). Clinical trials and late-stage drug development for Alzheimer's disease: an appraisal from 1984 to 2014. *Journal of Internal Medicine* 275 (3): 251–283.

14 Valkova, P. and Pohanka, M. (2021). Novel trends in electrochemical biosensors for early diagnosis of Alzheimer's disease. *International Journal of Analytical Chemistry* 2021: 1–13.

15 Passeri, E., Elkhoury, K., Morsink, M. et al. (2022). Alzheimer's disease: treatment strategies and their limitations. *International Journal of Molecular Sciences* 23 (22): 120–139.

16 Tysnes, O.-B. and Storstein, A. (2017). Epidemiology of Parkinson's disease. *Journal of Neural Transmission* 124 (8): 901–905.

17 Jankovic, J. and Tan, E.K. (2020). Parkinson's disease: etiopathogenesis and treatment. *Journal of Neurology, Neurosurgery, and Psychiatry* 91 (8): 795–808.

18 Elsworth, J.D. (2020). Parkinson's disease treatment: past, present, and future. *Journal of Neural Transmission* (Vienna, Austria: 1996) 127 (5): 785–791.

19 Lang, A.E. and Espay, A.J. (2018). Disease modification in Parkinson's disease: current approaches, challenges, and future considerations. *Movement Disorders Official Journal of the Movement Disorder Society* 33 (5): 660–677.

20 Hulisz, D. (2018). Amyotrophic lateral sclerosis: disease state overview. *The American Journal of Managed Care* 24 (15 Suppl): S320–S326.

21 Oskarsson, B., Gendron, T.F., and Staff, N.P. (2018). Amyotrophic lateral sclerosis: an update for 2018. *Mayo Clinic Proceedings* 93 (11): 1617–1628.

22 Al-Chalabi, A., Jones, A., Troakes, C. et al. (2012). The genetics and neuropathology of amyotrophic lateral sclerosis. *Acta Neuropathologica* 124 (3): 339–352.

23 Brown, R.H. and Al-Chalabi, A. (2017). Amyotrophic lateral sclerosis. *New England Journal of Medicine* 377 (2): 162–172.

24 Stoker, T.B., Mason, S.L., Greenland, J.C. et al. (2022). Huntington's disease: diagnosis and management. *Practical Neurology* 22 (1): 32–41.

25 Pandey, M. and Rajamma, U. (2018). Huntington's disease: the coming of age. *Journal of Genetics* 97 (3): 649–664.

26 Johnson, E.B. and Gregory, S. (2019). Huntington's disease: brain imaging in Huntington's disease. *Progress in Molecular Biology and Translational Science* 165: 321–369.

27 Hui, C., Tadi, P., and Patti, L. (2022). *Ischemic Stroke. StatPearls*. Treasure Island (FL): StatPearls Publishing. Copyright ©, StatPearls Publishing LLC.

28 Unnithan, A.K.A., J, M.D., and Mehta, P. (2022). *Hemorrhagic Stroke. StatPearls*. Treasure Island (FL): StatPearls Publishing. Copyright ©, StatPearls Publishing LLC.

29 Verma, N. (2016). Review on phytoconstituents and their mode of extractions: an overview. *Research Journal of Chemical and Environmental Sciences* 4 (2): 8–15.

30 Densie, W. (2013). Phytochemicals role in good health. *The Magazine for Nutrition Professionals* 15: 1–10.

31 Archivio, D., Filesi, M., Di Benedetto, C. et al. (2007). Polyphenols, dietary sources and bioavailability. *Annali-IstitutoSuperiore Di Sanita* 43: 348.

32 Scalbert, A., Johnson, I.T., and Saltmarsh, M. (2005). Polyphenols: antioxidants and Beyond. *The American Journal of Clinical Nutrition* 81: 215S–217S.

33 Marrugat, J., Covas, M.I., Fitó, M. et al. (2004). Effects of differing phenolic content in dietary olive oils on lipids and LDL oxidation. *European Journal of Nutrition* 43: 140–147.

34 Behl, T., Bungau, S., Kumar, K. et al. (2020). Pleotropic effects of polyphenols in cardiovascular system. *Biomedicine & Pharmacotherapy* 130: 110714.

35 Rasouli, H., Farzaei, M.H., and Khodarahmi, R. (2017). Polyphenols and their benefits: a review. *International Journal of Food Properties* 20 (2): 1700–1741.

36 Shahidi, F. and Yeo, J. (2018). Bioactivities of phenolics by focusing on suppression of chronic diseases: a review. *International Journal of Molecular Sciences* 19 (6): 1573.

37 Rashmi, H.B. and Negi, P.S. (2020). Phenolic acids from vegetables: a review on processing stability and health benefits. *Food Research International* 10: 92–98.

38 Ozcan, T., Akpinar-Bayizit, A., Yilmaz-Ersan, L., and Delikanli, B. (2014). Phenolics in human health. *International Journal of Chemical Engineering and Applications* 5 (5): 393.

39 Gupta, S., Bishnoi, J.P., Kumar, N. et al. (2018). Terminalia arjuna (Roxb.) Wight &Arn.: competent source of bioactive components in functional food and drugs. *The Pharma Innovation Journal* 7 (3): 223–231.

40 Singla, R.K., Dubey, A.K., Garg, A. et al. (2019). *Natural Polyphenols: Chemical Classification, Definition of Classes, Subcategories, and Structures*. Oxford University Press.

41 Teixeira, J., Gaspar, A., Garrido, E.M. et al. (2013). Hydroxycinnamic acid antioxidants: an electrochemical overview. *BioMed Research International* 4: 112–136.

42 Baniwal, P., Mehra, R., Kumar, N. et al. (2021). Cereals: functional constituents and its health benefits. *Pharma Innovation* 10 (3): 01–7.

43 Yadav, D., Kumar, H., Kumar, A. et al. (2016). Optimization of polyphenolic fortification of grape peel extract in stirred yogurt by response surface methodology. *Indian Journal of Dairy Science* 69 (1): 124–144.

44 Harborne, J.B. (ed.) (2013). *The Flavonoids: Advances in Research Since 1980*. 621. New York, NY: Springer. https://doi.org/10.1007/978-1-4899-2913-6.

45 Ballard, C.R. and Junior, M.R.M. (2019). Health benefits of flavonoids. In: *Bioactive Compounds*, 185–201. Elsevier. https://doi.org/10.1016/b978-0-12-814774-0.00010-4.

46 Sandu, M., Bîrsă, L.M., and Bahrin, L.G. (2017). Flavonoids–small molecules, high hopes. *Acta Chemica Iasi* 25 (1): 6–23.

47 Nikolić, I.L., Savić-Gajić, I.M., Tačić, A.D., and Savić, I.M. (2017). Classification and biological activity of phytoestrogens: a review. *Advanced Technologies* 6 (2): 96–106.

48 Zaheer, K. and Humayoun, A.M. (2017). An updated review of dietary isoflavones: nutrition, processing, bioavailability and impacts on human health. *Critical Reviews in Food Science and Nutrition* 57 (6): 1280–1293.

49 Calderón-Oliver, M. and Ponce-Alquicira, E. (2018). Fruits: a source of polyphenols and health benefits. In: *Natural and Artificial Flavouring Agents and Food Dyes* (ed. A.M. Grumezescu and A.M. Holban), 189–228. Academic Press, Elsevier. https://doi.org/10.1016/B978-0-12-811518-3.00007-7.

50 Wrolstad, R.E. (2004). Anthocyanin pigments—bioactivity and coloring properties. *Journal of Food Science* 69 (5): C419–C425.

51 Welch, C.R., Wu, Q., and Simon, J.E. (2008). Recent advances in anthocyanin analysis and characterization. *Current Analytical Chemistry* 4 (2): 75–101.

52 Krga, I. and Milenkovic, D. (2029). Anthocyanins: from sources and bioavailability to cardiovascular-health benefits and molecular mechanisms of action. *Journal of Agricultural and Food Chemistry* 67 (7): 1771–1783.

53 Cutrim, C.S. and Cortez, M.A.S. (2018). A review on polyphenols: classification, beneficial effects and their application in dairy products. *International Journal of Dairy Technology* 71 (3): 564–578.

54 Kumar, H., Choudhary, N., Varsha, K.N., and Suman, S.R. (2014). Phenolic compounds and their health benefits: a review. *Journal of Food Technology Research* 2: 46–59.

55 Popa, V.I., Dumitru, M., Volf, I., and Anghel, N. (2008). Lignin and polyphenols as allelochemicals. *Industrial Crops and Products* 27 (2): 144–149.

56 Radko, L. and Cybulski, W. (2007). Application of silymarin in human and animal medicine. *Journal of Pre-Clinical and Clinical Research* 1 (1): 125–146.

57 Shen, T., Wang, X.N., and Lou, H.X. (2009). Natural stilbenes: an overview. *Natural Product Reports* 26 (7): 916–935.

58 Chou, Y.C., Ho, C.T., and Pan, M.H. (2018). Stilbenes: chemistry and molecular mechanisms of anti-obesity. *Current Pharmacology Reports* 4 (3): 202–209.

59 Baur, J.A. and Sinclair, D.A. (2006). Therapeutic potential of resveratrol: the in vivo evidence. *Nature Reviews Drug Discovery* 5 (6): 493–506.

60 Nandagopal, A. and Siddiqui, K. (2019). Role of nutraceuticals in neurodegenerative diseases. *Acta PharmaceuticaSuecica* 57 (4): 117–131.

61 Defelice, S.L. (1995). The nutraceutical revolution – its impact on food industry R&D. *Trends in Food Science and Technology* 6: 59–61.

62 Feng, Y., Wang, X.P., Yang, S.G. et al. (2009). Resveratrol inhibits betaamyloid oligomeric cytotoxicity but does not prevent oligomer formation. *NeuroToxicology* 30 (6): 986–995.

63 Huang, T.C., Lu, K.T., Wo, Y.Y. et al. (2011). Resveratrol protects rats from Aβ induced neurotoxicity by the reduction of iNOS expression and lipid peroxidation. *PLoS ONE* 6 (12): Article IDe29102.

64 Frozza, R.L., Bernardi, A., Paese, K. et al. (2010). Characterization of trans-resveratrol-loaded lipid-core nanocapsules and tissue distribution studies in rats. *Journal of Biomedical Nanotechnology* 6 (6): 694–703.

65 Ushikubo, H., Watanabe, S., Tanimoto, Y. et al. (2012). 3, 3′, 4′, 5, 5′ Pentahyroxyflavone is a potent inhibitor of amyloid β fibril formation. *Neuroscience Letters* 513 (1): 51–56.

66 Gong, E.J., Park, H.R., Kim, M.E. et al. (2011). Morin attenuates tau hyperphosphorylation by inhibiting GSK3β. *Neurobiology of Disease* 44 (2): 223–230.

67 Devi, L. and Ohno, M. (2012). 7, 8-dihyroxyflavone, a small molecule TrkB agonist, reverses memory deficits and BACE1 elevation in amouse model of Alzheimer's disease. *Neuropsychopharmacology* 37 (2): 434–444.

68 Wang, J., Pfleger, C.M., Friedman, L. et al. (2010). Potential application of grape derived polyphenols in huntington's disease. *Translational Neuroscience* 1 (2): 95–100.

69 Sternberg, Z., Chadha, K., Lieberman, A. et al. (2008). Quercetin and interferon-β modulate immune response(s) in peripheral blood mononuclear cells isolated from multiple sclerosis patients. *Journal of Neuroimmunology* 205: 142–147.

70 Herges, K., Millward, J.M., Hentschel, N. et al. (2016). Neuroprotective effect of combination therapy of glatiramer acetate and epigallocatechin-3-gallate in neuroinflammation. *PLoS ONE* 6 (10): Article ID e25456.

71 Davis, J.M., Murphy, E.A., Carmichael, M.D., and Davis, B. (2009). Quercetin increases brain and muscle mitochondrial biogenesis and exercise tolerance. *American Journal of Physiology* 296 (4): 1071–1077.

72 Xiang, L., Sun, K., Lu, J. et al. (2011). Anti-aging effects of phloridzin, an apple polyphenol, on yeast via the SOD and Sir2 genes. *Bioscience, Biotechnology and Biochemistry* 75 (5): 854–858.

73 Hong, K.S., Park, J.I., Kim, M.J. et al. (2012). Involvement of SIRT1 in hypoxic down-regulation of c-Myc and β-catechin and hypoxic precondition effect of polyphenol. *Toxicology and Applied Pharmacology* 259 (2): 210–218.

74 Park, J.W., Hong, J.S., Lee, K.S. et al. (2010). Green tea polyphenol (-)-epigallocatechin gallate reduces matrix metalloproteinase-9 activity following transient focal cerebral ischemia. *Journal of Nutritional Biochemistry* 21 (11): 1038–1044.

75 Panickar, K.S., Polansky, M.M., and Anderson, R.A. (2009). Green tea polyphenols attenuate glial swelling and mitochondrial dysfunction following oxygen-glucose deprivation in cultures. *Nutritional Neuroscience* 12 (3): 105–113.

76 Khan, M.M., Ahmad, A., Ishrat, T. et al. (2009). Rutin protects the neural damage induced by transient focal ischemia in rats. *Brain Research* 1292: 123–135.

77 Gelderblom, M., Leypoldt, F., Lewerenz, J. et al. (2012). The flavonoid fisetin attenuates postischemic immune cell infiltration, activation and infract size after transient cerebral middle artery occlusion in mice. *Journal of Cerebral Flow and Blood Metabolism* 32 (5): 835–843.

78 Tu, X.K., Yang, W.Z., Liang, R.S. et al. (2011). Effect of baicalin on matrix metalloproteinase-9 expression and blood brain barrier permeability following focal cerebral ischemia in rats. *Neurochemical Research* 36 (11): 2022–2028.

79 Cao, Y., Mao, X., Sun, C. et al. (2011). Baicalin attenuates global cerebral ischemia/reperfusion injury in gerbils via anti-oxidative and anti-apoptotic pathways. *Brain Research Bulletin* 85 (6): 396–402.

80 Cui, L., Zhang, X., Yang, R. et al. (2010). Baicalein is neuroprotective in ratMCAOmodel: role of 12/15-lipoxygenase, mitogen-activated protein kinase and cytosolic phospholipase A2. *Pharmacology Biochemistry and Behavior* 96 (4): 469–475.

81 Tu, X.K., Yang, W.Z., Shi, S.S. et al. (2011). Baicalin inhibits TLR2/4 signaling pathway in rat brain following permanent cerebral ischemia. *Inflammation* 34 (5): 463–470.

82 Ho, D.J., Calingasan, N.Y., Wille, E. et al. (2010). Resveratrol protects against peripheral deficits in amouse model of Huntington's disease. *Experimental Neurology* 225: 74–84.

83 Jin, F., Wu, Q., Lu, Y.F. et al. (2008). Neuroprotective effect of resveratrol on 6–OHDA-induced Parkinson's disease in rats. *European Journal of Pharmacology* 600: 78–82.

84 Khan, M.M., Ahmad, A., Ishrat, T. et al. (2010). Resveratrol attenuates 6-hydroxydopamine-induced oxidative damage and dopamine depletion in rat model of Parkinson's disease. *Brain Research* 1328: 139–151.

85 Filomeni, G., Graziani, I., Zio, D.D. et al. (2010). Neuroprotection of kaempferol by autophagy in models of rotenone-mediated acute toxicity: possible implications for Parkinson's disease. *Neurobiology of Aging* 24: 85–91.

86 Vauzour, D., Corona, G., and Spencer, J.P.E. (2018). Caffeic acid, tyrosol and p-coumaric acid are potent inhibitors of 5-S-cysteinyldopamine induced neurotoxicity. *Archives of Biochemistry and Biophysics* 501: 106–111.

87 Tai, K.K. and Truong, D.D. (2010). (-)-Epigallocatechin-3-gallate (EGCG), a green tea polyphenol, reduces dichlorodiphenyltrichloroethane (DDT)-induced cell death in dopaminergic SHSY-5Y cells. *Neuroscience Letters* 482 (3): 183–187.

88 Kim, H.G., Ju, M.S., Shim, J.S. et al. (2010). Mulberry fruit protects dopaminergic neurons in toxin-induced Parkinson's disease models. *The British Journal of Nutrition* 104: 8–16.

89 Maher, P., Dargusch, R., Bodai, L. et al. (2011). Erk activation by the polyphenols fisetin and resveratrol provides neuroprotection in multiple models of Huntington's disease. *HumanMolecular Genetics* 20: 261–270.

90 Kumar, P. and Kumar, A. (2010). Protective effect of hesperidin and naringin against 3-nitropropionic acid induced Huntington's like symptoms in rats: possible role of nitric oxide. *Behavioural Brain Research* 206: 38–46.

91 Hickey, M.A., Zhu, C., Medvedeva, V. et al. (2003). Improvement of neuropathology and transcriptional deficits in CAG 140 knockin mice supports a beneficial effect of dietary curcumin in Huntington's disease. *Molecular Neurodegeneration* 4: 7–12.

92 Ehrnhoefer, D.E., Duennwald, M., Markovic, P. et al. (2006). Mitochondrial dysfunction and tau hyperphosphorylation in Ts1Cje, a mouse model for down syndrome. *Human Molecular Genetics* 18: 2752–2762.

93 Duraˇckov´a, Z. (2010). Some current insights into oxidative stress. *Physiological Research* 59: 459–469.

94 Sanmukhani, V., Satodia, J., Trivedi, J. et al. (2014). Efficacy and safety of curcumin in major depressive disorder: a randomized controlled trial. *Phytotherapy Research* 28 (4): 579–585.

95 Chen, W.Q., Zhao, X.L., Hou, Y. et al. (2009). Protective effects of green tea polyphenols on cognitive impairments induced by psychological stress in rats. *Behavioural Brain Research* 202: 71–76.

96 Zhu, W.L., Shi, H.S., Wei, Y.M. et al. (2012). Green tea polyphenols produce antidepressant-like effects in adult mice. *Pharmacological Research* 65: 74–80.

97 Vignes, M., Maurice, T., Lant´e, F. et al. (2006). Anxiolytic properties of green tea polyphenol (-)-epigallocatechin gallate (EGCG). *Brain Research* 1110: 102–115.

98 Zhang, Q., Yang, H., Wang, J. et al. (2013). Effect of green tea on reward learning in healthy individuals: a randomized, double-blind, placebo-controlled pilot study. *Nutrition Journal* 12: 1–7.

99 Hou, Y., Aboukhatwa, M.A., Lei, D.L. et al. (2010). Anti-depressant natural flavonols modulate BDNF and beta amyloid in neurons and hippocampus of double TgAD mice. *Neuropharmacology* 58: 911–920.

100 Dreiseitel, G., Korte, P., Schreier, P. et al. (2009). Berry anthocyanins and their aglycons inhibit monoamine oxidases A and B. *Pharmacological Research* 59: 306–311.

101 Messaoudi, M., Bisson, J.F., Nejdi, A. et al. (2008). Antidepressant-like effects of a cocoa polyphenolic extract in Wistar-Unilever rats. *Nutritional Neuroscience* 11: 269–276.

102 Sathyapalan, T., Beckett, S., Rigby, A.S. et al. (2010). High cocoa polyphenol rich chocolate may reduce the burden of the symptoms in chronic fatigue syndrome. *Nutrition Journal* 9 (55): 87–95.

103 Coghill, D., Bonnar, S., Duke, S. et al. (2009). *Child and Adolescent Psychiatry*. New York, NY, USA: Oxford University Press.

104 Yoon, S.Y., Pe˜na, I.D., Kim, S.M. et al. (2013). Oroxylin A improves attention deficit hyperactivity disorder-like behaviors in the spontaneously hypertensive rat and inhibits reuptake of dopamine in vitro. *Archives of Pharmacal Research* 36: 134–140.

105 Loftis, J.M., Wilhelm, C.J., and Huckans, M. (2013). Effect of epigallocatechin gallate supplementation in schizophrenia and bipolar disorder: an 8-week, randomized, double-blind, placebo controlled study. *Therapeutic Advances in Psychopharmacology* 3: 21–27.

106 Dietrich-Muszalska, K.B., Olas, B., and Rabe- Jablo´nska, J. (2012). Epicatechin inhibits human plasma lipid peroxidation caused by haloperidol in vitro. *Neurochemical Research* 37: 557–562.

107 Dietrich-Muszalska, A., Kopka, J., and Kontek, B. (2014). Polyphenols from berries of Aronia melanocarpa reduce the plasma lipid peroxidation induced by Ziprasidone. *Schizophrenia Research and Treatment* 7: 45–52.

108 Zhang, W.F., Tan, Y.L., Zhang, X.Y. et al. (2011). Extract of Ginkgo biloba treatment for tardive dyskinesia in schizophrenia: a randomized, doubleblind, placebo-controlled trial. *Journal of Clinical Psychiatry* 72: 615–621.

109 Suresh, P. and Raju, A.B. (2013). Antidopaminergic effects of leucine and genistein on shizophrenic rat models. *Neurosciences* 18: 235–241.

110 Li, H., Yan, Z., Zhu, J. et al. (2011). Neuroprotective effects of resveratrol on ischemic injury mediated by improving brain energy metabolism and alleviating oxidative stress in rats. *Neuropharmacology* 60: 252–258.

111 Li, C., Yan, Z., Yang, J. et al. (2010). Neuroprotective effects of resveratrol on ischemic injury mediated by modulating the release of neurotransmitter and neuromodulator in rats. *Neurochemistry International* 56: 495–500.

112 Yang, X., Li, Z., Wang, N. et al. (2015). Curcumin-encapsulated polymeric micelles suppress the development of colon cancer in vitro and in vivo. *Scientific Reports* 18; 5 (1): 1–5.

113 Aguilar, Z.P. (2013). Types of nanomaterials and corresponding methods of synthesis. *Nanomaterials for Medical Applications* 50: 15–29.

114 Yang, B., Dong, Y., Wang, F., and Zhang, Y. (2020). Nanoformulations to enhance the bioavailability and physiological functions of polyphenols. *Molecules* 25 (20): 4613.

115 Mohanty, C., Acharya, S., Mohanty, A.K. et al. (2010). Curcumin-encapsulated MePEG/PCL diblock copolymeric micelles: a novel controlled delivery vehicle for cancer therapy. *Nanomedicine* 5 (3): 433–449.

116 Khonkarn, R., Mankhetkorn, S., Hennink, W.E., and Okonogi, S. (2011). PEG-OCL micelles for quercetin solubilization and inhibition of cancer cell growth. *European Journal of Pharmaceutics and Biopharmaceutics* 79 (2): 268–275.

117 Šmejkalová, D., Muthný, T., Nešporová, K. et al. (2017). Hyaluronan polymeric micelles for topical drug delivery. *Carbohydrate Polymers* 20; 156: 86–96.

118 Lapteva, M., Mondon, K., Möller, M. et al. (2014). Polymeric micelle nanocarriers for the cutaneous delivery of tacrolimus: a targeted approach for the treatment of psoriasis. *Molecular Pharmaceutics* 11 (9): 2989–3001.

119 Chavoshy, F., Zadeh, B.S., Tamaddon, A.M., and Anbardar, M.H. (2020). Delivery and anti-psoriatic effect of silibinin-loaded polymeric micelles: an experimental study in the psoriatic skin model. *Current Drug Delivery* 17 (9): 787–798.

120 Casiraghi, A., Franzè, S., Selmin, F. et al. (2017). Investigation of the effect of different emulsifiers on the transdermal delivery of EGCG entrapped in a polymeric micelle system. *Planta Medica* 83 (05): 405–411.

121 Jiang, W., Wang, Q., Cui, D. et al. (2023). Metal-polyphenol network coated magnetic hydroxyapatite for pH-activated MR imaging and drug delivery. *Colloids and Surfaces B: Biointerfaces* 1 (222): 113076.

122 Guo, Y., Sun, Q., Wu, F.G. et al. (2021). Polyphenol-containing nanoparticles: synthesis, properties, and therapeutic delivery. *Advanced Materials* 33 (22): 2007356.

123 Park, S., Cha, S.H., Cho, I. et al. (2016). Antibacterial nanocarriers of resveratrol with gold and silver nanoparticles. *Materials Science and Engineering: C* 58: 1160–1169.

124 Qiu, C., McClements, D.J., Jin, Z. et al. (2020). Wang J. Resveratrol-loaded core-shell nanostructured delivery systems: cyclodextrin-based metal-organic nanocapsules prepared by ionic gelation. *Food Chemistry* 317: 126328.

125 Product brochures, indena. Available at http://www.indena.com/pages/brochures.php (Accessed on 21st April 2011).

126 Pandey, S. and Patel, K. (2010). Phytosomes: technical revolution in phytomedicine. *International Journal of Pharmtech Research* 2: 627–631.

127 Rudrapal, M., Mishra, A.K., Rani, L. et al. (2022). Nanodelivery of dietary polyphenols for therapeutic applications. *Molecules* 27: 8706. https://doi.org/10.3390/molecules27248706.

128 Verma, P. and Pathak, K. (2010). Therapeutic and cosmeceutical potential of ethosomes: an overview. *Journal of Advanced Pharmacy Technology Research* 1: 274–282.

129 Paliwal, S., Tilak, A., Sharma, J. et al. (2019). Flurbiprofen loaded ethosomes transdermal delivery of anti-inflammatory effect in rat model. *Lipids Health Disease* 518: 133.

130 Kayal, E.M., Nasr, M., Elkheshen, S., and Mortada, N. (2019). Colloidal (-)-epigallocatechin-3-gallate vesicular systems for prevention and treatment of skin cancer: a comprehensive experimental study with preclinical investigation. *European Journal of Pharmaceutical Sciences* 137: 104972.

131 Faisal, W., Soliman, G.M., and Hamdan, A. (2016). Enhanced skin deposition and delivery of voriconazole using ethosomal preparations. *Journal of Liposome Research* 28: 14–21.

132 Palle, S. and Neerati, P. (2018). Improved neuroprotective effect of resveratrol nanoparticles as evinced by abrogation of rotenone-induced behavioral deficits and oxidative and mitochondrial dysfunctions in rat model of Parkinson's disease. *Naunyn-Schmiedeberg's Archives of Pharmacology* 391: 445–453. https://doi.org/10.1007/s00210-018-1474-8.

133 Frozza, R.L., Bernardi, A., Paese, K. et al. (2010). Characterization of trans-resveratrol-loaded lipid-corenanocapsules and tissue distribution studies in rats. *Journal of Biomedical Nanotechnology* 6: 694–703. https://doi.org/10.1166/jbn.2010.1161.

134 Frozza, R.L., Bernardi, A., Hoppe, J.B. et al. (2013b). Neuroprotective effects of resveratrol against Ab administration in rats are improved by lipid-core nanocapsules. *Molecular Neurobiology* 47: 1066–1080. https://doi.org/10.1007/s12035-013-8401-2.

135 Lu, X., Ji, C., Xu, H. et al. (2009). Resveratrol-loaded polymeric micelles protect cells from Abeta-induced oxidative stress. *International Journal of Pharmaceutics* 375: 89–96. https://doi.org/10.1016/j.ijpharm.2009.03.021.

136 Loureiro, J., Andrade, S., Duarte, A. et al. (2017). Resveratrol and grape extract-loaded solid lipid nanoparticles for the treatment of Alzheimer's disease. *Molecules* 22: E277. https://doi.org/10.3390/molecules22020277.

137 Wang, Y., Xu, H., Fu, Q. et al. (2011). Protective effect of resveratrol derived from Polygonum cuspidatum and its liposomal form on nigral cells in parkinsonian rats. *Journal of the Neurological Sciences* 304: 29–34. https://doi.org/10.1016/j.jns.2011.02.025.

138 Da Rocha Lindner, G., Bonfanti, S.D., Colle, D. et al. (2015). Improved neuroprotective effects of resveratrol-loaded polysorbate 80-coated poly(lactide) nanoparticles in MPTP induced Parkinsonism. *Nanomedicine* 10: 1127–1138. https://doi.org/10.2217/nnm.14.165.

139 Shytle, R.D., Bruce, B.D.B., Tampa, F.L. et al. (2008). *Composition of polyphenols and methods of use*. WO2008/005577 A3.

140 Beijing, Wehand Bio Pharmaceutical Co ltd, Institute of Materia Medica of CAMS (2019). *Flavone polyphenol medicine self-emulsifying composition, preparation method thereof, medicine composition and application*. Beijing Wehand Bio Pharmaceutical Co ltd Institute of Materia Medica of CAMS. CN111803632A.

141 Ishak, K.A., Mohamad, A.M.S., and Ahmed, N. (2017). chapter 9: nanodelivery systems for nutaceuticals application. In: *Nanotechnology Applications in Food, Academic Press* (ed. E.O. Alexandra and M.G. Alexandru), 179–202. Elsevier. https://doi.org/10.1016/b978-0-12-811942-6.00009-1. ISBN 9780128119426.

12

Nanodelivery of Polyphenols as Nutraceuticals for Neurological Disorders

*Arpita Paul, Madhusmita Gogoi, and Kamaruz Zaman**

Department of Pharmaceutical Sciences, Faculty of Science and Engineering, Dibrugarh University, Dibrugarh, Assam, India
**Corresponding author*

Neurological disorders (NDs) affect not only the brain and spinal cord but also all the nerves connecting them. NDs are one of the leading causes of death and disability worldwide [1]. The prevalence of NDs is greater in low- and middle-income countries than in high income countries [2]. Every year approximately 10 million people across the globe suffer from NDs, and this number is expected to increase in the near future [3]. In NDs, damage to the neurons occur either by injury, accumulation of abnormal proteins, gene mutation, or an increase in reactive oxygen species (ROS). The most prevalent NDs include Alzheimer's disease (AD), Parkinson's disease (PD), multiple sclerosis (MS), amyotrophic lateral sclerosis (ALS), Huntington's disease (HD), ischemic stroke, epilepsy, and neuropathic pain (NP). These neurological conditions share the same structural and mechanical abnormalities at the cellular, molecular, and functional levels, such as oxidative damage, inflammation, synaptic loss, and neuronal apoptosis, eventually leading to disruption of neuronal signal transduction pathways [4–6]. These disorders can manifest with symptoms such as motor dysfunction, cognitive impairment, difficulty in speech, and pain [7]. NDs are heterogeneous; therefore, a single therapy is inefficient for treating the pathological condition, which explains the failure of current monotherapeutic treatments and encourages multi-target approach [8]. Moreover, clinical trials on drugs for CNS disorders are challenging because of the complexity of the nervous system and impermeability of the blood–brain barrier (BBB) [9]. Polyphenols can provide greater therapeutic effect than monotherapy with synthetic drugs because of their pleiotropic mechanism of action. Furthermore, in vitro and preclinical studies demonstrated polyphenols were effective in epidemiology. However, poor solubility, low bioavailability, and restricted entry of polyphenols across the BBB limits their application in the management of NDs.

Recently, nanodelivery systems can encapsulate polyphenols, thereby protecting them from enzymatic degradation, enhancing their absorption, bioavailability, and transportation to target organs, and altering cell signaling pathways [10]. Most of the nanodelivery systems comprise biodegradable and biocompatible polymers, and then they are formulated into different nanostructures. The nanodelivery systems, such as nanospheres, nanocapsules, nanoemulsions, solid lipid nanoparticles, cyclodextrins, liposomes, and micelles, have revolutionized the delivery of therapeutic substances across the nervous system [11].

Polyphenols: Food, Nutraceutical, and Nanotherapeutic Applications, First Edition. Edited by Mithun Rudrapal.

12.1 Polyphenols and Their Role in Neurological Disorders (NDs)

Polyphenols are a large group of natural compounds biosynthesized by plants as secondary metabolites with chemical characteristics related to phenolic substances. In particular, the well-known class of phenolic systems, "polyphenols", is characterized by the presence of at least two phenyl rings and one or more hydroxyl groups, resulting in a large number of heterogeneous compounds. Consequently, polyphenols can be divided into numerous subclasses or simply classified as flavonoids and non-flavonoids depending on the number of phenol units in their molecular structure, substituent groups, and/or the type of linkage between the phenol units [12].

NDs are a heterogeneous group of disorders characterized by the dysfunction and/or progressive loss of post-mitotic neuronal cells in the CNS or PNS that represent a critical problem regarding human suffering and the economic burden on the healthcare system. Cognitive decline, dementia, motor irregularities, sleep difficulties, behavioral, and psychosocial disorders are the main clinical symptoms of neurodegeneration [5]. Common neurodegenerative diseases include AD, PD, and HD, and the illnesses share similar cellular and molecular mechanisms, such as the buildup of abnormal, misfolded, and aggregated proteins, mitochondrial dysfunction, inflammation, the accumulation of oxidative stress, impaired neuronal transport, impairment of the autophagic process, and changes in proteasome activity. The imbalance between the generation of ROS and antioxidant defenses leads to oxidative stress. It causes mitochondrial dysfunction, genomic instability, and oxidative damage to DNA, proteins, and lipids, in addition to impairing DNA repair pathways. The high oxygen consumption needed to meet the brain's high energy requirements specifically leads to excessive ROS production, high levels of polyunsaturated fatty acids in neuronal membranes that make them more susceptible to oxidation, and insufficient antioxidant defense mechanisms. These factors make the brain more vulnerable to ROS injury than other organs [7].

A diet rich in polyphenols lowers the cellular oxidative stress and is a successful technique for preventing NDs. Several different mechanisms, including interaction with the hypoxia-inducible factor 1-alpha (HIF-1 alpha) pathway, modulation of the expression of genes that protect against oxidative stress, regulation of ROS by interacting with oxidative pathways, and scavenging metal ions to prevent free radical damage, are used by polyphenols to exert their antioxidant activity. Numerous polyphenols have the capacity to chelate metal ions (Fe^{2+}, Zn^{2+}, and Cu^{2+}) that accumulate in specific brain regions of ND patients and lead to oxidative stress. For instance, these compounds seem to function by reducing the amount of iron and its accumulation by complexing transition divalent metal ions. One of the ways that polyphenol compounds have neuroprotective effects is by reducing the formation of beta amyloid aggregates and/or fibrils [13]. Numerous in vitro studies have revealed that curcumin, resveratrol, and epigallocatechin gallate (EGCG) are beneficial in the direct disruption of β-pleated sheets [14]. Particularly, polyphenols seem to bind to different surface areas of the β-sheet structure, leading to a variety of outcomes, including the formation of quick and non-toxic oligomers. The anti-inflammatory property of polyphenols exerts protective action against neurodegeneration. These compounds predominantly influence the expression of pro-inflammatory genes, such as nitric oxide synthase, lipoxygenase, cyclooxygenase, chemokines, and numerous cytokines, by the nuclear factor kappa-light-chain-enhancer of activated B cells and mitogen-activated protein kinase signaling [13]. Table 12.1 represents a list of polyphenols used in NDs.

Polyphenols also interact with other pathways that are either directly or indirectly connected to the neurodegenerative process. They play an important role in the signaling pathways that control cell growth, survival, apoptosis, and autophagy. The phosphorylation state and expression levels of proteins involved in the signaling pathways for phosphoinositide 3-kinase, Akt/protein kinase B, tyrosine kinases, and protein kinase C are altered by polyphenols, which in this case, affect cellular function [34].

Table 12.1 Polyphenols used in the prevention and treatment of NDs.

Nds	Active Polyphenol	Study Model	References
AD	Resveratrol	Neuro-2a (N2a) cells	[15]
	Rosmarinic acid	C57Bl/6 J mice	[16]
	Rutin	Tau-P301S mic	[17]
	Luteolin	Sprague–Dawley rats	[18]
PD	Resveratrol	SN4741 cells	[19]
	Myricetin	*Drosophila* model	[20]
	Chrysin	C57BL/6 mice	[21]
	Ellagic acid	6-hydroxydopamine (6-OHDA)-induced rat	[22]
HD	Quercetin	Wistar rats	[23]
	Resveratrol and fisetin	*Drosophila* expressing mutant Httex1 and the R6/2 mouse model	[24]
	Naringin	Wistar rats	[25]
	Hesperidin	Wistar rats	[26]
ALS	Wedelolactone and gallic acid	Wistar rats	[27]
	Resveratrol	SOD1- G93A mice	[28]
	Epigallocatechin-3-gallate	SOD1- G93A mice	[29]
	7,8-dihydroxyflavone	SOD1- G93A mice	[30]
MS	Resveratrol	C57/Bl6 mice	[31]
	Naringenin	C57/Bl6 mice	[32]
	Curcumin	C57BL/6 mice	[33]

Some studies reported that in PD, hesperidin and naringenin act by decreasing the ROS level and increasing the level and activity of GSH; additionally, ferulic acid and pelargonidin act by decreasing the oxidative stress. In vitro studies reported that silymarin in MS acts by increasing the expressions of STAT5, JAK3, and FOXP3, as well as the levels of TGF-β. Many studies have shown that gallic acid, quercetin, apigenin, and naringenin act by various mechanisms in AD [35].

The major limitation of polyphenols is their poor bioavailability. Polyphenols have a relatively low bioavailability because of extrinsic factors (such as limited stability in the gastrointestinal system, substantial phase I and phase II metabolism, and quick elimination) and intrinsic factors (such as chemical structure, molecular weight, and low hydrosolubility). The cell membrane permeability of polyphenols is another factor that influences their bioavailability [36].

12.2 Nanocarriers for Brain Targeting

Nanotechnology offers a promising approach to treat CNS disorders through drug delivery, leveraging the small size of biodegradable and biocompatible nanoparticles that can cross the BBB. The surface of these nanoparticles can also be tailored to enhance the compatibility with the drug being loaded. Nanoparticles feature a two-fold composition with the first part safeguarding the drug from degradation and targeting specific brain cells, crossing the BBB and releasing the drug at a specific pH. The second part constitutes the nano-engineered complex. With the ability to deliver drugs to specific sites, the BBB crossing mechanism is a major advantage of nanoparticles. Example nanodelivery strategies adopted for delivering drugs across the CNS are shown in Figure 12.1, including the nanodelivery of polyphenols across the BBB through major pathways.

12.2.1 Solid Lipid Nanoparticles (SLNs)

Solid lipid nanoparticles (SLNs) are a desirable colloidal drug carrier system that comprises spherical solid lipid particles in the nanometer range, dispersed in water or an aqueous surfactant solution. SLNs have a solid hydrophobic core with a phospholipid coating, and the solid core can hold the drug in a solid high-melting fat matrix with the hydrophobic end of the phospholipid chains embedded. SLNs have the potential to transport either lipophilic and hydrophilic drugs or diagnostics [37]. SLNs delivery offers a novel approach to delivering drugs into the brain by addressing solubility, permeability, and toxicity issues, and has benefits over traditional invasive methods. Their high physical stability is another advantage. However, the use of polymeric micro/nanoparticles for drug delivery is limited because of the cytotoxicity of the polymers and difficulty in large-scale production. In contrast, solid lipids are a well-known matrix material for drug delivery and can be produced inexpensively and simply through high pressure homogenization or microemulsion technologies [38].

12.2.2 Polymeric Nanoparticles

Polymeric nanoparticles are a promising option for CNS drug delivery because of their controlled release, programmable size (10–1000 nm), biocompatibility, and non-toxicity. They can also be modified with ligands to increase binding to endothelial cell receptors and enhance transcytosis. Polymeric nanoparticles have a longer circulation duration, biodegradability, and the potential for

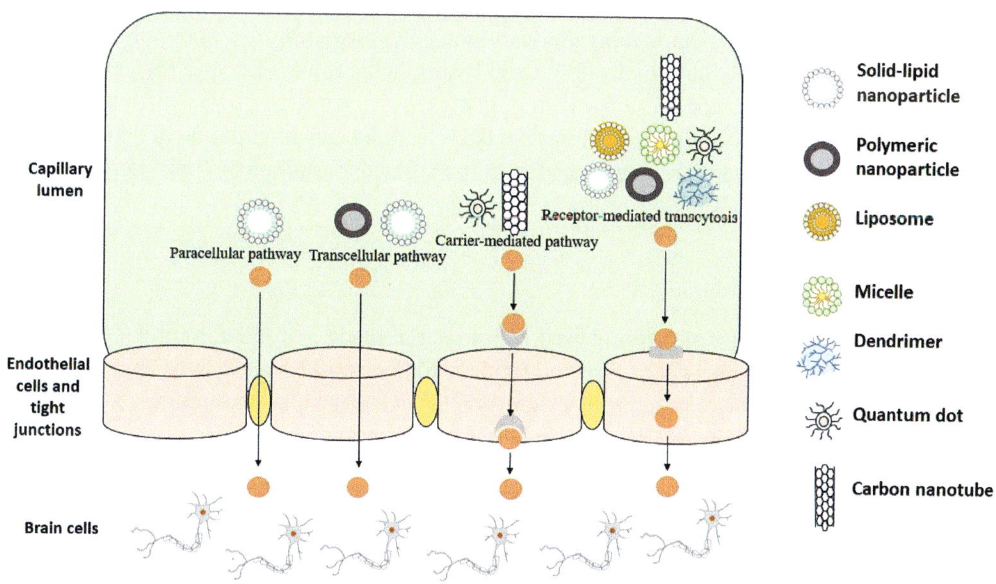

Figure 12.1 Delivery of polyphenols with the help of nanocarriers across the BBB into the CNS. BBB: Blood Brain Barrier; CNS: Central Nervous System.

sustained, targeted, and protected drug delivery via cellular absorption and activation of the polymeric lattice. They have the versatility to deliver a wide range of medications through hydrophobic, hydrophilic, electrostatic, and covalent interactions [39]. Natural polysaccharides (hyaluronic acid, chondroitin sulfate, and chitosan) and synthetic polymers (polylactic acid (PLA), polyglycolic acid (PGA), and poly lactic-co-glycolic acid (PLGA)) are extensively used in the formulation of polymeric nanoparticles [40].

12.2.3 Liposomes

Liposomes are spherical vesicles made of lipid bilayers that can entrap both hydrophilic and hydrophobic compounds. They are widely used as drug delivery systems to improve the safety and efficacy of therapeutics. They also have potential in neurological applications as they can cross the BBB and deliver therapeutic and diagnostic agents to the brain. Liposomes can reach the brain through various paths such as adsorption-mediated transcytosis, receptor-mediated endocytosis, and disruption of the BBB [41].

12.2.4 Micelles

Micelles are nanocarriers with a size of 5–50 nm that form spontaneously from self-assembling amphiphilic molecules. They have a hydrophobic core and a hydrophilic surface, making them useful for delivering poorly water soluble and lipophilic compounds. Micelles can penetrate the BBB through endocytosis and/or transcytosis, and their penetration can be enhanced by attaching specific ligands or through external forces [42].

12.2.5 Dendrimers

Dendrimers are nanoscale artificial macromolecules with a highly branched, globular structure comprising an initiator core, branched repeat units, and functional terminal groups. Commonly used dendrimer materials include polyamidoamine, polypropylenimine, and polyaryl ether. Dendrimers can encapsulate both hydrophilic and hydrophobic molecules, and they are used as nanocarriers for various therapeutic and imaging agents. They can overcome the BBB and are used in the treatment of CNS disorders. The cellular uptake is facilitated by modulation of tight junction proteins, and specific ligands can be conjugated to the surface for enhanced brain targeting and transport across the BBB [43].

12.2.6 Carbon Nanotubes

Carbon nanotubes are a type of carbon-based nanomaterial made by rolling graphite sheets into tubes with nanoscale diameters. They can be single-walled or multi-walled with open or closed ends. Carbon nanotubes have unique properties such as high electrical conductivity, mechanical strength, and thermal resistance, making them attractive for various applications in medicine and biotechnology. These include drug delivery, gene therapy, tissue engineering, and biosensing. Carbon nanotubes can be functionalized with specific chemical compounds to modify their physical and biological properties, making them useful as nanocarriers. However, they cannot cross the BBB via passive diffusion and they require conjugation with compounds to enable active transport to the brain for use in neuro-nanomedicine [44].

12.2.7 Inorganic Nanoparticles

Inorganic NPs, including metals, semiconductors, and metal oxides, have unique properties that make them attractive for biomedical applications. Their performance and functionality can be improved by adjusting the size, shape, composition, structure, and porosity. Silver NPs (AgNPs), iron oxide NPs (IONPs), and titanium dioxide NPs (TiO_2 NPs) are mainly used in disease diagnosis through bioimaging. Gold NPs (AuNPs) and SiO_2 NPs have been used as nanocarriers to target the CNS [45].

12.2.8 Quantum Dots

Quantum dots are zero-dimensional nanomaterials known for their exceptional optical and electrical properties. They are used in medicine and biology for drug delivery, targeted cancer therapy, bioimaging, cell labeling, and cell tracking. To target the brain and cross the BBB, quantum dots need to be functionalized, and the main mechanism for reaching the brain parenchyma is through carrier-mediated transport [46].

12.3 Nanodelivery of Polyphenols in Neurological Disorders (NDs)

Nanotechnology provides a new method to address the problems associated with NDs. Table 12.2 lists various treatment strategies for NDs using nanotechnology.

12.3.1 Solid Lipid Nanoparticles (SLNs)

Puerarin, an isoflavone from the roots of *Pueraria lobata* (Willd.), is beneficial in PD and AD; however, it has poor water aqueous solubility (0.46 mg/mL) that hinders its use. However, SLN-puerarin has better absorption, shorter Tmax and three fold higher bioavailability when compared with a puerarin suspension [58].

12.3.2 Polymeric Nanoparticles

The ability of resveratrol (Res) to protect against PD in mice was investigated using Res-loaded polysorbate 80 (PS80)-coated poly (lactide) nanoparticles. The effects were compared with those of bulk Res in C57BL/6 mice. The Res-loaded nanoparticles showed significant neuroprotective effects against the harmful effects of 1-methyl-4-phenyl-1,2,3,6-tetrahydropyridine (MPTP), unlike bulk Res [51].

12.3.3 Liposomes

A study investigated the effectiveness of using phosphatidic acid (PA) and apolipoprotein E (ApoE) to modify the surface of quercetin (QT) and rosmarinic acid (RA)-loaded liposomes to penetrate the BBB and treat β-amyloid (Aβ1-42)-induced neurotoxicity in an AD model. The results showed that incorporating ApoE or Tween 80 into the liposomes improved their ability to target the BBB, Aβ1-42, neurons, and reduced oxidative toxicity. However, Tween 80 had slightly higher cytotoxicity than ApoE. It was further observed that increasing the proportion of PA in the liposomes

Table 12.2 Treatment strategies for NDs using polyphenol-based nanoformulations.

Nds	Nanocarrier Type	Active Polyphenol	Study Model	Outcomes	References
AD	Liposomes	Quercetin and rosmarinic acid	Sprague–Dawley rats	Improved drug delivery, release and activity	[47]
	Micelles	Curcumin	Sprague–Dawley rats	Better antioxidant activity, enhanced bioavailability, distribution, and retention	[48]
	Carbon nanotube	Berberine	Wistar rats	Improved drug absorption and better performance	[49]
	Carbon dot	Curcumin	PC12 cells	Enhanced cellular uptake	[50]
PD	Polymeric nanoparticle	Resveratrol	C57BL/6 mice	Enhanced neuroprotection against MPTP	[51]
	Nanocrystal	Quercetin	6-OHDA-induced rat	Greater bioavailability	[52]
	Liposomes	Gallic acid, catechin, and epicatechin	SH-SY5Y-derived neurons	Excellent distribution	[53]
HD	Selenium nanoparticles	———	*Caenorhabditis elegans* transgenic model	Reduced neuronal death and relieved behavioral dysfunction, protecting from damage in stress conditions	[54]
	Solid lipid nanoparticles	Curcumin	Wistar rats	Reduced ROS, improved neuromotor coordination	[55]
MS	Polymeric nanoparticles	Curcumin	Lewis rat	Efficient therapeutic effect	[56]
ALS	Micelles	Curcumin	Adipose tissue sample	Reduced toxicity	[57]

increased the particle size and improved Aβ1-42 targeting. The addition of ApoE or Tween 80 also slowed the drug release and increased drug activity by reducing AchE activity and lipid peroxidation in rats with AD [47].

12.3.4 Micelles

To increase bioavailability and enhance delivery of curcumin in the brain of patients with AD, a group of researchers formulated cocrystals of curcumin and incorporated them in a micellar nanocarrier system for efficient delivery directly to the brain via the nose. The results of the study showed that curcumin cocrystals and their micelles had a much lower IC_{50} than regular curcumin, leading to improved antioxidant performance. Furthermore, in animal studies, the bioavailability of curcumin cocrystal micelles was 1.7 times greater, with increased brain distribution and longer retention [48].

12.3.5 Dendrimers

Res is effective in several NDs. However, the limitations of Res, such as low oral bioavailability, poor solubility, and instability, can be overcome by encapsulating it in water-soluble poly(amidoamine) dendrimers, which improves its solubility, stability, and overall effectiveness. Dendrimers not only enhance the properties of Res, they also have the potential to promote the solubility, stability, and controlled delivery for improved bioavailability and efficacy [59].

12.3.6 Carbon Nanotubes

A study investigated the use of berberine-loaded multiwalled carbon nanotubes (BRB-MWCNTs) coated with polysorbate and phospholipid as a potential treatment for AD. Results showed improved drug absorption in rats and better performance in memory tests compared with pure berberine. The coated MWCNTs also showed potential in reducing AD symptoms by maintaining normal biochemical levels in the brain tissue [49].

12.3.7 Inorganic Nanoparticles

A study combined the Aβ absorption property of selenium nanoparticles with Res to form Res@SeNPs. The in vitro evaluation showed that Res@SeNPs provide a synergistic effect on Cu^{2+}-induced Aβ42 aggregation and ROS generation, and effectively protect neuron cells (PC-12) from Aβ42-Cu^{2+} induced cell death. Therefore, the results suggest that the combination of Res and SeNPs is more effective in reducing Aβ42 toxicity than Res alone, making it a promising approach for long-term AD treatment [60].

12.3.8 Quantum Dots

A new delivery system, a Fe_3O_4 carbon dot (CD) nanocomposite, was used to load curcumin (CUR-Fe_3O_4@CDs) for the treatment of AD. The CDs have a natural fluorescence and are biocompatible with low toxicity. In vitro studies have shown that CUR-Fe_3O_4@CDs are highly specific towards Aβ42, thereby preventing aggregation and countering toxic effects on neuronal cells (PC-12). These findings suggest that CUR-Fe_3O_4@CDs could be a promising candidate for the treatment of AD [50].

12.4 Conclusions and Future Perspectives

The improvement in lifespan has led to a rise in the prevalence of NDs that require new treatment methods to improve a patient's symptoms and quality of life. The CNS is protected by various barriers, so drugs must pass the BBB for successful treatment. Phytochemical-based nanocarriers offer advantages such as safety, eco-friendliness, low toxicity, affordability, and the ability to control particle size and morphology. These nanocarriers can improve the delivery of phytotherapeutic compounds to the CNS, increase drug penetration into the brain, and prevent drug aggregation in the brain. Although many preclinical studies have shown the therapeutic effects of these nanocarriers in NDs, more research is needed to address safety concerns. The long-term clinical efficacy of these nanocarriers in neurological medicine also needs to be evaluated. Designing nanocarriers with specific targeting to different brain cells and for each type of ND is important to maximize their effectiveness.

References

1 Feigin, V.L., Vos, T., Nichols, E. et al. (2020). The global burden of neurological disorders: translating evidence into policy. *Lancet Neurology* 19 (3): 255–265.

2 World Bank Country and Lending Groups – World Bank Data Help Desk (2022). Available from: https://datahelpdesk.worldbank.org/knowledgebase/articles/906519-world-bank-country-and-lending-groups (accessed on 01 April 2023).

3 Makkar, R., Behl, T., Bungau, S. et al. (2020). Nutraceuticals in neurological disorders. *International Journal of Molecular Sciences* 21 (12): 1–19.

4 Dias, B.D.C., Jovanovic, K., Gonsalves, D. et al. (2011). Structural and mechanistic commonalities of amyloid-β and the prion protein. *Prion* 5 (3): 126–137. http://doi.org/104161/pri5317025.

5 Peden, A.H. and Ironside, J.W. (2012). Molecular pathology in neurodegenerative diseases. *Current Drug Targets* 13 (12): 1548–1559.

6 Holmes, B.B. and Diamond, M.I. (2012). Cellular mechanisms of protein aggregate propagation. *Current Opinion in Neurology* 25 (6): 721–726.

7 Paul, A., Guha, N., and Zaman, M.K. (2022). Nanotherapeutics of phytoantioxidants for CNS disorders. *Phytoantioxidants and Nanotherapeutics* 449–476. https://doi/full/10.1002/9781119811794.ch20.

8 Frautschy, S.A. and Cole, G.M. (2010). Why pleiotropic interventions are needed for alzheimer's disease. *Molecular Neurobiology* 41 (2–3): 392–409.

9 Lingineni, K., Belekar, V., Tangadpalliwar, S.R., and Garg, P. (2017). The role of multidrug resistance protein (MRP-1) as an active efflux transporter on blood-brain barrier (BBB) permeability. *Molecular Diversity* 21 (2): 355–365.

10 Squillaro, T., Schettino, C., Sampaolo, S. et al. (2018). Adult-onset brain tumors and neurodegeneration: are polyphenols protective? *Journal of Cellular Physiology* 233 (5): 3955–3967.

11 Kim, J., Ahn, S.I., and Kim, Y.T. (2019). Nanotherapeutics engineered to cross the blood-brain barrier for advanced drug delivery to the central nervous system. *Journal of Industrial and Engineering Chemistry* 73: 8.

12 Tsao, R. (2010). Chemistry and biochemistry of dietary polyphenols. *Nutrients* 2 (12): 1231.

13 Moradi, S.Z., Jalili, F., Farhadian, N. et al. (2021). Polyphenols and neurodegenerative diseases: focus on neuronal regeneration. *Critical Reviews in Food Science and Nutrition* 62 (13): 3421–3436. https://doi.org/101080/1040839820201865870.

14 Ebrahimi, A. and Schluesener, H. (2012). Natural polyphenols against neurodegenerative disorders: potentials and pitfalls. *Ageing Research Reviews* 11 (2): 329–345.

15 He, X., Li, Z., Rizak, J.D. et al. (2017). Resveratrol attenuates formaldehyde induced hyperphosphorylation of tau protein and cytotoxicity in N2a cells. *Frontiers in Neuroscience* 10: 598.

16 Hase, T., Shishido, S., Yamamoto, S. et al. (2019). Rosmarinic acid suppresses Alzheimer's disease development by reducing amyloid β aggregation by increasing monoamine secretion. *Scientific Reports* 9 (1): 1–13.

17 Sun, X.Y., Li, L.J., Dong, Q.X. et al. (2021). Rutin prevents tau pathology and neuroinflammation in a mouse model of Alzheimer's disease. *Journal of Neuroinflammation* 18 (1): 1–14.

18 Facchinetti, R., Valenza, M., Bronzuoli, M.R. et al. (2020). Looking for a treatment for the early stage of alzheimer's disease: preclinical evidence with co-ultramicronized palmitoylethanolamide and Luteolin. *International Journal of Molecular Sciences* 21 (11).

19 Zeng, W., Zhang, W., Lu, F. et al. (2017). Resveratrol attenuates MPP+-induced mitochondrial dysfunction and cell apoptosis via AKT/GSK-3β pathway in SN4741 cells. *Neuroscience Letters* 637: 50–56.

20 Ara, G., Afzal, M., Jyoti, S. et al. (2017). Effect of myricetin on the transgenic Drosophila model of Parkinson's disease. *Bulletin of Faculty of Pharmacy* 55 (2): 259–262.

21 Guo, B., Zheng, C., Cai, W. et al. (2016). Multifunction of chrysin in parkinson's model: anti-neuronal apoptosis, neuroprotection via activation of MEF2D, and inhibition of monoamine oxidase-B. *Journal of Agricultural and Food Chemistry* 64 (26): 5324–5333.

22 Baluchnejadmojarad, T., Rabiee, N., Zabihnejad, S. et al. (2017). Ellagic acid exerts protective effect in intrastriatal 6-hydroxydopamine rat model of Parkinson's disease: possible involvement of ERβ/Nrf2/HO-1 signaling. *Brain Research* 1662: 23–30.

23 Sandhir, R. and Mehrotra, A. (2013). Quercetin supplementation is effective in improving mitochondrial dysfunctions induced by 3-nitropropionic acid: implications in Huntington's disease. *Biochimica et Biophysica Acta* 1832 (3): 421–430.

24 Maher, P., Dargusch, R., Bodai, L. et al. (2011). ERK activation by the polyphenols fisetin and resveratrol provides neuroprotection in multiple models of Huntington's disease. *Human Molecular Genetics* 20 (2): 261–270.

25 Gopinath, K. and Sudhandiran, G. (2012). Naringin modulates oxidative stress and inflammation in 3-nitropropionic acid-induced neurodegeneration through the activation of nuclear factor-erythroid 2-related factor-2 signaling pathway. *Neuroscience* 227: 134–143.

26 Menze, E.T., Tadros, M.G., Abdel-Tawab, A.M. et al. (2012). Potential neuroprotective effects of hesperidin on 3-nitropropionic acid-induced neurotoxicity in rats. *Neurotoxicology* 33 (5): 1265–1275.

27 Maya, S., Prakash, T., and Goli, D. (2018). Evaluation of neuroprotective effects of wedelolactone and gallic acid on aluminium-induced neurodegeneration: relevance to sporadic amyotrophic lateral sclerosis. *European Journal of Pharmacology* 835: 41–51.

28 Schiaffino, L., Bonafede, R., Scambi, I. et al. (2018). Acetylation state of RelA modulated by epigenetic drugs prolongs survival and induces a neuroprotective effect on ALS murine model. *Scientific Reports.* 8 (1).

29 Xu, Z., Chen, S., Li, X. et al. (2006). Neuroprotective effects of (-)-epigallocatechin-3-gallate in a transgenic mouse model of amyotrophic lateral sclerosis. *Neurochemical Research* 1 (10): 1263–1269.

30 Korkmaz, O.T., Aytan, N., Carreras, I. et al. (2014). 7,8-Dihydroxyflavone improves motor performance and enhances lower motor neuronal survival in a mouse model of amyotrophic lateral sclerosis. *Neuroscience Letters* 566: 286.

31 Fonseca-Kelly, Z., Nassrallah, M., Uribe, J. et al. (2012). Resveratrol neuroprotection in a chronic mouse model of multiple sclerosis. *Frontiers in Neurology* MAY:84.

32 Wang, J., Qi, Y., Niu, X. et al. (2018). Dietary naringenin supplementation attenuates experimental autoimmune encephalomyelitis by modulating autoimmune inflammatory responses in mice. *The Journal of Nutritional Biochemistry* 54: 130–139.

33 ELBini-Dhouib, I., Manai, M., Neili, N.E. et al. (2022). Dual mechanism of action of curcumin in experimental models of multiple sclerosis. *International Journal of Molecular Sciences* 23 (15).

34 Ramassamy, C. (2006). Emerging role of polyphenolic compounds in the treatment of neurodegenerative diseases: a review of their intracellular targets. *European Journal of Pharmacology* 545 (1): 51–64.

35 Aryal, S., Skinner, T., Bridges, B. et al. (2020). The pathology of Parkinson's disease and potential benefit of dietary polyphenols. *Molecules* 25 (19).

36 Squillaro, T., Cimini, A., Peluso, G. et al. (2018). Nano-delivery systems for encapsulation of dietary polyphenols: an experimental approach for neurodegenerative diseases and brain tumors. *Biochemical Pharmacology* 154: 303–317.

37 Blasi, P., Giovagnoli, S., Schoubben, A. et al. (2007). Solid lipid nanoparticles for targeted brain drug delivery. *Advanced Drug Delivery Reviews* 59 (6): 454–477.

38 Kaur, I.P., Bhandari, R., Bhandari, S. et al. (2008). Potential of solid lipid nanoparticles in brain targeting. *The Journal of Controlled Release* 127 (2): 97–109.

39 Mitchell, M.J., Billingsley, M.M., Haley, R.M. et al. (2020). Engineering precision nanoparticles for drug delivery. *Nature Reviews. Drug Discovery* 20 (2): 101–124.

40 Zhang, T.T., Li, W., Meng, G. et al. (2016). Strategies for transporting nanoparticles across the blood-brain barrier. *Biomaterials Science* 4 (2): 219–229.

41 Montesinos, R.N. and Montesinos, R.N. (2017). Liposomal drug delivery to the central nervous system. *Liposomes*, https://www.intechopen.com/chapters/56546 (accessed on 03 April 2023).

42 Wang, X.J., Gao, Y.P., Lu, N.N. et al. (2016). Endogenous polysialic acid based micelles for calmodulin antagonist delivery against vascular dementia. *ACS Applied Materials and Interfaces* 8 (51): 35045–35058.

43 Srinageshwar, B., Peruzzaro, S., Andrews, M. et al. (2017). PAMAM dendrimers cross the blood-brain barrier when administered through the carotid artery in C57BL/6J mice. *International Journal of Molecular Sciences* 18 (3).

44 Alshehri, R., Ilyas, A.M., Hasan, A. et al. (2016). Carbon nanotubes in biomedical applications: factors, mechanisms, and remedies of toxicity. *Journal of Medicinal Chemistry* 59 (18): 8149–8167.

45 Teleanu, D.M., Chircov, C., Grumezescu, A.M. et al. (2019). Neuronanomedicine: an up-to-date overview. *Pharmaceutics* 11 (3).

46 Abdellatif, A.H., Tawfeek, H.M., Younis, M.A. et al. (2022). Biomedical applications of quantum dots: overview, challenges, and clinical potential. *International Journal of Nanomedicine* 17: 1951.

47 Kuo, Y.C., Chen, I.Y., and Rajesh, R. (2018). Use of functionalized liposomes loaded with antioxidants to permeate the blood–brain barrier and inhibit β-amyloid-induced neurodegeneration in the brain. *Journal of the Taiwan Institute of Chemical Engineers* 87: 1–14.

48 Desai, P.P. and Patravale, V.B. (2018). Curcumin cocrystal micelles-multifunctional nanocomposites for management of neurodegenerative ailments. *Journal of Pharmaceutical Sciences* 107 (4): 1143–1156.

49 Lohan, S., Raza, K., Mehta, S.K. et al. (2017). Anti-Alzheimer's potential of berberine using surface decorated multi-walled carbon nanotubes: a preclinical evidence. *International Journal of Pharmaceutics* 530 (1–2): 263–278.

50 Kuang, Y., Zhang, J., Xiong, M. et al. (2020). A novel nanosystem realizing curcumin delivery based on Fe3O4@carbon dots nanocomposite for Alzheimer's disease therapy. *Frontiers in Bioengineering and Biotechnology* 8: 1335.

51 Rocha Lindner, G.D., Santos, D.B., Colle, D. et al. (2015). Improved neuroprotective effects of resveratrol-loaded polysorbate 80-coated poly(lactide) nanoparticles in MPTP-induced Parkinsonism. *Nanomedicine* 10 (7): 1127–1138. https://doi.org/10.2217/nnm.14.165.

52 Ghaffari, F., Moghaddam, A.H., and Zare, M. (2018). Neuroprotective effect of quercetin nanocrystal in a 6-hydroxydopamine model of parkinson disease: biochemical and behavioral evidence. *Basic and Clinical Neuroscience* 9 (5): 317.

53 Marino, A., Battaglini, M., Desii, A. et al. (2021). Liposomes loaded with polyphenol-rich grape pomace extracts protect from neurodegeneration in a rotenone-based in vitro model of Parkinson's disease. *Biomaterials Science* 9 (24): 8171–8188.

54 Cong, W., Bai, R., Li, Y.F. et al. (2019). Selenium nanoparticles as an efficient nanomedicine for the therapy of Huntington's disease. *ACS Applied Materials and Interfaces* 11 (38): 34725–34735.

55 Sandhir, R., Yadav, A., Mehrotra, A. et al. (2014). Curcumin nanoparticles attenuate neurochemical and neurobehavioral deficits in experimental model of Huntington's disease. *Neuromolecular Medicine* 16 (1): 106–118.

56 Mohajeri, M., Sadeghizadeh, M., Najafi, F. et al. (2015). Polymerized nano-curcumin attenuates neurological symptoms in EAE model of multiple sclerosis through down regulation of inflammatory and oxidative processes and enhancing neuroprotection and myelin repair. *Neuropharmacology* 99: 156–167.

57 Tripodo, G., Chlapanidas, T., Perteghella, S. et al. (2015). Mesenchymal stromal cells loading curcumin-INVITE-micelles: a drug delivery system for neurodegenerative diseases. *Colloids Surf B Biointerfaces* 125: 300–308.

58 Luo, C.F., Hou, N., Tian, J. et al. (2013). Metabolic profile of puerarin in rats after intragastric administration of puerarin solid lipid nanoparticles. *International Journal of Nanomedicine* 8: 933–940.

59 Chauhan, A.S. (2015). Dendrimer nanotechnology for enhanced formulation and controlled delivery of resveratrol. *Annals of the New York Academy of Sciences* 1348 (1): 134–140.

60 Li, J., Gao, R., Zhao, D. et al. (2017). Separation and preparation of xanthochymol and guttiferone E by high performance liquid chromatography and high speed counter-current chromatography combined with silver nitrate coordination reaction. *Journal of Chromatography A* 1511: 143–148. https://pubmed.ncbi.nlm.nih.gov/28688721.

13

Addressing Antimicrobial Resistance by Repurposing Polyphenolic Phytochemicals with Novel Antibacterial Potential

Mithun Rudrapal[1,], Biswatrish Sarkar[2], Prashanta Kumar Deb[2], Atul R. Bendale[3], and Akhil Nagar[4]*

[1] Department of Pharmaceutical Sciences, School of Biotechnology and Pharmaceutical Sciences, Vignan's Foundation for Science, Technology & Research, Guntur, India
[2] Department of Pharmaceutical Sciences and Technology, Birla Institute of Technology, Ranchi, Jharkhand, India
[3] Mahavir Institute of Pharmacy, Nashik, Maharashtra, India
[4] Department of Pharmaceutical Chemistry, R. C. Patel Institute of Pharmaceutical Education and Research, Shirpur, Maharashtra, India
*Corresponding author

13.1 Introduction

Infectious diseases caused by pathogenic microbes (bacteria, fungi, viruses, protozoa) are one of the leading causes of morbidity and mortality across the globe. According to the World Health Organization (WHO), it has been estimated that after 2050, 10 million people will die annually because of microbial infections [1], which can be attributed to the increasing emergence of antimicrobial resistance (AMR) and multi-drug resistance (MDR) [2–4]. AMR and MDR are some of the greatest challenges to global public health today. AMR is currently responsible for over 700,000 deaths annually worldwide. As AMR continues to increase, the effective treatment of drug-resistant infectious diseases is a serious threat. In recent years, there have been several outbreaks of deadly infectious diseases, including Ebola, Zika, swine flu (H1N1 influenza virus), antibiotic-resistant bacterial infections, *Candida auris* infections, *Plasmodium falciparum* malaria, dengue, and coronavirus diseases-2019 (COVID-19). AMR has emerged as one of the dominant public health problems of the 21st century that threatens the effective prevention and treatment of a wide range of infections caused by bacteria, viruses, fungi, and parasites. The emergence of AMR combined with limited therapeutic efficacy and clinical usefulness of existing antimicrobial agents, including antibacterial antibiotics, poses a significant challenge to public health [5]. These issues urgently necessitate the discovery of novel and effective antimicrobial chemotherapeutics that would fight against MDR pathogenic organisms. The medical need for novel antimicrobial drugs/therapies globally is undeniable, and an effort to discover and develop novel antimicrobial drugs is, therefore, urgent. However, modern drug discovery and development is a lengthy and arduous process that inevitably struggles to deliver new therapies in a timely manner. The lengthy process and high cost associated with research and development often culminates in high failure rates because of safety and toxicity issues, among many other sources of attrition. Thus, strategies to improve the efficiency of drug development are urgently needed to enable effective drugs to enter the clinic. One creative strategy for the identification of novel antimicrobials that is gaining momentum is drug repurposing [6, 7]. In recent years, interest in new approaches to drug research

Polyphenols: Food, Nutraceutical, and Nanotherapeutic Applications, First Edition. Edited by Mithun Rudrapal.

and development, especially for antimicrobial drug discovery, has brought drug repurposing to the forefront as a promising strategy. Faced with the threat of antimicrobial resistance, options for treating infectious diseases are increasingly limited. Thus, drug repurposing is gaining traction as a strategic approach in antimicrobial research with the potential to reduce costs and expedite approval timelines. Drug repurposing efforts (in vitro and in vivo techniques) led to uncovering novel antimicrobial agents to treat microbial infections [8]. Bioactive plant-based products or dietary phytochemicals/polyphenols are a potential approach for effective drug development against infections caused by multi-drug resistant bacteria. Dietary plants are a rich source of polyphenolic compounds, including flavonoids (e.g., catechins, epigallocatechin gallate (EGCG), quercetin, apigenin, luteolin, genistein, etc.), which act as potential antioxidants (radical scavenging agents) in counteracting various pathological conditions in infectious illnesses because of their antimicrobial, anti-inflammatory, and antioxidant properties.

13.2 Antimicrobial Resistance (AMR)

During the last few decades, the incidence of microbial infections has increased dramatically. Continuous deployment of antimicrobial drugs in treating infectious diseases has led to the emergence of resistance among various microorganisms (e.g., bacteria, fungi, viruses, and parasites). Nearly all effective antimicrobial agents have created high levels of MDR with enhanced morbidity and mortality; thus, they are referred to as "super bugs" [5, 8]. MDR is defined as insensitivity or resistance of a microorganism to the administered antimicrobial medicines (which are structurally unrelated and have different molecular targets) despite earlier sensitivity to the medicine. Although the development of MDR is a natural phenomenon, the inappropriate use of antimicrobial drugs, inadequate sanitary conditions, inappropriate food handling, and poor infection prevention and control practices contribute to the emergence of and spread of MDR.

An extensive increase in the number of immunocompromised conditions, such as HIV infections, diabetic patients, chemotherapy for cancer treatments, individuals who have undergone organ transplantation, and intensive care for pre-term newborns, makes the body an easy target for hospital acquired (nosocomial) infectious diseases, thereby contributing to the further spread of MDR [9 11]. According to the WHO, microorganisms (bacteria, fungi, viruses, and parasites) that have developed resistance to antimicrobial drugs may lead to ineffective treatment, resulting in persistence and spreading of infection. Some serious bacterial infections include *Escherichia coli*, *Pseudomonas*, tuberculosis, pneumonia, methicillin-resistant *Staphylococcus aureus* (MRSA), meningitis, gonorrhea, syphilis, *Salmonella*, and MDR-Acinetobacter infections. Studies indicate very high rates of resistance in bacteria such as *E. coli* against cephalosporin and fluoroquinolones antibiotics, *Klebsiella pneumoniae* against cephalosporin and carbapenems, *S. aureus* against methicillin, *Streptococcus pneumonia* against penicillin, non-typhoidal *Salmonella* against fluoroquinolones, *Shigella* species against fluoroquinolones, *Neisseria gonorrhoeae* against cephalosporin, and *Mycobacterium tuberculosis* against rifampicin, isoniazid, and fluoroquinolone, causing common infections (such as urinary tract infections, pneumonia, and blood stream infections) [12–14] (Table 13.1). *Staphylococcal* species, most notably *S. epidermidis* and *S. aureus*, cause 60%–70% of all infections, and numerous outbreaks of MRSA have been reported [15].

The list of WHO priority drug-resistant pathogens and their current antibiotics of choice for treatment include carbapenem-resistant *Acinetobacter baumannii* (colistin, carbapenems, sulbactam, rifampin, and tigecycline), carbapenem-resistant *P. aeruginosa* (ticarcillin-clavulanate, ceftazidime, aztreonam, imipenem, ciprofloxacin and colistin, vancomycin-resistant), *Enterococcus*

Table 13.1 Multi-drug resistant (MDR) bacteria and associated diseases.

Bacteria	Diseases Caused By The Bacteria	Drugs Resistant To The Bacteria
Escherichia coli	Urinary tract infections and blood stream infections	Cephalosporins, fluoroquinolones
Klebsiella pneumoniae	Pneumonia and urinary tract infections	Cephalosporins, carbapenems
Staphylococcus aureus	Wound and blood stream infections	Methicillin
Streptococcus pneumoniae	Pneumonia, meningitis, otitis	Penicillin
Nontyphoidal Salmonella	Food borne diarrhea, blood stream infections	Fluoroquinolones
Shigella species	Diarrhea (bacillary dysentery)	Fluoroquinolones
Neisseria gonorrhoeae	Gonorrhea	Cephalosporins
Mycobacterium tuberculosis	Tuberculosis	Rifampicin, isoniazid, fluoroquinolone

faecium (streptogramin, linezolid, daptomycin, oritavancin, and tigecycline), methicillin-resistant *S. aureus* (vancomycin, trimethoprim-sulfamethoxazole, clindamycin, linezolid, tetracyclines, and daptomycin), fluoroquinolone-resistant *campylobacter* spp. (erythromycin, ciprofloxacin, and fluoroquinolones), and clarithromycin-resistant *Helicobacter pylori* (amoxicillin, esomeprazole, rabeprazole, omeprazole, metronidazole, levofloxacin, and clarithromycin) [14, 15].

Antibiotics have emerged as a firsthand strategy over the years to combat pathogenic organisms. Over several decades, bacteria causing common or severe infections have developed resistance to each new antibiotic coming to market. The WHO has long recognized the need for an improved and coordinated global effort to contain AMR. The impact of antibiotic resistance in terms of mortality and the public health cost is quite difficult to estimate, and there are few studies addressing this issue. The US Center for Disease Control and Prevention (CDC) conservatively estimated that, in the US, more than two million people every year are affected with antibiotic-resistant infections, with about 23,000 dying because of the infection. The antibiotics seemed to induce different kinds of stress in the bacterium and host, where the immune response towards oxidative stress (OS) is one of the defense functions. Investigations indicate that several antibiotic functions to increase OS. Thus, a bacterium may protect itself from both the host immune response and antibiotic therapy by increasing its antioxidant capacities. Generally, there are three main modes of action for antimicrobial drugs:

- *By disruption of bacterial cell wall biosynthesis*: Penicillin and other β-lactam antibiotics target cell wall synthesis by inhibiting transpeptidation.
- *By interfering with bacterial protein synthesis*: Antibiotics, such as tetracycline, can accumulate in the cytosol and bind to 30S ribosomal subunits to block key RNA interactions and thereby stop the peptide chains formation.
- *By blocking bacterial DNA replication and repair*: Quinolones target DNA-topoisomerase complexes to inhibit the cell division process of microbes.

Various mechanisms contribute to the development of MDR. The main mechanisms of resistance include enzymatic inactivation, limiting uptake of a drug, modification of a drug target, inactivation of a drug, and active efflux of a drug (Figure 13.1)

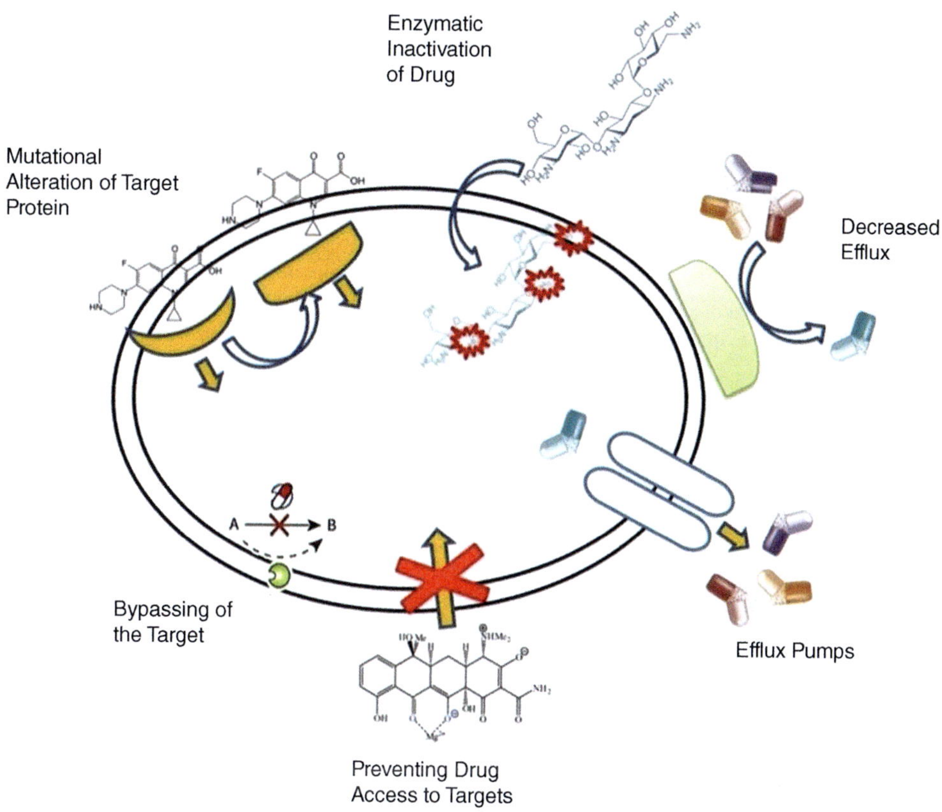

Figure 13.1 Mechanisms of developing multi-drug resistance (MDR) in bacterial cells.

13.3 Cellular Oxidative Stress (OS) and Microbial Infections

13.3.1 Oxidative Stress (OS) and Infectious Diseases

Pathogenesis of diseases because of microbial invasion in the human body occurs because of microbial toxins or various other mechanisms. Apart from virulence of pathogenic microorganisms or microbial cytotoxins, increased reactive oxygen species (ROS) and reactive nitrogen species (RNS) generated by activated immune cells create highly cytotoxic conditions (cytokine storm) that directly damage target organs in the context of unbalanced inflammatory responses. Where free radicals protect against invading microorganisms, they also cause tissue damage resulting in further inflammation and exacerbation of the disease. Thus, a dual role of OS can be found in human infectious diseases [16–18]. Expression of various enzymatic factors contributes to the generation of OS during the process of microbial invasions. Table 13.2 summarizes the role of a few enzymes that serve as biomarkers for inflammation involving OS.

Microorganisms are often equipped with the production of antioxidant peptides themselves to escape the action of ROS/RNS generated by immune cells. For example, *Mycobacterium tuberculosis*, the causative organism of tuberculosis, releases mycothiol, and mycothiol-dependent reductase mycoredoxin-1 to survive within the macrophages [19]. Viral infections can severely impair the oxidative homeostasis. During chronic viral hepatitis, OS has been found to cause liver fibrosis,

Table 13.2 Enzymes responsible for oxidative stress (OS) in microbial infections.

Sl. No.	Enzyme	Origin	Mechanism Of Action
1	Myeloperoxidase	Activated neutrophils and within lysosomes of cells	• Has viricidal and bactericidal properties • Converts hypochlorous acid from hydrogen peroxide and chloride ($H_2O_2 + Cl^- \rightarrow ClO^- + H_2O$) • Microorganisms captured within lysosomes are killed by the action of hypochlorous acid along with lysozymes
2	NADPH oxidase	Activated neutrophils	• Catalyzes oxidation of NADPH with concurrent reduction of oxygen to a superoxide ($NADPH^+ 2O_2$, $NADP^+ + H^+ + 2O_2^-$) • Expressed during the inflammatory reaction to kill captured pathogens • Similar function as myeloperoxidase
3	Inducible nitric oxide synthase (iNOS)	Macrophages and leukocytes	• Catalyzes the formation of nitric oxide (NO) via oxidation of L-arginine to N-hydroxy-L-arginine and subsequent oxidation into L-citruline and NO • Involved in several processes such as cellular proliferation, apoptosis, and angiogenesis • NO reacts with peroxynitrite in the presence of superoxide • Peroxynitrite is essential in macrophages for the destruction of pathogens • Expression of iNOS and peroxynitrite is induced by stress-activated protein kinases JNK/SAPK and p38MAP kinase

cirrhosis, cancer, and metabolic dysfunction. The depletion of cellular antioxidant defenses, such as ascorbic acid and GSH, an increase in malonaldehyde, along with the expression of catalase, are causative factors behind OS [17]. HIV-induced OS contributes to the neurodegenerative complications observed in AIDS patients [18]. OS is commonly implicated in the case of acute respiratory viral infections. Oxidative burst and hypoxemia have been recently associated with SARS-CoV-2/COVID-19 infections. Despite the overwhelming evidence of the role of OS in microbial infection and the associated diseases, the impact of most infectious agents on host redox systems is not adequately investigated and requires separate attention [18].

13.3.2 Oxidative Stress-induced Microbial Infections

ROS plays a large role in immune cells by killing phagocytosed microorganisms or involving immune and inflammatory signaling cascades. Inflammation is an integral part of immune defense that is characterized by infiltration of cells, such as monocytes, macrophages, lymphocytes, neutrophils, and plasma cells, and the release of ROS by these cells leading to tissue injury. Neutrophils, such as polymorphonuclear neutrophils, generate ROS by activating NADPH oxidase. Besides causing direct OS to remove pathogens, ROS also stimulates inflammatory pathways and molecular signaling [20]. Figure 13.2 depicts the various ROS-induced cellular oxidative damages observed in microbial/bacterial infections. Our understanding of inflammatory signaling indicates that an array of pathways is activated with recognition of molecular patterns derived from pathogens and oxidative damages. The immune receptors are equipped to sense pathogen associated molecular

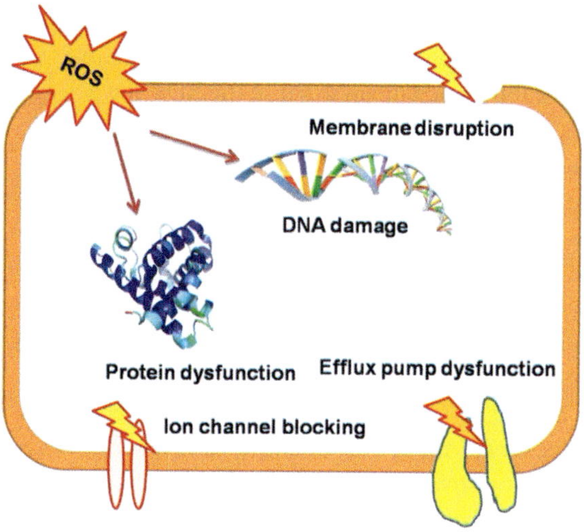

Figure 13.2 Reactive oxygen species (ROS)/oxidative stress (OS)-mediated cellular damages in bacterial infections.

patterns (PAMPs), pathogenic microbes or danger-associated molecular patterns (DAMPs), and endogenous stress, triggering downstream signaling cascades. These multiprotein complexes, which assemble upon sensing PAMPs or DAMPs (called inflammasomes), in turn initiate a caspase signaling pathway that further leads to cell injury and death. Inflammasome-induced cell injury and apoptosis contributes to the pathogenesis of Alzheimer's and Parkinson's diseases, as well as neurodegenerative and metabolic disorders [20, 21].

Microorganisms can possess the ability to modulate the ROS formation in the host and thereby increase the infectivity. By inducing or inhibiting oxygen radicals, pathogens reduce ROS-mediated host responses to infection and facilitate colonization. The following are a few examples used by the pathogens to evade the host redox defense system to cause persistent infections in human:

- Organisms such as *Francisella tularensis*, *Anaplasma phagocytophilum*, and *Chlamydia trachomatis* modulate with assembly of NADPH oxidase (NOX2) by averting incorporation of gp91phox, p22phox, and p47/p67phox (the key component of major cellular ROS-producing enzyme) into phagosomes that engulf the microorganisms. This process also avoids activation of the immune sensors, which may trigger the inflammatory cascade against pathogens [22–24].
- Microorganisms can also regulate the intercellular ROS levels. In neutrophils, microorganisms form exopolysaccharide, which inhibits ROS production and directly scavenges free oxygen radicals.
- *Entamoeba histolytica* uses ROS molecules as a secondary signaling messenger to activate NOX2 and induce apoptosis and host cell death, and in turn helps the spread of pathogens [25].
- Several viruses mediate ROS accumulations within the host cells to stimulate the apoptotic signal and subsequent cell death, which promotes the spread of virions and persisting viral infection [26].

13.3.3 Oxidative Stress/Reactive Oxygen Species (OS/ROS) and Antimicrobial Resistance (AMR)

Physiologically, ROS are generated spontaneously as a byproduct of the respiratory cycle the same time ROS generation is usually supplemented by redox-cycling agents, disinfectants, and antibiotics, which contributes to their antimicrobial bioactivity. Antibacterial agents (like disinfectants and antiseptics), which cause oxidative damage to invading pathogens, is considered an effective bactericidal agent [27]. Prokaryotes have developed their own inducible antioxidant defense network against toxic effects of these ROS. There is a subtle mechanism to sense the ROS and activate the expression of effectors that detoxify the ROS.

Irrespective of the targets and drug-target interactions, all antibacterials use a common mechanism of inactivation to stimulate the production of lethal doses of hydroxyl radicals via the Fenton reaction. Iron misregulation by the superoxide-mediated oxidation of iron–sulfur clusters generates hydroxyl radicals [28]. Investigations have proved that OS contributes to bactericidal antibiotic-mediated cell death such as in aminoglycoside antibiotics [29]. These studies indicate that OS is involved in the antibiotic resistance within pathogens and the exposure of bacteria to antibiotics may alter the antioxidant defense system and redox mechanisms in cells.

The frontline of drug resistance in bacteria is offered by a cell wall comprising a lipopolysaccharide–phospholipid bilayer, which passively resists entry of antibiotics and other antibacterial drugs. This can be illustrated by the susceptibility of gram-positive bacteria towards different antibiotics because of their absence of outer membranes. Apart from this possible reduction in the affinity of receptor sites, the substitution of alternative pathways, production of detoxifying enzymes, and activation of an efflux pump to inhibit cellular antibiotic uptake are different mechanisms that further contributes to the drug resistance of bacteria. Similarly, genetic modification introduced by plasmid conjugation, phage-based transduction, and horizontal gene transfer are principal mechanisms by which bacteria induce resistance against different antibacterial agents [30].

13.3.4 ROS/OS Pathways and Therapeutic Interventions for Antimicrobial/Antibacterial Agents

Current antimicrobial drugs are characterized on the basis of their ability to kill bacteria or inhibit growth. The mechanism of action of these drugs ranges from inhibition of the cellular processes essential for the biosynthesis of proteins, RNA, DNA, cell wall, and folic acid during the logarithmic growth phase. However, pathogenic bacteria are continuously developing drug resistance mechanisms against these antibacterial agents. Hence, in recent years, the focus has been exploring alternative biochemical pathways as a target space for the development of antibacterial classes [31].

Manipulating the metabolic flux, which is a hallmark of cancer metabolism, as a therapeutic intervention could be adapted to the development of novel antibacterial agents. Oncogenic cells uptake more glucose without changing mitochondrial oxidative phosphorylation, a phenomenon known as aerobic glycolysis or the Warburg effect. Modulation of the oncogenic mammalian pyruvate kinase (PYK) isoform 2 (M2PYK) to the M1PYK isoform (expressed in healthy cells) leads to the reversal of the Warburg effect and a reduction in tumor proliferation [31, 32]. In recent times, similar induction of tricarboxylic acid (TCA) cycle flux has been revealed within enveloped viruses (such as human cytomegalovirus), which has led to the development of novel antivirals like fatty acid biosynthesis inhibitors [33]. Many classical antibacterial methods have been developed against the backdrop of inhibition of the folate pathway in bacteria. Folate is an essential cofactor and is an important precursor for the biosynthesis of purines, pyrimidines, and amino acids. Antimicrobials, such as trimethoprim, selectively inhibit microbial dihydrofolate reductases (DHFR) to deplete the cellular pool of precursors of DNA, RNA, and proteins, thereby impacting bacteriostasis [34, 35]. Prontosil, the first synthetic antimicrobial agent, targets dihydropteroate synthase (DHPS), another such folic acid biosynthetic pathway enzyme. A combination DHFR–DHPS inhibitor was used as the first-line therapy in the prophylaxis and treatment of HIV-associated secondary pneumonia caused by *Pneumocystis jirovecii*. Similarly, sulfanilamides act by competitive inhibition at the active site of dihydropteroate synthase. Para-amino salicylic acid (PASA), the antituberculosis drug, also functions by inhibiting the folate pathway [31].

Augmenting the OS pathway and inducting ROS formation have emerged as potential antimicrobial targets recently. ROS have been involved in antibiotic-mediated lethality. For example, bactericidal antibiotics can generate ROS in mammalian cells through mitochondrial dysfunction [36]. A major class of antibiotics, such as β-lactams (e.g. ampicillin), quinolones (e.g. norfloxacin), and aminoglycosides (e.g., kanamycin), can induce ROS generation that can inhibit the vital function of oxygen-respiring cells [28, 29]. To increase the antibiotic-mediated lethality, the TCA cycle pathway can be targeted, which is a major source of $OH^•$ radicals. Antimicrobial agents enhance the generation of $O_2^{•-}$, which then is converted to H_2O_2 by SOD, and finally $OH^•$ radicals are formed. Unlike the other radicals, namely $O_2^{•-}$ and H_2O_2 that are eradicated by SOD and catalase and peroxidase, there is no enzyme to detoxify $OH^•$. Thus, $OH^•$ exerts major cytotoxic effects by damaging protein, lipid, and nucleic acid that also become the hallmark of bactericidal effects [37].

It has been further suggested that increasing the production of ROS may potentiate the antibacterial activity of bactericidal antibiotics. Inhibiting target genes like succinate dehydrogenase by carboxin increases the endogenous ROS formation and accumulation [36]. Similarly, if genes encoding antioxidant defense systems are impaired in bacterial cells, then it could be easily targeted and killed by ROS-generating antibacterials. Such genes as OxyR (confers resistance during increased H_2O_2 conditions), KatA (virulence factor in *P. aeruginosa*), and IscR (transcriptional regulator of an iron–sulfur assembly and full activity of KatA and H_2O_2 resistance) can be targeted to compromise the virulence of the pathogens. For example, avicin can chemically modulate OxyR in *E. coli* by avicinylation [38]. Pharmacophores capable of ROS generations, but effective in non-bacterial systems, can also be repurposed as antibacterial agents. Naphthoquinone derivatives, such as menadione, menaquinone, and alkannin, have shown promise as antibacterial, antifungal, and antiprotozoal agents [36].

13.4 Polyphenolic Phytochemicals as Antimicrobial Agents

13.4.1 Dietary Polyphenolics and Flavonoids

Polyphenols are the major group of naturally occurring secondary metabolites that exist in the plant kingdom. They are abundant in various plant parts, including fruits, the flowers and leaves of herbs, and terrestrial plants. More than 8,000 phenolic compounds of diverse structural arrangements have been reported from the plant kingdom [39]. Polyphenols are essentially biosynthesized by plants for a defense mechanism against microbes, environmental stress, and other predators [40]. These phytoconstituents are often found in plants as a conjugate with one or more sugar moieties and are termed glycosides. Chemically, they contain one or more phenolic rings with multiple hydroxyl groups on aromatic rings comprising many substitutions and structural diversity [41]. Because of the presence of multiple hydroxyl groups, most of these classes of compounds exhibit strong antioxidants and are well known as free radical scavengers. Polyphenols also exhibit wide ranges of biological activities, such as antioxidant, hepatoprotective, antibacterial, anticancer, antidiabetic, and antihypertensive behaviors, depending on their structural features [42].

Chemically, polyphenols can be broadly divided into two major categories, flavonoids and non-flavonoids, depending on the number and arrangement of different phenolic subunits and the linkage of hydroxyl moieties to the phenolic skeleton [41]. The flavonoid class of compounds can be subdivided into subclasses based on the degree of oxidation of the heterocyclic ring including flavones, flavanones, isoflavones, flavonones, flavanols, flavonols, dihydroflavonols, flavandiols, chalcones, dihydrochalcones, aurones, anthocyanidins, proantocyanidins, biflavanoids, and neoflavonoids [39]. They are richly available in various plant pigments. The non-flavonoid class is divided into four different subclasses: (1) phenolic acids, (2) stilbenes, (3) lignans, and (4) coumarins [41].

Phenolic acids are a dominant category under the non-flavonoid class of polyphenols and are further subdivided into hydroxybenzoic acids (C1–C6 backbone) and hydroxycinnamic acids (C3–C6 backbone), and structurally characterized by a carboxylic acid group linked to the phenolic ring [42]. They generally exist in plants either in free form or esterified form. They also often exist as a conjugate with a sugar moiety and proteins, and hydrolysable on acid or alkali treatment. Many foods and beverages, such as wine, tea, coffee chocolate, vegetables, whole grains, and fruits, contain hydroxycinnamic acid in very high concentrations [43].

Stilbenes are biosynthesized by plants during external influences such as infection or injury. They contain a C6–C2–C6 backbone and structurally represent a 1,2-diphenylethylene nucleus and exist either in the monomeric or oligomeric form. Resveratrol is a naturally occurring important bioactive compound that falls under this category [41, 44].

Stilbenes, a coumarin type of polyphenol, also synthesize and accumulate in the plant tissues because of abiotic stress and microbial attacks. Stilbenes comprise a 1,2-benzopyrone skeleton (α-chromone) and they also frequently exist in prenylated form. Coumarin cores are often used as a template in the synthesis of various pharmacologically important novel compounds [41, 45].

Lignans are a comparatively less abundant class of phenolic compounds structurally characterized by a dibenzylbutane skeleton. These types of compounds are generally found in higher plants (gymnosperms, angiosperms, pteridophytes, etc.). Often, they are found in the plant material in bound form and are difficult to extract [41, 45]. Table 13.3 depicts polyphenolic compounds, including flavonoids and their dietary sources.

13.4.2 Chemistry of Plant Polyphenolics/ Flavonoids

In plant-derived polyphenolic compounds, flavonoids comprise the largest group with approximately 10,000 natural analogues. These are hydroxylated phenolic substances that are synthesized by plants in response to microbial infection. They often exist as bright colored (yellow to red) pigments in plants and microbes [49]. The structural framework of flavanoid compounds comprises a benzo-γ-pyrone ring system (C6–C3–C6 backbone). Structurally, they are characterized as C15 compounds and comprise two phenolic (C6) rings that are linked by a bridge of heterocyclic pyrone rings. Two phenolic rings are denoted as A and B rings, whereas, connecting heterocyclic rings are considered as a C ring in the structural skeleton [41, 49]. The basic structural skeletons/backbones of various types of flavonoids classes are represented in Figure 13.3.

The chemical nature and biological potential of flavonoids depends on their structural class, degree of hydroxylation, other functional group substitutions, conjugations, and degree of polymerization that categorize them in different subclasses [50]. The different categories of flavonoids varies in the degree of oxidation and arrangement of substitution of the C ring, while different compounds within a class differ in the pattern of substitution of the A and B rings. Generally, the B ring binds at Position 2 of the C ring; however, it can also be attached at Position 3 or 4. Ring B can adopt different structural features and the three rings can undergo glycosylation and hydroxylation [48]. Flavonoids generally contain three or more –OH groups that are linked to their structural skeleton and exert a wide array of structural configurations. Additionally, in nature they are frequently found in glycosylated form with multiple sugar units, chemically termed as flavanoid glycosides. Many of them are hydrolysable by acid treatment and the non-sugar part is called a glycon [49, 51].

Table 13.3 Classification of polyphenolic compounds/flavonoids and their dietary sources with examples [43, 46–48].

Class Of Polyphenols	Subclass	Prototype Compounds	Major Dietary Sources
Flavonoids	Flavone	Baicalein, apigenin, luteolin, chrysin	Herbal tea, fenugreek, onion, garlic, pepper, citrus fruits, green leafy vegetables
	Isoflavone	Genistein, daidzein, biochanin A, glycitein	Chickpea, pea nut, dairy products, egg, meat, seafood, soy products, legumes
	Flavonol	Rutin, quercetin, myricetin, fisetin	Tea, chocolate, cocoa, onions, scallions, kale, broccoli, apples, berries,
	Flavanonol	Taxifolin, aromadendrin, engeletin	Milk thistle seeds, citrus fruits
	Flavanol	(−)- epicatechin, (+)-catechin, (−)-epigallocatechin gallate (EGCG), theaflavins	Cocoa, chocolate, tea, grapes, apples
	Chalcone	Butein, xanthoangelol, 4-hydroxyderricin, cardamomin, isoliquiritigenin, isosalipurposide	Tomatoes, licorice, shallots, bean sprouts
	Flavanone	Hesperidin, hesperetin, naringin, naringenin, eriodictyol	Pomegranate, citrus fruits, tomatoes, grape fruit
	Dihydrochalcones	Phlorizin, aspalathin, nothofagin	Apples and apple products, rooibos tea
	Anthocyanidins	Cyanidin, peonidin, delphinidin, petunidin, pelargonidin, malvidin	Red wines, cherries, red grapes, berries, flowers, oranges, black soy beans, hibiscus sp., purple/black rice, onions, red potatoes, purple cabbage
	Proantocyanidins	Procyanidin B1, procyanidin B2, procyanidin B3	Berries, cherries, red grapes, red wines, flowers, oranges, black soy beans, banana, cocoa, and apricot, cereals such as sorghum and barley
Non-flavonoids	Phenolic acids	Caffeic acid, sinapic acid, shikimic acid, gallic acid, protocatechuic acid, ferulic acid, p-coumaric acid,	Green tea, citrus fruits, kiwi, coffee, berries, apples, rice bran, passion fruit, cherries, mangoes, wheat, corn flours
	Stilbenes	Resveratrol	Grapes (skin), mulberries, peanuts, red wine
	Lignans	Silymarin, sylibin, sesamin, syringaresinol, ecoisolariciresinol, matairesinol, medioresinol, pinoresinol, laricircsino	Flaxseed, soy beans, broccoli, cabbage, milk thistle, apricots, strawberries, etc.
	Coumarins	Dicumarol, osthole	Cinnamon, green tea, carrot, bison grass

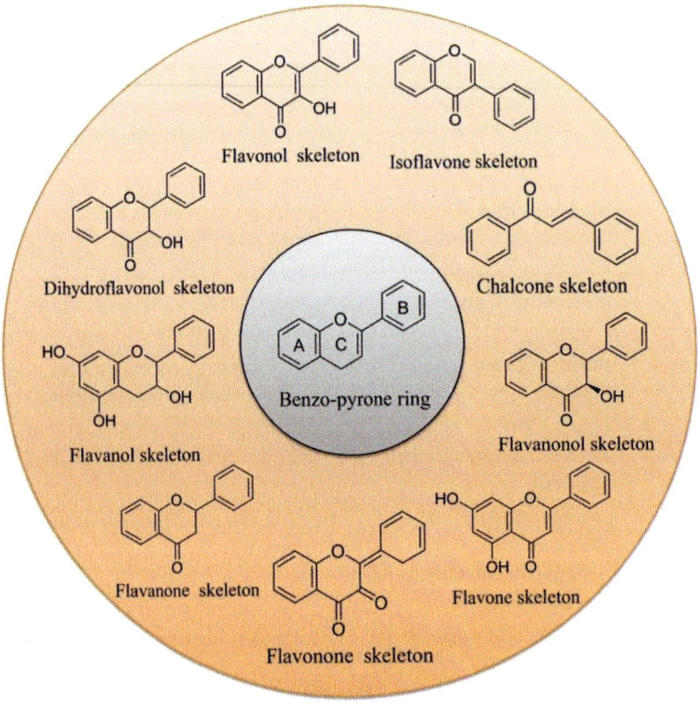

Figure 13.3 Basic structural skeletons/backbones of flavonoids.

The flavanoid ring system (benzo-γ-pyrone ring), which contains a double bond between the C-2 and C-3 positions with a ketone group at C-4 of the C ring denoted as flavones, whereas, in flavanones, the C ring is saturated containing a double bond between the C-2 and C-3 positions of the ring. These are also called dihydroflavones [48, 50]. Flavonol has an additional OH group at the C-3 position of the C ring, which makes them more polar than flavone and flavanones. Additionally, they exist in dihydroflavonol form, which has a double bond between the C3–C4 positions. Generally, glycosylation of flavonols occurs in the C-3 position of the C ring [41, 50]. Flavan-3-ols or flavanols contain a hydroxyl group at the C-3 position of the C ring. Unlike other flavanoids, isoflavonoids contain a B ring attached at the C-3 position of the C ring, whereas, in other flavanoids, the B ring is attached at the C-4 position [51].

Anthocyanidins and anthocyanins are bright colored (blue, red, or purple pigments) flavanoid compounds found in flowers, fruits, and leaves. These are positively charged compounds containing flavylium cations, and they often occur as chloride salts [41, 52]. Anthocyains comprise one or more sugar moieties in the C-3 position of the C ring. Frequently, these compounds are found in the plants as a conjugate with phenolic acids and other organic acids. The deglycosylated forms of anthocyanins are called anthocyanidins. Variation in the color of the anthocyanin compounds is reliant on the pH acylation and methylation of –OH groups attached to the A and B ring, and also the pH of the environment [52].

Proanthocyanidins are the dimer or trimer of flavanols in condensed form, also known as condensed tannins. Based on the interflavanic linkages, they can be divided as type A (C2–*O*–C7 or C2–*O*–C5 bonding) or type B (C4–C6 or C4–C8). They are often produced from flavanol-rich materials during fermentation [51]. Open C rings containing flavanoids are categorized as chalcones. Chalcone compounds exert a common chemical scaffold of 1,3 diaryl-2-propen-1-one, which is also known as a chalconoid [53]. Figure 13.4 represents the structures of plant-derived polyphenolic compounds, including flavonoids.

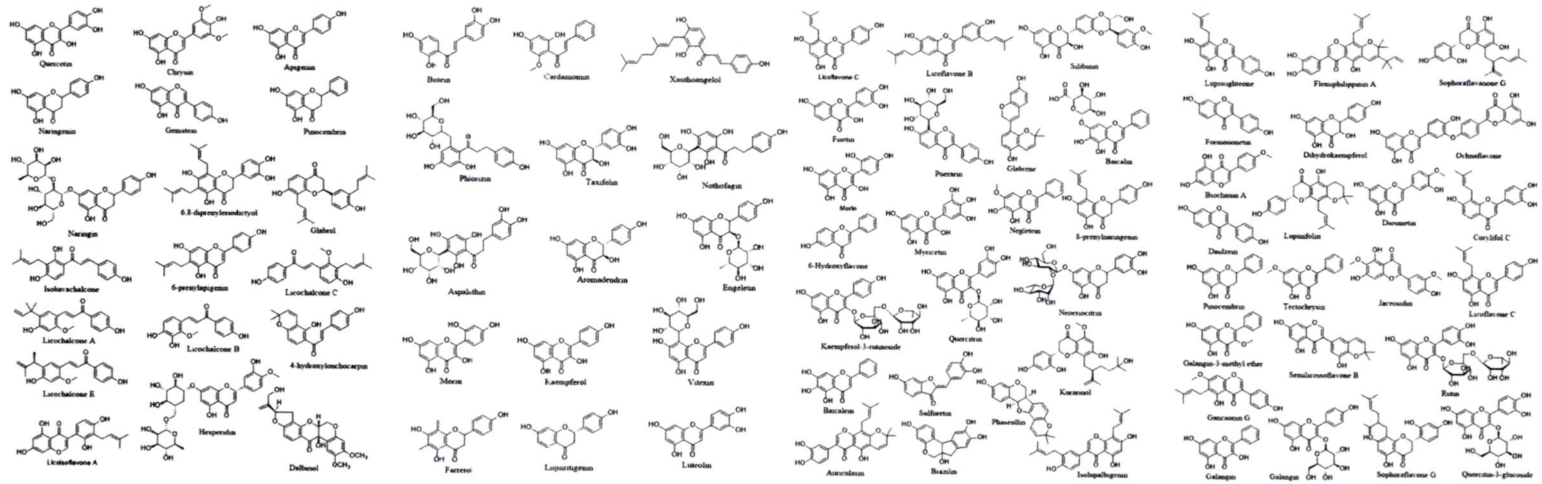

Figure 13.4 Structures of plant-derived polyphenols and flavonoids.

13.4.3 Biological Importance and Antioxidant Potential of Flavonoids

Flavonoids are essential to plants for survival during infections, predator attacks, and environmental stresses such as drought. Colors of the different plant parts are dependent on the type of flavonoids present in the tissue. Flavonoids are also responsible for the plant's communication with insects and microbes as a chemical messenger [49].

Consumption of flavonoids is proven to have various potential therapeutic benefits against human disease. A flavonoid-rich diet and supplements are attracting tremendous attention worldwide because of their wide array of positive ameliorative roles in human disease and healthy aging [50]. Flavanoid-rich herbal supplements are often found to be significantly effective in the management of hypertension, diabetes, and obesity, along with several other complications of metabolic syndromes. Additionally, flavanoid-rich food also helps heal infections and wounds faster, and improve the immune system [47, 50, 51]. Scientific validations indicate that flavonoids have remarkable potential in the prevention and management of several diseases. A variety of flavanoid molecules have a wide range of therapeutic benefits including antioxidant, anticancer, antibacterial, antiviral, antifungal, hepatoprotective, cardioprotective, antidiabetic, analgesic, and anti-inflammatory characteristics [47, 54].

Flavonoids exert excellent antioxidant effects by scavenging free radicals or chelating with metal ions. These properties are because of the multiple hydroxyl groups (–OH) present in the structure [55]. Flavonoids also activate the Nrf2-HO antioxidant pathway, which gives signals to produce endogenous antioxidant enzymes to maintain the redox balance in the human body during various OS conditions [56]. In addition to strong antioxidant properties, flavanoids also possess impressive potential towards the protection of DNA damage and mitochondrial death [57]. The approximate intake of 100 mg/day of total flavonoids in the daily diet may reduce the risk of all causes of mortality and cardiovascular disease by 6% and 4%, respectively [51]. Recent studies demonstrated that polyphenols like flavonoids improve gut-microbial health and maintain healthy gastrointestinal functions [58]. The free radical scavenging action of dietary flavonoids is displayed in Figure 13.5.

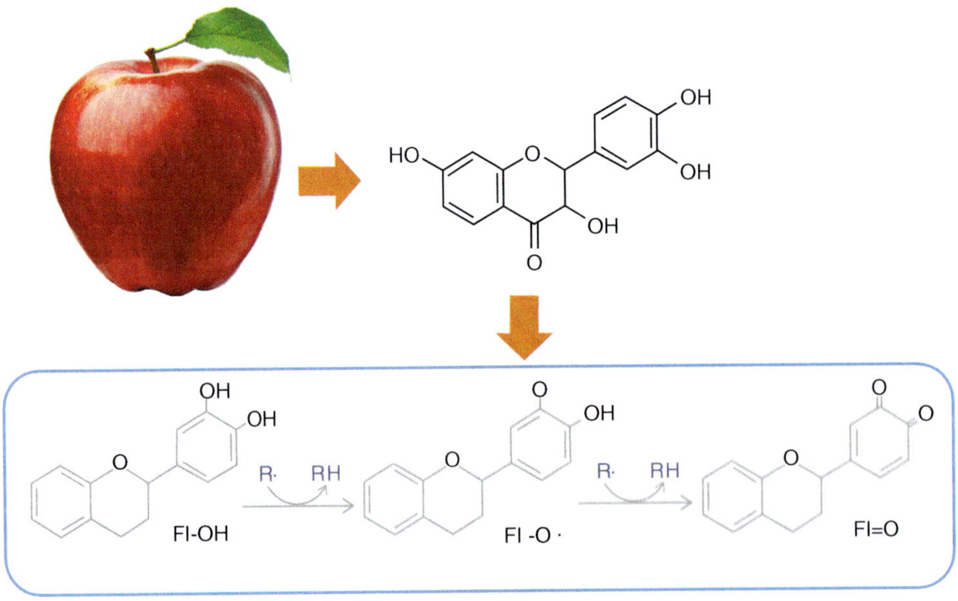

Figure 13.5 Free radical scavenging action of dietary flavonoids. Credit: JAVIER / Adobe Stock.

13.4.4 Antimicrobial/Antibacterial Effectiveness of Flavonoids

Flavonoids exhibit a broad spectrum of antimicrobial actions through different mechanisms that are often similar to those of conventional antibiotics and thus could be of importance in the augmentation of antimicrobial therapeutics. The antimicrobial properties of various flavonoids, along with their effectiveness against specific microbial strains and the mechanisms of action, are summarized in Table 13.4.

The specific mechanisms of antimicrobial/antibacterial actions are discussed below:

1) **Bacterial cell membrane disruption:** The integrity of the bacteria plasma membrane is a prerequisite for their essential cellular processes, including peptidoglycan biosynthesis, osmoregulation, cellular respiration, transport processes, and crosslinking of peptidoglycan, and therefore, the targeting of the bacteria cell membrane to interrupt the functions with chemical agents creates opportunities to kill them [61, 65]. A study shows that flavonoids can interact with the polar and non-polar heads of a lipid bilayer of a bacterial cell membrane as they hold

Table 13.4 Flavonoids and their antibacterial activity with mechanisms of action [59–67].

Flavonoids	Source	Bacterial Strain Against which They are Effective	Minimum Inhibitory Concentration (MIC)	Mechanism(s) Of Antibacterial Action
Kaempferol	*Maluspumila*	*S. aureus*	20 μg/ml	Inhibition of bacterial toxin;
		V. harveyi	6.25–100 μg/ml	Quorum sensing inhibition
Quercetin	*Anthemis cotula*	*S. aureus*	50 μg/ml	Inhibition of bacterial toxin; inhibition of sortase
Vitexin	*Vitex negundo*	*S. aureus*	252 μg/ml	Inhibition of biofilm formation; Quorum sensing inhibition
Farrerol	*Rhododendron hainanense*	*S. aureus*	0.5–8 μg/ml	Quorum sensing inhibition
Taxifolin	*Combretum albiflorum*	*P. aeruginosa*	4 mM	Inhibition of cell envelope synthesis
Luteolin	*Cecropia pachystachya*	*S. aureus*	12.5–50 μg/ml	Inhibition of biofilm formation; Quorum sensing inhibition
Chrysin	Honey	*S. aureus*	2–16 μg/ml	Quorum sensing inhibition

(Continued)

Table 13.4 (Continued)

Flavonoids	Source	Bacterial Strain Against which They are Effective	Minimum Inhibitory Concentration (MIC)	Mechanism(s) Of Antibacterial Action
Apigenin	*Citrus sinensis*	*E. coli* *S. mutans* *V. harveyi*	6.25–100 μg/ml 1.33 μg/ml 6.25–100 μg/ml	Inhibition of biofilm formation; Quorum sensing inhibition; Fatty acid synthase (FAS-II) inhibition
Naringenin	Bitter orange	*E. coli* *V. harveyi* *S aureus* *P. aeruginosa*	1–16 μg/ml 6.25–100 μg/ml 4 mM	Inhibition of bacterial toxin; Inhibition of biofilm formation; Quorum sensing inhibition; Membrane disruption
Naringin	Bitter orange	*Y. enterocolitica*	200 μg/ml	Quorum sensing inhibition
Hesperidin	Bitter orange	*Y. enterocolitica*	200 μg/ml	Quorum sensing inhibition
Dalbinol	*Amorpha fruticosa*	*P. aeruginosa*	6.25–25 μM	Inhibition of biofilm formation
Genistein	*Malusdo mestica*	*E. coli*	50 μg/ml	Inhibition of biofilm formation
Liquiritigenin	*Glycyrrhiza glabra*	*S. aureus*	4–32 μg/ml	Quorum sensing inhibition
Pinocembrin	*Apismellifera*	*S. aureus*	1–16 μg/ml	Quorum sensing inhibition
6,8-diprenyleriodictyol	*Dorstenia sp.*	*S. aureus* MSSA1	0.5 μg/ml	Depolarization of cell membrane; inhibition of DNA, RNA and protein synthesis
Isobavachalcone	*Dorstenia sp.*	*S. aureus* MSSA1	2 μg/ml	RNA and protein synthesis
4-hydroxylonchocarpin	*Dorstenia sp.*	*S. aureus* MSSA1	16 μg/ml	RNA and protein synthesis
6-prenylapigenin	*Dorstenia sp.*	*S. aureus* MSSA1	4 μg/ml	RNA and protein synthesis
Glabrol	*Glycyrrhiza glabra*	MSSA ATCC29213 MRSA T144	2 μg/ml	Disruption of the membrane permeability

Table 13.4 (Continued)

Flavonoids	Source	Bacterial Strain Against which They are Effective	Minimum Inhibitory Concentration (MIC)	Mechanism(s) Of Antibacterial Action
Licochalcone A	*Glycyrrhiza glabra*	MSSA ATCC29213 MRSA T144	2–4 μg/ml	ns
Licochalcone B	*Glycyrrhiza glabra*	MRSA T144	16 μg/ml	ns
Licochalcone C	*Glycyrrhiza glabra*	MSSA ATCC29213 MRSA T144	4 μg/ml	ns
Licochalcone D	*Glycyrrhiza glabra*	MSSA ATCC29213 MRSA T144	16–32 μg/ml	ns
Licochalcone E	*Glycyrrhiza glabra*	MSSA ATCC29213 MRSA T144	4 μg/ml	ns
Glabrene	*Glycyrrhiza glabra*	MSSA ATCC29213 MRSA T144	16 μg/ml	ns
Licoisoflavone A	*Glycyrrhiza glabra*	MSSA ATCC29213 MRSA T144	32 μg/ml	ns
Licoflavone B	*Glycyrrhiza glabra*	MSSA ATCC29213 MRSA T144	16–32 μg/ml	ns
Licoflavone C	*Glycyrrhiza glabra*	MSSA ATCC29213 MRSA T144	32–34 μg/ml	ns
Puerarin	*Radix puerariae*	*S. aureus*	2–16 μg/ml	Inhibition of bacterial toxin
Silibinin	*Silybum marianum*	*S. aureus*	4–32 μg/ml	Inhibition of bacterial toxin
Baicalin	*Scutellaria baicalensis*	*L. monocytogenes* *P. aeruginosa*	0.5–128 μg/ml 0.2–200 μM	Inhibition of bacterial toxin; Inhibition of biofilm formation
Fisetin	*Hymenaea courbaril*	*L. monocytogenes*	0.88–28 μM	Inhibition of bacterial toxin; FAS-I synthase inhibitor
Morin	*Morus alba*	*P. aeruginosa*	750 μg/ml	Inhibition of bacterial toxin; Coagulase inhibitor; Membrane disruption; FAS-I synthase inhibitor

(Continued)

Table 13.4 (Continued)

Flavonoids	Source	Bacterial Strain Against which They are Effective	Minimum Inhibitory Concentration (MIC)	Mechanism(s) Of Antibacterial Action
6-Hydroxyflavone	*Barleria prionitis*	*S. aureus*	50 μg/ml	Inhibition of bacterial toxin
Myricetin	*Vitis vinifera*	*Streptococcus agalactiae*	0.44–1.22 mM	Inhibition of bacterial toxin; FAS-I synthase inhibitor
Negletein	*Scutellaria oblonga*	*S. aureus*	32 μg/ml	Inhibition of biofilm formation
8-prenylnaringenin	*Humulus lupulus*	*S. aureus*	100 μg/ml	Inhibition of biofilm formation
Alopecurone A, D	*Saphora alopecuroides*	*S. epidermidis*	1.56–100 μg/ml	Inhibition of biofilm formation
Neoeriocitrin	*Citrus bergamia*	*V. harveyi*	6.25–100 μg/ml	Inhibition of biofilm formation
Neohesperidin	Bitter orange	*Y. enterocolitica*	200 μg/ml	Inhibition of biofilm formation; Quorum sensing inhibition
Kaempferol-3-rutinoside	*Saphora japanica*	*S. mutans*	60.7 μg/ml	Sortase enzyme inhibition
Kurarinol	*Saphora flavescens*	*S. aureus*	25–100 μg/ml	Sortase enzyme inhibition
Quercitrin	*Rhus verniciflua*	*S. mutans*	16–256 μg/ml	Sortase enzyme inhibition
Baicalein	*Scutellaria baicalensis*	*P. aeruginosa* *H. pylori*	0.2–200 μM 0.42–1 mM	Inhibition of biofilm formation; Urease inhibitor
Genistein	*Malusdo mestica*	*E. coli*	50 μg/ml	Inhibition of biofilm formation
Dalbinol	*Amorpha fruticosa*	*P. aeruginosa*	6.25–25 μM	Inhibition of biofilm formation
Amoradicin, amoricin, amorisin, isoamoritin	*Amorpha fructosa*	*C. perfringens*	1–100 μM	Neuraminidase inhibitor
Auriculasin, flemingsin, flemiphilippinin E	*Flemingia philippinensis*	*C. perfringens*	0.05–80 μg/ml	Neuraminidase inhibitor
Erysubin E, neorautenol, phaseollin, cristacarpin	*Erythrina abyssinica*	*C. perfringens* *V. cholerae*	0.1–100 μM	Neuraminidase inhibitor
Sulfuretin	*Dalbergia odorifera*	*R. solanacearum*	50 mg/mL	ns
Hematoxylin	*Haematoxylon brasiletto*	*S. aureus*	4.57 mM	ns

Table 13.4 (Continued)

Flavonoids	Source	Bacterial Strain Against which They are Effective	Minimum Inhibitory Concentration (MIC)	Mechanism(s) Of Antibacterial Action
Brazilin	*Haematoxylon brasiletto*	*S. aureus*	4.83 mM	ns
Isolupalbigenin	*Erythrina poeppigiana*	MRSA	1.56–3.13 μg/ml	ns
Lupiwighteone	*Ficustikoua* Bur.	*P. infestans*	90.365 μg/ml	ns
Formononetin	*Dalbergia frutescens*	*G. lamblia*	0.03–0.01 μg/ml	ns
Biochanin A	*Cassia fistula*	*T. cruzi*	18.32 μg/ml	ns
Daidzein	Apple fruit	*C. violaceum*	50–200 μg/ml	ns
Flemiphilippinin A	*Flemingia philippinensis*	*C. perfringens*	0.05–80 μg/ml	ns
Erythribyssin L	*Erythrina abyssinica*	*C. perfringens* *V. cholera*	0.1–100 μM	ns
Conrauiflavonol	*Ficus conraui*	*E. coli*	64 μg/ml	ns
Quercetin 3-O-methyl ether	*Cistus laurifolius*	*H. pylori*	3.9 μg/ml	ns
Ericoside	*Erica mannii*	*E. coli*	64 μg/ml	ns
3-O-methyldiplacol	*Paulownia tomentosa*	*S. aureus*	2.4 μg/ml	ns
Taxifolin-7-O-α-l-rhamnopyranoside	*Hypericum japonicum*	*S. aureus*	32 μg/ml	ns
Dihydrokaempferol	*Commiphora pedunculata*	*S. aureus*	625 μg/ml	ns
Sepicanin A	*Artocarpus sepicanus*	MRSA	2.9 μM	ns
Mimulone	*Paulownia tomentosa*	*S. aureus*	4.9 μM	ns
Sophoraflavanone G	*Sophora alopecuroides*	*S. epidermidis*	3.1–12.5 μg/ml	ns
Ochnaflavone	*Ochna pretoriensis*	*P. aeruginosa*	31.3 μg/ml	ns
Lupinifolin	*Mundulea sericea*	*S. aureus*	0.5 μg/ml	ns
Pinocembrin	*Cryptocarya chinensis*	*M. tuberculosis*	3.5 μg/ml	ns
Kurarinol	*Sophora flavescens*	*S. aureus*	107.7 μM	ns
Quercetin 3-O-β-D-rutinoside	*Marrubium globosum*	*P. vulgaris*	320 μg/ml	ns
Piliostigmol	*Piliostigma reticulatum*	*E. coli*	2.57 μg/ml	ns
Elatoside A, B	*Epimedium elatum*	*S. typhi*	20 mM	ns

(Continued)

Table 13.4 (Continued)

Flavonoids	Source	Bacterial Strain Against which They are Effective	Minimum Inhibitory Concentration (MIC)	Mechanism(s) Of Antibacterial Action
Entadanin	*Entada abyssinica*	*S. typhi*	1.56 μg/ml	ns
Quercetin-3-O-rutinoside	*Calotropisprocera*	*S. aureus*	19.5 mM	ns
Myricetrin-3-O-rhamnoside	*Croton menyharthii*	*B. cereus*	30–250 μg/ml	ns
Astragalin	*Garcinia preussii*	*S. aureus*	128 μg/ml	ns
Galangin-3-methyl ether	*Alpinia calcarata*	*S. aureus*	62.5 μg/ml	ns
Dorsmanin G	*Dorstenia mannii*	*P. aeruginosa*	8.0 μg/ml	ns
Techtochrysin	*Scutellaria oblonga*	*E. faecalis*	24–32 μg/ml	ns
Semilicoisoflavone B	*Glycyrrhiza uralensis*	*E. faecium*	32 μM	ns
Diosmetin	*Sophora moorcroftiana*	*S. aureus*	8 μg/ml	ns
Atocarpin	*Artocarpus anisophyllus*	*P. putida*	450 μg/ml	ns
Licoflavone	*Retama raetam*	*E. coli*	7.5 μg/ml	ns
Erysubin D	*Erythrina subumbrans*	*T. aureus*	50 μg/ml	ns
Amentoflavone	*Dorstenia barteri*	*B. megaterium*	3 μg/ml	ns
Jaceosidin	*Centaurea diluta*	MRSA	16.32 μg/ml	ns
Corylifol C	*Psoralea corylifolia*	*S. epidermis* *S. aureus*	0.147 mM 16 μg/ml	ns
Gancaonin Q	*Dorstenia angusticornis*	*B. subtilis*	2.44 μg/ml	ns
Amentoflavone	*Dorstenia barteri*	*B. cereus* *B. megaterium*	3 μg/ml	ns
Erysubin F	*Erythrina subumbrans*	*T. aureus*	50 μg/ml	ns
Cycloartocarpesin	*Morus mesozygia*	*P. aeruginosa*	156 μg/ml	ns
Licoflavone C	*Retama raetam*	*E. coli*	7.5 μg/ml	ns
Psiadiarabin	Saudi Arabian propolis	*M. marinum*	61.9 μg/ml	ns
Techtochrysin	*Scutellaria oblonga*	*E. coli* *E. faecalis* *B. subtilis*	24–32 μg/ml	Inhibition of biofilm formation
Gancaonin G, semilicoisoflavone B	*Glycyrrhiza uralensis*	*E. faecium*	32 μm	ns
5-Carbomethoxymethyl-4′,7 dihydroxyflavone	*Selaginella moellendorffii*	*E. coli*	25 μg/ml	ns

Table 13.4 (Continued)

Flavonoids	Source	Bacterial Strain Against which They are Effective	Minimum Inhibitory Concentration (MIC)	Mechanism(s) Of Antibacterial Action
Gnaphaliin A	*Achyrocline satureioides*	*S. aureus*	128 μg/ml	ns
Galangin	*Propel*	*S. aureus*	50 μg/ml	ns
Piliostigmol	*Piliostigma reticulatum*	*E. coli*	2.57 μg/ml	ns
Quercetin-3-O-rutinoside	*Calotropis procera*	*B. subtilis*	80 μg/ml	ns
Rutin	*Litchi chinensis*	*S. aureus* *E. coli* *S. dysenteriae*	62.5 μg/ml	Membrane disruption; Quorum sensing inhibition
3-Cinnamoyltribuloside	*Heritiera littoralis*	*M. madagascariense*	80–160 μg/ml	ns
Astragalin	*Garcinia preussii*	*S. aureus*	128 μg/ml	ns
Quercitin-3-glucoside	*Scutellaria oblonga*	*S. aureus*	32 μg/ml	Inhibition of biofilm formation
Quercetin-3-O-β-D-glucopyranoside	*Maytenus buchananii*	*S. aureus*	16 μg/ml	ns
Galangin-3-methyl ether	*Alpinia calcarata*	*S. aureus*	62.5 μg/ml	ns
Entadanin	*Entada abyssinica*	*S. typhimurium*	1.56 μg/ml	ns
Sophoraflavanone G	*Sophora flavescens*	*S. aureus*	1 μg/ml	ns
Mimulone; 3′-O-methyl-5′-hydroxydiplacone; Diplacone; 3′-O-methyl-5′-O-methyldiplacone	*Paulownia tomentosa*	*E. faecalis* *B. subtilis*	2 μg/ml	ns
Abyssione-V 4′-O-methyl ether	*Erythrina caffra*	*E. coli* *S. aureus*	3.9–62 μg/ml	ns
Sophoraflavone G	*Sophora alopecuroides*	*S. epidermidis*	3.1–12.5 μg/ml	ns
Taxifolin-7-O-α-L-rhamnopyranoside	*Hypericum japonicum*	*S. aureus*	32 μg/ml	ns
3′-O-methyldiplacol	*Paulownia tomentosa*	*B. cereus* *B. subtilis* *S. epidermidis*	2–4 μg/ml	ns

MIC: Minimum inhibitory concentration.
ns: not specified
MRSA: Methicillin-resistant *S. aureus*
MSSA: Methicillin-susceptible *S. aureus*

both polar and non-polar residues in their structure. The non-polar groups of flavonoid molecules can adhere to the inner lipophilic region of the bacterial membrane; on the contrary, polar head groups of the lipid bilayer interact by formation of the hydrogen bonds with the hydrophilic residues of flavanoids in the bacterial membrane surface [64, 65]. For example, flavonoids, such as catechin, cause disruption of the bacterial membrane through binding with the lipid bilayer following production of ROS and leakage of potassium ions in MRSA [68]. Mirzoeva et al. reported that quercetin decreases the proton motive force and increases the membrane permeability in *S. aureus*. Other flavonoids, such as naringenin, rutin, morin, and rhamnetin, also exhibit membrane disruption as an antimicrobial mechanism [69].

2) **Inhibition of quorum sensing (QS):** The chemical mediated communication system in certain bacteria helps them communicate with each other using small diffusible signaling molecules to build a protective barrier, known as "quorum sensing" (QS). These signaling molecules are produced by certain virulence genes at high cell density and serve a potential role in the colonization of bacterial species in the host. Moreover, QS facilitates bacteria to track the environment for other bacteria and to help each survive during environmental stresses that create a resistance mechanism [65, 70]. Interruption of this QS network by interfering with signal molecules or altering the signal generation and receiving mechanisms is known as "quorum quenching", which is a successful strategy to develop antimicrobial agents [61]. Similarly, conventional antibiotics of many plant-derived and synthetic flavanoids exhibit anti-QS potential to control microbial infection. Flavonoids, such as quercetin, kaempferol, rutin, apigenin, sinensetin neohesperidin, neoeriocitrin, naringin, and naringenin, exhibit inhibition of bacterial QS [65, 71, 72]. Some examples of flavonoids with anti-QS potential are given in Table 13.2.
3) **Inhibition of cell envelope synthesis:** Fatty acid synthase type II (FAS-II) in gram-negative bacteria is essential for membrane development by generation of acyl-ACP and β-hydroxyacyl-ACP. These are the two essential components for the biosynthesis of membrane phospholipids and lipopolysaccharides, where acyl-ACP molecules are required to generate patidic acid, which is a precursor to all cytoplasmic membrane phospholipids. Peptidoglycan is the essential component of a gram-positive bacterial cell wall. Flavanoids can inhibit both the FAS-II enzyme, as well as peptidoglycan, and alter the cell envelope synthesis [61, 64, 65]. Flavanoids, such as sakuranetin, quercetin, and apigenin, inhibit β hydroxyacyl-ACP dehydrase in *H. pylori* with IC_{50} values of 0.1 μM, 2.7 μM, and 2.5 μM, respectively [65, 73]. Flavanoids, such as eriodictyol, naringenin, and taxifolin, were found in 3-ketoacyl-ACP synthase in *Enterococcus faecalis*. Catechins inhibit the biosynthesis of the bacteria cell wall by interaction with peptidoglycan. Many flavanoids (quercetin, morin, fisetin, myricetin) also inhibit the synthesis of mycolic acid, which is a distinct component of the mycobacterium cell wall [65, 74].
4) **Inhibition of nucleic acid synthesis:** Flavonoids also exhibit inhibition of topoisomerases, such as DNA gyrase, helicase, and DHFR, which leads to the inhibition of the synthesis of the nucleic acid necessary for bacterial DNA replication [64, 65]. A study reported by Ohemeng et al. showed the DNA gyrase inhibitor potential of apigenin, quercetin, and 3,6,7,30,40-pentahydroxyflavone [75]. Genistein was reported to counter the growth of *Vibrio harveyi* by stabilization of the topoisomerase II–DNA cleavage complex. A literature study revealed that EGCG can alter pyrimidines and purine synthesis by inhibiting DHFRs at a concentration of 5–80 μM. Structurally similar flavonoids, such as luteolin, and related flavonoids, such as morin and myricetin, were found to counter the DnaB and RecBCD helicase/nuclease of *E. coli*. Flavanoids also demonstrated antimicrobial action by acting as intercalating agents, and for example, inhibit the nucleic acid synthesis of EGCG, robinetin, and myricetin [64, 76].

5) **Inhibition of biofilm formation:** A large colony of microbial cells connects each other and adheres to a solid surface (human tissue and medical devices) by the formation of a biofilm, which is a robust extracellular matrix comprising polymeric substances, mainly mucilage. It becomes challenging to eradicate bacteria after the formation of biofilm because of the protective barrier of the mature biofilm, which contains water channels to transport oxygen and nutrients to the bacteria inside the biofilm [65]. Interactions with the membrane surface to kill the bacteria are established as one of the antimicrobial mechanisms of flavanoids either by inhibition of the biofilm formation or disruption of the biofilm extracellular polysaccharide matrix of the biofilm. Evidence showed that phloretin, catechin, EGCG, apigenin, kaempferol, and rutin inhibit biofilm formation in various strains of biofilm producing bacteria [64, 65, 77].
6) **Inhibition of the electron transport chain and ATP synthesis:** Electron transport chain reactions are essential to the transfer of electrons and protons across the bacteria membrane, which leads to ATP synthesis for survival. A variety of flavanoids inhibit ATP synthase by binding with the polyphenol binding site in the bacterial cytoplasmic membrane and alter the energy production and use [61, 64]. Flavonoid licochalcone isolated from *Glycyrrhiza inflata* inhibited oxygen consumption. Other flavonoids, such as isobavachalcone and 6-prenylapigenin isolated from *Dorstenia* spp., exhibited bacteria membrane depolarization on *S. aureus* [64]. Some flavanoids are also reported to show inhibition of F1F0 ATPase, which plays an important role in ATP synthesis. For example, baicalein, silibinin, epicatechin, and morin inhibit F1FO ATPase in *E. coli* [65].
7) **Antibacterial mechanism of flavonoid–metal complexes:** Flavanoids are well known for their metal chelation properties because of the hydroxyl and keto groups present in the skeleton [65]. Complexation with metal ions makes the flavonoid structure more stable than the parent structure and also the affinities to various intracellular proteins. Flavanoids can also bind with bacterial enzymes containing transition metal ions [64]. Numerous flavanoids containing metallic nanoparticles exhibited a potent antimicrobial effect compared with their parent structure. One such well-studied complex is the quercetin complex with Mn, Co, Cd, and Hg, which has a potent bactericidal effect against *Klebsiella pneumonia*, *S. aureus*, and *Bacillus cereus*. However, the actual mechanism of antibacterial action of the flavonoid–metal complexes has not been proven [7n].
8) **Miscellaneous mechanisms of action:** Apart from the above mentioned mechanisms, some flavanoid molecules, such as morin, scutellarin, and baicalin, also exhibit antibacterial action by inhibiting bacterial virulent enzymes such as hyaluronidases, protease, urease, sortase, neuraminidase, and coagulase [64]. Some flavanoids have been reported for their potential to counter bacterial endotoxins, which plays a pivotal role in the development of infections in the host cell [64, 65]. In addition, a variety of flavanoids exhibited potential to arrest the swarming motility and hamper the bacterial movement, altering the function of various surface proteins present in the pili and capsule [64].

Figure 13.6 illustrates various mechanisms of antibacterial actions of dietary flavonoids.

13.4.5 Flavonoids as Antimicrobial Potentiators

Bacterial cells can develop the ability to remove antimicrobial drugs and small molecules from the cell by an efflux pump and this phenomenon is known to be a major contributor of antimicrobial drug resistance [79]. Flavonoids show efflux pump inhibition potential in various drug resistant

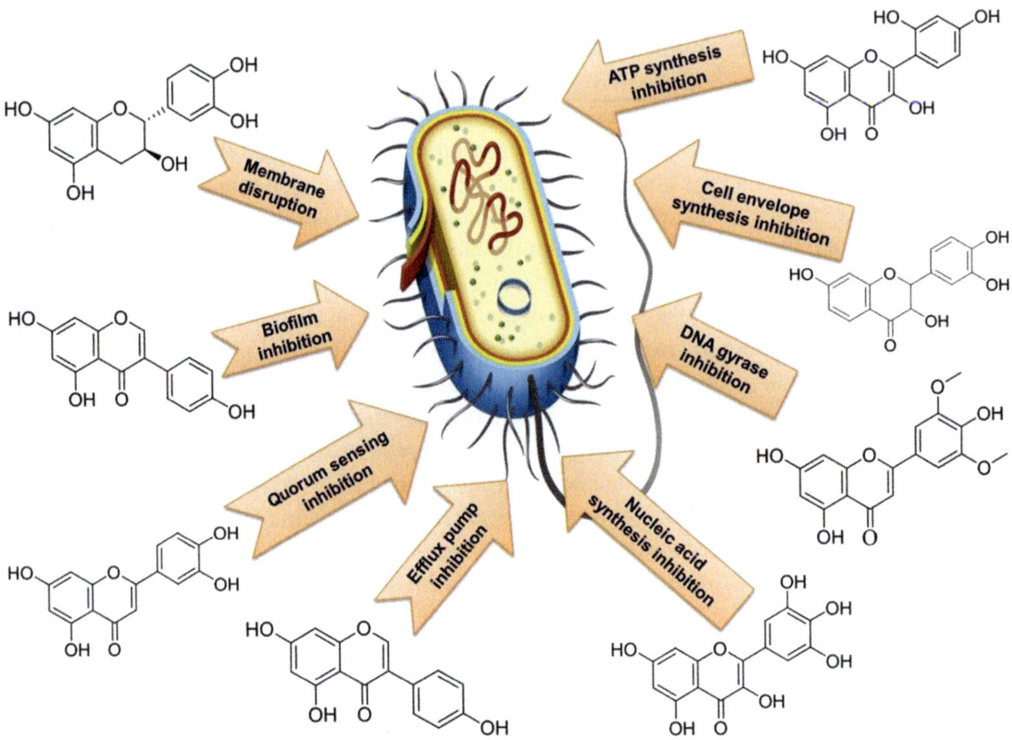

Figure 13.6 Mechanisms of the antibacterial action of dietary flavonoids.

strains by binding with the efflux pump transporter proteins. Flavonoids eventually block the drug efflux from bacteria by altering the function of transporters such as the ATP-binding cassette (ABC) transporter, multi-antimicrobial and toxic compound extrusion (MATE) transporter, small multidrug resistance (SMR) transporter, resistance-nodulation cell division (RND) transporter, and major facilitator (MFS) transporters [80]. Evidently, flavanoid silibinin co-administered with ciprofloxacin blocks the efflux pump of MRSA and potentiates antibiotic action. Similarly, morin, quercetin, and luteolin are also reported to inhibit the efflux pump of MRSA via leakage of potassium. Another study reported by Wang et al. [81] declared that silybin obtained from milk thistle seed enhances the efficiency in combination with ciprofloxacin by inhibition of the efflux pump.

It is obvious to have increased OS and inflammation via generation of ROS because of the host immune response during bacterial infections. Increased OS may lead to the vulnerability of the infection and also trigger the malfunction of the cellular metabolism [82]. Flavonoids are well known for their modulatory effect against OS in the human body by scavenging a variety of free radicals and chelating metallic ions [55]. Numerous antibacterial drugs kill the bacteria by activation of the ROS pathways, whereas, a mild amount of ROS is proven to be beneficial to the microorganism for their signaling mechanism. In such cases, flavonoids may act as a ROS scavenger to break the intracellular signals of bacterial cells and hamper the biochemical processes. Similarly, many existing antibacterial agents' flavonoids also possess antibacterial action by generation of ROS [64, 82]. Therefore, the antibacterial action of flavanoids and the relation with the antioxidant property seems controversial and inconclusive.

13.5 Repurposing Plant Polyphenolics/Flavonoids as Novel Antibacterial Agents

13.5.1 Drug Repurposing Strategy and Natural Product-based Drug Discovery

Drug repurposing, repositioning, or redirecting are common terms used to describe the process of generating novel clinical opportunities for known approved drugs, whether through new indications or new commercial opportunities for already marketed drugs. Repurposing is a common practice in the pharmaceutical sector [83]. The drug repurposing strategy has been employed as an alternative to conventional techniques in the drug discovery and development program. Drug repurposing primarily allows for an accelerated drug development process with less expenses than the typical route for de novo drug development. The repurposing strategy eases the complex drug discovery process because it bypasses many of the discovery and preclinical stages with the availability of pharmacokinetic, pharmacodynamic, and toxicity profiles of drug molecules under investigation [81, 84]. Recently, different chemical (PubChem, ChEMBL, DrugBank, etc.) and biological/protein (PDB) databases have been developed with the aim of in silico screening of plant-derived natural products. In silico screening of bioactive phytochemicals using computational tools and techniques with further experimental (in vitro and in vivo) assays accelerates the discovery of lead/drug molecules.

13.5.2 Repurposing Plant Polyphenolics/Flavonoids as Antibacterial Therapeutics

Plant-derived compounds or phytochemicals, including polyphenols and flavonoids, play a dominant role as antimicrobials, antibacterials, antifungals, and antivirals. In the current scenario, the problem of emerging MDR bacteria is posing a global medical threat and is continuously challenging the scientific community. The reduction in efficacy and the increase in toxicity of the synthetic drugs is further aggravating the problem [85, 86]. Many flavonoids obtained from plant/dietary sources possess a broad-spectrum antibacterial effectiveness as already described. Some of the potent antibacterial flavonoids include apigenin, luteolin, quercetin, rutin, myricetin, taxifolin, kaempferol, (–)-epicatechin, EGCG, isorhamnetin, and genistein. In view of their promising antimicrobial potential, plant polyphenols or flavonoids can be effectively repurposed as antimicrobial therapeutics to combat MDR microbials, particularly bacterial infections.

13.5.3 Synergistic Antibacterial Action of Flavonoids with Existing Antibiotics

Synergistic combinations of naturally occurring molecules with already existing synthetic drugs are gaining global interest. Synergy is defined as a therapeutic strategy where two or more compounds are given together in a combination that exerts higher efficacy and fewer side effects than the individual molecules [87]. Synergistic interactions of the flavanoids with existing antibiotics are one of the emerging strategies to fight various MDR strains. Combinations of flavonoids with antibacterial drugs may improve or facilitate the interaction of an antimicrobial agent in a multi-targeted basis to combat drug resistance. Moreover, better safety can be expected because multiple molecules are used together in comparatively lower concentrations [88]. Flavanoids exhibit immense potential as resistance modifying agents in antimicrobial drug therapy. Evidently, flavanoids counter the resistant mechanism by increasing drug permeability via bacterial efflux pump inhibition. Moreover, many flavanoids also neutralize the bacterial enzymes, such as

Table 13.5 Synergistic action of flavonoids with existing antibiotic regimens [55, 64, 65, 79].

Sl. No.	Flavonoids	Antibiotics	Strain
1	Flavone	Vancomycin, oxacillin	VISA ATCC 700699
2	Baicalein	Tetracycline, penicillin, amoxicillin, cloxacillin	MRSA PPSA *S. aureus* DMST 20651
3	Rutin + Morin	Methicillin	MRSA
4	Apigenin	Ampicillin, ceftriaxone	MRSA
5	Luteolin	Ampicillin, cephradine, ceftriaxone, imipenem, methicillin, ceftazidime	MRSA ATCC 43300 *S. pyogenes* DMST
6	Diosmetin	Streptomycin, ciprofloxacin	*S. aureus* 1199B, RN4220 *S. aureus* EMRSA-15
7	Galangin	Cloxacillin, amoxicillin	*S. aureus* DMST 20651 *E. coli*
8	Genistein	Norfloxacin, Ciprofloxacin	*S. aureus* 1199B, RN4220 S. *aureus* EMRSA-15
9	(−)-Epicatechin, myrecetin	Isoniazid	*M. smegmatis*
10	Morin	Ampicillin	MRSA ATCC 3359, MRSA DPS-1
11	Quercetin	Cloxacillin, ceftriaxone	*S. aureus* DMST 20651 MRSA ATCC 43300
12	EGCG	Tetracycline, oxacillin	MRSA6975, MRSA3202
13	3-Arylideneflavanones	Oxacillin	*S. aureus* A3
14	Quercetin + luteolin	Imipenem	MRSA clinical isolates
15	Quercetin, rutin, myricetin, taxifolin, kaempferol, isorhamnetin	Isoniazid, gentamycin	*Mycobacterium* spp.
16	Tiliroside	Norfloxacin, ofloxacin, ciprofloxacin, lomefloxacin	*S. aureus*

β-lactamase and penicillinases, which degrades the antibacterial compounds and maintains the drug concentration to kill the bacterial cells [64, 65, 79]. Table 13.5 presents the combined or additive effect of antibacterial drugs with various flavonoids.

13.6 Conclusion

The emergence of MDR and XDR pathogens is a serious global health concern. No potent single chemotherapeutic drug/dosage regimen is currently available clinically that can effectively halt/prevent AMR. Hence, plant-based antimicrobials may be an alternative option to identify suitable curative or preventive agents that can fight MDR pathogens. Most of the chemically synthesized antibacterials and antibiotics can cause serious adverse side effects and are very expensive. Therefore, there is increasing attention towards the use of medicinal plants/dietary plants as alternative sources of antimicrobial agents that can effectively control antibiotic-resistant bacterial infections. Owing to the promising antibacterial potential, dietary polyphenolic compounds/flavonoids may be developed as future antibacterial leads/drugs candidates for the treatment of MDR infections. However, these plant-derived polyphenolic compounds should be investigated for further development into more potential and safe antimicrobial agents.

References

1 Jones, K.E., Patel, N.G., Levy, M.A. et al. (2008). Global trends in emerging infectious diseases. *Nature* 451 (7181): 990–993.
2 Singh, S., Singh, S.K., Chowdhury, I., and Singh, R. (2017). Understanding the mechanism of bacterial biofilms resistance to antimicrobial agents. *Open Microbiology Journal* 11: 53–62.
3 González-Lamothe, R., Mitchell, G., Gattuso, M. et al. (2009). Plant antimicrobial agents and their effects on plant and human pathogens. *International Journal of Molecular Sciences* 10 (8): 3400–3419.
4 Anand, U., Jacobo-Herrera, N., Altemimi, A., and Lakhssassi, N. (2019). A comprehensive review on medicinal plants as antimicrobial therapeutics: potential avenues of biocompatible drug discovery. *Metabolites* 9 (11): 258.
5 Sua, S., Yina, P., Li, J. et al. (2020). *In vitro* and *in vivo* anti-biofilm activity of pyran derivative against *staphylococcus aureus* and pseudomonas aeruginosa. *Journal of Infection and Public Health* 13: 791–799.
6 Jalli, N., Santhi, S.K.V., Hnamte, S. et al. (2019). Antioxidant, anti-quorum sensing and anti-biofilm potential of ethanolic leaf extract of Phryniumcapitatum and Dryptesindica. *Asian Pacific Journal of Tropical Biomedicine* 9 (8): 323–332.
7 Roya, R., Tiwari, M., Donelli, G., and Tiwari, V. (2018). Strategies for combating bacterial biofilms: a focus on anti-biofilm agents and their mechanisms of action. *Virulence* 9 (1): 522–554.
8 Ong, K.S., Mawanga, C.I., Daniel-Jambuna, D. et al. (2018). Current anti-biofilm strategies and potential of antioxidants in biofilm control. *Expert Review of Anti-Infective Therapy* 16 (11): 855–864.
9 Khameneh, B., Iranshahy, M., Soheili, V., and Bazzaz, B.S. (2019). Review on plant antimicrobials: a mechanistic viewpoint. *Antimicrobial Resistance and Infection Control* 8 (1): 1–28.
10 Holmes, A.H., Moore, L.S., Sundsfjord, A. et al. (2016). Understanding the mechanisms and drivers of antimicrobial resistance. *The Lancet* 387 (10014): 176–187.
11 Hofer, U. (2019). The cost of antimicrobial resistance. *Nature Reviews Microbiology* 17 (1): 3.
12 Wu, C.P., Ohnuma, S., and Ambudkar, S.V. (2011). Discovering natural product modulators to overcomemultidrug resistance in cancer chemotherapy. *Current Pharmaceutical Biotechnology* 12 (4): 609–620.
13 Pinner, R.W., Lynfield, R., Hadler, J.L. et al. (2015). Cultivation of an adaptive domestic network for surveillance and evaluation of emerging infections. *Emerging Infectious Diseases* 21 (9): 1499–1509.
14 Hadler, J.L., Vugia, D.J., Bennett, N.M., and Moore, M.R. (2015). Emerging infections program efforts to address health equity. *Emerging Infectious Diseases* 21 (9): 1589–1594.
15 Li, X.Z. and Nikaido, H. (2004). Efflux-mediated drug resistance in bacteria. *Drugs* 64: 159–204.
16 Pizzino, G., Irrera, N., Cucinotta, M. et al. (2017). Oxidative stress: harms and benefits for human health. *Oxidative Medicine and Cellular Longevity* 2017: 8416763.

17 Pohanka, M. (2013). Role of oxidative stress in infectious diseases. A review. *Folia Microbiologica* 58 (6): 503–513.

18 Ivanov, A.V., Bartosch, B., and Isaguliants, M.G. (2017). Oxidative stress in infection and consequent disease. *Oxidative Medicine and Cellular Longevity 2017: 3496043.*

19 Van Laer, K., Buts, L., Foloppe, N. et al. (2012). Mycoredoxin-1 is one of the missing links in the oxidative stress defense mechanism of mycobacteria. *Molecular Microbiology* 86 (4): 787–804.

20 Chatterjee, S. (2016). Oxidative Stress, Inflammation, and Disease. In: *Oxidative Stress and Biomaterials* (eds. T. Dziubla, and D.A. Butterfield), 35–58. Elsevier.

21 Lamkanfi, M. and Dixit, V.M. (2014). Mechanisms and functions of inflammasomes. *Cell* 157: 1013–1022.

22 Carlyon, J.A., Latif, D.A., Pypaert, M. et al. (2004). Anaplasma phagocytophilum uti-lizes multiple host evasion mechanisms to thwart NADPH oxidase-mediated killing during neutrophil infection. *Infection and Immunity* 72 (8): 4772–4783.

23 Boncompain, G., Schneider, B., Delevoye, C. et al. (2010). Production of reactive oxygen species is turned on and rapidly shut down in epithelial cells infected with chlamydia trachomatis. *Infection and Immunity* 78 (1): 80–87.

24 McCaffrey, R.L. and Allen, L.A.H. (2006). *Francisella tularensis* LVS evades killing by human neutrophils via inhibition of the respiratory burst and phagosome escape. *Journal of Leukocyte Biology* 80 (6): 1224–1230.

25 Sim, S., Yong, T.S., Park, S.J. et al. (2005). NADPH oxidase-derived reactive oxygen species-mediated activation of ERK1/2 is required for apoptosis of human neutrophils induced by entamoeba histolytica. *The Journal of immunology* 174 (7): 4279–4288.

26 Yang, T.-C., Lai, -C.-C., Shiu, S.-L. et al. (2010). Japanese encephalitis virus down-regulates thioredoxin and induces ROS-mediated ASK1-ERK/p38 MAPK activation in human promonocyte cells. *Microbes and Infections* 12 (8–9): 643–651.

27 Kim, S.J., Kim, H.S., and Seo, Y.R. (2019). Understanding of ROS-inducing strategy in anticancer therapy. *Oxidative Medicine and Cellular Longevity* 2019: 5381692.

28 Dwyer, D.J., Kohanski, M.A., Hayete, B., and Collins, J.J. (2007). Gyrase inhibitors induce an oxidative damage cellular death pathway in Escherichia coli. *Molecular Systems Biology* 3 (1): 91.

29 Kohanski, M.A., Dwyer, D.J., Hayete, B. et al. (2007). A common mechanism of cellular death induced by bactericidal antibiotics. *Cell* 130 (5): 797–810.

30 Marrakchi, M., Liu, X., and Andreescu, S. (2014). Oxidative stress and antibiotic resistance in bacterial pathogens: state of the art, methodologies, and future trends. In: *Advancements of Mass Spectrometry in Biomedical Research* (A.G. Woods and C.C. Darie ed.), 483–498. Cham: Springer.

31 Murima, P., McKinney, J.D., and Pethe, K. (2014). Targeting bacterial central metabolism for drug development. *Chemico-Biological* 21 (11): 1423–1432.

32 Anastasiou, D., Poulogiannis, G., Asara, J.M. et al. (2011). Inhibition of pyruvate kinase M2 by reactive oxygen species contributes to cellular antioxidant responses. *Science* 334 (6060): 1278–1283.

33 Munger, J., Bennett, B.D., Parikh, A. et al. (2008). Systems-level metabolic flux profiling identifies fatty acid synthesis as a target for antiviral therapy. *Nature Biotechnology* 26 (10): 1179–1186.

34 Bushby, S.R. and Hitchings, G.H. (1968). Trimethoprim, a sulphonamide potentiator. *British Journal of PharmacologyCemotheraphy* 33 (1): 72–90.

35 Kwon, Y.K., Higgins, M.B., and Rabinowitz, J.D. (2010). Antifolate-induced depletion of intracellular glycine and purines inhibits thymineless death in E. coli. *ACS Chemical Biology* 5 (8): 787–795.

36 Kim, S.Y., Park, C., Jang, H.J. et al. (2019). Antibacterial strategies inspired by the oxidative stress and response networks. *Journal of Microbiology* 57 (3): 203–212.

37 Dwyer, D.J., Kohanski, M.A., and Collins, J.J. (2009). Role of reactive oxygen species in antibiotic action and resistance. *Current Opinion in Microbiology* 12 (5): 482–489.
38 Haridas, V., Kim, S.O., Nishimura, G. et al. (2005). Avicinylation (thioesterification): a protein modification that can regulate the response to oxidative and nitrosative stress. *The Proceedings of the National Academy of Sciences* 102 (29): 10088–10093.
39 Tresserra-Rimbau, A., Lamuela-Raventos, R.M., and Moreno, J.J. (2018). Polyphenols, food and pharma. Current knowledge and directions for future research. *Biochemical Pharmacology* 156: 186–195.
40 Ştefan, M. and Bîrsă, M.L. (2019). Flavonoids–An amazing group of compounds with potent antimicrobial properties. *Memoirs of the Scientific Sections of the Romanian Academy* 42.
41 Durazzo, A., Lucarini, M., Souto, E.B. et al. (2019). Santini A. Polyphenols: a concise overview on the chemistry, occurrence, and human health. *Phytotherapy Research* 33 (9): 2221–2243.
42 Tsao, R. (2010). Chemistry and biochemistry of dietary polyphenols. *Nutrients* 2 (12): 1231–1246.
43 Liu, K., Luo, M., and Wei, S. (2019). The bioprotective effects of polyphenols on metabolic syndrome against oxidative stress: evidences and perspectives. *Oxidative Medicine and Cellular Longevity* 2019: 6713194.
44 Shen, T., Wang, X.N., and Lou, H.X. (2009). Natural stilbenes: an overview. *Natural Product Reports* 26 (7): 916–935.
45 Pereira, D.M., Valentão, P., Pereira, J.A., and Andrade, P.B. Phenolics: from chemistry to biology. *Molecules* 14 (6): 2201–2211.
46 Kumar, S. and Pandey, A.K. (2013). Chemistry and biological activities of flavonoids: an overview. *The Scientific World Journal* 2013: 162750.
47 Panche, A.N., Diwan, A.D., and Chandra, S.R. (2016). Flavonoids: an overview. *Journal of Nutritional Science* 5.
48 Cook, N.C. and Samman, S. (1996). Flavonoids—chemistry, metabolism, cardioprotective effects, and dietary sources. *The Journal of Nutritional Biochemistry* 7 (2): 66–76.
49 Santos, E.L., Bhlns, M., Ferriani, A.P., and Teixeira, S.D. (2017). *Flavonoids: Classification, Biosynthesis and Chemical Ecology*. 6. 482. InTech.
50 Silva, A.S., Reboredo-Rodríguez, P., Süntar, I. et al. (2020). Evaluation of the status quo of polyphenols analysis: part I—phytochemistry, bioactivity, interactions, and industrial uses. *Comprehensive Reviews in Food Science and Food Safety* 19 (6): 3191–3218.
51 Corcoran, M.P., McKay, D.L., and Blumberg, J.B. (2012). Flavonoid basics: chemistry, sources, mechanisms of action, and safety. *Journal of Nutrition in Gerontology and Geriatrics* 31 (3): 176–189.
52 Khoo, H.E., Azlan, A., Tang, S.T., and Lim, S.M. (2017). Anthocyanidins and anthocyanins: colored pigments as food, pharmaceutical ingredients, and the potential health benefits. *Food and Nutrition Research* 61 (1): 1361779.
53 Zhuang, C., Zhang, W., Sheng, C. et al. (2017). Chalcone: a privileged structure in medicinal chemistry. *Chemical Reviews* 117 (12): 7762–7810.
54 Patel, J.M. (2008). A review of potential health benefits of flavonoids. *Lethbridge Undergraduate Research Journal* 3 (2): 1–6.
55 Amic, D., Davidovic-Amic, D., Beslo, D. et al. (2007). SAR and QSAR of the antioxidant activity of flavonoids. *Current Medicinal Chemistry* 14 (7): 827–845.
56 Velagapudi, R., El-Bakoush, A., and Olajide, O.A. (2018). Activation of Nrf2 pathway contributes to neuroprotection by the dietary flavonoid tiliroside. *Molecular Neurobiology* 55 (10): 8103–8123.
57 Romanova, D., Vachalkova, A., Cipak, L. et al. (2001). Study of antioxidant effect of apigenin, luteolin and quercetin by DNA protective method. *Neoplasma* 48 (2): 104–107.

58 Espley, R.V., Butts, C.A., Laing, W.A. et al. (2014). Dietary flavonoids from modified apple reduce inflammation markers and modulate gut microbiota in mice. *The Journal of Nutrition* 144 (2): 146–154.

59 Xie, Y., Yang, W., Tang, F. et al. (2015). Antibacterial activities of flavonoids: structure-activity relationship and mechanism. *Current Medicinal Chemistry* 22 (1): 132–149.

60 Tagousop, C.N., Ekom, S.E., Ngnokam, D., and Voutquenne-Nazabadioko, L. (2018). Antimicrobial activities of flavonoid glycosides from Graptophyllumgrandulosum and their mechanism of antibacterial action. *BMC Complementary and Alternative Medicine* 18 (1): 1–0.

61 Cushnie, T.T. and Lamb, A.J. (2005). Antimicrobial activity of flavonoids. *International Journal of Antimicrobial Agents* 26 (5): 343–356.

62 Dzoyem, J.P., Hamamoto, H., Ngameni, B. et al. (2013). Antimicrobial action mechanism of flavonoids from Dorstenia species. *Drug Discoveries and Therapeutics* 7 (2): 66–72.

63 Wu, S.C., Yang, Z.Q., Liu, F. et al. (2019). Antibacterial effect and mode of action of flavonoids from licorice against methicillin-resistant Staphylococcus aureus. *Frontiers in Microbiology* 10: 2489.

64 Górniak, I., Bartoszewski, R., and Króliczewski, J. (2019). Comprehensive review of antimicrobial activities of plant flavonoids. *Phytochemistry Reviews* 18 (1): 241–272.

65 Biharee, A., Sharma, A., Kumar, A., and Jaitak, V. (2020). Antimicrobial flavonoids as a potential substitute for overcoming antimicrobial resistance. *Fitoterapia* 7: 104720.

66 Das, M.C., Das, A., Samaddar, S. et al. (2018). Vitexin alters Staphylococcus aureus surface hydrophobicity to interfere with biofilm formation. *BioRxiv* 301473.

67 Daglia, M. (2012). Polyphenols as antimicrobial agents. *Current Opinion in Biotechnology* 23 (2): 174–181.

68 Reygaert, W.C. (2014). The antimicrobial possibilities of green tea. *Frontiers in Microbiology* 5: 434.

69 Mirzoeva, O.K., Grishanin, R.N., and Calder, P.C. (1997). Antimicrobial action of propolis and some of its components: the effects on growth, membrane potential and motility of bacteria. *Microbiological Research* 152 (3): 239–246.

70 Rasmussen, T.B. and Givskov, M. (2006). Quorum-sensing inhibitors as anti-pathogenic drugs. *Indian Journal of Medical Microbiology* 296 (2–3): 149–161.

71 Paczkowski, J.E., Mukherjee, S., McCready, A.R. et al. (2017). Flavonoids suppress Pseudomonas aeruginosa virulence through allosteric inhibition of quorum-sensing receptors. *The Journal of Biological Chemistry* 292 (10): 4064–4076.

72 Zhang, L.H. (2003). Quorum quenching and proactive host defense. *Trends in Plant Science* 8 (5): 238–244.

73 Zhang, L., Kong, Y., Wu, D. et al. (2008). Three flavonoids targeting the β-hydroxyacyl-acyl carrier protein dehydratase from Helicobacter pylori: crystal structure characterization with enzymatic inhibition assay. *Protein Sci* 17 (11): 1971–1978.

74 Li, B.H. and Tian, W.X. (2004). Inhibitory effects of flavonoids on animal fatty acid synthase. *Journal of Biochemistry* 135 (1): 85–91.

75 Ohemeng, K.A., Schwender, C.F., Fu, K.P., and Barrett, J.F. (1993). DNA gyrase inhibitory and antibacterial activity of some flavones (1). *Bioorganic and Medicinal Chemistry Letters* 3 (2): 225–230.

76 Russo, P., Del Bufalo, A., and Cesario, A. (2012). Flavonoids acting on DNA topoisomerases: recent advances and future perspectives in cancer therapy. *Current Medicinal Chemistry* 19 (31): 5287–5293.

77 Asahi, Y., Noiri, Y., Miura, J. et al. (2014). Effects of the tea catechin epigallocatechin gallate on P orphyromonasgingivalis biofilms. *Journal of Applied Microbiology* 116 (5): 1164–1171.

78 Brown, A.K., Papaemmanouil, A., Bhowruth, V. et al. (2007). Flavonoid inhibitors as novel antimycobacterial agents targeting Rv0636, a putative dehydratase enzyme involved in Mycobacterium tuberculosis fatty acid synthase II. *Microbiology* 153 (10): 3314–3322.

79 Seukep, A.J., Kuete, V., Nahar, L.et al. (2020). Plant-derived secondary metabolites as the main source of efflux pump inhibitors and methods for identification. *Journal of Pharmaceutical Analysis* 10 (4): 277–290.

80 Chambers, C.S., Viktorová, J., Řehořová, K.et al. (2019). Defying multidrug resistance! Modulation of related transporters by flavonoids and flavonolignans. *Journal of Agricultural and Food Chemistry* 68 (7): 1763–1779.

81 Farha, M.A. and Brown, E.D. (2019). Drug repurposing for antimicrobial discovery. *Nature Microbiology* 4 (4): 565–577.

82 Miroshnichenko, A.G., Perfil'Yev, V.Y., Lysenko, I.V., and Zhernakova, N.I. (2017). Interaction between some antibiotics and antioxidants. *Research Results in Pharmacology* 3 (4).

83 Gyawali, R. and Ibrahim, S.A. (2014). Natural products as antimicrobial agents. *Food Control* 46: 412–429.

84 Rudrapal, M. and Khairnar, S.J. (2020). Drug repurposing: an emerging approach in drug discovery. In: *Drug Repurposing – Hypothesis, Molecular Aspects and Therapeutic Applications* (ed. A. Dekebo). IntechOpen.

85 Barbieri, R., Coppo, E., Marchese, A. et al. (2017). Phytochemicals for human disease: an update on plant-derived compounds antibacterial activity. *Microbiological Research* 196: 44–68.

86 Davidson, P.M. and Juneja, V.K. (1990). Antimicrobial agents. *Food Additives* 83–137.

87 Wagner, H. and Ulrich-Merzenich, G. (2009). Synergy research: approaching a new generation of phytopharmaceuticals. *Phytomedicine* 16 (2–3): 97–110.

88 Ayaz, M., Ullah, F., Sadiq, A. et al. (2019). Synergistic interactions of phytochemicals with antimicrobial agents: potential strategy to counteract drug resistance. *Chemico-Biological Interactions* 308: 294–303.

14

Toxicity of Polyphenols Consumed as Food and Nutraceuticals

Remedies through Nanotherapeutic Approaches

Anyanwu O. Gabriel and Ogah F. Chinenye*

Department of Biochemistry, Bingham University, Karu, Nasarawa, Nigeria.
** Corresponding author*

Polyphenols are a vast group of phytochemicals abundant in nature, with at least 10,000 distinct compounds. These phytochemicals are observed in diverse sources, including medicinal plants, foods, nutraceuticals, and spices, and are known for their bioactive properties [1, 2]. Plants produce these compounds as secondary metabolites because of biotic or abiotic factors [3], where they play multiple functions, such as acting as volatile chemicals to attract or repel the same or different species, providing protection against oxidative stress in environments with high UV light, serving as pigments to attract or deter other species, or acting as poisonous substances to defend against predators, all of which are essential for their survival [4]. In nature, polyphenols are found in cereals, beverages, fruits, and vegetables. Cherries, apples, pears, berries, and grapes are examples of fruits that have a polyphenol content of 200–300 mg/100 g of wet weight [5].

Polyphenols are food constituents that are highly beneficial to humans because of their ability to prevent or combat various diseases. They have significant anti-inflammatory, anticarcinogenic, antioxidant, hypolipidemic, antimicrobial, antihypertensive, neuroprotective, and calorigenic properties [1].

Polyphenols are categorized by one or more hydroxyl substituents and at least two phenyl rings [6]. These compounds can be grouped as either non-flavonoids or flavonoids, or sorted into various subcategories in terms of the quantity of phenol units present in their molecular architecture, the type of functional groups, as well as how the phenol units are linked [7].

Flavonoids have a fundamental diphenyl propane structure, where phenolic rings are typically connected by a heterocyclic pyran ring [8]. The differences between these compounds depend on whether there is an alkene bond connecting C2 and C3 and whether an oxogroup is formed by C4 on the connecting ring [9]. The subclass includes flavanones, anthocyanins, flavanols, isoflavones, anthocyanidins, flavonols, and flavones [2].

Non-flavonoids have a fundamental structure of a solitary aromatic ring. Examples of non-flavonoid compounds are stilbenes, phenolic acids, and lignans [9]. The polyphenol amides are polyphenols with an amide group. Notable examples include capsaicinoids and avenanthramides. In addition, there are other subcategories that are not presently recognized, including curcuminoids, alkylphenols, hydroxybenzaldehydes, furanocumarins, tyrosol, and hydroxy benzoketones [10].

Polyphenols: Food, Nutraceutical, and Nanotherapeutic Applications, First Edition. Edited by Mithun Rudrapal.

Several polyphenols have been found to have therapeutic properties, with varying mechanisms of action. Turmeric, grapes, and red onion are examples of sources of these polyphenols that include curcumin, resveratrol, and quercetin. These compounds have been found to have antiobesity properties [11]. They also activate antioxidant pathways in the body, which helps to prevent oxidative damage by inducing the synthesis of enzymes that are capable of preventing or reducing oxidative damage [12], as well as providing a series of health benefits as a result of their multiple beneficial properties.

Dietary sources of polyphenols undergo significant biotransformation by metabolic reactions within enterocytes, the liver, and gut microbiome, which results in their limited absorption when taken orally. Although polyphenols are poorly absorbed when ingested, they still demonstrate notable biological effects, resulting in a paradoxical situation of high bioactivity despite low bioavailability. Recently, polyphenol metabolites have gained interest because of their comparable or greater intrinsic biological activity compared with the original compounds [13].

To meet the required daily allowance of these classes of phytochemicals, polyphenols have been formulated into nutraceuticals to augment the insufficiency of the dietary sources. However, the safety of some supplements may be influenced by the method used to extract them, even if the supplements are derived from plants that are safe to consume. For example, a slimming supplement made from a hydroalcoholic extract of tea buds was taken off the market because of instances of severe hepatotoxicity that were linked to the product [13]. As a result, polyphenols have now been loaded in nanomaterials to reduce their toxicity or ensure they get to their target site when used for therapeutics.

14.1 Consumption of Polyphenols/Food Polyphenols

Polyphenols are major constituents in various food types, with spices taking the lead with the most polyphenols per 100 g. Amongst all spices, cloves have the greatest concentration of polyphenols at 15,188 mg/100 g [14]. Pérez-Jiménez et al. [14] reported rich dietary sources of polyphenols. Spices, fruits, and seeds had the highest concentration of polyphenols among the dietary sources reported. The values were 654–15,188 mg/100 g for spices, 560–1,756 mg/100 g for fruits, and 466–1,528 mg/100 g for seeds [14].

14.2 Polyphenols as Food

A diet rich in polyphenols can provide protection against chronic diseases, such as diabetes, neurological disorders, diseases that affect the circulatory system, as well as cancer, by interfering with various biological activities, including cell proliferation, enzymatic activity, signal transduction pathways, and cellular redox potential [15]. A few polyphenols found in food sources are discussed.

14.2.1 Resveratrol

Resveratrol is a non-flavonoid polyphenol synthesized by plants to protect against harsh conditions [15] with a large quantity found in certain fruits, including grapes with 19–508 mg/g (1 mg/L in the skin and 20 mg/L in wine) [16, 17], berries (between 32.5 μg/g and 0.77 μg/g), and 5.1 μg/g in peanuts (91 μg/g in its shells and 0.3 μg/g in peanut butter) [15]. Other sources include fruits and vegetables such as cucumber, Surinam cherry, banana, leech, guava, peach, potato, apple, passion fruit, pineapple, pear, pistachio, and rhubarb [15].

Resveratrol is beneficial against various pathologies as well as redox reactions where it acts as an antioxidant against the reactive oxygen species (ROS) generated in the body [18]. It has anticancer, anti-inflammatory, antidiabetic, antiaging, neuroprotective, and cardiovascular protective properties [19–23].

14.2.2 Curcumin

Curcumin is the main bioactive compound found in the rhizome of turmeric, which accounts for 0.58% to 3.14% of its dry weight [24, 25]. Turmeric has a long record of use as a spice, dye, and herbal remedy; in combination with other spices, it forms a staple ingredient used for cooking in some parts of the world [26]. According to several sources, curcumin is a significant pleiotropic agent that exhibits various activities including hypoglycemic, anti-inflammatory, wound healing, antioxidant, antitumor, and antimicrobial effects [27–30], with a recommended daily intake of 0–3 mg/kg body weight [31].

14.2.3 Quercetin

Quercetin is a flavonoid polyphenol found in food sources such as pepper (32.59 mg/100 g), lettuce (40.27 mg/100 g), tomato (4.56 mg/100 g), onion (12.65–17.22 mg/100 g), black chokeberry (8.90 mg/100 g), apple (2.47 mg/100 g), and broccoli (4.25 mg/100 g). They add to the approximate nutritional consumption of quercetin of 5–40 mg daily, with the possibility of higher daily intakes of 200–500 mg if there is a significant consumption of fruits and vegetables with a high amount of quercetin [15]. Nevertheless, the primary issue related to the consumption of quercetin is its limited oral bioavailability because of its low solubility in water, weak chemical and metabolic stability, and restricted ability to pass through membranes.

An increasing amount of research indicates that quercetin has promising characteristics such as antibacterial, antioxidant, anticancer, anti-inflammatory, neuroprotective, and cardiovascular protective properties. Other studies have also reported that quercetin has the potential to reduce the ROS level *in vitro* and its metabolites quercetin trisulfate, quercetin-sulfoglucuronide, and quercetin-glucuronide were also effective in ex vivo studies. Others reported the cytotoxic effects of various quercetin metabolites. In a study by Kusaczuk et al. [32], the effects of quercetin on human glioblastoma cell lines were investigated. They found that quercetin caused apoptotic death in one cell line and necrotic death in another, suggesting that quercetin could be a possible treatment for brain tumors [32].

14.2.4 Genistein

Genistein, primarily present in soy beans (Glycine max), is one of the most prevalent isoflavones at a concentration of 322 mg/g [33]. While soy beans are the primary source of genistein in a diet, there are also other food options that could supply this isoflavone. Currants (2.167 mg/g), raisins (1.458 mg/g), mango (0.212 mg/g), dried cooked prunes (0.663 mg/g), peanuts (0.158 mg/g), passion fruit (0.4 mg/g) (34), and even quinoa seeds (0.4–4.1 mg/g) were identified to contain significant concentrations of genistein [34].

14.2.5 Ellagitannins

Ellagitannins are a type of hydrolyzable tannin that can be found in many different plant species and they are significant in terms of their nutritional value for humans [35]. Ellagitannins typically contain a hexahydroxydiphenoyl group that is connected to a sugar molecule, often glucose, through esterification. Ellagitannins exhibit low stability when they polymerize and are hydrolyzed. The product of ellagitannin hydrolysis, ellagic acid, is not very soluble in water. In addition to being present in ellagitannins, free-form ellagic acid or derivatives of ellagic acid resulting from methylation, methoxylation, and glycosylation of hydroxyl groups can also be found in plant vacuoles [35]. Fruits, such as raspberry, blackberry, pomegranate, and blueberry, nuts such as pecans, walnuts, and hazelnuts, and specific medicinal plants, such as oak and geranium, are the primary dietary sources of ellagic acid and its precursors [36, 37].

The amount of ellagic acid in various foods was estimated by breaking down their ellagitannin content. Among the fruits in the *Rosaceae* family, blackberries were found to contain the highest concentration of ellagic acid at 150 mg/100 g. Raspberries had a concentration of 47–270 mg/100 g, making them another significant source of ellagic acid. Strawberries have been identified as a significant source, with a concentration of 31–81 mg/100 g [35]. The overall concentration of ellagic acid was 59 mg/100 g in walnuts and 33 mg/100 g in pecans [35]. Ellagic acid has been linked to several health benefits, including acting as an antioxidant [38, 39], antiproliferative [40], anti-inflammatory [15, 39], antiatherogenic [41], liver protective [15, 42–44], antiviral [15], neuroprotective [45, 46], cardioprotective [47], antidiabetic [48, 49], anti-epileptic [50], and antimicrobial [51] agent.

14.2.6 Proanthocyanidins

Proanthocyanidins, i.e., compacted tannins, belong to one of the most common groups of polyphenols found in plants, wherein they function to protect plants from biotic and abiotic stressors as well as play essential roles in the growth and development of various parts of their host plant [52]. Rich food sources include cinnamon with a total proanthocyanidins (TP) content of 8,108.2 mg/100 g, cereals and beans with a TP content of 8.1 mg/100 g in black beans to 3,965.4 mg/100 g in sorghum, fruits with a TP content of 4.0 mg/100 g in banana to 663.7 mg/100 g in chokeberries, and nuts with a TP content of 8.7 mg/100 g in cashews to 500.7 mg/100 g in hazelnuts [53]. Proanthocyanidins are also beneficial to humans where they serve various functions in alleviating diseases [52].

14.3 Toxicity of Polyphenols

The phrase "Everything is toxic" has been reoccurring in recent times. This includes every food or drink we ingest that is tagged safe to those that are actually toxic. The toxic effect observed with safe foods/drinks is mostly related to its dosage and/or in combination with other compounds. When taken at a safe dose (in combination or alone), no toxic effect is observed. Polyphenols may not exhibit a toxic effect; however, their low oral bioavailability, poor solubility, susceptibility to degradation, or breakdown during exposure to either processing conditions and/or interactions with the gastrointestinal (GI) tract produce toxic metabolites after they are metabolized. The toxic impact commonly observed with polyphenols can be attributed to these metabolites. Some studies have reported toxic/side effects of consuming dietary or supplemental forms of polyphenol compounds.

14.4 Polyphenols in Toxic Foods

Polyphenols are found in both beneficial and toxic plants, and their effects on humans depend on the circumstance. Some toxic plants contain high levels of polyphenols, as shown in Table 14.1, which are beneficial as anti-inflammatory, anticancer, and antioxidant agents [54]. However, in

Table 14.1 Polyphenols in toxic plants/foods.

Plant	Botanical Name	Polyphenols	References
Black elderberry	*Sambucus nigra*	Anthocyanins, cyanidin-3-*o*-sambubioside, hyperoside, isoquercetin, quercetin, rutoside, epicatechin-3-*o*-gallate	[55]
Bitter almond	*Prunus amygdalus*	Kaempferol, vanillic acid, ferulic acid, rosmarinic acid, naringenin, hydroxycinnamic acid, myricetin, caffeic acid	[56]
Red kidney beans	*Phaseolus vulgaris L*	Delphinidin 3-o-glucosyl glucoside, kaempferol, pelargonidin 3,5-o-diglucoside, petunidin 3-o-6′′-acetyl-glucoside	[57]
Castor beans	*Ricinus communis*	Ellagic acid, catechin, gallic acid	[58]
Rhubarb leaves	*Rheum rhabarbarum L*	Anthocyanins, gallic acid, catechin, Epicatechin, myreticin, quercetin, isorhamnetin	[59]
Poison ivy	*Toxicodendron radicans*	Quercetin and kaempferol	
Yew	*Taxus cuspidata*	Catechin, taxiresinol, secoisolariciresinol, isotaxiresinol, isolariciresinol, epicatechin	[59]

some cases, polyphenols can worsen the toxic effects of the compounds. Therefore, it is crucial to be cautious when consuming toxic foods and seek guidance from a medical professional. Overall, the relationship between polyphenols and toxic foods is complex and context-dependent.

14.5 Toxicity of Polyphenolic Compounds

Polyphenols may exhibit various beneficial properties; however, these benefits are hampered or reduced by their metabolism in the gut, preventing them from reaching their target. This creates toxic metabolites that confer toxic potentials to the polyphenol, which is discussed in this section, and the toxic effects of some polyphenols are highlighted in Table 14.2.

14.5.1 Genistein

Genistein belongs to the subclass of flavonoids, isoflavone, that is mostly found in soy beans and is known to be a phytoestrogen [60]. Genistein is metabolized into 6′-hydro-o-demethylangolesin, o-demethylangolesin, and equol. These metabolites are responsible for the toxic effect observed with isoflavones.

Numerous studies have indicated that isoflavones can be helpful in avoiding or managing hormone-related conditions such as osteoporosis, cancer, and cardiovascular disease (CVD) [61]. However, this bioactive phytoestrogen has a huge effect on hormones. The phytoestrogens cause premature activation of the hypothalamic–pituitary–gonadal axis (responsible for the development and regulation of the reproductive systems) in children before puberty, bringing about breast development in girls, and increased testicular size in boys, with an advancement of skeletal age and accelerated linear growth [60]. Basically, genistein makes children grow faster than they

Table 14.2 Effect of polyphenols in animal and human subjects.

Polyphenol	Models	Effect	References
Resveratrol	1,000 mg/day in humans	Onset and progression of cardiovascular diseases	[18, 75]
	0.3 & 1.0 g/kg/day in rats	Hepatotoxicity	[78]
	3,000 mg/kg body weight in rats	Renal toxicity	[77]
Curcumin	25 mg/kg in dogs	Increased bile secretions	[82; 106]
	2,000 mg in humans	Hepatotoxicity	[84]
	100 mg/kg in rats	Ulcerogenic effect	[81]
Quercetin	500–2,000 mg/kg in mice	Hepatotoxicity	[89]
	1,900 mg/kg in rats	Renal tubular tumors and reduced body weight	[86]
	@50 mg/kg co-administration with gigoxin in pigs	Death	[107]
Catechin (cyanidanol drug)	2 g/day in humans	Autoimmune cytopenia	[97]
EGCG	1,500 mg/kg in mice	Hepatotoxicity	[95]
Pol 60	30 mg/kg Pol 60 co-administration with 2.5 mg/kg ebulin f in mice	Death	[94]
Flavokawain B	12.5 μM zebra fish embryos	Teratogenic effect	[100]
Caffeic acid	5 and 150 mg/kg/day	Reproductive and developmental toxicity	[108]
NDGA	2% NDGA in rats	Cystic nephropathy	[102]
	50 mg/kg in mice	Increased plasma ALT levels	[104]
	75 mg/kg in mice	Liver damage	[103]
Genistein	50 & 100 mM primary neuronal culture	Neurotoxicity	[62]
	400–480 mg in humans	Hepatotoxicity	[109]

should. Estrogens are known to have the ability to protect neurons from excitotoxicity. This was investigated in a study by Jin et al. to determine the protective potential of genistein (phytoestrogens) in a primary rat neuronal culture [62]. Despite having estrogen-like activity, genistein fails to protect the neurons, instead inducing neurotoxicity in the primary neuronal culture at concentrations of 50 mM and 100 mM. The reason for this toxicity was the blocking of the gamma-aminobutyric acid (GABA) receptor, which caused an increased excitation by glutamate, ultimately resulting in cellular damage [62].

14.5.2 Resveratrol

Resveratrol (RE) is a non-flavonoid polyphenol whose function is to protect its host plant against harsh conditions [15]. RE is sourced from grapes, red wine, berries, banana, guava, pineapple, apple, peanuts, potato, and cucumber [15]. RE is rapidly metabolized after absorption following the phase I and II reactions, producing metabolites that result in its beneficial properties, as well its toxic effects [63].

Although there has been reports indicating how well RE is tolerated by humans [64, 65], other studies reported its toxicity in both *in vitro* and *in vivo* studies [66]. For example, in high doses, RE has been found to inhibit the cytochrome P450 and interact with various drugs, which can reduce its effectiveness [67]. Furthermore, prolonged consumption of RE can cause thyroid disruption and act as a goitrogen [19, 68], in addition to other toxic side effects such as high-dose-associated hermetic and pro-oxidant effects [69–74]. The toxic effect of RE is a result of oxidation to its *o*-quinone metabolite that confers a pro-oxidant property to RE where it alters the oxidation/reduction status of endothelial human cells, especially in obese individuals, resulting in the beginning and advancement of CVDs at a high dose of 1,000 mg/day [18, 75]. In addition, RE induces oxidative stress affecting male reproductive functions at an increasing dose of 2–20 mg/kg of body weight [76]. RE has also been shown to be associated with renal toxicity in rats at a high dose of 3000 mg trans-RE/ kg body weight daily for four weeks [77], resulting in an increase in BUN, creatinine, alkaline phosphatase, alanine aminotransferase (ALT), total bilirubin, and albumin. RE also decreases the hemoglobin, hematocrit, and RBC counts with an increase in kidney weight as well as an increase in incidence and severity of nephropathy [77]. RE also results in hepatotoxicity by showing an abnormal expression of liver genes in female rats treated with RE administered at doses of 0.3–3.0 g/kg daily for 28 days [78]. In a study by Hebbar et al. [78], genes responsible for the first phase of drug metabolism were inhibited at doses of 0.3 g/kg/day and 1.0 g/kg/day, while the expression of other genes (cytochrome P450 reductase, manganese superoxide dismutase, thiosulfate sulphur transferase, and quinone oxidoreductase) increased in a manner that was dependent on the dose.

14.5.3 Curcumin

Curcumin is also a flavonoid polyphenol that is the active ingredient in turmeric at 0.58% to 3.14% its dry weight. Turmeric has been used as a herbal and dietary supplement for ailments, such as a digestive disorders, arthritis, and liver conditions, owing to its antioxidant and anti-inflammatory properties [79]. The recommended maximum amount of curcumin that can be safely consumed daily, also known as the acceptable daily intake (ADI), is 3 mg/kg/day [80]. Reports have shown that curcumin in high concentration and in conjunction with other compounds has a toxic effect.

Curcumin produces gastric ulceration in albino rats when administered orally at a high dose of 100 mg/kg daily for six days [81]. Additionally, parenteral administration of 25 mg/kg curcumin in anaesthetized dogs caused a maximal increase of bile secretions in nearly all cases [82]. Authors also noted that while the drug caused a reduction in the amount of solids present in the bile, the overall quantity of cholesterol, bile salts, and bilirubin expelled increased [82]. Curcumin's poor absorption was greatly enhanced by piperine by more than 2,000% [83], increasing curcumin's bioavailability and resulting in hepatotoxicity following an oral ingestion of 2,000 mg curcumin supplement for 90 days [84].

14.5.4 Quercetin

Quercetin is a flavonoid polyphenol belonging to the flavonol subgroup that is found in onions, apples, black tea, broccoli, and grapes. Studies have identified possible adverse effects of quercetin, such as lowered levels of glutathione, an elevation in the discharge of lactate dehydrogenase, and an increase in the cytosolic [illegible] Ca^{2+} concentration in a rat lung epithelial cell line. In another study, the co-administration of quercetin and digoxin, a cardiac glycoside, in pigs enhanced the absorption of digoxin, resulting in an increased serum concentration at a 40 mg/kg dose while causing instant death of the animal at 50 mg/kg [85]. Furthermore, the consumption of extremely large amounts of quercetin (1,000 ppm and 40,000 ppm) over an extended period (58 weeks and 104 weeks) caused carcinogenic effects in male rats, resulting in treatment-related injuries in the intestine, bladder, and kidneys [86–88]. Quercetin was delivered at a dose of approximately 40–1,900 mg/kg/day to determine the toxicity and carcinogenicity of the polyphenol [86], and no negative effects were observed initially. However, towards the end of the two-year study, the group treated with high-dose quercetin showed reduced body weight gain. Additionally, quercetin exhibited a carcinogenic effect in male rats, resulting in primarily benign tumors in the renal tubular epithelium [86]. In another study, quercetin was hepatotoxic at 500–2,000 mg/kg body weight, causing an elevation of biomarkers of hepatotoxicity, lipid peroxidation, a decrease in glutathione levels, and the expression of stress-regulated genes [89].

14.5.5 Green Tea Polyphenols

Up to 30% of the dry weight of green tea is polyphenols, including phenolic acids, flavandiols, flavanols, and flavonoids [90]. Numerous polyphenols from green tea are flavonols, commonly known as catechins [91], which can also be found in red wine, broad beans, apricots, black grapes, and strawberries [92]. Four types of catechins have been identified in green tea: epigallocatechin, epicatechin-3-gallate, epicatechin, and epigallocatechin gallate (EGCG) [93, 94].

When given to mice orally either as a one-time intake of 1,500 mg/kg or in smaller doses, high amounts of EGCG were hepatotoxic. This results in a significant increase in the plasma ALT level, with increases of 138-fold and 184-fold observed in the two different administration methods, respectively. Additionally, the mice experienced an 85% reduction in survival rate because of the increased hepatotoxic response [95]. In another report by Gandolfo et al., autoimmune cytopenia was observed in five patients who received 2 g/day cyanidanol (a drug form of catechin) for 4–36 months. Autoimmune cytopenia refers to a set of different yet interrelated conditions characterized by the destruction of hematologic cell lineages, such as erythrocytes, platelets, and leukocytes, because of an immune response [96]. The suspension of the medication resulted in normalization of the hematological values in all five patients [97].

Polyphenon 60 (Pol 60), another type of polyphenol found in green tea, was also reported to be toxic in mice when co-administered with ebulin f found in dwarf elder [94]. The co-administration of oral Pol 60 at 30 mg/kg body weight and intraperitoneal ebulin f at 2.5 mg/kg body weight [94] resulted in a decrease in the mice survival rate by 70% following the initial recorded fatality on Day 2 and the last fatality on Day 11. Notably, independent administration did not trigger lethal toxicity.

14.5.6 Polyphenols From Kava (*Piper Methysticum*)

Piper methysticum is a plant indigenous to the South Pacific islands [98] with kavalactones and three additional kavachalcones (flavokawain A, flavokawain B, and flavokawain C), all of which are lactone derivatives [99, 100]. Kava contains chalcones, which are precursors of flavonoids. The chalcones are characterized by the absence of a closed C ring. While traditionally used by islanders, kava has been sold in the Western world as a mood enhancer and anxiolytic [101]. Despite having some beneficial attributes, the kavachalcone derivative flavokawain B is toxic to zebrafish embryos, causing a teratogenic effect at a dose of 12.5 μM [100].

14.5.7 Nordihydroguaiaretic Acid (NDGA)

Nordihydroguaiaretic acid (NDGA) is a constituent of the chaparral plant (*Larrea tridentate*) and a type of lignan present in significant quantities (up to 10% of the plant's dry weight) in the plant's twigs and leaves, as well as the major lignin (a non-flavonoid) in chaparral. NDGA in tea has mainly been used to alleviate joint pain, allergic reactions, muscle cramps, and eradicate parasites. It has been used topically to lessen inflammation and discomfort and to foster the recovery of small injuries. Despites its abundant benefits, studies have demonstrated that NDGA can cause cystic nephropathy in rats fed a 2% concentration of NDGA for a period of 6 weeks [102]. Intraperitoneal administration of NDGA also proved fatal to mice, with 75 mg/kg being the lethal dose that caused elevated levels of serum ALT, an indication of liver damage [103]. In addition, CD-1 mice exhibited a four-fold increase in plasma ALT levels 24 hours following an intraperitoneal administration of 50 mg/kg NDGA [104]. A report of NDGA toxicity in humans has also been recorded in a study where 13 out of 18 patients showed increased liver injury biomarkers, jaundice, histopathological features of hepatocellular necrosis, cholestasis, cholangitis, and cirrhosis after the consumption of a NDGA supplement at doses of 400–480 mg at least four times daily [105].

14.6 Strategies to Overcome Toxicities

14.6.1 Increasing Polyphenol Bioavailability

Polyphenol toxicity arises mainly after being orally consumed because of their potential to exhibit poor oral bioavailability. Upon consumption, polyphenols are considered xenobiotics by the body, much like pharmaceuticals. As a result, they must overcome numerous barriers during digestion and absorption, including chemical and extensive enzymatic modification, to reach the location(s) where they exert their effects [110]. When consumed in high concentration, polyphenols may be metabolized by the liver into non-toxic and/or toxic metabolites that may/may not be excreted. These toxic metabolites confer toxic potential to the polyphenols. Therefore, increasing the bioavailability of polyphenols may result in a reduction of polyphenol toxicity.

Bioavailability describes the rate and amount of a compound that was absorbed that reaches its target site in the body. One way that bioavailability can reduce the toxicity of polyphenols is by limiting the amount of the polyphenol that is absorbed into the bloodstream. This can be achieved by designing specific synergistic interactions (e.g., via encapsulation) with compounds or other polyphenols. This approach may be helpful in improving the absorption and distribution of polyphenols to their target organs, which may increase their beneficial effects while reducing their potential toxicity. By limiting the amount of the compound that is absorbed, promoting their metabolism, and enhancing their delivery, bioavailability can help reduce the potential toxicity of polyphenols while maximizing their health benefits.

14.6.2 Modulating Gut Metabolic Processes

The metabolism of polyphenols from food within the GI tract can limit their bioavailability and reduce their beneficial effects. Gut esterase enzymes and gut monoamine oxidase enzymes, from both bacterial and host sources, are the primary mechanisms responsible for deactivating polyphenols and generating toxic metabolites. Inhibition of these enzymes may allow polyphenols to remain in the GI tract long enough to be absorbed more effectively. This can be achieved by conjugating polyphenols with gut esterase or MAO inhibitors, which can block the enzyme activities thereby increasing the absorption of the polyphenol of interest [110].

14.6.3 Modulating Polyphenols to Act as a Substrate for ABC Transporters

Polyphenols can be both substrates for and regulators of ABC transporter function, which can affect the bioavailability [110]. For example, the flavonoid quercetin has limited access to the brain; however, it has anxiolytic potential by decreasing the corticotropin releasing factor, a neuropeptide associated with anxiety and fear [111]. A study found that the combination of the flavonoid quercetin and PSC833 (Pgp inhibitor) did not affect its entry into the brain; however, a combination with GF120918 (BCRP inhibitor) increased quercetin entry into the brain 20-fold [112]. The data suggest that quercetin penetrates the blood–brain barrier (BBB) epithelial cells, likely through osmosis because of its hydrophobicity. However, when the BCRP efflux transporter recognizes quercetin, it is exclusively exported out of the cells and into the gut lumen, which reduces its availability in BBB epithelial cells.

Similarly, resveratrol also has low oral bioavailability because of the activity of the BCRP efflux pump [110]. However, it is a substrate for the MRP3 transporter, which increases its plasma levels by pumping it into the blood side of gut epithelia [110]. Therefore, while BCRP causes reduced oral bioavailability by pumping resveratrol out and into the gut lumen, MRP3 is involved in increasing its plasma levels by pumping resveratrol into the blood side of the gut epithelia. Co-consuming polyphenols with a polyphenol ABC transporter inhibitor can increase its bioavailability to the target tissue. Additionally, some polyphenols can increase the expression or function of other ABC efflux transporters, enhancing the influx of substrates into the bloodstream thereby contributing to the synergy between the substances.

14.6.4 Encapsulation of Dietary Polyphenol

This technique involves the coating of bioactive compounds with a barrage of material or entrapping them inside carriers or shells [113]. This technique can improve the water dispersibility of polyphenols, making them more easily incorporated into aqueous-based beverages and foods [113]. Encapsulation also provides protection against the harsh stomach and intestinal environment [114], which can improve polyphenol bioavailability and eliminate toxicity. Various delivery systems have been reported for encapsulation, among these, nanoencapsulation is the preferred method for delivering polyphenols because it has proven bioavailability and improves stability.

14.7 Remedies of Toxicity

14.7.1 Remedies of the Toxicity of Polyphenols Through Nanotherapeutic Approaches

Polyphenols can function as curative agents for a variety of illnesses; however, because of their low absorption and bioavailability, their beneficial effects are very limited.

Progress in delivering polyphenols to tissues of interest revealed the potential for using nanoparticles to increase the efficiency while reducing the degradability. The use of nanoformulations has been beneficial in shielding polyphenols from deterioration, boosting their absorption, and reducing their toxicity [115]. Incorporating polyphenols into nanostructures stimulates dendritic cells, leading to a heightened immune response in the body [116]. These methods of delivering substances could be used to convey polyphenols, which are typically insoluble and lipophilic [115].

Nanoformulations have been demonstrated to boost the bioavailability of polyphenols in addition to preserving their stability within the physiological environment through *in vitro* and *in vivo* studies, all while achieving a sustained release [117]. Various nanoparticles have been introduced with the aim of enhancing polyphenol bioavailability [118].

Although polyphenols have been shown to be beneficial therapeutic agents against various diseases, their vulnerability to environmental conditions, their low solubility, low bioavailability, poor permeability, instability, and rapid release makes their application in medicine and food greatly limited [119]. Nanoformulations can be used as polyphenol delivery agents to enhance the effectiveness of polyphenols by preventing their degradation and improving their intracorporal interactions. This can be achieved by encapsulating phenolic compounds in nanocarriers, which can enhance bioavailability, provide targeted release, as well as protect the bioactive molecules [120]. Various inorganic, organic, or organic/inorganic hybrid nanomaterials with distinct physical and chemical properties have been produced. These materials, ranging from 1 nm to 100 nm in size, have larger surface areas and have been created for different biomedical applications such as disease sensing, imaging, and therapy [121]. A new type of nanomaterial has been developed that shows high performance while increasing polyphenol biocompatibility [122]. A few of these nanoformulations for loading and delivering polyphenols are discussed.

14.7.2 Self Nanoemulsifying Drug Delivery Systems (SNEDDS)

In 2011, Li et al. [123] created nanoemulsifying drug delivery systems (SNEDDS) containing the extract of persimmon leaf that has quercetin in abundance. This enhanced the bioavailability of plasma quercetin by 1.5 fold when given orally to beagle dogs at a concentration of 5.3 mg/kg body weight.

14.7.3 Solid Liquid Nanoparticles (SLNs)

Quercetin-encapsulated solid liquid nanoparticles (SLNs) were made and administered at 50 mg/kg body weight [123], which caused a 5-fold increase in the oral bioavailability of quercetin and also exhibited sustained release of the polyphenols compared with free quercetin. Encapsulation or incorporation of resveratrol into the lipid compartment of SLN nanoparticles can increase its solubility in water by more than 100 times [124]. SLNs also improved the hydrophilicity and chemical stability of curcumin [125], while also exhibiting a sustained release pattern [126].

14.7.4 Nanostructured Lipid Carriers (NLCs)

Nano-quercetin was developed and administered intravenous and orally at 10 mg/kg and 50 mg/kg body weight, respectively. The study showed that quercetin–NLCs improved the bioavailability of the total quercetin concentration in plasma by 2.0 fold and 2.8 fold after oral and intravenous administration when compared with a free quercetin solution [124]. Resveratrol was also encapsulated in the lipid compartment of NLCs in a study by Frozza et al. This resulted in a more than 100 times improvement in the solubility of resveratrol with increased concentration in tissues by more than two times than with free resveratrol and also better GI safety [127].

14.7.5 Nanoemulsions

Nanoemulsions are systems of two immiscible liquids where the internal phase is distributed within the external phase. Owing to their large surface area, small size, stability, and high optical clarity, polyphenolic compounds, such as curcumin and EGCG, have been enclosed within a nanoemulsion to improve drug bioavailability. Curcumin loaded into a nanoemulsion was found to be the most effective for GI absorption and had an enhanced oral bioavailability of more than two-fold [128].

Encapsulating EGCG in an oil/water nanoemulsion resulted in significantly greater antitumor activity *in vitro* when compared with free EGCG [129]. Another study explored the effects of nanoemulsion encapsulation on the biological activity, epithelial permeability, and physical and chemical properties of EGCG. The study found that the EGCG–nanoemulsion was approximately three times more bio-accessible, and the intestinal permeability of EGCG increased significantly when compared with unencapsulated catechins [130]. Nanoemulsions have good stability, a sustained release pattern, and can improve the chemical stability and aqueous solubility of polyphenolic compounds.

14.7.6 Phytosome

A phytosome is a stable complex created through electrostatic interaction between plant extracts (mostly polyphenols) and phospholipids (primarily phosphatidylcholine) [131]. These phytosomes are more bio-available than the pure extracts because of the presence of phosphatidylcholines, which helps the plant extracts flow through the body [132]. This is confirmed in many studies that reported the benefits of the polyphenol formulations compared with using the natural product alone.

In a study that used a straightforward approach to produce a complex of rosmarinic acid and phospholipids (RA-PLC), the findings indicated that RA-PLC had better membrane permeability, increased antioxidant properties, and 1.2 times greater biological utilization than natural RA [119]. The bioavailability of curcumin has been shown to increase with the use of curcumin–phospholipid complexes, with better pharmacokinetics, as well as better hepatoprotection [133, 134]. When administered to rats, the amount of drug content in the plasma of the curcumin–phospholipid group was five times greater than that of the free curcumin group [135]. A different study demonstrated that curcumin–phospholipid complexes have a longer half-life than unbound drugs [136].

Another study revealed the use of a complex formed between polyphenols and phospholipids in the treatment of cancer [137]. According to the research, both silybin and silybin-phosphatidylcholine caused a reduction in the HER2 expression of SKBR3breast cancer cells. However, the silybin-phospholipid multifaceted compound showed a stronger ability to impede the growth of cancerous cells than silybin alone [137].

14.7.7 Nanoformulations That are Protein Based

A protein-based system for delivering drugs has excellent functional characteristics and high nutritional value, which is why it is increasingly being developed in the food industry.

14.7.7.1 Casein-derived Nanoparticles

Casein is the main protein found in milk, at approximately 80% of the composition. Casein has both hydrophobic and hydrophilic domains and can come together on its own in the presence of calcium phosphate, forming a colloid that measures 50–500 nm in diameter [138]. Because of its exceptional properties, such as self-assembly, water binding, emulsification, and surface activity [139], casein is an ideal choice for drug delivery systems, particularly for polyphenol delivery.

Encapsulating curcumin in β-casein has been shown to increase its solubility by 2,500-fold and enhance its cytotoxicity to the human leukemia cell line (K-562). The casein–curcumin complex also exhibited much greater antioxidant action than unbound curcumin [140]. Luo et al. [141] prepared sodium-rutin caseinate/pectin complex nanoparticles and discovered that the process of heating increased the speed of forming nanoparticles and the encapsulation of rutin. Pectin's existence delayed the hydrolysis of sodium caseinate, which facilitated the regulated discharge of rutin in the digestive system [141]. Although encapsulated quercetin and curcumin results in a high packaging efficiency and solubility in nanoparticles, the polyphenol–casein nanoparticles were found to be more toxic to MCF-7 human breast cancer cells than molecules that were not loaded [142].

14.7.7.2 Gelatin Nanoparticles

Gelatin nanoparticles have been used for delivering various drugs, including polyphenols, with successful outcomes. Polyphenolic compounds, such as tannic acid, catechin, curcumin, and EGCG, have been enclosed in gelatin nanoparticles and a polyelectrolyte shell to improve the gel's stability and regulate the release of polyphenols [143]. Gelatin nanoparticles were used to encapsulate EGCG, which was released over a span of up to eight hours with maintained biological efficacy.

14.7.7.3 Whey Protein Nanoparticles

Whey protein, obtained from whey (a secondary product from making cheese), comprises several proteins such as bovine serum albumin, α-lactalbumin, lactoferrin, immunoglobulins, and β-lactoglobulin (β-lg) [144]. Whey protein is a suitable choice for encapsulating and transporting compounds like polyphenols because of its nutritional value, low cost, and versatility, as well as its ability to resist pepsin, making it advantageous for oral delivery of polyphenols and various compounds [145]. Shpigelman et al. [146] used thermally altered lactoglobulin to transport EGCG, and they observed that the connection coefficient between preheated protein and EGCG was about four fold greater than that of the unmodified protein. Encapsulation of EGCG with lactoglobulin nanoparticles greatly preserved the antioxidant activity of EGCG as well as slowed the degradation rate of EGCG by 3.2 fold compared with the unbound EGCG in eight days [146]. Li et al. used β-lg and a nanoemulsion as a carrier to transport curcumin, and the findings indicated that the pH stability, water solubility, and permeability was enhanced drastically. The complex of curcumin and β-lg was found to be resilient to pepsin, but sensitive to trypsin [147].

14.8 Safety Considerations of the Toxicity of Polyphenols

Polyphenols are phytochemicals found in various plant-based foods. Although generally safe for exhibiting antioxidant and anti-inflammatory properties, in addition to other benefits, certain polyphenols have exhibited toxic effects at elevated dosage. Some safety considerations of polyphenol toxicity are:

1. Processing: The processing and preparation of foods high in polyphenols can affect their toxicity. For example, overcooking or boiling certain vegetables may decrease the bioavailability of polyphenols or even convert them into toxic compounds.
2. Potentially toxic effects: Some polyphenols have been found to have toxic effects in animal studies. For example, the consumption of catechins found in green tea may cause autoimmune cytopenia at a dose of 2 g/day when taken in the drug form cyanidanol [97], and quercetin can be toxic to human and animal cells at high concentrations.

3. Dose-dependent effects: Like many natural compounds, the effects of polyphenols are often dose-dependent. While low to moderate doses of polyphenols are generally considered safe and may confer health benefits, taking a high amount or using for an extended period may elevate the probability of negative consequences.
4. Interaction with medications: Some polyphenols can interact with medications and affect their effectiveness. For example, resveratrol can interfere with blood-thinning medications, such as warfarin, and increase the risk of bleeding [148].
5. Sensitivity: Certain individuals may be more susceptible to the effects of polyphenols than others, particularly those with pre-existing medical conditions. For example, resveratrol has been shown to modify the redox condition of human endothelial cells in overweight people, resulting in the development and advancement of CVDs. As a result, obese individuals may need to limit their intake of foods high in these polyphenols [71].

Overall, while polyphenols have potential health benefits, it is important to consider potential safety concerns and to consult with a healthcare provider before making significant changes to your diet or supplement regimen.

14.9 Conclusion

The consumption of polyphenols through dietary sources and nutraceuticals has gained significant popularity over time because of their possible health benefits. Nevertheless, the excessive consumption of polyphenols can lead to toxicity; therefore, the need to develop novel therapeutic strategies to counter this problem is greatly required. In order to overcome this toxicity, nanotechnology-based approaches have yielded promising results at mitigating the toxic effects of polyphenols by improving their bioavailability, targeting specific tissues, and reducing their degradability and toxicity, thus increasing their beneficial attributes in the target tissues. Therefore, the use of nanotherapeutic approaches can improve the safety and effectiveness of polyphenols, making them a more viable option for preventive and therapeutic interventions. More investigations in this area are needed to explore the full prospects of nanotherapeutic strategies in polyphenol toxicity, as well as the potential toxicity of nanotherapies, and to develop appropriate strategies for risk assessment and management.

14.10 Future Perspectives

While the advantages of polyphenols have been extensively researched in diverse pathologies, there is growing trepidation regarding the potential toxicity of polyphenols as well as their metabolites, particularly in the context of high-dose supplementation or chronic exposure. Studies on the toxicity of polyphenols are expected to concentrate on a few essential domains. First, there is a need for more comprehensive studies of the toxicokinetics and toxicodynamics of different polyphenols, including their absorption, metabolism, distribution, and excretion in the body, in addition to their mechanisms of toxicity at both the cellular and molecular level.

Second, there is a need for a better understanding of compound–compound interactions, i.e., between polyphenols and other dietary components as in the case of curcumin and piperine, as well as the impact of individual genetic variations on polyphenol metabolism and toxicity. Third, there is a need for more standardized methods for assessing polyphenol toxicity *in vitro* and *in vivo*, including the utilization of appropriate dose range exposure durations and endpoints.

Finally, there is a need for more research on the potential long-term effects of persistent polyphenol exposure, most importantly in susceptible groups such as expectant mothers, babies, and the elderly. There is therefore a need for regulatory agencies to closely monitor polyphenols with known toxicity to ensure that concentrations in herbal preparations are maintained at a low concentration and are safe for consumption.

References

1. Tresserra-Rimbau, A., Lamuela-Raventos, R.M., and Moreno, J.J. (2018). Polyphenols, food and pharma. Current knowledge and directions for future research. *Biochemical Pharmacology* 156: 186–195.
2. Truzzi, F., Tibaldi, C., Zhang, Y. et al. (2021). An overview on dietary polyphenols and their biopharmaceutical classification system (BCS). *International Journal of Molecular Sciences* 22 (11): 5514.
3. Piccolella, S., Crescente, G., Candela, L. et al. (2019). Nutraceutical polyphenols: new analytical challenges and opportunities. *Journal of Pharmaceutical and Biomedical Analysis* 175: 112774.
4. Köhler, H., Contreras, R.A., Pizarro, M. et al. (2017). Antioxidant responses induced by UVB radiation in Deschampsia antarctica Desv. *Frontiers in Plant Science* 8: 921.
5. Pandey, K.B. and Rizvi, S.I. (2009). Plant polyphenols as dietary antioxidants in human health and disease. *Oxidative Medicine and Cellular Longevity* 2 (5): 270–278.
6. Han, X., Shen, T., and Lou, H. (2007). Dietary polyphenols and their biological significance. *International Journal of Molecular Sciences* 8 (9): 950–988.
7. Cardona, F., Andrés-Lacueva, C., Tulipani, S. et al. (2013). Benefits of polyphenols on gut microbiota and implications in human health. *The Journal of Nutritional Biochemistry* 24 (8): 1415–1422.
8. Khoddami, A., Wilkes, M.A., and Roberts, T.H. (2013). Techniques for analysis of plant phenolic compounds. *Molecules* 18 (2): 2328–2375.
9. Singla, R.K., Dubey, A.K., Garg, A. et al. (2019). *Natural Polyphenols: Chemical Classification, Definition of Classes, Subcategories, and Structures*, 102. 1397–1400. Oxford University Press.
10. Shahidi, F. (2009). Nutraceuticals and functional foods: whole versus processed foods. *Trends in Food Science & Technology* 20 (9): 376–387.
11. Anyanwu, G.O., Kolb, A.F., and Bermano, G. (2020). Antiobesity functional leads and targets for drug development. In: *Phytochemicals as Lead Compounds for New Drug Discovery* (ed. C. Egbuna, S. Kumar, J.C. Ifemeje, S.M. Ezzat, S. Kaliyaperumal), 143–160. Elsevier.
12. González-Reyes, S., Guzmán-Beltrán, S., Medina-Campos, O.N. et al. (2013). Curcumin pretreatment induces Nrf2 and an antioxidant response and prevents hemin-induced toxicity in primary cultures of cerebellar granule neurons of rats. *Oxidative Medicine and Cellular Longevity* 2013.
13. Mennen, L.I., Walker, R., Bennetau-Pelissero, C. et al. (2005). Risks and safety of polyphenol consumption. *The American Journal of Clinical Nutrition* 81 (1): 326S–329S.
14. Pérez-Jiménez, J., Neveu, V., Vos, F. et al. (2010). Identification of the 100 richest dietary sources of polyphenols: an application of the phenol-explorer database. *European Journal of Clinical Nutrition* 64 (3): S112–S120.
15. Luca, S.V., Macovei, I., Bujor, A. et al. (2020). Bioactivity of dietary polyphenols: the role of metabolites. *Critical Reviews in Food Science and Nutrition* 60 (4): 626–659.
16. Vincenzi, S., Tomasi, D., Gaiotti, F. et al. (2013). Comparative study of the resveratrol content of twenty-one Italian red grape varieties. *South African Journal of Enology and Viticulture* 34 (1): 30–35.
17. Di Donna, L., Taverna, D., Indelicato, S. et al. (2017). Rapid assay of resveratrol in red wine by paper spray tandem mass spectrometry and isotope dilution. *Food Chemistry* 229: 354–357.

18 Shaito, A., Posadino, A.M., Younes, N. et al. (2020). Potential adverse effects of resveratrol: a literature review. *International Journal of Molecular Sciences* 21 (6): 2084.

19 Giuliani, C., Iezzi, M., Ciolli, L. et al. (2017). Resveratrol has anti-thyroid effects both *in vitro* and *in vivo*. *Food and Chemical Toxicology* 107: 237–247.

20 He, Y., Zeng, H.-Z., Yu, Y. et al. (2017). Resveratrol improves prostate fibrosis during progression of urinary dysfunction in chronic prostatitis. *Environmental Toxicology and Pharmacology* 54: 120–124.

21 Mallebrera, B., Maietti, A., Tedeschi, P. et al. (2017). Antioxidant capacity of trans-resveratrol dietary supplements alone or combined with the mycotoxin beauvericin. *Food and Chemical Toxicology* 105: 315–318.

22 Yin, X., Yu, J., Kong, Q. et al. (2017). Mechanism of isomers and analogues of resveratrol dimers selectively quenching singlet oxygen by UHPLC-ESI-MS2. *Food Chemistry* 237: 1101–1111.

23 Dytrtová, J.J., Straka, M., Bělonožníková, K. et al. (2018). Does resveratrol retain its antioxidative properties in wine? Redox behaviour of resveratrol in the presence of Cu (II) and tebuconazole. *Food Chemistry* 262: 221–225.

24 Tayyem, R.F., Heath, D.D., Al-Delaimy, W.K. et al. (2006). Curcumin content of turmeric and curry powders. *Nutrition and Cancer* 55 (2): 126–131.

25 Aggarwal, B.B., Surh, Y.-J., and Shishodia, S. (2007). *The Molecular Targets and Therapeutic Uses of Curcumin in Health and Disease*. New York: Springer Science & Business Media.

26 Esatbeyoglu, T., Huebbe, P., Ernst, I.M. et al. (2012). Curcumin—from molecule to biological function. *Angewandte Chemie International Edition* 51 (22): 5308–5332.

27 Crețu, E., Trifan, A., Vasincu, A. et al. (2012). Plant-derived anticancer agents-curcumin in cancer prevention and treatment. *Revista Medico-Chirurgicala a Societatii de Medici si Naturalisti din Iasi* 116 (4): 1223–1229.

28 Prasads, T. (2014). AGGARWALBB. *Recentdevelopmentsindelivery, Bioavailability, Absorption and metabolismofcurcumin: Thegoldenpigmentfromgoldenspice* 46 (1): 2–18.

29 Hewlings, S.J. and Kalman, D.S. (2017). Curcumin: a review of its' effects on human health. *Foods* 6: 10–92.

30 Kocaadam, B. and Şanlier, N. (2017). Curcumin, an active component of turmeric (Curcuma longa), and its effects on health. *Critical Reviews in Food Science and Nutrition* 57 (13): 2889–2895.

31 Authority, E.F.S. (2014). Refined exposure assessment for curcumin (E 100). *EFSA Journal* 12 (10): 3876.

32 Kusaczuk, M., Krętowski, R., Naumowicz, M. et al. (2022). A preliminary study of the effect of quercetin on cytotoxicity, apoptosis, and stress responses in glioblastoma cell lines. *International Journal of Molecular Sciences* 23 (3): 1345.

33 Mujić, I., Šertović, E., Jokić, S. et al. (2011). Isoflavone content and antioxidant properties of soybean seeds. *Croatian Journal of Food Science and Technology* 3 (1): 16–20.

34 Lutz, M., Martínez, A., and Martínez, E.A. (2013). Daidzein and Genistein contents in seeds of quinoa (Chenopodium quinoa Willd.) from local ecotypes grown in arid Chile. *Industrial Crops and Products* 49: 117–121.

35 Landete, J. (2011). Ellagitannins, ellagic acid and their derived metabolites: a review about source, metabolism, functions and health. *Food Research International* 44 (5): 1150–1160.

36 Heilman, J., Andreux, P., Tran, N. et al. (2017). Safety assessment of Urolithin A, a metabolite produced by the human gut microbiota upon dietary intake of plant derived ellagitannins and ellagic acid. *Food and Chemical Toxicology* 108: 289–297.

37 Pattanayak, R., Basak, P., Sen, S. et al. (2017). An insight to the binding of ellagic acid with human serum albumin using spectroscopic and isothermal calorimetry studies. *Biochemistry and Biophysics Reports* 10: 88–93.

38 Ortiz-Ruiz, C.V., Berna, J., Tudela, J. et al. (2016). Action of ellagic acid on the melanin biosynthesis pathway. *Journal of Dermatological Science* 82 (2): 115–122.

39 Verotta, L., Panzella, L., Antenucci, S. et al. (2018). Fermented pomegranate wastes as sustainable source of ellagic acid: antioxidant properties, anti-inflammatory action, and controlled release under simulated digestion conditions. *Food Chemistry* 246: 129–136.

40 Qiu, Z., Zhou, B., Jin, L. et al. (2013). *In vitro* antioxidant and antiproliferative effects of ellagic acid and its colonic metabolite, urolithins, on human bladder cancer T24 cells. *Food and Chemical Toxicology* 59: 428–437.

41 Mele, L., Mena, P., Piemontese, A. et al. (2016). Antiatherogenic effects of ellagic acid and urolithins *in vitro*. *Archives of Biochemistry and Biophysics* 599: 42–50.

42 Bharathi, E., Jagadeesan, G., and Vijayakumar, M. (2014). Hepato-ameliorative effect of hesperidin and ellagic acid on mercuric chloride intoxicated rats. *Biomedicine & Aging Pathology* 4 (1): 17–21.

43 García-Niño, W.R. and Zazueta, C. (2015). Ellagic acid: pharmacological activities and molecular mechanisms involved in liver protection. *Pharmacological Research* 97: 84–103.

44 Aslan, A., Gok, O., Erman, O. et al. (2018). Ellagic acid impedes carbontetrachloride-induced liver damage in rats through suppression of NF-kB, Bcl-2 and regulating Nrf-2 and caspase pathway. *Biomedicine & Pharmacotherapy* 105: 662–669.

45 Baluchnejadmojarad, T., Rabiee, N., Zabihnejad, S. et al. (2017). Ellagic acid exerts protective effect in intrastriatal 6-hydroxydopamine rat model of Parkinson's disease: possible involvement of ERβ/Nrf2/HO-1 signaling. *Brain Research* 1662: 23–30.

46 Firdaus, F., Zafeer, M.F., Anis, E. et al. (2018). Ellagic acid attenuates arsenic induced neuro-inflammation and mitochondrial dysfunction associated apoptosis. *Toxicology Reports* 5: 411–417.

47 Warpe, V.S., Mali, V.R., Arulmozhi, S. et al. (2015). Cardioprotective effect of ellagic acid on doxorubicin induced cardiotoxicity in wistar rats. *Journal of Acute Medicine* 5 (1): 1–8.

48 Nankar, R.P. and Doble, M. (2017). Hybrid drug combination: anti-diabetic treatment of type 2 diabetic Wistar rats with combination of ellagic acid and pioglitazone. *Phytomedicine* 37: 4–9.

49 Yin, P., Yang, L., Xue, Q. et al. (2018). Identification and inhibitory activities of ellagic acid-and kaempferol-derivatives from Mongolian oak cups against α-glucosidase, α-amylase and protein glycation linked to type II diabetes and its complications and their influence on HepG2 cells' viability. *Arabian Journal of Chemistry* 11 (8): 1247–1259.

50 Dhingra, D. and Jangra, A. (2014). Antiepileptic activity of ellagic acid, a naturally occurring polyphenolic compound, in mice. *Journal of Functional Foods* 10: 364–369.

51 Abuelsaad, A.S., Mohamed, I., Allam, G. et al. (2013). Antimicrobial and immunomodulating activities of hesperidin and ellagic acid against diarrheic Aeromonas hydrophila in a murine model. *Life Sciences* 93 (20): 714–722.

52 Yu, D., Huang, T., Tian, B. et al. (2020). Advances in biosynthesis and biological functions of proanthocyanidins in horticultural plants. *Foods* 9 (12): 1774.

53 Gu, L., Kelm, M.A., Hammerstone, J.F. et al. (2004). Concentrations of proanthocyanidins in common foods and estimations of normal consumption. *The Journal of Nutrition* 134 (3): 613–617.

54 Gomaa, E.Z. (2013). *In vitro* antioxidant, antimicrobial, and antitumor activities of bitter almond and sweet apricot (Prunus armeniaca L.) kernels. *Food Science and Biotechnology* 22: 455–463.

55 Vlachojannis, J., Cameron, M., and Chrubasik, S. (2010). A systematic review on the sambuci fructus effect and efficacy profiles. *Phytotherapy Research: An International Journal Devoted to Pharmacological and Toxicological Evaluation of Natural Product Derivatives* 24 (1): 1–8.

56 Keser, S., Demir, E., and Yilmaz, O. (2014). Some bioactive compounds and antioxidant activities of the Bitter Almond Kernel (Prunus dulcis var. amara). *Journal of the Chemical Society of Pakistan* 36 (5).

57 Ganesan, K. and Xu, B. (2017). Polyphenol-rich dry common beans (Phaseolus vulgaris L.) and their health benefits. *International Journal of Molecular Sciences* 18 (11): 2331.

58 Luzardo-Ocampo, I., Vasco-Leal, J., Cuellar-Nuñez, M.L. et al. (2022). Castor bean (Ricinus communis L.) polyphenolic extracts exhibited anti-inflammatory effects on LPS-stimulated RAW 264.7 macrophages. *Current Developments in Nutrition* 6 (Supplement_1): 309-309.

59 Kalisz, S., Oszmiański, J., Kolniak-Ostek, J. et al. (2020). Effect of a variety of polyphenols compounds and antioxidant properties of rhubarb (Rheum rhabarbarum). *Lwt* 118: 108775.

60 Kim, J., Kim, S., Huh, K. et al. (2011). High serum isoflavone concentrations are associated with the risk of precocious puberty in Korean girls. *Clinical Endocrinology* 75 (6): 831–835.

61 Cassidy, A. (1996). Physiological effects of phyto-oestrogens in relation to cancer and other human health risks. *Proceedings of the Nutrition Society* 55 (1B): 399–417.

62 Jin, Y., Wu, H., Cohen, E.M. et al. (2007). Genistein and daidzein induce neurotoxicity at high concentrations in primary rat neuronal cultures. *Journal of Biomedical Science* 14 (2): 275–284.

63 Tomé-Carneiro, J., Larrosa, M., González-Sarrías, A. et al. (2013). Resveratrol and clinical trials: the crossroad from *in vitro* studies to human evidence. *Current Pharmaceutical Design* 19 (34): 6064–6093.

64 Cottart, C.H., Nivet-Antoine, V., Laguillier-Morizot, C. et al. (2010). Resveratrol bioavailability and toxicity in humans. *Molecular Nutrition & Food Research* 54 (1): 7–16.

65 Salehi, B., Mishra, A.P., Nigam, M. et al. (2018). Resveratrol: a double-edged sword in health benefits. *Biomedicines* 6 (3): 91.

66 Calabrese, E.J., Mattson, M.P., and Calabrese, V. (2010). Resveratrol commonly displays hormesis: occurrence and biomedical significance. *Human & Experimental Toxicology* 29 (12): 980–1015.

67 Detampel, P., Beck, M., Krähenbühl, S. et al. (2012). Drug interaction potential of resveratrol. *Drug Metabolism Reviews* 44 (3): 253–265.

68 Giuliani, C., Bucci, I., Di Santo, S. et al. (2014). Resveratrol inhibits sodium/iodide symporter gene expression and function in rat thyroid cells. *PLoS One* 9 (9): e107936.

69 Dey, A., Guha, P., Chattopadhyay, S. et al. (2009). Biphasic activity of resveratrol on indomethacin-induced gastric ulcers. *Biochemical and Biophysical Research Communications* 381 (1): 90–95.

70 Dai, Z., Li, Y., Quarles, L. et al. (2007). Resveratrol enhances proliferation and osteoblastic differentiation in human mesenchymal stem cells via ER-dependent ERK1/2 activation. *Phytomedicine* 14 (12): 806–814.

71 Posadino, A.M., Giordo, R., Cossu, A. et al. (2019). Flavin oxidase-induced ROS generation modulates PKC biphasic effect of resveratrol on endothelial cell survival. *Biomolecules* 9 (6): 209.

72 Guha, P., Dey, A., Chatterjee, A. et al. (2010). Pro-ulcer effects of resveratrol in mice with indomethacin-induced gastric ulcers are reversed by l-arginine. *British Journal of Pharmacology* 159 (3): 726–734.

73 Rocha, K., Souza, G., Ebaid, G.X. et al. (2009). Resveratrol toxicity: effects on risk factors for atherosclerosis and hepatic oxidative stress in standard and high-fat diets. *Food and Chemical Toxicology* 47 (6): 1362–1367.

74 Gadacha, W., Ben-Attia, M., Bonnefont-Rousselot, D. et al. (2009). Resveratrol opposite effects on rat tissue lipoperoxidation: pro-oxidant during day-time and antioxidant at night. *Redox Report* 14 (4): 154–158.

75 Mankowski, R., You, L., Buford, T. et al. (2020). Higher dose of resveratrol elevated cardiovascular disease risk biomarker levels in overweight older adults–A pilot study. *Experimental Gerontology* 131: 110821.

76 Ranawat, P., Khanduja, K., and Pathak, C. (2014). Resveratrol-an ingredient of red wine abrogates the reproductive capacity in male mice. *Andrologia* 46 (6): 650–658.

77 Crowell, J.A., Korytko, P.J., Morrissey, R.L. et al. (2004). Resveratrol-associated renal toxicity. *Toxicological Sciences* 82 (2): 614–619.

78 Hebbar, V., Shen, G., Hu, R. et al. (2005). Toxicogenomics of resveratrol in rat liver. *Life Sciences* 76 (20): 2299–2314.

79 Padmaja, S. and Raju, T. (2004). Antioxidant effect of curcumin in selenium induced cataract of Wistar rats. *Indian journal of Experimental Biology* 42: 601–603.

80 Additives, E.P.o.F.and Food, N.S.a.t. (2010). Scientific opinion on the re-evaluation of curcumin (E 100) as a food additive. *EFSA Journal* 8 (9): 1679.

81 Prasad, D., Gupta, B., Srivastava, R. et al. (1976). Studies on ulcerogenic activity of curcumin. *The Indian Journal of Physiology and Pharmacology* 20: 92.

82 Ammon, H.P. and Wahl, M.A. (1991). Pharmacology of Curcuma longa. *Planta Medica* 57 (01): 1–7.

83 Shoba, G., Joy, D., Joseph, T. et al. (1998). Influence of piperine on the pharmacokinetics of curcumin in animals and human volunteers. *Planta Medica* 64 (04): 353–356.

84 Sohal, A., Alhankawi, D., Sandhu, S. et al. (2021). Turmeric-induced hepatotoxicity: report of 2 cases. *International Medical Case Reports Journal* 14: 849.

85 Wang, Y.-H., Chao, P.-D.L., Hsiu, S.-L. et al. (2004). Lethal quercetin-digoxin interaction in pigs. *Life Sciences* 74 (10): 1191–1197.

86 Dunnick, J.K. and Halley, J.R. (1992). Toxicity and carcinogenicity studies of quercetin, a natural component of foods. *Toxicological Sciences* 19 (3): 423–431.

87 Boots, A.W., Li, H., Schins, R.P. et al. (2007). The quercetin paradox. *Toxicology and Applied Pharmacology* 222 (1): 89–96.

88 Pamukcu, A.M., Yalçiner, Ş., Hatcher, J.F. et al. (1980). Quercetin, a rat intestinal and bladder carcinogen present in bracken fern (Pteridium aquilinum). *Cancer Research* 40 (10): 3468–3472.

89 Singh, P., Sharma, S., and Rath, S.K. (2022). A versatile flavonoid Quercetin: study of its toxicity and differential gene expression in the liver of mice. *Phytomedicine Plus* 2 (1): 100148.

90 Chacko, S.M., Thambi, P.T., Kuttan, R. et al. (2010). Beneficial effects of green tea: a literature review. *Chinese Medicine* 5 (1): 1–9.

91 Braicu, C., Ladomery, M.R., Chedea, V.S. et al. (2013). The relationship between the structure and biological actions of green tea catechins. *Food Chemistry* 141 (3): 3282–3289.

92 Cabrera, C., Artacho, R., and Giménez, R. (2006). Beneficial effects of green tea—a review. *Journal of the American College of Nutrition* 25 (2): 79–99.

93 Sano, M., Tabata, M., Suzuki, M. et al. (2001). Simultaneous determination of twelve tea catechins by high-performance liquid chromatography with electrochemical detection. *Analyst* 126 (6): 816–820.

94 Rojo, M.Á., Garrosa, M., Jiménez, P. et al. (2020). Unexpected toxicity of green tea polyphenols in combination with the Sambucus RIL Ebulin. *Toxins* 12 (9): 542.

95 Lambert, J.D., Kennett, M.J., Sang, S. et al. (2010). Hepatotoxicity of high oral dose (−)-epigallocatechin-3-gallate in mice. *Food and Chemical Toxicology* 48 (1): 409–416.

96 Teachey, D.T. and Lambert, M.P. (2013). Diagnosis and management of autoimmune cytopenias in childhood. *Pediatric Clinics* 60 (6): 1489–1511.

97 Gandolfo, G.M., Girelli, G., Conti, L. et al. (1992). Hemolytic anemia and thrombocytopenia induced by cyanidanol. *Acta Haematologica* 88 (2–3): 96–99.

98 Rouse, J. (1998). Kava: a South Pacific herb for anxiety, tension and insomnia. *Clinical Nutrition and Insights* 96: 3900–3905.

99 Bilia, A.R., Gallori, S., and Vincieri, F.F. (2002). Kava-kava and anxiety: growing knowledge about the efficacy and safety. *Life Sciences* 70 (22): 2581–2597.

100 Mohd Sakeh, N., Md Razip, N.N., Mohd Ma'in, F.I. et al. (2020). Melanogenic inhibition and toxicity assessment of flavokawain A and B on B16/F10 melanoma cells and zebrafish (Danio rerio). *Molecules* 25 (15): 3403.

101 d Estes, J., Stolpman, D., and Olyaei, A. (2003). High prevalence of potentially hepatotoxic herbal supplement use in patients with fulminant hepatic failure. *Archives of Surgery* 138 (8): 852–858.

102 Evan, A.P. and Gardner, K.D., Jr (1979). Nephron obstruction in nordihydroguaiaretic acid-induced renal cystic disease. *Kidney International* 15 (1): 7–19.

103 Lambert, J.D., Zhao, D., Meyers, R.O. et al. (2002). Nordihydroguaiaretic acid: hepatotoxicity and detoxification in the mouse. *Toxicon* 40 (12): 1701–1708.

104 Galati, G. (2003). *Dietary flavonoid/polyphenolic reactive metabolites and their biological properties.* [Doctoral thesis, University of Toronto]. TSpace Repository https://hdl.handle.net/1807/121254.

105 Sheikh, N.M., Philen, R.M., and Love, L.A. (1997). Chaparral-associated hepatotoxicity. *Archives of Internal Medicine* 157 (8): 913–919.

106 Jentzsch, K., Gonda, T., and Holler, H. (1959). Paper chromatographic and pharmacological investigations on Curcuma pigments. *Pharmaceutica Acta Helvetiae* 34: 181–188.

107 Wang, H., Provan, G.J., and Helliwell, K. (2004). Determination of rosmarinic acid and caffeic acid in aromatic herbs by HPLC. *Food Chemistry* 87 (2): 307–311.

108 Liu, Y., Qiu, S., Wang, L. et al. (2019). Reproductive and developmental toxicity study of caffeic acid in mice. *Food and Chemical Toxicology* 123: 106–112.

109 Shaikh, J., Ankola, D., Beniwal, V. et al. (2009). Nanoparticle encapsulation improves oral bioavailability of curcumin by at least 9-fold when compared to curcumin administered with piperine as absorption enhancer. *European Journal of Pharmaceutical Sciences* 37 (3–4): 223–230.

110 Scheepens, A., Tan, K., and Paxton, J.W. (2010). Improving the oral bioavailability of beneficial polyphenols through designed synergies. *Genes & Nutrition* 5 (1): 75–87.

111 Kosari-Nasab, M., Shokouhi, G., Ghorbanihaghjo, A. et al. (2019). Quercetin mitigates anxiety-like behavior and normalizes hypothalamus–pituitary–adrenal axis function in a mouse model of mild traumatic brain injury. *Behavioural Pharmacology* 30 (2 and 3): 282–289.

112 Youdim, K.A., Dobbie, M.S., Kuhnle, G. et al. (2003). Interaction between flavonoids and the blood–brain barrier: *in vitro* studies. *Journal of Neurochemistry* 85 (1): 180–192.

113 Garavand, F., Jalai-Jivan, M., Assadpour, E. et al. (2021). Encapsulation of phenolic compounds within nano/microemulsion systems: a review. *Food Chemistry* 364: 130376.

114 Rudrapal, M., Maji, S., Prajapati, S.K. et al. (2022). Protective effects of diets rich in polyphenols in cigarette smoke (CS)-induced oxidative damages and associated health implications. *Antioxidants* 11 (7): 1217.

115 Murthy, N.T.V., Agrahari, V., and Chauhan, H. (2021). Polyphenols against infectious diseases: controlled release nano-formulations. *European Journal of Pharmaceutics and Biopharmaceutics* 161: 66–79.

116 Greish, K., Mathur, A., Bakhiet, M. et al. (2018). Nanomedicine: is it lost in translation? *Therapeutic Delivery* 9 (4): 269–285.

117 Yang, -Q.-Q., Wei, X.-L., Fang, Y.-P. et al. (2022). Nanochemoprevention with therapeutic benefits: an updated review focused on epigallocatechin gallate delivery. *Critical Reviews in Food Science and Nutrition* 60 (8): 1243–1264.

118 Choi, G., Kim, T.-H., Oh, J.-M. et al. (2018). Emerging nanomaterials with advanced drug delivery functions; focused on methotrexate delivery. *Coordination Chemistry Reviews* 359: 32–51.

119 Yang, B., Dong, Y., Wang, F. et al. (2020). Nanoformulations to enhance the bioavailability and physiological functions of polyphenols. *Molecules* 25 (20): 4613.

120 Esfanjani, A.F. and Jafari, S.M. (2016). Biopolymer nano-particles and natural nano-carriers for nano-encapsulation of phenolic compounds. *Colloids and Surfaces B: Biointerfaces* 146: 532–543.

121 Anand, S., Sowbhagya, R., Ansari, M.A. et al. (2022). Polyphenols and their nanoformulations: protective effects against human diseases. *Life* 12 (10): 1639.

122 Guo, Y., Sun, Q., Wu, F.G. et al. (2021). Polyphenol-containing nanoparticles: synthesis, properties, and therapeutic delivery. *Advanced Materials* 33 (22): 2007356.

123 Li, W., Yi, S., Wang, Z. et al. (2011). Self-nanoemulsifying drug delivery system of persimmon leaf extract: optimization and bioavailability studies. *International Journal of Pharmaceutics* 420 (1): 161–171.

124 Wang, S., Su, R., Nie, S. et al. (2014). Application of nanotechnology in improving bioavailability and bioactivity of diet-derived phytochemicals. *The Journal of Nutritional Biochemistry* 25 (4): 363–376.

125 Mohanty, C. and Sahoo, S.K. (2010). The *in vitro* stability and *in vivo* pharmacokinetics of curcumin prepared as an aqueous nanoparticulate formulation. *Biomaterials* 31 (25): 6597–6611.

126 Kakkar, V., Singh, S., Singla, D. et al. (2011). Exploring solid lipid nanoparticles to enhance the oral bioavailability of curcumin. *Molecular Nutrition & Food Research* 55 (3): 495–503.

127 Frozza, R.L., Bernardi, A., Paese, K. et al. (2010). Characterization of trans-resveratrol-loaded lipid-core nanocapsules and tissue distribution studies in rats. *Journal of Biomedical Nanotechnology* 6 (6): 694–703.

128 Yu, H. and Huang, Q. (2012). Improving the oral bioavailability of curcumin using novel organogel-based nanoemulsions. *Journal of Agricultural and Food Chemistry* 60 (21): 5373–5379.

129 Ru, Q., Yu, H., and Huang, Q. (2010). Encapsulation of epigallocatechin-3-gallate (EGCG) using oil-in-water (O/W) submicrometer emulsions stabilized by ι-carrageenan and β-lactoglobulin. *Journal of Agricultural and Food Chemistry* 58 (19): 10373–10381.

130 Bhushani, J. and Karthik, P. (2016). Anandharama krishnan C. nanoemulsion based delivery system for improved bioaccessibility and Caco-2 cell monolayer permeability of green tea catechins. *Food Hydrocoll* 56: 372–382.

131 Bhingardeve, D., Patil, S., Patil, R. et al. (2014). Phytosome-valuable phyto-phospholipid carriers. *Journal of Current Pharma Research* 5 (1): 1386.

132 Rossi, R., Basilico, F., Rossoni, G. et al. (2009). Liquid chromatography/atmospheric pressure chemical ionization ion trap mass spectrometry of bilobalide in plasma and brain of rats after oral administration of its phospholipidic complex. *Journal of Pharmaceutical and Biomedical Analysis* 50 (2): 224–227.

133 Mirzaei, H., Shakeri, A., Rashidi, B. et al. (2017). Phytosomal curcumin: a review of pharmacokinetic, experimental and clinical studies. *Biomedicine & Pharmacotherapy* 85: 102–112.

134 Allam, A.N., Komeil, I.A., Fouda, M.A. et al. (2015). Preparation, characterization and *in vivo* evaluation of curcumin self-nano phospholipid dispersion as an approach to enhance oral bioavailability. *International Journal of Pharmaceutics* 489 (1–2): 117–123.

135 Marczylo, T.H., Verschoyle, R.D., Cooke, D.N. et al. (2007). Comparison of systemic availability of curcumin with that of curcumin formulated with phosphatidylcholine. *Cancer Chemotherapy and Pharmacology* 60 (2): 171–177.

136 Zhang, J., Tang, Q., Xu, X. et al. (2013). Development and evaluation of a novel phytosome-loaded chitosan microsphere system for curcumin delivery. *International Journal of Pharmaceutics* 448 (1): 168–174.

137 Mahmoodi, N., Motamed, N., Paylakhi, S.H. et al. (2015). Comparing the effect of Silybin and Silybin Advanced™ on viability and HER2 expression on the human breast cancer SKBR3 cell line by no serum starvation. *Iranian Journal of Pharmaceutical Research: IJPR* 14 (2): 521.

138 Fox, P. and Brodkorb, A. (2008). The casein micelle: historical aspects, current concepts and significance. *International Dairy Journal* 18 (7): 677–684.

139 Livney, Y.D. (2010). Milk proteins as vehicles for bioactives. *Current Opinion in Colloid & Interface Science* 15 (1–2): 73–83.

140 Esmaili, M., Ghaffari, S.M., Moosavi-Movahedi, Z. et al. (2011). Beta casein-micelle as a nano vehicle for solubility enhancement of curcumin; food industry application. *LWT-food Science and Technology* 44 (10): 2166–2172.

141 Luo, Y., Pan, K., and Zhong, Q. (2015). Casein/pectin nanocomplexes as potential oral delivery vehicles. *International Journal of Pharmaceutics* 486 (1–2): 59–68.

142 Ghayour, N., Hosseini, S.M.H., Eskandari, M.H. et al. (2019). Nanoencapsulation of quercetin and curcumin in casein-based delivery systems. *Food Hydrocolloids* 87: 394–403.

143 Shutava, T.G., Balkundi, S.S., Vangala, P. et al. (2009). Layer-by-layer-coated gelatin nanoparticles as a vehicle for delivery of natural polyphenols. *ACS Nano* 3 (7): 1877–1885.

144 O'Mahony, J.A., Fox, P.F., and Kelly, A.L. (2013). Indigenous Enzymes of Milk. In: *Advanced Dairy Chemistry: 1A: Proteins: Basic Aspects*, 4e. (ed. P.L.H. McSweeney and P.F. Fox), 337–385. Springer US.

145 Chen, L. and Subirade, M. (2005). Chitosan/β-lactoglobulin core–shell nanoparticles as nutraceutical carriers. *Biomaterials* 26 (30): 6041–6053.

146 Shpigelman, A., Israeli, G., and Livney, Y.D. (2010). Thermally-induced protein–polyphenol co-assemblies: beta lactoglobulin-based nanocomplexes as protective nanovehicles for EGCG. *Food Hydrocolloids* 24 (8): 735–743.

147 Li, M., Cui, J., Ngadi, M.O. et al. (2015). Absorption mechanism of whey-protein-delivered curcumin using Caco-2 cell monolayers. *Food Chemistry* 180: 48–54.

148 Chiba, T., Kimura, Y., Suzuki, S. et al. (2016). Trans-resveratrol enhances the anticoagulant activity of warfarin in a mouse model. *Journal of Atherosclerosis and Thrombosis* 23 (9): 1099–1110.

15

Nanodelivery of Food Polyphenols for Nutraceutical Applications

Aamena Yusuf[1], *Tanisha Ambawat*[1], *Sneha Dokhale*[2], *Shine Devarajan*[1], *and Samiksha Garse*[1,*]

[1] *School of Biotechnology and Bioinformatics, D Y Patil Deemed to be University, Navi Mumbai, Maharashtra, India.*
[2] *B. K. Birla College, Birla College Campus Rd, Kalyan, Maharashtra, India.*
* *Corresponding author*

15.1 Introduction

Polyphenols are plant-based, non-nutrient natural components also termed as secondary metabolites. These are included in plant-based daily diets [1]. The substances are generated through the shikimate/phenylpropanoid and/or polyketide pathways that have several phenolic units but no nitrogen-based functionalities are referred to as polyphenols [2]. Most plants naturally contain polyphenols, which have a complex and unique chemical structure resulting in a wide range of biological properties [3, 4]. According to epidemiological studies and meta-analyses, consuming diets high in plant polyphenols over the long term may provide some protection against the advancement of neurological diseases, diabetes, osteoporosis, and cardiovascular diseases, as shown in Figure 15.1 [3].

Many polyphenolic substances have limited oral bioavailability, which constrains their use in nutraceuticals. Green bio-based nanocarriers are ideal for encapsulating, storing, and packing polyphenols, thus enhancing their bioavailability [2, 4, 5].

In a variety of sectors, bio-based polymers have arisen as a viable alternative to petroleum-based polymers [6]. Proteins are versatile bio-based polymers with different functional qualities such as emulsification, amphiphilicity, gelation, and foaming [7]. Polysaccharides are also bio-based polymers with a diverse variety of origins, and their distinctive structure and physiological activity make them suitable as ingredients for the creation of a variety of nanocarriers and the transport of polyphenolic chemicals [8]. Lipid-based nanocarriers are also some of the most important methods for targeting such transportation in food and nutrition due to their excellent biocompatibility and biodegradability [9]. The introduction of several marketed modified polyphenol products sets the groundwork for additional study and development of novel polyphenol health nutrition products [5]. The label "Nutraceutical" was a blend of "Nutrition" and "Pharmaceutical", by Dr. Stephen DeFelice [10]. The definition of a nutraceutical given by DeFelice is "a food (or part of a food) that provides medical or health benefits, including disease

Polyphenols: Food, Nutraceutical, and Nanotherapeutic Applications, First Edition. Edited by Mithun Rudrapal.

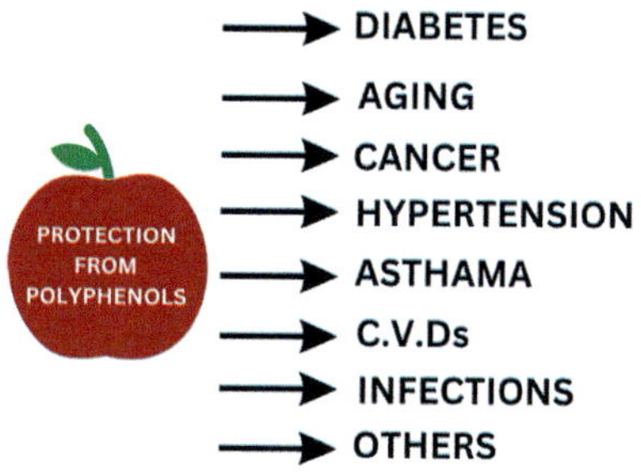

Figure 15.1 Health benefits of polyphenols.

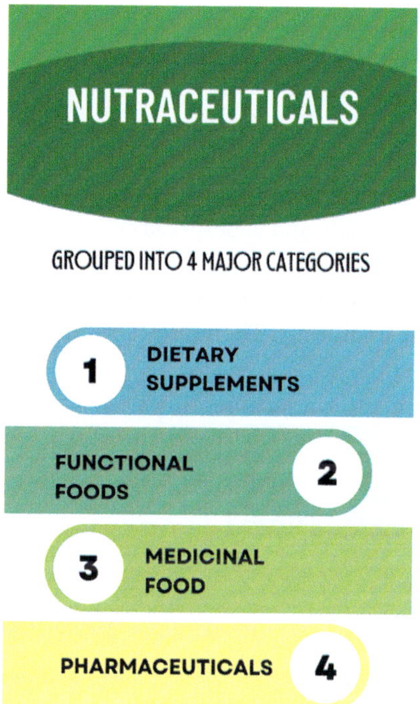

Figure 15.2 Classification of nutraceuticals.

prevention and treatment" [11]. Any food-sourced product that offers notable wellness advantages in addition to the essential nutritional content included in meals is referred to as a nutraceutical. Nutraceuticals are grouped into four major categories, as shown in Figure 15.2.

Hippocrates once said, "Let food be your medicine"; this is the fundamental tenet of nutraceuticals. Nutraceuticals signaled a new age in food health and medicine, as people become more mindful of the relationship between critical illnesses and eating behavior [11]. Natural honey is one of the best nutraceutical products because it contains antioxidants and enzymes that are necessary for digestion [12]. Because of their potential to increase bioavailability, component solubility, and stability, these have been widely employed in nanotechnology [13]. Nanotechnology is enhancing the whole industry of food, from production to processing, storage, and consumption. Nanoparticles (NPs) for edible delivery are physiologically consistent small materials ranging in size from 1 nm to 100 nm and they are generated using various ways that produce important physio-chemical properties. These features of NPs make them useful in an array of domains [13]. This chapter discusses the nanodelivery of food polyphenols for nutraceuticals with improved solubility, bioavailability, efficiency, encapsulation, and prolonged and focused drug delivery.

15.2 Polyphenols: Classification, Health Benefits, Bioavailability

Polyphenols are phenolic hydroxyl group-containing organic molecules that are found abundantly in natural flora. They have been shown to benefit human health immensely because of their antioxidant properties, adherence affinity, and anticancer properties. They also show great promise in the bioimaging, biomedical, and therapeutic industry by the preparation and modification of nanoassemblies [4, 14]. These compounds, which are classified as phenolic acids, flavonoids, anthocyanins, and tannins, possess complicated structures, and use phenolic rings as their core monomer. Polyphenols from tea and coffee are bioavailable; approximately 30% of these compounds are absorbed in circulation in humans, and regular consumption of these compounds results in improvements in biomarkers for oxidative stress [4, 14, 15]. Grosso revealed that 75% of polyphenols come from beverage and chocolate consumption in a population-specific analysis within Poland [16].

15.2.1 Classification of Polyphenols

Food polyphenols are a diverse group of organic substances from the kingdom Plantae, the majority of which have two phenyl rings and one or more hydroxyl groups [17]. The classification of the same is given in Figure 15.3. There are currently 8,000 polyphenolic chemicals known, and more than 4,000 of them are flavonoids [18]. Flavonoids and phenolic acids are the two main classes of the diverse collection of phenolic chemicals known as polyphenols; phenolic acids are split into two types, hydroxycinnamic and hydroxybenzoic acids. Flavonoids are colored chemicals that are further divided into flavones, flavanols, flavanones, flavonols, and isoflavones [19]. They can be found either unconjugated (as an aglycone) or conjugated with substances, including lipids, glucose, carboxylic acid, amines, and organic acids. Polyphenols are divided into phenolic acids, phenolic aldehydes (vanillin, salicylaldehyde, syringaldehyde, etc.), flavonoids, iso-flavonoids, tannins (hydrolyzable and condensed tannins), lignans, and lignins based on the structural variation [20, 21].

Dietary polyphenols exhibit a wide variety of structural types, from straightforward compounds like monomers and oligomers, to complex polymers with high densities. High-molecular-weight (>500) structures are referred to as tannins because of their capacity to interact with proteins. Condensed tannins stand out among their contribution to food quality and widespread occurrence in flora [18].

However, because of their extremely unstable nature, most phenolic compounds are quickly converted into their reaction products when the plant cells are in disarray. Therefore, it is difficult

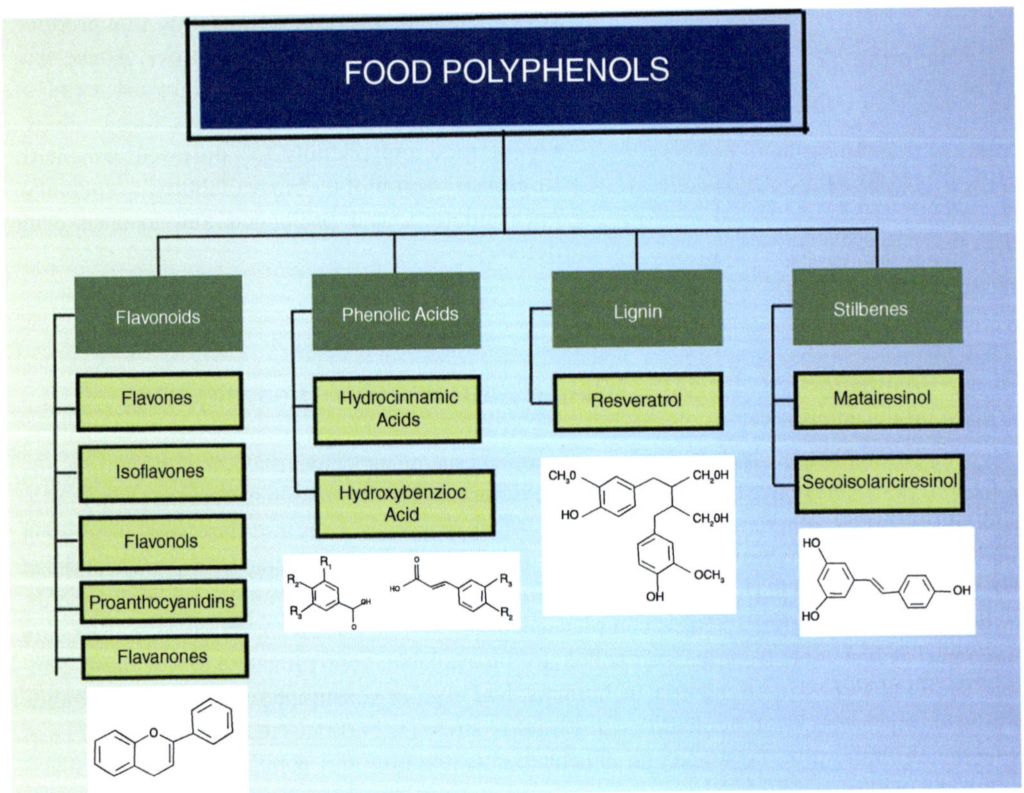

Figure 15.3 Classification of food polyphenols.

to incorporate food polyphenols into food compositions [19, 22]. Black tea has a well-established history of producing certain tannin-like substances via enzymatic oxidation. It has been shown that several chemical processes involving anthocyanins or flavanols occur during the aging of red wine [17, 19, 21–25].

15.2.1.1 Phenolic Acids

Phenolics are classified as aromatic rings with OH groups connected as non-flavonoid polyphenols. Benzoic acid, which has seven carbon atoms, and cinnamic acid, which has nine carbon atoms, are the two main characteristics that set phenolic acids apart [17]. Because of their multiple health benefits for humans, including their anti-inflammatory, anticancer, and antimicrobial qualities, they are essential dietary components [23].

15.2.1.2 Flavonoids

Flavonoids are a component of the vibrant colors found in herbs, fruits, and vegetables [26]. They can be further separated into flavonones, flavones, flavonols, flavan-3-ols, anthocyanidins, and isoflavones depending on the variation in the C-ring [27]. Because of the pattern and intensity of modifications of hydroxyl groups, methoxy groups, and glycan molecules, each of these subgroups differs in certain ways. Flavonoids include substances such as quercetin, naringenin, catechin, daidzein, and cyanidin-glucoside [28, 29]. They can exist in several modified forms in addition to being free glycosidic conjugates and aglycones [30–32]. Additional categories of flavonoids include flavones, flavonols, flavonones, isoflavones, and anthocyanins [33].

Flavanones are the most therapeutically important flavonoids that are present in all citrus fruits. Hesperiden, narigenin, and eriodictyol are examples of a few flavonones. Anthocyanins are the most significant subclass of flavonoids and naturally occurring pigments [34–36]. They are two phenyl benzopyrylium salts that have had their polyhydroxy and polymethoxy derivatives glycosylated. Black currants, red and merlot grapes, as well as raspberries, cranberries, strawberries, blueberries, and bilberries all have them in the outermost layers. They could potentially replace synthetic dyes and natural colors in the food companies [34–46].

15.2.1.3 Stilbenes

The non-flavonoid class of stilbenes comprises two phenyl rings bridged by a two-carbon methylene group. They are free isomers (cis and trans), with two aromatic rings, called A and B, in glycosylated forms. Two hydroxyl groups are present at the m-position of ring A, whereas methoxy and hydroxyl groups are substituted at several locations in ring B [47]. Stilbenes, such as resveratrol, are often produced by berries, grapes, and nuts [48].

15.2.1.4 Lignans and Lignins

Another non-flavonoid molecule is lignan, which is made up of two C6–C3 units linked together at positions 8 and 8′. Because the lignan C9 and C9′ positions are substituted in a variety of ways, they have a wide range of structural forms categorized such as dibenzylbutane, furan, and aryltetralin [47]. Pulses, seeds, and vegetable oils are the principal sources of lignans because they lack an abundance of glycosylated structures. Lignin is an aromatic biopolymer created when peroxidase enzymes oxidize p-hydroxycinnamyl alcohol monomers in a phenolic reaction [49].

15.2.1.5 Tannins

Tannins are a class of high molecular weight, water-soluble phenolics. They can also be separated into condensed tannins and hydrolyzable tannins. Galltannins and ellagitannins are two types of hydrolyzable tannins. Phlorotannin, a kind of tannin that makes up 25% of the subcellular structure of brown algae, has recently attracted interest from the food, feed, and pharmaceutical industries [50]. As functional molecules with a variety of anti-activities towards cancer, microbes, hypertensive, diabetic, and inflammatory effects, phlorotannins have shown a lot of promise [51]. These phenolic compounds could be exploited as antioxidants in the food industry because they have excellent superoxide and free radical scavenging capabilities. They inhibit lipase in the pancreas that is responsible for the breakdown of dietary fats, and was eventually used as a weight loss component [52].

15.2.2 Health Benefits

Polyphenols are secondary metabolites present in the kingdom Plantae. They provide significant health advantages because of their immunomodulatory, antibacterial, anti-inflammatory, and antioxidant capabilities [1, 2]. Polyphenols found in fruits, vegetables, tea, cocoa, and other foods are good for human health. For instance, there is a significant link between cocoa flavan-3-ols and a reduced risk of diabetes, myocardial infarction, and stroke. Additionally, dietary polyphenols help reduce lipid profiles, bodily inflammation, insulin resistance, and blood pressure [4, 5]. Resveratrol, a stilbene, and the flavonoid quercetin have been connected to cardiovascular health. It is possible that the capacity of dietary polyphenols has therapeutic effects that stem from a two-way relationship with the gut microbiota. This is mostly because polyphenols are known to improve human health by changing the composition of the gut microbiota [19]. Polyphenols are transformed into bioactive molecules with therapeutic benefits by the gut flora. In this chapter, we go through polyphenols' antioxidant, anticytotoxicity, anti-inflammatory, antihypertensive, and antidiabetic properties [21].

15.2.2.1 Antioxidant Activity

Polyphenols are functional and bioactive phytochemicals that have stirred interest from various scientists, nutritionists, and consumers because of their health benefits. One such health benefit is their antioxidant property. Typically, oxidative stress is induced by low levels of antioxidant synthesis or high levels of reactive oxygen species (ROS), such as OH radicals and H_2O_2, during various metabolic processes leading to chemical reactions triggering cell and tissue damage [53–55]. Moreover, diseases like atherosclerosis, inflammation, cancer, diabetes, arthritis, and even neurological disorders are caused by oxidative damage to cells [54–56]. Utilizing dietary plant polyphenols and other naturally occurring antioxidants from plant-based foods can help the body fight off oxidative stress and its harmful effects. Because of their essential role in disease prevention, these phytochemicals rank among the most important natural antioxidants used by humans [57]. Because of their antioxidant properties and ability to scavenge ROS, phenolic compounds have the potential to both treat and prevent several degenerative disorders [53]. According to research, eating foods high in polyphenols, such as fish and other seafood, which are natural antioxidants, can help prevent lipid oxidation as well as morbidity brought on by degenerative diseases [58, 59]. Another important factor influencing the antioxidant capabilities of these bioactive substances is the amount and position of the hydroxyl groups in phenols [60]. This group has a conjugation impact leading to reduction in the hydrogen ion's binding capacity [53]. Therefore, free radical generation can be stopped by phenolic substances. The free radicals receive an electron transfer through this mechanism, and after the radicals are neutralized, they become more stable. Thus, the domino effect will be broken [17]. However, there is a direct correlation between the class of plant species chosen and the kind of solvent used in the extraction processes and the potential antioxidant activity of phenolic compounds [61]. By using several mechanistic pathways, the polyphenols can exhibit antioxidant actions that reduce the generation of ROS [62]. According to research, polyphenols can interact with ROS by giving unpaired electrons to free radicals or hydrogen atoms, which in turn produces stable phenolic oxygen radicals. Therefore, these bioactive substances can remove the byproducts of free radicals [53]. Polyphenols, including tocopherols and flavonoids, are typically regarded as important main antioxidant components. By using electron transfer processes, their aromatic amines can prevent autoxidation [63].

15.2.2.2 Antihypertensive Activity

Although hypertension is one of the major global causes of cardiovascular illnesses, its origins are not well understood with no overt symptoms [63, 64]. To avoid the adverse effects of hypertension, prevention and treatment have become essential. Dietary modifications are the main strategy for controlling high blood pressure. The most effective non-pharmacological treatments include reducing sodium intake, adding potassium supplements, increasing physical activity, and losing weight [65]. Receptor blockers, angiotensin II receptor antagonists (ARBs), angiotensin-transferase inhibitors, calcium antagonists, and diuretics are only a few of the synthetic and chemical drugs that the World Health Organization (WHO) prescribes for the treatment of hypertension [66]. Although these medications have an efficient antihypertensive effect, they can also cause vertigo, coughing, ankle edema, high blood cholesterol levels, and sodium and water retention. Furthermore, these medications cannot be used for a long period to treat artery illness or reduce the symptoms of hypertension [66, 67]. Endothelial dysfunction has reportedly been identified as one of the symptoms of hypertension disease. Thus, the combination of endothelium-dependent vasodilatation with oxidative damage serves as a defining characteristic of endothelial dysfunction. In actuality, this dysfunction is discovered prior to changes in the composition or texture of the artery walls, and as a result, this dysfunctional state leads to the onset and development of cardiovascular disorders [67, 68]. Several studies have confirmed that phenolic compounds and their derivatives can enhance the performance of vascular endothelial cells through numerous methods, including by inhibiting the production of pro-oxidant enzymes [68]. For example, the COXs and NADPH oxidase may aid in enhancing endothelial function, slowing the aging of the vascular system, and preventing hypertension [69]. Additionally, the phenolic compounds can help lower the blood pressure by activating the mTORC2-Rictor survival pathway and suppressing the production of mTOR signaling proteins [70]. Additionally, multiple research studies [71–74] suggest that polyphenols may reduce the activation of the angiotensin system, and increase endothelial relaxation by modifying EDH, oxidative stress, and inflammatory response in order to enhance the angiotensin system and generate the vasodilation effect [69, 71, 73, 74]. Because of their role in lowering systolic blood pressure in spontaneously hypertensive rats, polyphenols and their derivatives have been shown to have a significant antihypertensive effect in other in vivo research studies [75, 76]. Additionally, polyphenols effectively lower blood pressure through a variety of methods of action. Polyphenol therapy for hypertension has been demonstrated to improve endothelial function and relax vascular tissues by inhibiting ACE and related pathways [64]. It has been shown that polyphenols can reduce the activity of metalloproteinases, especially those that cause hypertension [64, 77].

15.2.2.3 Antimicrobial Activity

Microorganisms are to blame for several diseases and fatalities. However, overuse of synthetic medications has increased the prevalence of resistant pathogenic forms of bacteria as well as antimicrobial resistance. Consequently, the need for naturally occurring antibacterial chemicals has increased [78–80]. Because of their numerous antimicrobial effects against a huge microbial population, interest in polyphenols has increased. They also help address a wide spectrum of beneficial microbes [79, 81].

The mode of action of polyphenols includes the disruption of the lipid membrane and alteration of the structure causing a leak of the cellular components, disturbing intracellular functions, and affecting the hydroxylation and oxidation levels [79, 80, 82]. The antimicrobial activity is given in Figure 15.4. The antibacterial activity of polyphenols has been improved by new extraction and encapsulation techniques [83–85]. Additionally, polyphenols have the capacity to give antibacterial activity, changing the permeability of cells and destroying the composition of cells as a result [63]. It is important to highlight that the hydroxyl group of phenolic polyphenols contributes significantly to the death of bacterial cells. As a result, the density gradient across the plasma membrane is reduced when the OH of the polyphenolic compound contacts the bacterial cell wall and it reduces the ATP pool, which causes microbial cells to die [63]. In a case study, the antibacterial potential of phenolics from Japanese apricots against Enterobacteria was examined [86]. The examined microorganisms were more resistant to the antibacterial activity of phenols; however, only at comparatively greater doses (1,250–5,000 g/mL). The outcomes of the chemical examination revealed the presence of chlorogenic acid and hydroxycinnamic acid derivatives [87]. On vero E6 cells, the treatment dramatically reduced the MTT assay at concentrations of 80 mg/mL and 100 g/mL. Similarly, the application reduced the viral titer in the supernatant of infected treated cells and prevented virus reproduction for the 24 hours that were observed. Additionally, some studies have documented that polyphenols have a unique capacity to shield humans against disease such as oral bacteria and non-communicable viral diseases [88–93]. Polyphenols also have anticancer properties and antiviral properties against COVID-19 [19].

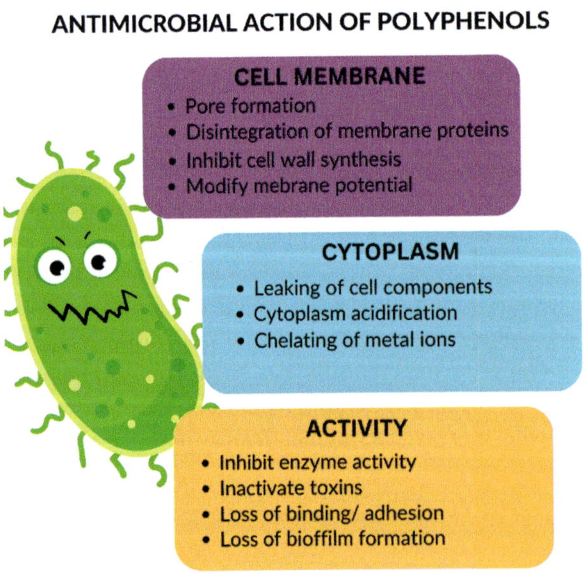

Figure 15.4 Antimicrobial action of polyphenols.

External Factors	Environmental factors and food availability
Food Processing related factors	Thermal treatment, homogenization, cooking and methods of culinary preparation, storage
Food related factors	Food Matrix, Presence of positive or negative factors for absorption
Interaction with other compounds	Bonds with proteins or polyphenols with similar mechanism of absorption
Polyphenols related factors	Chemical structure, amount introduced and concentration in diet
Host-related Factors	Intestinal factors and systemic factors

Figure 15.5 Factors affecting bioavailability.

15.2.3 Bioavailability

Bioavailability refers to the extent to which a substance or a compound is made completely available to its intended biological destinations. Most polyphenolic compounds reveal low bioavailability [9]. The reasons are usually linked to its sub-par bioaccessibility. Factors include its interaction with the food matrix, metabolic processes, and food processing, which are mediated by the intestine, liver, and microbiota, and can influence the bioavailability of polyphenolic compounds [94]. Polyphenols have powerful antioxidant, immunomodulatory, anticancer, prebiotic, and anti-inflammatory properties; however, more research is needed to confirm their gastroprotective activity. Natural polyphenols are anticarcinogenic and recommended for CRC chemoprevention and treatment. Polyphenols may have pharmacological activity, particularly antitumor activity [22]. Nanomedicine has the potential to improve tumor treatment by compensating for the low bioavailability of polyphenols. Polyphenols have immunomodulatory consequences on macrophages and may be utilitarian in the treatment and prevention of autoimmune diseases [95, 96]. The limited oral bioavailability of several polyphenolic substances limits their utility in nutraceuticals. The silver lining is that green bio-based nanocarriers are ideal for encasing, protecting, and dispersing polyphenols, hence enhancing bioavailability. When used as nutritional treatments, bio-based polymers with higher biocompatibility, biodegradability, resource sustainability, and nutrient content, such as proteins and polysaccharides, are appropriate delivery vehicles for addressing the drawbacks of oral polyphenol utilization [13, 29]. Factors affecting bioavailability of polyphenols are listed in Figure 15.5.

15.3 Nanocarriers and Nanodelivery Methods

15.3.1 Nanocarriers

15.3.1.1 Solid Lipid Nanoparticles

Various nanocarrier and nanodelivery systems are tabulated in Figure 15.7. Solid lipid nanoparticles (SLNs) are colloidal nanocarriers of submicron size (50–1,000 nm) comprising lipids compatible with human physiology. The SLNs are disseminated in an aqueous solution with surfactants. Ultrasonication and microfluidization, often known as high-pressure homogenization, are two techniques used to create SLNs [97]. SLNs feature a large surface, a large pharmaceutical payload, and improved interfacial contact because of their submicron size. Problems facing SLNs, which include nanoemulsions, microemulsions, polymer NPs (refer to Figure 15.6), and liposomes, are ineffective site-of-action targeting and uncontrolled release profiles [98]. A drug payload can be delivered using one of two methods. The drug can be attached to the polymeric component's surface or embedded into the polymeric core. One of the primary benefits of SLNs is that they trap the hydrophobic drug in a state of equilibrium without using hazardous chemical solvents. The size-related features of SLNs, as well as their drug integration capabilities, are additional benefits [99]. It is easier to set up large-scale manufacturing, which has better bioavailability and fewer negative effects. Differences in nucleic acid sizes and charge can impact lipid packing and the SLN structure. The molar proportion and fatty composition of the manufacturing techniques used to encapsulate nucleic acids in the systems must also be fine-tuned [98]. Additionally, surface-attached ligands (such as folate and transferrin) can be included in the outer layer of the SLNs to tailor them to recognize and bind to particular cell receptors.

Recently, two microfluidic mixing methods based on fast mixing have been developed. Staggered herringbone mixing, a continuous flow method called microfluidic hydrodynamic focusing, is the most widely applied method for producing lipid nanoparticles (LNPs) in a repeatable and scalable manner [99]. By using this method, there is better control over the mixing process and it is quicker. All of these approaches allow for the rapid mixing of a liquid layer, including nucleic acid and lipid elements, producing large nucleic acid encapsulation [98]. Certain LNPs can be generated by precisely controlling microfluidic operational parameters.

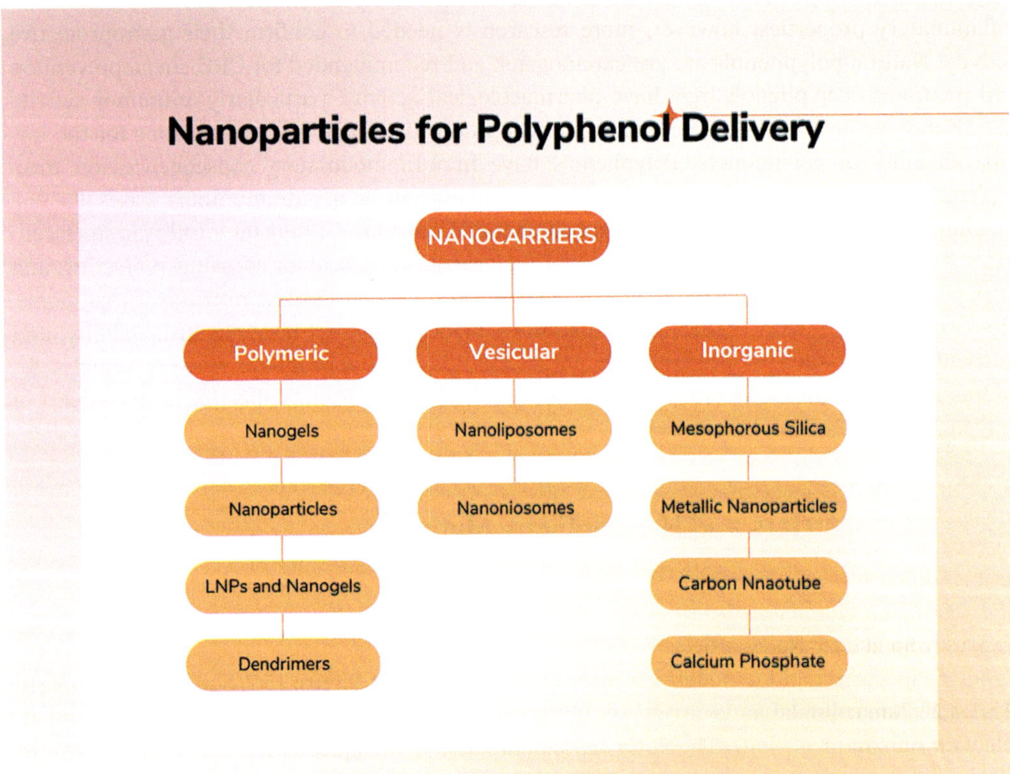

Figure 15.6 Nanoparticles for polyphenol delivery.

Delivery Systems	Types of Carriers	Materials	Major Outcomes
		Zein	Improvements in performance and efficiency from natural sources.
	Nanoparticles	Soy Protein	A variety of food ingredients having significant nutritional value and health impacts.
	Nanogels	Rice Protein	Complexes with high value added for proteins.
	Nanofilms	Ferritin	Natural protein for storing iron having a hollow for bioactive nutrients.
Protein Based	Nanofibres	Albumin	Secure and well-tolerated in Humans.
	Nanoemulsions	Gliadin	Eco-friendly manufacturing process, strong biocompatibility, sustainable materials
	Nanoparticles	Caesin	Have both hydrophilic and hydrophobic ability.
		Whey Protein	Emulsification, Gelation and Hydrophobic active ingredient binding.
	Nanoparticles	Starch	Variety of raw material sources, non-toxic, biocompatible, perfect for nanocarrier material.
	Nanogels	Cellulose	Extraction of natural fibre, Lot of crystals, High Young's Modulus
Carbohydrate Based	Nanofilms	Lignin	Hydrophobic and Hydrophilic interactions between Amphiphilic Nanoparticles and Polyphenols
	Nanofibres	Marine Polysaccharide	Versatility and Biocompatibility in transport of Polyphenols to colon to treat diseases
	Nanoemulsions	Glycogen	Mol. Wt. > Renal Threshold, Renal clearance is the only option of biodegradation
	Nanoparticles		
	Solid Lipid		
Lipid Based	Nanoparticles	Lipid Compounds	Enhances Intestine solubility, Boosts Surface Area to Mass Ratio, and strengthens active molecule resistance to environmental and enzymatic static stress
	Nano emulsion		
	Liposomes		

Figure 15.7 Common bio-based nanocarriers and nanodelivery systems.

To generate a nanoemulsion, at least two immiscible liquids must be present, one of which is water and another is a fat. The procedures used to manufacture these nanoformulations are mechanical or chemical [97]. Hydrophobic chemicals and emulsifiers mix to generate nanoemulsion droplets, whereas, in a mechanical process, large emulsion droplets are homogenized under tremendous pressure to form nanodroplets. The size and shape of the produced emulsion droplets are significant characteristics that set nanoemulsion apart from a traditional emulsion; the size ranges from 20 nm to 200 nm [97, 98, 100].

15.3.1.2 Nanocrystals

Nanocrystals are colloidal materials with a dimension of 20–850 nm that comprises drug molecules distributed in various phases [101]. These nanocrystals, which can also be created chemically or mechanically, benefit from fewer nanoscale particles and a large surface area to maintain contact with the dissolving phase. Two advantages of nanocrystals over conventional dosage forms are enhanced saturation solubility and increased drug loading with faster dissolving rates [101].

15.3.1.3 Nanopolymerosomes (NPS)

Nanopolymerosomes (NPS) are aqueous-cored polymer forms that range in size from 10 nm to 1 mm because of amphiphilic copolymers. The programmable features of NPS, which may be used to create synthetic organelles or drug delivery systems, allow for a wide range of biological benefits. They are created in the same way as polymeric NPs [102]. Because of their customizable qualities, several biodegradable and stimulation-responsive polymers are used in NPS to encapsulate drugs and improve release behaviors. NPS are effective in incorporating both lipophilic and hydrophilic pharmacological chemicals, namely a polymeric membrane bilayer and solvent core. NPS have a broader variety of use than nanostructured lipid carriers (NLCs), a more persistent vesicular composition, and a regulated vesicular content [103, 104].

15.3.1.4 Liposomes

Liposomes are phospholipid bilayer sphere-like vesicles that have a watery core that are biodegradable. They develop when phospholipids self-assemble [62, 105]. Their structure allows hydrophilic drug molecules to be loaded into the inner aqueous core and lipophilic medicinal compounds to be loaded into the surrounding phospholipid bilayer. Liposomes can be created using several techniques, including solvent injections, reversed-phase evaporation, and thin-film hydration. Depending on their size and the quantity of phospholipid bilayers they contain, these vesicles are categorized as unilamellar, multilamellar, or multivesicular [62, 105–107]. According to studies, liposomal polyphenols have therapeutic effects. Curcumin liposomes have higher antioxidant activity than uncomplexed curcumin. Researchers created a liposomal formulation of quercetin and observed that it had improved solubility, bioavailability, and antitumor effectiveness in vivo [108–111].

15.3.1.5 Ethosomes

These vesicular transporters are phospholipid-based and transport a sizable amount of ethanol. Touitou and colleagues originally developed these vesicles to effectively transmit active drugs through the epidermis [95]. The ethosome is a non-intrusive transport system. Because ethosomes contain more ethanol, they charge the skin's surface and make it easier to penetrate, better enabling medicinal medications to permeate the skin's underlying tissues for systemic circulation. Mechanical dispersion, a cold method, hot method, and other procedures are used to create ethosomes [112, 113]. A useful method to improve in situ stability, skin permeability for polyphenols, bioavailability, and therapeutic efficacy has been proposed by adding phytopolyphenols to ethosomes. Epigallocatechin-3-gallate (EGCG) was encapsulated in ethosomes, and researchers found that EGCG increased the antioxidant activity and photostability. However, the ethosomal formulation displayed greater in vivo skin targeting and efficacy against UVB radiation-induced skin inflammation [95, 114, 115].

15.3.1.6 Phytosomes

Phytosomes are phospholipid and herbal medicinal extract combination vesicles. Phospholipids can behave as active emulsifiers because of their polar head and lipid tail. The emulsifier characteristic of phytosomes increases the bioavailability of phytopolyphenols by facilitating their transition from the aqueous to the lipophilic environment of the cell membrane [116]. When compared with silybin alone, the composition of silybin phytosomes have better hepatoprotective and antioxidant capacities [117]. Ginkgo phytosomes have been linked to increased brain and vascular protection. Furthermore, researchers created a transdermal product containing rutin phytosomes and discovered that they had a significant effect on rheumatoid arthritis [118, 119].

15.3.1.7 Invasomes

Invasomes are phospholipid, ethanol, and terpene-based nanovesicular transdermal medication delivery devices. Invasomes promote drug permeability within the epidermal layers by eroding the stratum corneum's lipid packing [120]. Invasomes have a greater therapeutic effect and enhanced solubility as a medicinal carrier for phytopolyphenols. Scientists have created an invasive cream made of *Ocimum basilicum* to cure acne, and they have seen how well the medicine works by penetrating the epidermis [121].

15.3.1.8 Polymeric Nanocarriers

One of the most efficient and practical industrial methods for protecting and delivering phenolics is the application of polymeric nanoforms and natural NP-based carriers. The preparation, use, and characterization of polymeric-based nanocapsules and natural nanocarriers for phenolics have been conducted [122]. These include polymeric NPs, polymeric complex NPs, cyclodextrins, nanocaseins, nanocrystals, electrospun nanofibers, electrosprayed NPs, and nanosprayed dried particles [122]. NPs are submicron solid fragments that can be employed to nanoencapsulate substances that are bioactive. NPs, nanospheres, or nanocapsules can be obtained depending on the technique of preparation [123].

15.3.1.8.1 Nanocapsules

Systems that have a classic core–shell construction, where the medicine is contained within a liquid cavity on the inside and encased by a polymeric sheet or cavity on the outside [122, 124].

15.3.1.8.2 Nanospheres

Drugs can either be adsorbed on the surface of the NPs or trapped in the continuous polymeric network of solid matrix systems [83, 88].

15.3.1.8.3 Polymerosomes

Amphiphilic block copolymers self-assemble to create polymeric vesicles that have an aqueous inside and a polymer shell [103].

15.3.1.8.4 Dendrimers

Dendritic polymers have central cores, branching repeat units, and terminal groups in 3D, nanoscale, hyper-branched, well-defined, monodisperse topologies [56].

Nanodelivery systems are classified into two types: liquid and solid. The three types of fluid nanodelivery methods are nanoemulsions, nanoliposomes, and nanopolymersomes. Nanoemulsions are either emulsions or stabilized emulsions. There are three varieties of solid nanodelivery methods: LNPs, polymeric NPs, and nanocrystals [125].

15.3.2 Nanodelivery Systems

NP delivery systems are designed technologies that employ NPs to deliver therapies in a targeted and controlled manner. There are numerous effective drug delivery systems (Figure 15.8) that have been employed in recent years; nevertheless, there are still obstacles that must be addressed,

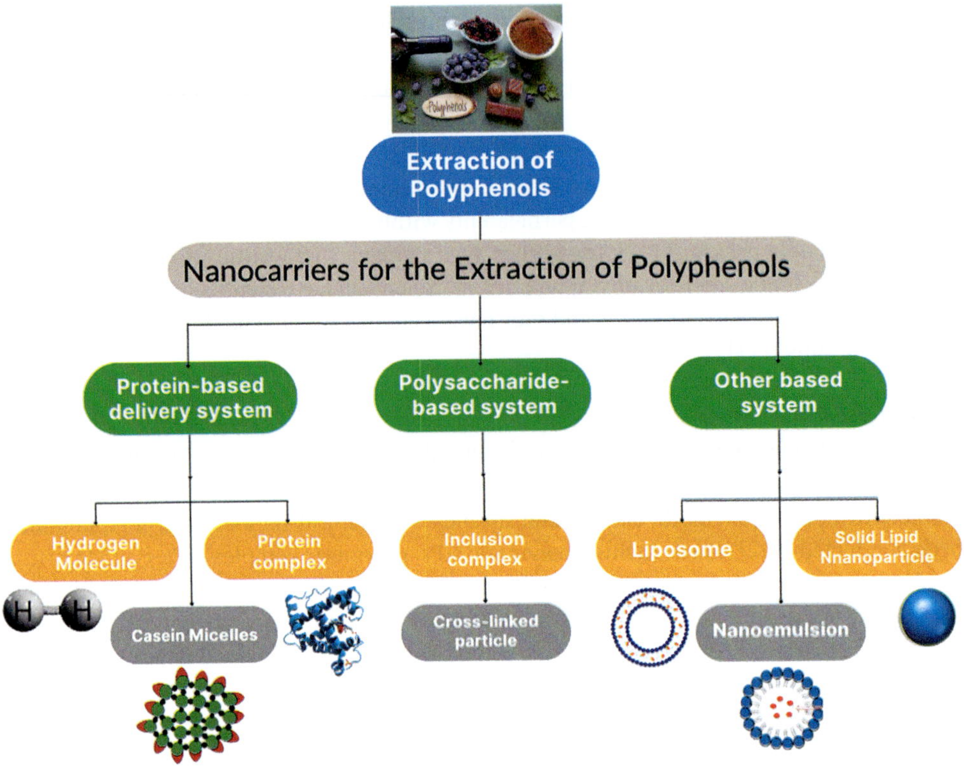

Figure 15.8 Process of extraction of polyphenols and their delivery systems. Credit: Danijela / Adobe Stock

and an improved technology must be created for successful drug delivery to its destinations. As a result, research into nano-based drug delivery systems is being conducted to improve drug delivery systems [125]. Richard Feynman first reported the idea of nanotechnology [126, 127].

There are two types of gene delivery strategies: viral and non-viral [128, 129]. There are potential side effects of existing gene delivery systems, resulting in the scope for further research on new gene delivery systems. Alternatively, non-viral gene delivery techniques have been developed using bio-polymers [129, 130]. It is widely recognized that NPs offer unique advantages for bioactive delivery, ranging from increased stability to regulated release, and targeting the bioactive for more substantial functioning [131]. Biologically active plant substances are widely recognized for their multiple medicinal properties; nevertheless, processing and storage instabilities restrict their bioavailability and bioaccessibility [132, 133]. Microfluidization has recently emerged as a transformative technique for creating methods of delivery with increased stability and accessibility of encapsulated plant compounds that are bioactive such as solid lipid nanocarriers, nanoemulsions, and liposomes.

The medical sector is intrigued by the possibility of nanodelivery systems created using microfluidization methods to regulate delivery with improved health benefits for the curing of numerous chronic illnesses. This study focuses on microfluidization-based nanodelivery technologies and their applications in the treatment of chronic illnesses [132, 134, 135]. Recently, microfluidization methods have been employed to develop nanodelivery systems with increased plant-based bioactive chemical resistance and bioavailability. Microfluidization is a high energy technique that operates on the dynamics of specifically engineered microchannels [132, 135]. Because of its capacity-constrained permeation enhancement, microneedle-mediated NPs have a tremendous potential and a wide range of applications [136, 137]. Nanodelivery/NP-based techniques have been developed to target tumor cells or the tumor microenvironment, and non-coding RNA-based medicines have the potential to be targeted therapies for cancer and other disorders [138–140]. Exosomes are a viable therapeutic nanodelivery platform because of their unique biological characteristics, stability, and composition. Delivering smart silencers in a secure and effective manner may be achieved by modifying synthetic NPs with exosome mimics [141–143]. The delivery systems are classified into four types:

1) Nanofabricated mode system
2) Carbohydrate mode system
3) Protein mode system
4) Lipid mode system

15.3.2.1 Nanofabricated Mode System

A common trend is the incorporation of functional food ingredients into food products to satisfy customer requests for a healthy lifestyle. Because of their low dissolution in water, unpleasant sensory nature, limited oral accessibility, and sensitivity to chemical degradation, these chemicals are detrimental to the food matrix. For food applications, the nanoencapsulation technology wraps bioactive substances in different nanofabricated delivery systems that are biocompatible and biodegradable [144–146].

15.3.2.2 Carbohydrate Mode System

Carbohydrate delivery systems feature outstanding characteristics such as biodegradability, abundance, and the ability to adapt to functioning [147]. Cyclodextrin is a carbohydrate-based delivery strategy that may be used to limit dietary bioactives that are less soluble, temperature sensitive, or chemically fragile. Nanofabrication processes are used to develop carbohydrate-mode delivery [147, 148].

15.3.2.3 Protein Mode System

Protein mode system techniques are popular because of their ability to incorporate both non-polar and polar bioactive substances. They can be produced via spray drying and electro-hydrodynamic techniques and they are mostly derived from plants, animals, and bacteria. Proteins from plants have received attention recently in nanotechnology to protect and control hydrophobic bioactive compounds, which has piqued the interest of the nutraceutical, food, and pharma sectors [149, 150]. Plant proteins are sustainable, eco-friendly, and energetic, which adds to their prospective function. To reduce the negative effects of employing raw material carriers, challenges must be overcome to enhance their technological efficiency and boundaries. Although this plant protein delivery method is limited and primarily used to transport lipophilic substances, further purification or extraction technologies are needed to evaluate other properties. Moreover, protein physicochemical molecular principles govern protein mode nanocarriers. Antisolvent precipitation, pH-driven gelation, and electrospray can all be used as preparation techniques [149–152].

15.3.2.4 Lipid Mode System

Lipid mode delivery systems, which were created for use in food applications, are the most efficient encapsulation techniques because they can include materials with different solubilities, enhance targeted delivery, and shield contents from free radicals [148, 153].

15.4 Polyphenol-based Nanodelivery

15.4.1 Nano-resveratrol

A polyphenolic compound called resveratrol is present in the skin of grapes and seeds, as well as in trace quantities in peanuts, plums, and apples. It defends herbal plants from microbes and fungi by functioning as a phytoalexin [121]. The redox properties of phenolic hydroxyl groups have a substantial impact on resveratrol's antioxidant ability. ROS can be eliminated by three of the OH groups found in resveratrol. Additionally, resveratrol activates the endogenous antioxidant system and the direct antioxidant pathway, which aids in cellular defense. The gene regulatory properties of these polyphenolic compounds can be connected to their antioxidant properties [154, 155]. Down regulating the expression of NADPH oxidase inhibited the formation of ROS. When resveratrol enhanced mitochondrial biogenesis, the tetrahydrobiopterin-synthesizing enzyme GTP cyclohydrolase-I was found to result in elevated antioxidant enzymatic levels and a decrease in mitochondrial superoxide [155].

Recently, a few nanoscale delivery systems with resveratrol have drawn interest because of possible medicinal uses. In the visceral organs of male Wistar rats, one study discovered that resveratrol-loaded nanocapsules had a two times higher bioavailability than free drug. When resveratrol is administered intravenously in its unprocessed state, the bioavailability of the medication was increased by six times using folic acid bound serum albumin-encapsulated resveratrol polymeric NPs [156, 157]. Resveratrol liposomes outperformed resveratrol alone in terms of solubility and chemical stability. Because resveratrol is tightly bound to lipid molecules, which enhances nano-resveratrol penetration as well as the sustained drug release effect, it has been observed that resveratrol SLNs could increase the penetration in keratocytes [99]. Additional research showed that the formation of tumor cells was more effectively suppressed in a mouse model of xenograft ovarian cancer when nano-resveratrol was combined with serum albumin. According to Shao et al., nano-resveratrol formulations based on methoxy-PEG-PCL showed more cytotoxic effects on tumor cells than the free drug because of the improved cellular uptake [158, 159].

15.4.2 Nano-curcumin

Curcumin, a polyphenolic molecule with medicinal qualities, has been used as a spice and nutritional supplement in Asian countries for a very long period. Rhizomes from the turmeric plant (*Curcuma longa*) are used to make curcumin. Notably, it has three structural components, two of which are phenolic groups and one of which is a diketone, all arranged in an aryl hydrocarbon skeleton. Curcumin has neurological effects in addition to anti-inflammatory and anticancer properties [111]. Among the ailments it is used to treat are age-related disorders, inflammation, atherosclerosis, oxidative stress, cardiovascular disease, type-2 diabetes, rheumatoid arthritis, and ocular problems.

Oral bioavailability of curcumin nanoformulations based on PLGA was 22 times greater than that of the raw state. In a cerebral ischemia rat model, curcumin-loaded SLNs exhibited a 16-fold increase in bioavailability [160]. Curcumin nanocapsules made with Eudragit RL100 polymer exhibit higher antioxidant capacity with reduced lipid peroxidation in dairy sheep milk. Researchers discovered that curcumin invasomes have more anti-inflammatory and antioxidant properties than other vesicular systems. Curcumin-loaded nanocomposites demonstrated better anti-inflammatory activity [124]. Studies have shown that curcumin has the strongest anti inflammatory effect of any polyphenol when delivered in an altered dosage form using a nanodelivery method [161]. The effectiveness of oral nano-curcumin formulation on individuals with intermediate COVID-19 was evaluated in open, non-randomized clinical studies. In a 14-day clinical trial, it was discovered that curcuminoids administered as nanomicelles significantly speed up the healing process for symptoms like myalgia, fever, and tachypnea. Nano-curcumin and omega-3 fatty acids cooperated to decrease the COX/iNOS mRNA gene in neuroinflammation, possibly providing migraine sufferers with symptomatic relief [124].

15.4.3 Nano-genistein

The flavonoid genistein is a potent antioxidant. Two enzymes, sodium oxide dismutase and catalase, are responsible for genistein's antioxidant properties. Numerous severe diseases, such as type-2 diabetes, osteoporosis, cancer, obesity, and neurodegeneration, have all been treated with it. However, because of its poor bioavailability, it has significant clinical limits, and some side effects, such as endocrine disruption and toxic consequences, have been recorded when used at larger doses [162]. These limitations have been successfully overcome by nanotechnology, with polymeric nanomicelles containing genistein exhibiting better plasma profiles and bioavailability than genistein alone [163]. When combined with lactoferrin, genistein NPs have also been demonstrated to significantly lower poly-comb protein expression and postpone the development of oral squamous cell carcinoma [164, 165].

15.4.4 EGCG-based Nanoforms

Green tea contains several important phytophenols, including (−)EGCG. Numerous pharmacological health benefits of EGCG have been reported, including functions as an antioxidant, tumor chemoprevention, improved heart health, weight loss, and defense against ionizing fallout [166, 167]. Nanoformulations are being studied for their potential to improve the stomach environment, systemic circulation transport capacity, and cancer targeting efficacy.

Nanoscale-based (lactide-co-glycolic acid) particles containing EGCGs were developed and studied for their anti-inflammatory capabilities because inflammation is a significant concern [166]. The anticancer and antiproliferative properties of EGCGs cause a marked arrest in the G1 phase of the cell cycle and the production of apoptosis. The G1 phase is greatly prolonged by the chemoprotective and antiproliferative effects of EGCG on cancer [166–168].

15.4.5 Nano-kaempferol

Kaempferol is an anti-inflammatory agent found in a variety of plants and fruits. It is used to treat a variety of conditions, including intervertebral disc degeneration, colitis, and fibroproliferative disorders. To improve patient compliance and effectiveness, the bioavailability, solubility, adverse drug reactions, and site-specific targeting must all be attained [115]. Poly(ethylene oxide)-poly(propylene oxide)-poly(ethylene oxide) NPs significantly decreased cancer cell viability because of their improved bioavailability compared with crude kaempferol [95]. When combined with paclitaxel, kaempferol-containing NLCs showed increased activity against MDA-MB 468 breast cancer cells. In A549 lung cancer cells, kaempferol gold nanoclusters with improved morphological properties and anticancer activity were investigated. Kaempferol NPs markedly decreased the levels of cardiac enzymes, vascular endothelial growth factor expression, oxidative stress, and increased heart tissue when compared with fluorouracil. These results were validated by histopathology studies [169]. A kaempferol nanomatrix was layer-by-layer tuned to increase the plasma and bone marrow content, boost the anabolic effect in osteopenic rats, and maximize bioavailability. These nanocarriers may increase the bioavailability of medications, accumulate in tumors, promote tumor cell uptake, combine therapeutic medicines with imaging methods, and enhance anticancer properties. Clinical studies have demonstrated the anti-inflammatory, type-2 diabetes, and cardiotonic benefits of the dietary supplement nano-kaempferol, making it a promising treatment for a range of diseases [96].

15.4.6 Naringenin-based Nanoforms

Citrus fruits, such as bergamot, tomatoes, and cherries, contain a substance called naringenin (NR). It has many pharmacokinetic properties, including antitumor, anti-inflammatory, and antioxidant properties [53, 56]. However, NR's therapeutic efficacy is limited by its strong hydrophobicity. NR can be added to nanodelivery vehicles, such as micelles, liposomes, SLN, nanosuspensions, and others, to overcome this issue. Sustained-release NR NPs with greater oral bioavailability, gastrointestinal tract absorption, and solubility were discovered to have increased anti-inflammatory effects in the Freund's adjuvant arthritis model [29, 41]. According to in vitro and in vivo experiments, NR-based Eudragit E100 Cationic Polymeric Nanoparticles have increased absorption and bioavailability by approximately 96%, which increased the anticancer potential by approx. 16%. A mechanistic method that paired NR with a PLGA doxorubicin nanoparticulate system demonstrated enhanced efficacy, a coactive effect, and decreased the toxicity. While an in vitro breast cancer study revealed more potent selective antitumor action, an in vivo tumor cell toxicity assay inhibited tumors in animal models [53, 56, 167]. Additionally, NR was effective in treating Parkinson's disease. After rats were administered an intranasal dose of an NR-vitamin E-loaded nanoemulsion, their behavioral activity returned to normal. The NR-loaded sulfobutylether–cyclodextrin/chitosan NPs were demonstrated to be a useful alternative for ocular administration of poorly soluble NR, with a sustained release and no irritating effects on the rabbit's eye. When Nile tilapia fish were exposed to NR NPs with an average size range of 165.1 nm, the oxidative stress created by cadmium was reduced [170], potentially by the increased antioxidant capacity and nano-NR bioaccumulation in liver and kidney cells. When NR is synthesized in the proper nanostructure, it can be used to treat a variety of ailments, including cancer, neurological disorders, liver diseases, ophthalmic disorders, inflammatory diseases, skin diseases, and diabetes. In a randomized, placebo-controlled clinical experiment, nano-NR had hepatoprotective effects in obese people, as well as secondary effects of decreased blood pressure and faster metabolism [170].

15.4.7 Apigenin-based Nanoform

Flavonoids are the most common type of polyphenol in plants, and apigenin (AG) is a highly potent bioactive substance. Using both conventional and nanodelivery methods, researchers have employed AG to treat conditions such as cancer, diabetes, Alzheimer's disease, dementia, and inflammatory illnesses [171]. The antidiabetic impact of AG-bilosomes was more effective than that of a basic AG dispersion, and an optimized formulation of AG bilosomes provided better release and penetration with a flux that was 4.49 times greater. A cancer cell with a high expression of CD44 receptors was the target of an AG nanoassembly with a high drug loading and entrapment efficacy [171, 172]. The formulation also offered an extended retention duration in the circulation and sustained release.

PLGA-loaded AG NPs were created, and their efficiency in preventing UV-induced skin cancer was assessed. By minimizing mitochondrial matrix edema brought on by greater carrier penetration in tissues, nano-AG showed strong anticancer potential. In rat studies, they also showed protective benefits against hepatocellular cancer [173]. Pharmacokinetics and biodistribution investigations have shown a considerable increase in AG in systemic circulation, showing the possibility for future patients with liver cancer. Nano-AG exhibits strong pharmacological activity against a variety of cancer types. SLN with a high AG content (80.44% drug encapsulation) and optimum particle size of approximately 161 nm was used to treat rheumatoid arthritis. In AG-loaded mucoadhesive SLN, a chitosan covering increased absorption and the antioxidant capacity [173]. Clinical experiments on nano-AG have revealed extraordinary effectiveness against several malignancies. Nanotechnology could be used to improve apigenin's solubility and bioavailability profile, which has shown promising outcomes against the growth of breast cancer cells [174].

15.4.8 Nano-theaflavins and Nano-thearubigins

The catechins in tea (*Camellia sinensis*) undergo enzymatic oxidation to produce theaflavins (TF) and thearubigins (TR), two naturally occurring polyphenolic substances. The environmentally friendly synthesis of nanoformulations uses these compounds [114, 115]. Tea leaves and green synthesis nanotechnology principles were used to create gold NPs, which have enhanced antioxidant and antibacterial properties. Silver NPs were also produced using green chemistry principles with TF and TE to increase their antibiotic-induced bactericidal activity against *Salmonella typhi* [16, 175].

To increase stability, absorption rate, intestinal epithelial cell targeting, and to stop TE and TF from being oxidized, chitosan-based NPs can be nanoencapsulated. Black tea leaf extract was used as a capping agent to stabilize the silver NPs made by electrolytic deposition using green nanotechnology [165]. A dose-dependent MTT experiment was conducted to examine their cancer resistant effectiveness against HeLa cervical carcinoma cells. Another work used tea extract that has potent antibacterial properties because of TE and TF to make stable gold and silver NPs. Polyelectrolyte-encapsulated 200 nm gelatin-based NPs were created using the layer-by-layer method [165].

Hepatocyte growth factor-induced breast cancer cells are strongly inhibited by polyphenols created from gelatinized NPs. Green nanotechnology was used to create gold NPs of tea polyphenols, which had significant anti-prostate and anti-breast cancer cell line action. Clinical studies have demonstrated the efficacy of TE and TF anti-inflammatory, antioxidant, anticancer, and anti-osteoporotic medications. Blood cholesterol levels were considerably reduced by green tea extracts [175, 176].

15.4.9 Quercetin Nanoforms

Quercetin (QT), a polyphenol with anti-inflammatory, anti-Alzheimer's, anti-arthritic, wound-healing, anti-ischemic, antihypertensive, antidiabetic, and antioxidant characteristics, has been optimized for use in a range of pharmaceutical applications [53, 56]. A nanotechnology-based formulation has shown considerable promise in the pharmaceutical area for improving numerous physicochemical and biological properties of QT. In conjunction with doxorubicin, nano-QT was used in the chemotherapeutic amelioration of apoptosis in cancer cell lines. An MTT assay was used to look for antiproliferative effects, while RT-PCR was used to look for gene targeting potential [110, 176]. A nano-QT was synthesized and subjected to characterization tests before being used in the quorum quenching of *Streptococcus* mutants using photodynamic therapy. This technique down regulated quorum-sensing system genes, eliminated microbial biofilm, and produced the greatest ROS. By enhancing biopharmaceutical properties and improving the cell membrane permeability, triphenylphosphonium-coated nano-QT was used to treat cerebral ischemia via ROS [106, 110]. By regulating mitochondrial delivery and significantly increasing QT absorption in the brain, oral treatment of nano-QT capsules lessened the severity of the histopathological changes. Nano-QT was better than QT in avoiding matrix metalloproteinase-9 and oxidative stress-induced gastric ulcers when it was examined for its capacity to halt mitochondrial damage in ethanol-induced gastric ulcer rat models [176].

The development of biopharmaceuticals is benefited by the enhanced QT bioavailability and carefully selected nanoformulation components of the nano-QT hydrogel. A clinical test using the nano-QT hydrogel on the skin wounds of 56 diabetic patients considerably sped up wound healing time as compared with a standard pharmacological treatment [177]. Intriguing pharmacological effects of nano-QT on humans have also been discovered in other clinical research.

15.5 Current Advances

Current research on polyphenols has revealed their antibacterial and antifungal properties. It is challenging to create a perfect, all-encompassing medicine because of their ubiquitous nature, complicated structures, high virulence, and intricate processes of infections [161]. Furthermore, these infections can avoid the negative effects of many treatments because of the development of drug resistance and genetic changes. Contrary to traditional drug formulations, polyphenols work through a variety of mechanisms by focusing on various cellular machinery and interfering with these microorganisms' main metabolic processes. Because of their variety of synergistic and immunomodulatory processes, polyphenols rank among the best nutritional supplements with the greatest ability to fight infections [178].

The systematic application of scientific knowledge to the operation and regulation of materials on the nanoscale is known as nanotechnology. Food nanotechnology is a cutting-edge, fascinating, and rapidly expanding topic with numerous applications in the food sector. It is connected to a wide range of fields [96]. Beneficial substances known as nutraceuticals are produced from nutrients, herbal products, dietary supplements, and genetically modified "designed foods". To enhance the delivery mechanism of natural bioactive compounds and nutraceuticals, these are nanofabricated. It not only increases efficacy and physicochemical stability, but also assures food quality [169].

Food biotechnology can use nanofabrication as a tactic to increase the effectiveness of biomolecules. Organic antioxidants with polyphenolic structures, such as curcumin and resveratrol, can prevent free radical damage, lessen lipid peroxidation, and inhibit DNA oxidation [179]. They may work synergistically or additively when combined. Many nanofood delivery approaches have combined curcumin and resveratrol to address their limited water solubility, bioavailability, and instability. There are significant challenges, including low entrapment efficiency, instability, a high rate of leakage, and a lack of safety. To increase the stability of food NPs, a new food hyalurasome was developed [51]. Hyaluronic acid (HA) is a polysaccharide polymer that occurs naturally and has antioxidant properties both in vitro and in vivo. Oligo-HA (oHA) is a low molecular weight HA with improved stability, bioavailability, and DPPH (2,2-diphenyl-1-picrylhydrazyl) radical scavenging activity compared wtih normal HA. It also possesses a variety of functional features, including immunostimulatory and antiangiogenic capabilities. Nutraceutical hyalurosome nano-food delivery systems (CRHs) were developed using advanced nanotechnology to increase the stability, bioavailability, and antioxidant activity of insoluble antioxidants [179, 180].

Bioactive compounds, which are dietary metabolites that prevent cancer, are present in fruits and vegetables. The stability and targeted delivery of biomolecules have improved with recent efforts to encapsulate bioactive components in nanodelivery systems [117, 169]. For their potential application in cancer therapy, a variety of nanodelivery techniques for bioactive substances, including polymeric NPs, SLNs, NLCs, liposomes, niosomes, and nanoemulsions, have been studied. In vivo models and cancer cell lines were used in recent human clinical studies and effectiveness analyses of the nanoformulations [131]. Because of the antiproliferative and pro-apoptotic characteristics of tumor cells, nanodelivery techniques were developed to increase the therapeutic effectiveness of bioactive compounds against a variety of cancers.

These substitutes were discovered and created to increase the effectiveness and security of new herbal treatments [133]. It has been demonstrated that polyphenolics, flavonoids, bioactive peptides, pigments, and essential fatty acids have medicinal or health benefits. Food science techniques, such as nanoencapsulation and nanofabricated delivery systems, enhance food quality and advance health [146]. Nanofabricated delivery methods based on lipids (solid and liquid), proteins, and carbohydrates are a few examples. Toxicology assessments need to be further investigated to guarantee the security of nanofabricated delivery systems, and advances in nanotechnology may play a crucial role in the creation of functional foods [148].

15.6 Challenges and Future Perspectives

Combining multiple administration methods can enhance therapeutic effectiveness. Despite significant improvements in human trials for gene delivery carriers, questions concerning how certain carriers are expected to target a particular nucleic acid to a particular specific cell type still persist [70]. Although CVnCoV two-dose vaccinations showed only 47% efficiency in preventing the disease, CureVac's CVnCoV mRNA LNP vaccine for COVID-19 was a potential option [19]. It employed a formulation similar to that of Pfizer and Moderna's successful vaccines. These results highlight the need to adapt the particle for the specific RNA sequence and the distinction between modified vs. unmodified mRNA payloads used by Pfizer, Moderna, and CureVac [128, 181]. Next generation gene delivery methods must consider material qualities, nucleic acid intracellular activity and alterations, and disease characteristics.

Artificial intelligence algorithms and state-of-the-art robotic high screening technologies are being developed to assess vast datasets of successful delivery vehicles. This primer focuses on several aspects of DNA-based delivery that incorporate NPs, emphasizing the most crucial characteristics that should be considered when developing delivery platforms and prospective production methods [70]. It looks at the analysis of the findings, explains the methods for characterizing the characteristics of nanomaterials, and helps infer possible biological impacts. The utilization of nucleic acid NPs in bioanalysis, nano-barcoding, gene silencing and editing, vaccines, and immunotherapy are only a few of their numerous significant applications. Data reproducibility and deposition are examined in relation to the field's limits and optimization [131]. It is essential to examine the toxicity of nano-coated materials and their various delivery mechanisms for bioactive substances and nutraceuticals. Although the use of nanofabricated materials in food packaging is expanding swiftly, there are still end-user regulatory and safety problems that need to be fully explored and addressed. There is no worldwide legal regulation in force, and many nations still lack governmental approval to assess the risk and safety of nanoencapsulated materials [128].

The full term for Steffen Foss Hansen's "React Now" technique is Registration, Evaluation, Authorization, Categorization, and Tools for Evaluating Nanomaterials Opportunities and

Weaknesses. To tackle food safety difficulties and successfully market nanofabricated programmable foods or nutraceuticals, organizations and businesses that work with nanofabricated materials must carefully analyze all of the aforementioned factors [181].

15.7 Conclusion

Secondary plant metabolites known as polyphenols have positive effects on human health and food preservation. Because of the expanding interest in and variety of biological functions of these products, the use of polyphenols as dietary supplements, antimicrobial medications, cosmetics, and natural food preservatives is a trend that promises to be successful in the market [4, 182, 183]. Polyphenols are secondary metabolites that have potential health benefits for humans as dietary sources of nutrition. When combined with nanotechnology-based drug delivery science, nutritional supplements, herbal medicines, and spices have the potential to boost biological function and overcome restrictions [154].

An encouraging development in the market is the use of polyphenols as alternatives to antibiotics, prescription medications, and natural food preservatives. However, barriers to moving these innovations to the industrial world persist [176]. The development of nanotechnology-based drug delivery systems has several problems, including obtaining multifunctional systems, scale-up methodologies, investigating targeting efficiency, meeting international criteria, and regulatory concerns for toxicity profiles and biocompatibility [184]. The supply chain cannot support the huge demand for polyphenols with the meager amount that is produced. As a result, extraction methods have been devised that enable production via extraction even when it comes from unconventional sources like organic waste. The development of novel methods, such as enzyme-assisted extraction, supercritical fluid, and high-voltage electric discharge, is necessary. The industrial uses are negatively impacted by the already low bioavailability of polyphenols as well as their interactions with other compounds. As a result, nanocarriers were used to increase the process efficiency [155].

NPs made from food macromolecules may improve the activity of polyphenols such as resveratrol, curcumin, and EGCG. They must endure the abrasive pH and environment created by digestive enzymes in the GI tract, retain the loaded polyphenols throughout oral administration, and then go to the small intestine, where the medicine is absorbed [83]. NP-based delivery systems can improve the bioavailability and stability of pharmaceuticals and bioactive substances through a variety of mechanisms. The method in which a substance interacts with the human body and its profile of absorption, distribution, metabolism, and excretion will depend on its physicochemical qualities, the behavior of the delivery system based on NPs, and morphological traits. The potential dangers of NPs to human health are unknown. Comprehensive research should be performed in this area [138].

References

1 Hano, C. and Tungmunnithum, D. (2020). *Plant polyphenols, more than just simple natural antioxidants: oxidative stress, aging and age-related diseases*. MDPI. 26.

2 Quideau, S., Deffieux, D., Douat-Casassus, C., and Pouységu, L. (2011). Plant polyphenols: chemical properties, biological activities, and synthesis. *Angewandte Chemie, International Edition in English* 50 (3): 586–621.

3 Graf, B.A., Milbury, P.E., and Blumberg, J.B. (2005). Flavonols, flavones, flavanones, and human health: epidemiological evidence. *Journal of Medicinal Food* 8 (3): 281–290.

4 Pandey, K.B. and Rizvi, S.I. (2009). Plant polyphenols as dietary antioxidants in human health and disease. *Oxidative Medicine and Cellular Longevity* 2: 270–278.

5 Zhang, Z., Li, X., Sang, S. et al. (2022). Polyphenols as plant-based nutraceuticals: health effects, encapsulation, nano-delivery, and application. *Foods* 11 (15): 2189.

6 Aziz, T., Ullah, A., Ali, A. et al. Manufactures of bio-degradable and bio-based polymers for bio-materials in the pharmaceutical field. *Journal of Applied Polymer Science* 139 (29): e52624.

7 Balla, E., Daniilidis, V., Karlioti, G. et al. Poly (lactic Acid): a versatile biobased polymer for the future with multifunctional properties—From monomer synthesis, polymerization techniques and molecular weight increase to PLA applications. *Polymers* 13 (11): 1822.

8 Ahmad, A., Gulraiz, Y., Ilyas, S., and Bashir, S. (2022). Polysaccharide based nano materials: health implications. *Food Hydrocolloids for Health* 100075.

9 Subramanian, P. (2021). Lipid-based nanocarrier system for the effective delivery of nutraceuticals. *Molecules* 26 (18): 5510.

10 Kalra, E.K. (2003). Nutraceutical-definition and introduction. *American Association of Pharmaceutical Scientists* 5 (3): 27–28.

11 Das, L., Bhaumik, E., Raychaudhuri, U., and Chakraborty, R. (2012). Role of nutraceuticals in human health. *Journal of Food Science and Technology* 49: 173–183.

12 Ajibola, A., Chamunorwa, J.P., and Erlwanger, K.H. (2012). Nutraceutical values of natural honey and its contribution to human health and wealth. *Nutrition & Metabolism* 9 (1): 1–12.

13 Arshad, R., Gulshad, L., Haq, I.U. et al. (2021). Nanotechnology: a novel tool to enhance the bioavailability of micronutrients. *Food Science & Nutrition* 9 (6): 3354–3361.

14 Fukushima, Y., Tashiro, T., Kumagai, A. et al. (2014). Coffee and beverages are the major contributors to polyphenol consumption from food and beverages in Japanese middle-aged women. *Journal of Nutritional Science* 3: e48.

15 Rahman, M.M., Rahaman, M.S., Islam, M.R. et al. (2021). Role of phenolic compounds in human disease: current knowledge and future prospects. *Molecules* 27 (1): 233.

16 Grosso, G., Stepaniak, U., Topor-Mądry, R. et al. (2014). Estimated dietary intake and major food sources of polyphenols in the polish arm of the HAPIEE study. *Nutrition* 30 (11–12): 1398–1403.

17 Tsao, R. (2010). Chemistry and biochemistry of dietary polyphenols. *Nutrients* 2 (12): 1231–1246.

18 Cheynier, V. (2005). Polyphenols in foods are more complex than often thought. *The American Journal of Clinical Nutrition* 81 (1): 223S–229S.

19 Rathod, N.B., Elabed, N., Punia, S. et al. (2023). Recent developments in polyphenol applications on human health: a review with current knowledge. *Plants* 12 (6): 1217.

20 Tsimogiannis, D. and Oreopoulou, V. (2019). Classification of phenolic compounds in plants. In: *Polyphenols in plants*, 263–284. Academic Press.

21 Vuolo, M., Lima, V., and Maróstica, M. (2019). *Chapter 2-Phenolic Compounds: Structure, Classification, and Antioxidant Power, Editor (S): Maira Rubi Segura Campos*. 33–50. Bioactive Compounds, Woodhead Publishing.

22 Abbas, M., Saeed, F., Anjum, F.M. et al. (2017). Natural polyphenols: an overview. *International Journal of Food Properties* 20 (8): 1689–1699.

23 Kumar, N. and Goel, N. (2019). Phenolic acids: natural versatile molecules with promising therapeutic applications. *Biotechnology Reports* 24: e00370.

24 Ziegler, V., Ferreira, C.D., Hoffmann, J.F. et al. (2018). Cooking quality properties and free and bound phenolics content of brown, black, and red rice grains stored at different temperatures for six months. *Food Chemistry* 242: 427–434.

25 Dias, R., Pereira, C.B., Pérez-Gregorio, R. et al. (2021). Recent advances on dietary polyphenol's potential roles in celiac disease. *Trends in Food Science & Technology* 107: 213–225.

26 Kanwal, Q., Hussain, I., Siddiqui, L.H., and Javaid, A. (2011). Antimicrobial activity screening of isolated flavonoids from Azadirachta indica leaves. *Journal of the Serbian Chemical Society* 76 (3): 375–384.

27 Wang, T.-Y., Li, Q., and Bi, K.-S. (2018). Bioactive flavonoids in medicinal plants: structure, activity and biological fate. *Asian Journal of Pharmaceutical Sciences* 13 (1): 12–23.

28 Dai, J. and Mumper, R.J. (2010). Plant phenolics: extraction, analysis and their antioxidant and anticancer properties. *Molecules* 15 (10): 7313–7352.

29 Kanwal, Q., Hussain, I., Siddiqui, L.H., and Javaid, A. (2011). Polyphenols, dietary sources and bioavailability. *Annali-IstitutoSuperiore di Sanita* 43 (4): 348.

30 Cuyckens, F. and Claeys, M. (2004). Mass spectrometry in the structural analysis of flavonoids. *Journal of Mass Spectrometry* 39 (1): 1–15.

31 He, X.-G. (2000). On-line identification of phytochemical constituents in botanical extracts by combined high-performance liquid chromatographic–diode array detection–mass spectrometric techniques. *Journal of Chromatography A* 880 (1–2): 203–232.

32 Stobiecki, M., Malosse, C., Kerhoas, L. et al. (1999). Detection of isoflavonoids and their glycosides by liquid chromatography/electrospray ionization mass spectrometry in root extracts of lupin (Lupinus albus). phytochemical analysis. *An International Journal of Plant Chemical and Biochemical Techniques* 10 (4): 198–207.

33 Jiang, N., Doseff, A.I., and Grotewold, E. (2016). Flavones: from biosynthesis to health benefits. *Plants* 5 (2): 27.

34 Pojer, E., Mattivi, F., Johnson, D., and Stockley, C.S. (2013). The case for anthocyanin consumption to promote human health: a review. *Comprehensive Reviews in Food Science and Food Safety* 12 (5): 483–508.

35 de Gaulejac, N.S.-C., Glories, Y., and Vivas, N. (1999). Free radical scavenging effect of anthocyanins in red wines. *Food Research International* 32 (5): 327–333.

36 Corradini, E., Foglia, P., Giansanti, P. et al. (2011). Flavonoids: chemical properties and analytical methodologies of identification and quantitation in foods and plants. *Natural Product Research* 25 (5): 469–495.

37 Lao, F. and Giusti, M.M. (2018). Extraction of purple corn (Zea mays L.) cob pigments and phenolic compounds using food-friendly solvents. *Journal of Cereal Science* 80: 87–93.

38 Khoo, H.E., Azlan, A., Tang, S.T., and Lim, S.M. (2017). Anthocyanidins and anthocyanins: colored pigments as food, pharmaceutical ingredients, and the potential health benefits. *Food & Nutrition Research* 61 (1): 1361779.

39 Khan, M.K. and Dangles, O. (2014). A comprehensive review on flavanones, the major citrus polyphenols. *Journal of Food Composition and Analysis* 33 (1): 85–104.

40 Panche, A.N., Diwan, A.D., and Chandra, S.R. (2016). Flavonoids: an overview. *Journal of Nutritional Science* 5: e47.

41 Hsiao, Y.-H., Ho, C.-T., and Pan, M.-H. (2020). Bioavailability and health benefits of major isoflavone aglycones and their metabolites. *Journal of Functional Foods* 74: 104164.

42 Alara, O.R., Abdurahman, N.H., and Ukaegbu, C.I. (2021). Extraction of phenolic compounds: a review. *Current Research in Food Science* 4: 200–214.

43 Bolca, S. (2014). Bioavailability of soy-derived isoflavones and human breast cancer. In: *Polyphenols in human health and disease*, 1241–1256. Academic Press.

44 Malla, A. and Ramalingam, S. (2018). Health perspectives of an isoflavonoid genistein and its quantification in economically important plants. In: *Role of Materials Science in Food Bioengineering*, 353–379. Academic Press.

45 Robards, K., Li, X., Antolovich, M., and Boyd, S. (1997). Characterisation of citrus by chromatographic analysis of flavonoids. *Journal of the Science of Food and Agriculture* 75 (1): 87–101.

46 Hollman, P.C.H. and Arts, I.C.W. (2000). Flavonols, flavones and flavanols–nature, occurrence and dietary burden. *Journal of the Science of Food and Agriculture* 80 (7): 1081–1093.

47 Singla, R.K., Dubey, A.K., Garg, A. et al. (2019). *Natural Polyphenols: Chemical Classification, Definition of Classes, Subcategories, and Structures*. 1397–1400. Oxford University Press.

48 Han, X., Shen, T., and Lou, H. (2007). Dietary polyphenols and their biological significance. *International Journal of Molecular Sciences* 8 (9): 950–988.

49 Pereira, D.M., Valentão, P., Pereira, J.A., and Andrade, P.B. (2009). *Phenolics: From chemistry to biology*. Molecular Diversity Preservation International. 2202–2211.

50 Koeberle, A. and Werz, O. (2014). Multi-target approach for natural products in inflammation. *Drug Discovery Today* 19 (12): 1871–1882.

51 Charoensiddhi, S., Abraham, R.E., Su, P., and Zhang, W. (2020). Seaweed and seaweed-derived metabolites as prebiotics. *Advances in Food and Nutrition Research* 91: 97–156.

52 Qin, Y. (2018). *Applications of Bioactive Seaweed Substances in Functional Food Products, in Bioactive Seaweeds for Food Applications*. 111–134. Elsevier.

53 Yan, Z., Zhong, Y., Duan, Y. et al. (2020). Antioxidant mechanism of tea polyphenols and its impact on health benefits. *Animal Nutrition* 6 (2): 115–123.

54 Inanli, A.G., Tümerkan, E.T.A., El Abed, N. et al. (2020). The impact of chitosan on seafood quality and human health: a review. *Trends in Food Science & Technology* 97: 404–416.

55 Kumarappan, C., Thilagam, E., and Mandal, S.C. (2012). Antioxidant activity of polyphenolic extracts of Ichnocarpus frutescens. *Saudi Journal of Biological Sciences* 19 (3): 349–355.

56 Lee, C.Y., Nanah, C.N., Held, R.A. et al. (2015). Effect of electron donating groups on polyphenol-based antioxidant dendrimers. *Biochimie* 111: 125–134.

57 Mihanfar, A., Nouri, M., Roshangar, L., and Khadem-Ansari, M.H. (2021). Polyphenols: natural compounds with promising potential in treating polycystic ovary syndrome. *Reproductive Biology* 21 (2): 100500.

58 Rathod, N.B., Ranveer, R.C., Benjakul, S. et al. (2021). Recent developments of natural antimicrobials and antioxidants on fish and fishery food products. *Comprehensive Reviews in Food Science and Food Safety* 20 (4): 4182–4210.

59 Saeed, N., Khan, M.R., and Shabbir, M. (2012). Antioxidant activity, total phenolic and total flavonoid contents of whole plant extracts Torilisleptophylla L. *BMC Complementary and Alternative Medicine* 12: 1–12.

60 Li, Q., Li, J., Duan, M. et al. (2022). Impact of food additive titanium dioxide on the polyphenol content and antioxidant activity of the apple juice. *LWT* 154: 112574.

61 Martins, N., Barros, L., and Ferreira, I.C. (2016). In vivo antioxidant activity of phenolic compounds: facts and gaps. *Trends in Food Science & Technology* 48: 1–12.

62 Emami, S., Azadmard-Damirchi, S., Peighambardoust, S.H. et al. (2016). Liposomes as carrier vehicles for functional compounds in food sector. *Journal of Experimental Nanoscience* 11 (9): 737–759.

63 Bensid, A., El Abed, N., Houicher, A. et al. (2022). Antioxidant and antimicrobial preservatives: properties, mechanism of action and applications in food–a review. *Critical Reviews in Food Science and Nutrition* 62 (11): 2985–3001.

64 Hügel, H.M., Jackson, N., May, B. et al. (2016). Polyphenol protection and treatment of hypertension. *Phytomedicine* 23 (2): 220–231.

65 Micucci, M., Bolchi, C., Budriesi, R. et al. (2020). Antihypertensive phytocomplexes of proven efficacy and well-established use: mode of action and individual characterization of the active constituents. *Phytochemistry* 170: 112222.

66 Cao, Y., Xie, L., Liu, K. et al. (2021). The antihypertensive potential of flavonoids from Chinese herbal medicine: a review. *Pharmacological Research* 174: 105919.

67 Lamb, S.A., Al Hamarneh, Y.N., Houle, S.K. et al. (2018). Hypertension Canada's 2017 guidelines for diagnosis, risk assessment, prevention and treatment of hypertension in adults for pharmacists: an update. *Canadian Pharmacists Journal/Revue Des Pharmaciens du Canada* 151 (1): 33–42.

68 Yamagata, K., Tagami, M., and Yamori, Y. (2015). Dietary polyphenols regulate endothelial function and prevent cardiovascular disease. *Nutrition* 31 (1): 28–37.

69 Oak, M.H., Auger, C., Belcastro, E. et al. (2018). Potential mechanisms underlying cardiovascular protection by polyphenols: role of the endothelium. *Free Radical Biology and Medicine* 122: 161–170.

70 Sanches-Silva, A., Testai, L., Nabavi, S.F. et al. (2020). Therapeutic potential of polyphenols in cardiovascular diseases: regulation of mTOR signaling pathway. *Pharmacological Research* 152: 104626.

71 An, P., Wan, S., Luo, Y. et al. (2022). Micronutrient supplementation to reduce cardiovascular risk. *Journal of the American College of Cardiology* 80 (24): 2269–2285.

72 Delgado, N.T., Rouver, W.N., and Dos Santos, R.L. (2020). Protective effects of pomegranate in endothelial dysfunction. *Current Pharmaceutical Design* 26 (30): 3684–3699.

73 Tropea, T., Renshall, L.J., Nihlen, C. et al. (2020). Beetroot juice lowers blood pressure and improves endothelial function in pregnant eNOS−/− mice: importance of nitrate-independent effects. *The Journal of Physiology* 598 (18): 4079–4092.

74 Valero, M.S., Nuñez, S., and Les, F. et al. (2022). The potential role of everlasting flower (Helichrysum stoechasMoench) as an antihypertensive agent: vasorelaxant effects in the rat aorta. *Antioxidants* 11 (6): 1092.

75 Chan, S.-L., Capdeville-Atkinson, C., and Atkinson, J. (2008). Red wine polyphenols improve endothelium-dependent dilation in rat cerebral arterioles. *Journal of Cardiovascular Pharmacology* 51 (6): 553–558.

76 López-Sepúlveda, R., Jiménez, R., Romero, M. et al. (2008). Wine polyphenols improve endothelial function in large vessels of female spontaneously hypertensive rats. *Hypertension* 51 (4): 1088–1095.

77 Botden, I.P., Langendonk, J.G., Meima, M.E. (2011). Daily red wine consumption improves vascular function by a soluble guanylyl cyclase-dependent pathway. *American Journal of Hypertension* 24 (2): 162–168.

78 Hyldgaard, M. (2020). Mechanisms of action, resistance, and stress adaptation. In: *Antimicrobials in Food* 735–784. CRC Press.

79 Othman, L., Sleiman, A., and Abdel-Massih, R.M. (2019). Antimicrobial activity of polyphenols and alkaloids in middle eastern plants. *Frontiers in Microbiology* 10: 911.

80 Zamuz, S., Munekata, P.E., Dzuvor, C.K. et al. (2021). The role of phenolic compounds against Listeria monocytogenes in food. *A review. Trends in Food Science & Technology* 110: 385–392.

81 López-Malo, A., Alzamora, S.M., Paris, M.J. et al. (2020). Naturally occurring compounds–Plant sources. *Antimicrobials in food* 527–594.

82 Freitas, I.R. and Cattelan, M.G. (2018). Antimicrobial and antioxidant properties of essential oils in food systems—An overview. *Microbial contamination and food degradation* 443–470.

83 Saini, A., Panwar, D., Panesar, P.S., and Bera, M.B. (2021). Encapsulation of functional ingredients in lipidic nanocarriers and antimicrobial applications: a review. *Environmental Chemistry Letters* 19: 1107–1134.

84 Ferrentino, G., Haman, N., Morozova, K. et al. (2021). Phenolic compounds extracted from spruce (Picea abies) by supercritical carbon dioxide as antimicrobial agents against gram-positive bacteria assessed by isothermal calorimetry. *Journal of Thermal Analysis and Calorimetry* 145: 3093–3103.

85 Dzah, C.S., Duan, Y., Zhang, H. et al. (2020). The effects of ultrasound assisted extraction on yield, antioxidant, anticancer and antimicrobial activity of polyphenol extracts: a review. *Food Bioscience* 35: 100547.

86 Mitani, T., Ota, K., Inaba, N. et al. (2018). Antimicrobial activity of the phenolic compounds of Prunus mume against Enterobacteria. *Biological and Pharmaceutical Bulletin* 41 (2): 208–212.

87 De Angelis, M., Della-Morte, D., Buttinelli, G. et al. (2021). Protective role of combined polyphenols and micronutrients against influenza A virus and SARS-CoV-2 infection in vitro. *Biomedicines* 9 (11): 1721.

88 Koch, W. (2019). Dietary polyphenols—important non-nutrients in the prevention of chronic noncommunicable diseases. *A Systematic Review. Nutrients* 11 (5): 1039.

89 Abenavoli, L., Larussa, T., Corea, A. et al. (2021). Dietary polyphenols and non-alcoholic fatty liver disease. *Nutrients* 13 (2): 494.

90 Margină, D., Ungurianu, A., Purdel, C. et al. (2020). Analysis of the intricate effects of polyunsaturated fatty acids and polyphenols on inflammatory pathways in health and disease. *Food and Chemical Toxicology* 143: 111558.

91 Rudrapal, M., Khairnar, S.J., Khan, J. et al. (2022). Dietary polyphenols and their role in oxidative stress-induced human diseases: Insights into protective effects, antioxidant potentials and mechanism (s) of action. *Frontiers in Pharmacology* 13: 283.

92 Li, Y., Jiang, X., Hao, J. et al. (2019). Tea polyphenols: application in the control of oral microorganism infectious diseases. *Archives of Oral Biology* 102: 74–82.

93 Montenegro-Landívar, M.F., Tapia-Quirós, P., Vecino, X. et al. (2021). Polyphenols and their potential role to fight viral diseases: an overview. *Science of the Total Environment* 801: 149719.

94 D'Archivio, M., Filesi, C., Varì, R. et al. (2010). Bioavailability of the polyphenols: status and controversies. *International Journal of Molecular Sciences* 11 (4). 1321–1342.

95 Faisal, W., Soliman, G.M., and Hamdan, A.M. (2018). Enhanced skin deposition and delivery of voriconazole using ethosomal preparations. *Journal of Liposome Research* 28 (1): 14–21.

96 D'Archivio, M., Filesi, C., Varì, R. et al. (2010). Bioavailability of the polyphenols: status and controversies. *International Journal of Molecular Sciences* 11 (4): 1321–1342.

97 Carlotti, M.E., Sapino, S., Ugazio, E. et al. (2012). Resveratrol in solid lipid nanoparticles. *Journal of Dispersion Science and Technology* 33 (4): 465–471.

98 Desai, J. and Thakkar, H. (2016). Effect of particle size on oral bioavailability of darunavir-loaded solid lipid nanoparticles. *Journal of Microencapsulation* 33 (7): 669–678.

99 Teskač, K. and Kristl, J. (2010). The evidence for solid lipid nanoparticles mediated cell uptake of resveratrol. *International Journal of Pharmaceutics* 390 (1): 61–69.

100 Ishkeh, S.R., Shirzad, H., Asghari, M. et al. (2021). Effect of chitosan nanoemulsion on enhancing the phytochemical contents, health-promoting components, and shelf life of raspberry (Rubus sanctus Schreber). *Applied Sciences* 11 (5): 2224.

101 Zhang, J., Huang, Y., Liu, D. et al. (2013). Preparation of apigenin nanocrystals using supercritical antisolvent process for dissolution and bioavailability enhancement. *European Journal of Pharmaceutical Sciences* 48 (4–5): 740–747.

102 Hashemi, M., Shamshiri, A., Saeedi, M. et al. (2020). Aptamer-conjugated PLGA nanoparticles for delivery and imaging of cancer therapeutic drugs. *Archives of Biochemistry and Biophysics* 691: 108485.

103 Rudrapal, M., Vallinayagam, S., Zothantluanga, J.H. et al. (2022). Nanophytomedicines: nature to medicines. In: *Applications of Nanotechnology in Drug Discovery and Delivery*, 71–93. Elsevier.

104 Khan, J., Rudrapal, M., Bhat, E.A. et al. (2021). Perspective insights to bio-nanomaterials for the treatment of neurological disorders. *Frontiers in Bioengineering and Biotechnology* 9: 724158.

105 Karn, P.R., Cho, W., and Hwang, S.-J. (2013). Liposomal drug products and recent advances in the synthesis of supercritical fluid-mediated liposomes. *Nanomedicine* 8 (9): 1529–1548.

106 Huang, Y.B., Tsai, M.J., Wu, P.C. et al. (2011). Elastic liposomes as carriers for oral delivery and the brain distribution of (+)-catechin. *Journal of Drug Targeting* 19 (8): 709–718.

107 Guimarães, D., Cavaco-Paulo, A., and Nogueira, E. (2021). Design of liposomes as drug delivery system for therapeutic applications. *International Journal of Pharmaceutics* 601: 120571.

108 El-Samaligy, M.S., Afifi, N.N., and Mahmoud, E.A. (2006). Evaluation of hybrid liposomes-encapsulated silymarin regarding physical stability and in vivo performance. *International Journal of Pharmaceutics* 319 (1–2): 121–129.

109 Coimbra, M., Isacchi, B., van Bloois, L. et al. (2011). Improving solubility and chemical stability of natural compounds for medicinal use by incorporation into liposomes. *International Journal of Pharmaceutics* 416 (2): 433–442.

110 Yuan, Z.P., Chen, L.J., Fan, L.Y. et al. (2006). Liposomal quercetin efficiently suppresses growth of solid tumors in murine models. *Clinical Cancer Research* 12 (10): 3193–3199.

111 Tønnesen, H.H., Smistad, G., Ågren, T., and Karlsen, J. (1993). Studies on curcumin and curcuminoids. XXIII: effects of curcumin on liposomal lipid peroxidation. *International Journal of Pharmaceutics* 90 (3): 221–228.

112 Paliwal, S., Tilak, A., Sharma, J. et al. (2019). Flurbiprofen loaded ethosomes-transdermal delivery of anti-inflammatory effect in rat model. *Lipids in Health and Disease* 18 (1): 1–15.

113 Verma, P. and Pathak, K. (2010). Therapeutic and cosmeceutical potential of ethosomes: an overview. *Journal of Advanced Pharmaceutical Technology & Research* 1 (3): 274.

114 Godin, B., Touitou, E., Rubinstein, E. et al. (2005). A new approach for treatment of deep skin infections by an ethosomal antibiotic preparation: an in vivo study. *Journal of Antimicrobial chemotherapy* 55 (6): 989–994.

115 El-Kayal, M., Nasr, M., Elkheshen, S., and Mortada, N. (2019). Colloidal (-)-epigallocatechin-3-gallate vesicular systems for prevention and treatment of skin cancer: a comprehensive experimental study with preclinical investigation. *European Journal of Pharmaceutical Sciences* 137: 104972.

116 Saneja, A., Panda, A.K., and Lichtfouse, E. (eds.) (2020). *Sustainable Agriculture Reviews 43: Pharmaceutical Technology for Natural Products Delivery Vol. 1 Fundamentals and Applications (Vol. 43). Springer Nature.*

117 Bijak, M. (2017). Silybin, a major bioactive component of milk thistle (Silybum marianum L. Gaernt.)—Chemistry, bioavailability, and metabolism. *Molecules* 22 (11): 1942.

118 Das, M.K. and Kalita, B. (2014). Design and evaluation of phyto-phospholipid complexes (phytosomes) of rutin for transdermal application. *Journal of Applied Pharmaceutical Science* 4 (10): 051–057.

119 Amin, T. and Bhat, S.V. (2012). A review on phytosome technology as a novel approach to improve the bioavailability of nutraceuticals. *International Journal of Advanced Research in Engineering and Technology* 1 (3): 43–57.

120 Nangare, S. and Dugam, S. (2020). Smart invasome synthesis, characterizations, pharmaceutical applications, and pharmacokinetic perspective: a review. *Future Journal of Pharmaceutical Sciences* 6 (1): 1–21.

121 Celik, I., Rudrapal, M., Yadalam, P.K. et al. (2022). Resveratrol and its natural analogues inhibit RNA dependant RNA polymerase (RdRp) of Rhizopus oryzae in mucormycosis through computational investigations. *Polycyclic Aromatic Compounds* 1–18.

122 Nagaich, U. (2018). Polymeric nanocapsules: an emerging drug delivery system. *Journal of Advanced Pharmaceutical Technology & Research* 9 (3): 65.

123 Sekhon, B.S. (2010). Food nanotechnology–an overview. *Nanotechnology, Science and Applications* 1–15.

124 Zanotto-Filho, A., Coradini, K., Braganhol, E. (2013). Curcumin-loaded lipid-core nanocapsules as a strategy to improve pharmacological efficacy of curcumin in glioma treatment. *European Journal of Pharmaceutics and Biopharmaceutics* 83 (2): 156–167.

125 Patra, J.K., Das, G., Fraceto, L.F. et al. (2018). Nano based drug delivery systems: recent developments and future prospects. *Journal of Nanobiotechnology* 16 (1): 1–33.

126 Lawal, S.K., Olojede, S.O., Faborode, O.S. et al. (2022). Nanodelivery of antiretroviral drugs to nervous tissues. *Frontiers in Pharmacology* 13: 1025160.

127 Bayda, S., Adeel, M., Tuccinardi, T. et al. (2019). The history of nanoscience and nanotechnology: from chemical–physical applications to nanomedicine. *Molecules* 25 (1): 112.

128 Mendes, B.B., Conniot, J., Avital, A. et al. (2022). Nanodelivery of nucleic acids. *Nature Reviews Methods Primers* 2 (1): 24.

129 Nayerossadat, N., Maedeh, T., and Ali, P.A. (2012). Viral and nonviral delivery systems for gene delivery. *Advanced Biomedical Research* 1.

130 Ramamoorth, M. and Narvekar, A. (2015). Non viral vectors in gene therapy-an overview. *Journal of Clinical and Diagnostic Research: JCDR* 9 (1): GE01.

131 Borel, T. and Sabliov, C. (2014). Nanodelivery of bioactive components for food applications: types of delivery systems, properties, and their effect on ADME profiles and toxicity of nanoparticles. *Annual Review of Food Science and Technology* 5: 197–213.

132 Ganesan, P., Karthivashan, G., Park, S.Y. et al. (2018). Microfluidization trends in the development of nanodelivery systems and applications in chronic disease treatments. *International Journal of Nanomedicine* 6109–6121.

133 Rein, M.J., Renouf, M., Cruz-Hernandez, C. et al. (2013). Bioavailability of bioactive food compounds: A challenging journey to bioefficacy. *British Journal of Clinical Pharmacology* 75 (3): 588–602.

134 Pingale, P., Kendre, P., Pardeshi, K., and Rajput, A. (2023). An emerging era in manufacturing of drug delivery systems: nanofabrication techniques. *Heliyon* 9 (3).

135 Ozturk, O.K. and Turasan, H. (2021). Applications of microfluidization in emulsion-based systems, nanoparticle formation, and beverages. *Trends in Food Science & Technology* 116: 609–625.

136 Jiang, X., Zhao, H., and Li, W. (2022). Microneedle-mediated transdermal delivery of drug-carrying nanoparticles. *Frontiers in Bioengineering and Biotechnology* 10: 121.

137 Ruan, S., Zhang, Y., and Feng, N. (2021). Microneedle-mediated transdermal nanodelivery systems: a review. *Biomaterials Science* 9 (24): 8065–8089.

138 Sabit, H., Abdel-Hakeem, M., Shoala, T. et al. (2022). Nanocarriers: a reliable tool for the delivery of anticancer drugs. *Pharmaceutics* 14 (8): 1566.

139 Jabir, N.R., Tabrez, S., Ashraf, G.M. et al. (2012). Nanotechnology-based approaches in anticancer research. *International Journal of Nanomedicine* 4391–4408.

140 Kara, G., Calin, G.A., and Ozpolat, B. (2022). RNAi-based therapeutics and tumor targeted delivery in cancer. *Advanced Drug Delivery Reviews* 182: 114113.

141 Kim, J., Li, S., Zhang, S., and Wang, J. (2022). Plant-derived exosome-like nanoparticles and their therapeutic activities. *Asian Journal of Pharmaceutical Sciences* 17 (1): 53–69.

142 Pei, W., Li, X., Bi, R. et al. (2021). Exosome membrane-modified M2 macrophages targeted nanomedicine: treatment for allergic asthma. *Journal of Controlled Release* 338: 253–267.

143 Sen, S., Xavier, J., Kumar, N. et al. (2023). Exosomes as natural nanocarrier-based drug delivery system: recent insights and future perspectives. *3 Biotech* 13 (3): 101.

144 El Khoury, D., Cuda, C., Luhovyy, B.L., and Anderson, G.H. (2012). Beta glucan: health benefits in obesity and metabolic syndrome. *Journal of Nutrition and Metabolism* 2012.

145 Galanakis, C.M. (2021). Functionality of food components and emerging technologies. *Foods* 10 (1): 128.

146 Lordan, S., Ross, R.P., and Stanton, C. (2011). Marine bioactives as functional food ingredients: potential to reduce the incidence of chronic diseases. *Marine Drugs* 9 (6): 1056–1100.

147 Qi, X., Tong, X., Pan, W. et al. (2021). Recent advances in polysaccharide-based adsorbents for wastewater treatment. *Journal of Cleaner Production* 315: 128221.

148 Puttasiddaiah, R., Lakshminarayana, R., Somashekar, N.L. et al. (2022). Advances in nanofabrication technology for nutraceuticals: new insights and future trends. *Bioengineering* 9 (9): 478.

149 Fathi, M., Donsi, F., and McClements, D.J. (2018). Protein-based delivery systems for the nanoencapsulation of food ingredients. *Comprehensive Reviews in Food Science and Food Safety* 17 (4): 920–936.

150 Gomes, A. and Sobral, P.J.D.A. (2021). Plant protein-based delivery systems: an emerging approach for increasing the efficacy of lipophilic bioactive compounds. *Molecules* 27 (1): 60.

151 Detsi, A., Kavetsou, E., Kostopoulou, I. et al. (2020). Nanosystems for the encapsulation of natural products: the case of chitosan biopolymer as a matrix. *Pharmaceutics* 12 (7): 669.

152 Malekzad, H., Mirshekari, H., Sahandi Zangabad, P. et al. (2018). Plant protein-based hydrophobic fine and ultrafine carrier particles in drug delivery systems. *Critical Reviews in Biotechnology* 38 (1): 47–67.

153 Mozafari, M.R., Flanagan, J., Matia-Merino, L., et al. (2006). Recent trends in the lipid-based nanoencapsulation of antioxidants and their role in foods. *Journal of the Science of Food and Agriculture* 86 (13): 2038–2045.

154 Hendawy, O.M. (2021). Nano-delivery systems for improving therapeutic efficiency of dietary polyphenols. *Alternative Therapies in Health & Medicine* 27.

155 Mishra, Y., Amin, H.I.M., Mishra, V. et al. (2022). Application of nanotechnology to herbal antioxidants as improved phytomedicine: an expanding horizon. *Biomedicine & Pharmacotherapy* 153: 113413.

156 Detoni, C.B., Souto, G.D., da Silva, A.L.M. et al. (2012). Photostability and skin penetration of different e-resveratrol-loaded supramolecular structures. *Photochemistry and Photobiology* 88 (4): 913–921.

157 Sanna, V., Siddiqui, I.A., Sechi, M., & Mukhtar, H. (2013). Resveratrol-loaded nanoparticles based on poly (epsilon-caprolactone) and poly (d, l-lactic-co-glycolic acid)–poly (ethylene glycol) blend for prostate cancer treatment. *Molecular Pharmaceutics* 10 (10): 3871–3881.

158 Shao, J., Li, X., Lu, X. et al. (2009). Enhanced growth inhibition effect of resveratrol incorporated into biodegradable nanoparticles against glioma cells is mediated by the induction of intracellular reactive oxygen species levels. *Colloids and Surfaces B: Biointerfaces* 72 (1): 40–47.

159 Guo, L., Peng, Y., Yao, J. et al. (2010). Anticancer activity and molecular mechanism of resveratrol–Bovine serum albumin nanoparticles on subcutaneously implanted human primary ovarian carcinoma cells in Nude mice. *Cancer Biotherapy & Radiopharmaceuticals* 25 (4): 471–477.

160 Ranjan, A.P., Mukerjee, A., Helson, L. et al. (2013). Efficacy of liposomal curcumin in a human pancreatic tumor xenograft model: inhibition of tumor growth and angiogenesis. *Anticancer Research* 33 (9): 3603–3609.

161 Wang, D., Wang, T., Zhang, Z. et al. (2022). Recent advances in the effects of dietary polyphenols on inflammation in vivo: potential molecular mechanisms, receptor targets, safety issues, and uses of nano-delivery system and polyphenol polymers. *Current Opinion in Food Science* 100921.

162 Tang, J., Xu, N., Ji, H. et al. (2011). Eudragit nanoparticles containing genistein: formulation, development, and bioavailability assessment. *International Journal of Nanomedicine* 2429–2435.

163 Zhang, H., Liu, G., Zeng, X. et al. (2015). Fabrication of genistein-loaded biodegradable TPGS-b-PCL nanoparticles for improved therapeutic effects in cervical cancer cells. *International Journal of Nanomedicine* 10: 2461.

164 Javed, Z., Khan, K., Herrera-Bravo, J. et al. (2021). Genistein as a regulator of signaling pathways and microRNAs in different types of cancers. *Cancer Cell International* 21 (1): 1–12.

165 Dev, A., Sardoiwala, M.N., Kushwaha, A.C. et al. (2021). Genistein nanoformulation promotes selective apoptosis in oral squamous cell carcinoma through repression of 3PK-EZH2 signalling pathway. *Phytomedicine* 80: 153386.

166 Ju, J., Hong, J., Zhou, J.N. et al. (2005). Inhibition of intestinal tumorigenesis in Apc min/+ mice by (−)-epigallocatechin-3-gallate, the major catechin in green tea. *Cancer Research* 65 (22): 10623–10631.

167 Gu, J.-W., Makey, K.L., Tucker, K.B. et al. (2013). EGCG, a major green tea catechin suppresses breast tumor angiogenesis and growth via inhibiting the activation of HIF-1α and NFκB, and VEGF expression. *Vascular Cell* 5: 1–10.

168 Shankar, S., Ganapathy, S., Hingorani, S.R., and Srivastava, R.K. (2008). *EGCG* inhibits growth, invasion, angiogenesis and metastasis of pancreatic cancer. *Frontiers in bioscience A Journal and Virtual Library* 13: 440–452.

169 Qian, Y.S., Ramamurthy, S., Candasamy, M. et al. (2016). Production, characterization and evaluation of kaempferol nanosuspension for improving oral bioavailability. *Current Pharmaceutical Biotechnology* 17 (6): 549–555.

170 Naeini, F., Namkhah, Z., Tutunchi, H. et al. (2021). Effects of naringenin supplementation in overweight/obese patients with non-alcoholic fatty liver disease: study protocol for a randomized double-blind clinical trial. *Trials* 22 (1): 1–9.

171 Ding, S.-M., Zhang, Z.H., Song, J. et al. (2014). Enhanced bioavailability of apigenin via preparation of a carbon nanopowder solid dispersion. *International Journal of Nanomedicine* 9: 2327.

172 Telange, D.R., Patil, A.T., Pethe, A.M. et al. (2017). Formulation and characterization of an apigenin-phospholipid phytosome (APLC) for improved solubility, in vivo bioavailability, and antioxidant potential. *European Journal of Pharmaceutical Sciences* 108: 36–49.

173 Cochran, D.B., Gray, L.N., Anderson, K.W., and Dziubla, T.D.(2016). Degradable poly (apigenin) polymer inhibits tumor cell adhesion to vascular endothelial cells. *Journal of Biomedical Materials Research Part B: Applied Biomaterials* 104 (7): 1438–1447.

174 Adel, M., Zahmatkeshan, M., Akbarzadeh, A. et al. (2022). Chemotherapeutic effects of Apigenin in breast cancer: preclinical evidence and molecular mechanisms; enhanced bioavailability by nanoparticles. *Biotechnology Reports* e00730.

175 Chamcheu, J.C., Siddhiqui, I.A., Adhami, V.M. et al. (2018). Chitosan-based nanoformulated (−)-epigallocatechin-3-gallate (EGCG) modulates human keratinocyte-induced responses and

alleviates imiquimod-induced murine psoriasiform dermatitis. *International Journal of Nanomedicine* 13: 4189.

176 Shan, Z., Nisar, M.F., Li, M. et al. (2021). Theaflavin chemistry and its health benefits. *Oxidative Medicine and Cellular Longevity* 2021.

177 Hussain, Y., Mirzaei, S., Ashrafizadeh, M. et al. (2021). Quercetin and its nano-scale delivery systems in prostate cancer therapy: paving the way for cancer elimination and reversing chemoresistance. *Cancers* 13 (7): 1602.

178 Joshi, V., Tambat, S., Bhori, M. et al. (2022). *Dietary Polyphenols in Bacterial and Fungal Infections. Dietary Polyphenols in Human Diseases*. 158–183. CRC Press.

179 Vieira, I.R.S. and Conte-Junior, C.A. (2022). Nano-delivery systems for food bioactive compounds in cancer: prevention, therapy, and clinical applications. *Critical Reviews in Food Science and Nutrition* 1–26.

180 Guo, C., Yin, J., and Chen, D. (2018). Co-encapsulation of curcumin and resveratrol into novel nutraceutical hyalurosomes nano-food delivery system based on oligo-hyaluronic acid-curcumin polymer. *Carbohydrate Polymers* 181: 1033–1037.

181 Piccolella, S., Crescente, G., Candela, L., and Pacifico, S. (2019). Nutraceutical polyphenols: new analytical challenges and opportunities. *Journal of Pharmaceutical and Biomedical Analysis* 175: 112774.

182 Dini, I. and Grumetto, L. (2022). Recent advances in natural polyphenol research. *Molecules* 27 (24): 8777.

183 Ofosu, F.K., Daliri, E.B.M., Elahi, F. et al. (2020). New insights on the use of polyphenols as natural preservatives and their emerging safety concerns. *Frontiers in Sustainable Food Systems* 4: 525810.

184 de Araújo, F.F., de Paulo Farias, D., Neri-Numa, I.A., and Pastore, G.M. (2021). Polyphenols and their applications: an approach in food chemistry and innovation potential. *Food Chemistry* 338: 127535.

16

Polyphenols in Food Products and Nutraceuticals

Bioavailability and Pharmacokinetic Issues

Priaunsh Dutt[1], *Yashika Shadija*[1], *Sneha Dokhale*[2], *Sweta Das*[2], *Shine Devarajan*[1], *and Samiksha Garse*[1,*]

[1] *School of Biotechnology and Bioinformatics, D Y Patil Deemed to be University, Navi Mumbai, Maharashtra, India*
[2] *B. K. Birla College, Birla College Campus Rd, Kalyan, Maharashtra, India*
**Corresponding author*

16.1 Introduction

Naturally occurring polyphenols, well known for their antioxidant properties, are obtained from plants and certain fungal species. These compounds cannot be synthesized by animals or humans, and hence need to be included in a dietary intake [1, 2]. A variety of primary and secondary substituent modifications (such as acylation or glycosylation) and the wide range of carbon backbone chains in this class of compounds contribute to the diversity of the phenolic compound structures [1, 3]. Depending on their chemical and physiological properties, the diverse group of these phenolic compounds are further categorized as discussed in this chapter [3]. The positive effect of polyphenols on human health has attracted the attention of many researchers recently. Furthermore, these effects have a direct connection between the health impact and phenolic component bioavailability in the host body [4]. Phenols have poor bioavailability because of their complex structure, thus they require certain strategies to increase it. This chapter provides a brief overview of polyphenols, their classes, pharmacokinetics, as well as the strategies to increase the bioavailability.

16.2 Polyphenols

A growing interest in polyphenols and their research has been observed in the past few years because of their beneficial properties in food products and nutraceuticals. Researchers and food manufacturers have been experimenting with food polyphenols to enhance their role in healthcare [5]. With antioxidant properties, foods of plant origin are a well-known rich source of polyphenols. Figure 16.1 highlights important polyphenols and their plant sources. The distribution of these secondary metabolites is nearly ubiquitous in nature. Polyphenols are an enormous class of phytochemicals having anti-inflammatory, antioxidant, antibacterial, and neuroprotective activities, among other health advantages [5, 6]. Because they are naturally available, these organic

Polyphenols: Food, Nutraceutical, and Nanotherapeutic Applications, First Edition. Edited by Mithun Rudrapal.

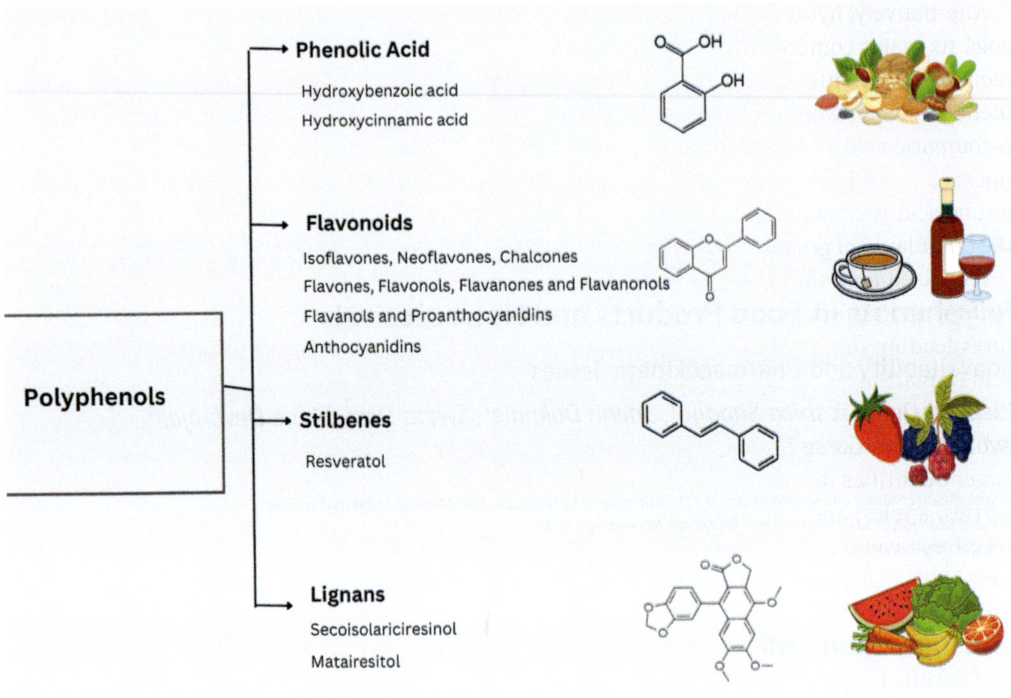

Figure 16.1 Classification of polyphenols.

compounds are in abundance in plants and their derivative products [7]. For example, 100 g fresh fruits or vegetables contain approximately 200–300 mg of polyphenols that contributes to approximately 1,193 mg ± 500 mg of polyphenols in an adult's daily dietary consumption [8, 9]. Moreover, plant derivates, such as chocolate, coffee, tea, and red wine, are rich in various forms of polyphenols [7]. Approximately 8,000 phenolic metabolites and roughly 4,000 flavonoids have been identified [10]. Because of the diversity and widespread distribution in plants, these compounds are broadly classified into phenols and polyphenols. These are further categorized as stilbenes and lignans based on the phenolic structure with varied hydroxyl (OH) side chains on the aromatic rings [4].

16.2.1 Phenolic Acids

Phenolic acids are compounds with COOH and OH groups on the first and sixth carbon skeleton, respectively. These are freely available in fruits and vegetables; however, they are present in a bound state in grains and seeds with a higher concentration in dried fruits [11–13]. Their two principal compounds are hydroxy derivatives of benzoic and cinnamic acids, which are non-phenolic in nature [14]. Compared with cinnamic acid, the concentration of hydroxybenzoic acid is less in plants, and moderately present in root vegetables, such as onion and black radish, with up to 10 mg/kg fresh weight [15]. Tannin-bound gallic acid is a complex compound with three hydroxy side chains of benzoic acid [16]. This tannin is a key component of tea leaves and red fruits, such as pomegranate and strawberries, at 4.5 g/kg fresh weight [17]

Alternatively, hydroxycinnamic acid (HCA) prevails in its free state compared with hydroxybenzoic. Its major components are p-coumaric (carrot) [18], caffeic acid (olive oil) [19], ferulic acid (commelinid plants) [20], and sinapic acid (spices) [21]. HCA and its derivates are widely distributed in the cell wall, roots, leave extract, and fruits. Carrots have 6.8 mg/100 g fresh weight of p-coumaric acid [18, 22], olives have 1.40–2.40 g/100 g dry weight of caffeic acid [19, 23], and mustard seeds have 12,500 μg/g dry weight of sinapic acid [21, 24]. The main dietary supply of ferulic acid is cereal grains, with the highest concentration 0.8–2 g/kg dry weight obtained from the outer layer of grains [25, 26].

Processed food that is either frozen, sterilized, or fermented is bound with the derivatives of cyclitol and tartaric acid. Cyclitols are glycosylated derivatives with two primary cyclic structures leading to quinic acid (cinchona bark) and shikimic acid (*Illicium* plants) [27]. Tartaric acid (tamarind) is an ester derivative with 12%–18% in the pulp and leaves of tamarind [28, 29]. Chlorogenic acid is created when quinic acid and caffeic acid interacts through a process of hydrolysis and is catalyzed by esterase. Although it is present in various fruit, it is present in larger quantities in coffee with approximately 70–350 mg/coffee cup [30]. Cherries, kiwis, blueberries, apples, and plums are fruits with the highest amount HCA, with approximately 0.5–2 g/kg fresh weight [31]. Compared with the other components of HCA, caffeic acid constitutes 70%–100% of its content in free and esterified form. Overall, considering the universality in fruits parts, the HCA concentration is greater in the ripened outer layer. However, the concentration is inversely proportional to the ripening of fruits and is directly proportional to the size of the fruits [31, 32].

16.2.2 Flavonoids

The class of phytonutrients, known as dietary polyphenols, includes a number of phenolic compounds properties of low molecular weight. These are called flavonoids, and approximately 6,000 are known [33]. The flavonoid basic backbone comprises three rings, two of which possess phenolic nature, called A and B, and the other ring is heterocyclic in nature with an oxygen atom embedded, called C [34]. These flavonoids are further subdivided based on the pattern of hydroxylation and the differences in the ring C composition [35]. Flavones and flavonols have a ketone composition with a phenyl (Ph) substitute and OH group, respectively [34]. Isoflavones and neoflavanoids are also extracted from plants with a phenyl ring attached at the third and fourth carbon atom, respectively. The commonality between these two flavonoids is that they both lack an OH group. For all flavonoids, ring B is joined by the second carbon of ring C [36]. Nevertheless, chalcones lack a heterocyclic ring, yet they fall under the flavonoids category [35]. Proanthocyanidins have multiple subunits of polyhydroxy flavan (3-ol and 4-ol) with a molecular weight of approximately 1–30 kDa [37].

16.2.2.1 Isoflavones, Neoflavones, and Chalcones

Ring B of isoflavones is connected to ring location C3, and they are typically found among plants of the leguminous family [36]. Because many cultures depend extensively on beans, particularly soy beans, isoflavones influence human health at a substantial rate. The two primary isoflavones found in soy are genistein and daidzein, in addition to glycetein, biochanin A, and formononetin [38, 39]. They are also in red clovers [40]. Most of these isoflavone aglycones are found as 7-O-glucosides and 6”-O-malonyl-7-O-glucosides. Even though neoflavonoids are rarely found in food plants, dalbergin is the most ubiquitous and widely dispersed neoflavone in the plant kingdom [41]. Open-ring chalcones are found in fruits such as apples [42], hops, and beers [43].

16.2.2.2 Flavones, Flavonols, Flavanones, and Flavanonols

In the plant kingdom, flavonoids are most prevalent as secondary metabolites. The largest group among all flavonoids are flavones and their 3-hydroxy derivatives flavonols, as well as their glycosides, methoxides, and other acylated products on all three rings [33]. There are at least 279 and 347 distinct glycosidic combinations in the two most popular flavonol aglycones, quercetin and kaempferol, respectively [44]. In the past 15 years, there have been more flavanones and their 3-hydroxy derivatives, flavanonols, also known as dihydroflavonols, discovered [45, 46]. Some flavanones, such as prenylated flavanones, furanoflavanones, pyranoflavanones, and benzylated flavanones, have distinctive patterns of substitution that yield numerous variants that are substituted. Citrus fruit taxifolin is a well-known flavanonol [45, 47].

16.2.2.3 Flavanols and Proanthocyanidins

Catechins, also known as flavanols or flavan-3-ols, are widely used. In contrast to the majority of flavonoids, flavanols lack the C4 carbonyl and the C2 and C3 double bonds. As a result, flavanols can have two chiral centers on their molecules (on C2 and C3) and hydroxylation at C3, resulting in four potential diastereoisomers. Epicatechin has a cis configuration, whereas catechin has a trans configuration. There are two stereoisomers of each of these two configurations: (+)-catechin, (−)-catechin, (+)-epicatechin, and (−)-epicatechin. The two isomers (+)-catechin and (−)-epicatechin are frequently present in food plants [36]. Many fruits contain flavanols; however, grapes, apples, and blueberries have the highest concentrations [42]. Condensed tannins are the conventional definition of proanthocyanidins. Strong antioxidants, known as flavanols and oligomers (2–7 monomeric unit compounds), have been linked to numerous potential health advantages. Oligomeric proanthocyanidins can have an A-type structure, in which monomers are connected by C2–O–C7 or C2–O–C5 bonding, or a B-type structure, in which C4–C6 or C4–C8 are frequently seen. This depends on the interflavanic connections. Procyanidin is a trimer [48].

16.2.2.4 Anthocyanidins

The pigments known as anthocyanins are dispersed in the cell sap of the upper layer of flowers and fruit, giving these a pigmented tint of pink, red, blue, or purple [49]. Depending on the pH, they can be found in many chemical forms that are both colored and uncolored. In plants, they are resistant to sunlight, pH level, and oxidation conditions, which are prone to trigger them to decay; although they are extremely unstable in the aglycone form (anthocyanidins) [50].

16.2.3 Stilbenes

Stilbenes are a subclass of phenylpropanoids that are differentiated by having a 1,2-diphenylethylene nucleus with two Ph rings bridged by a methylene group of two carbon atoms [51]. Stilbenes are rare in the human diet. The majority of stilbenes in plants function as phytoalexins, which are substances that are only processed against triggered infections or damage. The naturally occurring polyphenol stilbene resveratrol (3,4′,5-trihydroxystilbene) is one of the most studied [52]. Red wine contains a high concentration of resveratrol (0.3–7 mg aglycones/L and 15 mg glycosides/L) that is hypothesized to provide health benefits by acting as the representative stilbene molecule [4, 52–55].

16.2.4 Lignans

Lignans are diphenolic compounds with 2,3-dibenzylbutane content, synthesized by forming a dimer of two residues of cinnamic acid. Secoisolariciresinol is one of many lignans that have been suggested to act as a phytoestrogen [56]. The most abundant dietary source is linseed, with small concentrations of matairesinol and secoisolariciresinol (approx. 3.7 g/kg dry weight). Most fiber-rich plants, including vegetables (broccoli, carrots, and garlic), cereals (oats, barley), and legumes (soybeans, beans), contain lignans, and a diet packed with lignans may be advantageous to human health. Lignans have been found to have a variety of pharmacological effects, particularly when used as a cancer treatment [57].

16.3 Pharmacokinetic Properties of Polyphenols

Pharmacokinetics is the study of the administration of a substance in the body, and it mainly involves four stages: absorption, distribution, metabolism, and excretion. Polyphenols are useful components in dietary supplements and culinary preparations. It is crucial to study the pharmacokinetic stages in order to utilize the therapeutic effects of polyphenols in a more efficient and sustainable way.

16.3.1 Absorption

Polyphenols, in their free form, are present as esters, glycosides, or are polymeric in nature. These forms are not readily absorbed by the host. Thus, the compounds undergo certain enzymatic alterations to aid absorption. Several hydroxyl groups make up a significant portion of polyphenolic compounds, which are catalyzed by enzymes for methylation, glucuronidation, or sulfation. The small intestine only absorbs 5%–10% of the consumed polyphenol amount [58]. The remaining polyphenols can build up in the large intestine and be eliminated in the feces. Depending on the type of diet, the consumption of polyphenolic chemicals may differ greatly. Recent studies evaluated the presence of anthocyanin derivatives in the plasma of rat 6 minutes post administration. Such rapid appearance of this substance in the blood stream indicates the presence of polyphenol transporters in the stomach wall. Transporters, such as a bilitranslocase, may play a significant role in such activity. This organic anion transporter, which is also involved in the movement of anthocyanins and flavonoid aglycons, is found in the liver, kidney, vascular endothelium, and gastric epithelium [59, 60]. Polyphenols, such as ascorbic acid and chlorogenic acid, are absorbed in the small intestine via Na^+-dependent active transportation. Similarly, aglycone or glycoside links are better absorbed in host systems than polyphenol complexes formed with rhamnose. This was demonstrated using quercetin glycosides; for quercetin 4′-glucoside, maximum absorption occurs 0.5–0.7 hours after consumption, whereas it takes 6–9 hours for rutin (quercetin-3-rutinoside) [55].

16.3.2 Distribution

Bioactive ingredients must be bioavailable, which means they must be properly absorbed from the gut into the bloodstream and transported to the intended area inside the body. The small intestine and liver frequently conjugate phenolics throughout the absorption process. Three primary methods of conjugation used for polyphenols are methylation, sulfation, and glucuronidation [61].

16.3.3 Metabolism

Different polyphenolic substances possess different affinities for proline-rich proteins. (+)-catechin was shown to have a stronger affinity for salivary glands. Polyphenols are subjected to extremely acidic conditions in the stomach, which may affect their stability. However, research on a number of these substances, including resveratrol [62], quercetin [63], and catechin [59], showed stability at lower pH. Proanthocyanidins were also discovered to be stable at such pH levels [48, 60], despite research that suggested they would degrade to monomers in an acidic environment [64]. It is interesting to note that some research suggests phenolic compounds may be absorbed in the stomach. Experiments were carried out on laboratory animals to study the metabolism of these naturally occurring polyphenols. Malvidin-3-glucoside was detected in the plasma of rats approximately 6 minutes after anthocyanins were administered. The rapid appearance of this substance in the bloodstream is likely because the transporters are present in the stomach wall [65]. The transporter bilitranslocase would likely be responsible for carrying out such a function. This organic anion transporter, which is also involved in the movement of anthocyanins and flavonoid aglycons, is found in the liver, kidney, vascular endothelium, and gastric epithelium [66]. Additionally, the stomach may absorb phenolic acids such as dimethoxycinnamic, chlorogenic, gallic, caffeic, and p-coumaric acids [67–69].

The primary location of polyphenol glucuronidation, which is carried out by enzymes from the family of uridine diphosphate glucuronosyltransferases (UGT), is the small intestine. The UGT1A8 and UGT1A10 isoforms drive the glucuronidaton of flavonoids at the C5 and C7 positions of the A-ring [70]. High levels of O-methylated forms and O-methylated flavanol glucuronides are also found; catechol-O-methyltransferases are the enzymes involved in the production of these derivatives. Typically, at position $m^{-3'}$-O-, these enzymes methylate compounds that contain a catechol residue. S-adenosylmethionine is a donor of the methyl group [71].

Similar to other xenobiotics, phenolic substances go through reactions in the liver, including oxidation, reduction, hydrolysis, and hydration, which are catalyzed by phase I enzymes. The primary site of polyphenol absorption is the large intestine, and the colonic microbiota is crucial to the breakdown of these substances. Enzymes produced by *Clostridium orbiscindens, Eubacterium ramulus*, and *Enterococcus casseliflavus* that are capable of deglycosylating quercetin-3-glucoside and performing the fission of the C-ring in quercetin and naringenin have been found in humans [72]. The first step in the breakdown of ()-epicatechin is the fission of the C-ring, which produces 1-(3′,4′-dihydroxyphenyl)-3-(2″,4″,6″-trihydroxy)propan-2-ol that is then transformed into 5-(3′,4′)-dihydroxyphenyl-valerolactone. Next, the valerolactone ring undergoes oxidation to form 3-hydroxyphenylpropionic acid from 5-(3′,4′)-dihydroxyphenyl-valeric acid. This substance produces 3-hydroxyphenylacetic acid when it is oxidized. The galloyl moiety is removed via the breakdown of epigallocatechin gallate (EGCG) and epicatechin gallate (ECG) [73].

16.3.4 Excretion

To have an impact on certain tissues or organs, bioactive compounds must be bioavailable, which means they must be properly ingested from the gut into the bloodstream and transported to the right location inside the body. Phenolics frequently conjugate in the small intestine and then in the liver throughout the absorption process. Polyphenol metabolites can be eliminated by either the biliary or urine systems [74]. Large, highly conjugated intermediates are more inclined to be removed in bile than small conjugates, including monosulfates, which are preferentially excreted in urine. The relative amount of biliary and urine excretion in laboratory animals differs from one polyphenol to another. Genistein, EGCG, and eriodictyol appear to be primarily excreted through the biliary system [75, 76]. The biliary excretion of polyphenols in humans may differ greatly from that in rats because of the presence of the gall bladder; however, this has never been investigated. Intestinal bacteria have glucuronidases that can break up conjugated compounds released in bile, into free aglycones. Aglycones have the potential to be reabsorbed, causing enterohepatic cycling. Human studies have frequently been used to determine urinary excretion. The maximum plasma concentrations and total amount of metabolites discharged in urine are roughly connected [77]. Interestingly, genistein's efficient biliary excretion can be the reason the genistein plasma concentrations are typically greater than daidzein concentrations, despite daidzein's higher urinary excretion. Urinary recovery varies from 5.9% to 27% for caffeic and ferulic acids; however, it is between 0.5% and 6% for certain tea catechins, 2%–10% for red wine catechin, and up to 30% for chocolate epicatechin [77–80].

For other polyphenols, such as anthocyanins (0.005%–0.1% of intake), these percentages might be extremely low [81, 82]. However, the fact that anthocyanins have a variety of different molecular structures and that numerous potential metabolites can be produced may be the only reason for their low bioavailability [83]. Furthermore, because of analytical challenges, some metabolites may still be unknown. It has been demonstrated that when urine samples were frozen [84], all of the metabolites of strawberry anthocyanins were extremely unstable and substantially destroyed.

16.4 Bioavailability of Polyphenols

Bioavailability is a term borrowed from the study of pharmacology and it is used to define a fractional oral dose that enters the circulation from the prepared compound or derivative (i.e., parent ingredient or its active ingredient) [85]. In the case of polyphenols, bioavailability may be described as "the amount of polyphenolic content absorbed by the body and made biologically active in the metabolism of the given organism" [85, 86]. Any compound post-absorption is bioactive because of the metabolic consequences that the compound will have in the system it enters. The health benefits provided by polyphenols are dependent on the bioavailability of the compounds in the system. However, the bioavailability of polyphenols is adversely affected because of the low absorption of polyphenols in the human body, high rate of metabolism, and rapid excretion of these compounds. The majority of polyphenols ingested in humans are not found in the urine, indicating poor or no absorption through the gut barrier, or absorption and excretion in bile [87]. Understanding the bioavailability is of utmost importance for the health effects of polyphenols to be extensively studied.

16.4.1 Bioavailability of Different Types of Polyphenols

A wide spectrum of plant and plant-based sources are naturally available. The bioavailability of polyphenols with their respective sources and plasma concentration are discussed in Table 16.1 [32, 88].

16.4.2 Factors Affecting the Bioavailability of Polyphenols

There are numerous factors affecting the bioavailability of polyphenols, including external factors affecting the plants from which these polyphenolic compounds are derived. For example, the degree of ripeness affects the concentration of polyphenol content, where the phenolic acid concentration generally decreases with ripening while anthocyanins increases [31, 95]. Other factors include the interaction of polyphenols with other components present in the food such as proteins, fiber, carbohydrates, and fat. The intermolecular bonds formed between serum albumin and quercetin molecules might indicate the reason behind the relatively slow excretion of quercetin molecules from the body [87]. Some polyphenols associated with dietary fibers are partially bioavailable in the human body, even though the fibers delay absorption [96].

Table 16.1 Polyphenol sources and plasma concentration in humans.

Polyphenol		Source of Polyphenol	Concentration of Polyphenol in Plasma (μm)	References
Phenolic acid	Caffeic acid	Red wine (200 ml)	0.06	[89, 90]
	Chlorogenic acid	Coffee (200 ml)	0.5	
Flavanols	Quercetin	Onion	0.74	[91]
		Apple	0.30	
Catechins	-	Red wine (120 ml)	0.072	[78]
Isoflavones	Genistein	Soy milk and Soy proteins	0.74	[92]
Flavanones	Naringenin	Grape fruit juice	5.99	[93]
Anthocyanins	Cyanidin 3-glucoside	Orange juice (1 L)	0.002	[94]

In a host, the bioavailability and bioefficacy of polyphenols are dependent on the absorption of polyphenols through the gut. The absorption is mainly affected by the chemical and structural factors of the polyphenol compounds. Polyphenols are not present in their native form; however, they are present in their glycosylated, acylated, conjugated, or conjugated forms, which in turn affect the chemical structure of the polyphenol [97]. The degree of polymerization, molecular size, hydrophobicity, and hydrophilicity influence the bioavailability of the phenolic compounds. This creates difficulty in absorption of the compounds through the membrane. Isoflavones, flavanols, and flavanones are reported to have the highest bioavailability, while proanthocyanins and anthocyanidins are absorbed poorly by the body [98].

16.4.2.1 Interactions Between Polyphenols and the Food Matrix

The simultaneous intake of different kind of foods can have a variety of effects on the absorption of polyphenols in the host system. Bioaccessibility is a term defined as a fractional compound of the food matrix that has entered the gastrointestinal (GI) tract and is readily available to be absorbed by the intestine. The bioaccessibility of polyphenols is thus dependent on the food matrix with which it is ingested [99]. The food matrix influences the absorption of polyphenols because of the synergistic and antagonistic relationship between the compounds, and because of other properties of the matrix such as pH and temperature. There have been studies suggesting that polyphenol absorption in the host system may also occur because of inter-component interactions. A study carried out by Tulipani et al. in 2012 [100] found that the addition of an oil matrix significantly increased the absorption of tomato phenolics. It was suggested that that the matrix stimulated the re-absorption. Carbohydrates also help enhance the absorption of flavanols, as was reported by Schramm et al. in 2003 [101]. The study identified an increase in uptake of flavanols from cocoa by simultaneous consumption of carbohydrates. The study also found that protein and lipid factors do not play a major role in the absorption of flavanols. Compounds, such as ethanol, have been proven to improve anthocyanin bioavailability when included in the dietary matrix [102]. Thus, the selection of the right food matrix that has a synergistic effect on the absorption of polyphenols is extremely important for increasing the bioavailability and bioaccessibility of polyphenols.

16.4.2.2 Metabolic Activity in the Liver

Polyphenols that are bioaccessible in the GI tract undergo a series of reactions, which mainly include methylation, sulfation, and glucuronidation, prior to their transfer into the bloodstream [103]. These conjugation reactions mainly occur in the liver and represent a detoxification process that helps restrict any toxic effects that the compounds might have by facilitating the biliary and urinary elimination of the compound. Glucuronidation plays a major role in facilitating the excretion of compounds by increasing their molecular weight [104]. Methylation is carried out by the catechol-O-methyl transferase enzyme [105, 106], which shows its highest activity in the liver and kidneys. Sulfation is carried out by the sulfotransferase enzyme mostly in the liver [105, 107]. Glucuronidation is performed by the UGT enzyme and occurs first in the enterocytes before further processing in the liver [64, 108]. These extensive modifications help alter the biological activity of polyphenols, increasing the active metabolite production from the polyphenols and increasing the excretion of the polyphenols.

16.4.2.3 The Microbiome in the Gut

Numerous polyphenols do not get absorbed into the gut and thus enter the colon. The colon plays host to a significant number of microorganisms that help in the metabolism of polyphenols [97]. These microorganisms are exposed to two major types of polyphenols: dietary polyphenols that are not absorbed by the gut and those that are excreted through the liver. The dietary polyphenols are mostly in glycosidic form, while the later are found in their conjugated form [109]. The microorganisms help the metabolism by excreting extracellular enzymes, glycosidases, which help breakdown the glycoside bonds of the polyphenols with their sugar moiety. The polyphenols are broken down into aglycones that can then be absorbed by colonic cells in the host and thus help increase the bioavailability of the polyphenols [110, 111]. These reabsorbed aglycones are sent back into the liver for reconjugation and further processing.

16.4.3 Polyphenols Affecting the Bioavailability of Other Foods

The impact of polyphenols on the metabolism of carbohydrates has been studied for quite some time [99]. Carbohydrate digestion involves key enzymes such as amylase and α-glucosidases. Polyphenols inhibit the activity of these enzymes and hence slow the breakdown of carbohydrates. A study conducted by Bräunlich et al. [112] found that α-glucosidases are severely inhibited by polyphenols from the chokeberry and thus help in decreasing blood glucose levels. Polyphenolic compounds can be inhibitors even prior to absorption because α-glucosidase is an enzyme that is membrane bound and present in the epithelial cells of the intestine [102]. Thus, polyphenols that are natural antioxidants can also be used as potent antidiabetic agents.

Polyphenols found in black and green tea impact the emulsification and absorption of fats. Emulsification of fats is important for the breakdown into lipid droplets that are absorbed by the host post-enzymatic lipid digestion. Polyphenols have been shown to cause an increase in the size of lipid droplets while consequently decreasing the area of contact with the lipolytic enzyme and hence affecting the breakdown and digestion of fats and fatty acids [113].

Polyphenols, such as chlorogenic acid, also result in lower protein digestion. However, such effects were only observed with high polyphenol intake. Reducing the polyphenol intake showed no repercussions on the digestion of protein in the host system [114].

16.5 Strategies to Improve the Dietary Bioavailability of Polyphenols

The poor absorption of phenolic compounds in the host system severely impacts the bioaccessibility and bioefficacy of the compounds. Therefore, it is necessary to look at strategies to improve the bioavailability of phenolic compounds to help capitalize on the health benefits of polyphenol consumption. Polyphenol bioavailability may be increased via two fundamental pathways: modifying host mechanisms to ensure better adsorption of phenolic compounds or the use of strategies to modify the intrinsic factors of polyphenol resulting in poor absorption.

16.5.1 Modulation of the Metabolism

Improvising the bioavailability of any compound necessitates a thorough understanding of the metabolism of the compound [115]. For enhancing polyphenol bioavailability, metabolic process, such as absorption in the intestine [116], metabolic stabilization [117, 118], regulation of enzyme activity, and regulation of polyphenol transporters, can be used to increase bioaccessibility in the host system.

16.5.1.1 Modulation of Gut Microbiomes

Polyphenols enter the gut or are metabolized by the gut microbiota, and enzymes are released by the host. Polyphenols have been known to cause significant modulations in the gut microbiome [7]. Modulation of the gut microbiome for increasing the metabolism of polyphenols is mainly focused on the enzymes released by these microbes, which may result in degradation of the phenolic compound. Enzymes, such as esterases, cause significant breakdown of phenolic compounds such as chlorogenic acid. A strategy may be developed to increase the metabolism of polyphenols, where the compounds may be co-administered with an antibiotic that will result in a decrease in the esterases producing microbiota and consequently increase the stability of chlorogenic acid in the gut. However, this system has not been studied in depth and is currently only in use in the pharmacotherapy industry [115, 119].

16.5.1.2 Regulation of the Activity of Polyphenol Transporters

ATP-binding cassette (ABC) transporters are proteins that use the energy from the hydrolysis of the ATP molecule to help in the translocation of various molecules across membranes. Certain polyphenols act as substrates for these transporters. This results in a severe decrease in their bioavailability [115].

Studies conducted by Youdim et al. in 2004 shed light on quercetin, a phenolic molecule, which with co-administration with a breast cancer resistant protein (BCRP) inhibitor, resulted in a significant increase in its ability to cross the blood–brain barrier (BBB). This data may imply that the quercetin molecule has the ability to penetrate the BBB; however, on entry, the molecules are recognized as substrates to the BCRP transporter and pumped out [120].

Similar studies have shown the involvement of the BCRP efflux pump in the reduced bioavailability of resveratrol [65].

A strategy may be designed using an ABC transporter inhibitor and a polyphenol, which is a known substrate for the respective efflux pump. This will thus reduce the chances of the polyphenol being identified as the substrate and result in increased bioavailability of the polyphenol [115].

16.5.1.3 Regulation of Enzyme Activities

Polyphenols post-consumption react with various enzymes in the gut that are either of bacterial or host origin. This results in the deactivation of the phenolic compounds of the polyphenols.

For example, chlorogenic acid is found as an ester of caffeic or quininic acid. Esterase enzyme present in the gut breaks down the acid and hence deactivates it. Similarly, enzymes, such as monoamine oxidase enzymes [121], oxidize monoamine groups present on certain phenolic compounds, which alters the function or deactivates the phenolic compounds.

Use of enzyme inhibitors can help reduce the rate of breakdown of phenolic compounds, resulting in increased metabolic stability of the compound. This helps the compound remain longer in its native form in the GI tract, which consequently helps the absorption by passive diffusion [117].

16.5.2 Improving the Transport of Polyphenols

A high bioavailability does not indicate that the active compound, in this case a polyphenol, reaches the site of action. Therefore, improving the transport of polyphenols will have a significant impact on the bioaccessibility of the drug at the desired site [122]. This improvement in transport can be introduced by various methods such as nanoencapsulation, liposomal technology emulsion technologies, and other nanodelivery methods. These methods help improve the stability of polyphenol by reducing its interactions with other factors and hence enhance the transport of polyphenols through the cell membranes [97].

16.5.2.1 Nanodelivery

Nanodelivery is a field of science that is relatively new and rapidly developing. It uses nanovectors that deliver the active compound (drug) to the target organs in the human body. Nanotechnology in the context of polyphenols can be used to improve the transport of polyphenols as well as increase the dosage of polyphenols. Methods, such as nanoencapsulation, liposomal technology, spray drying, and emulsion technology, are of current and upcoming interest in the field of nanodelivery of polyphenols [122].

Nanoencapsulation techniques use a biopolymer for encapsulation of phenolic compounds and to increase their stability. This technique has been applied extensively to curcumin and it was observed that nanoencapsulated curcumin was present in higher plasma concentrations than native curcumin in the host system, suggesting enhanced bioavailability [123].

Another method similar to nanoencapsulation is liposome encapsulation, where colloidal particles are used to form a capsule. This method has been used for studies related to the delivery of quercetin and curcumin in an effort to increase their bioavailability. An increase in bioavailability was reported in comparison to the native form using oral administration in a rat model [124].

Spray drying technologies are another encapsulation method in which an active compound is atomized in hot gas to instantaneously create a powder. This technique was used to encapsulate the compound curcumin, and further studies observed an increase in the serum concentration of bread enriched with this encapsulated curcumin compared with bread enriched with curcumin in its native form [125].

Emulsion technologies are used for encapsulation of active compounds in aqueous solutions. These active compounds can then be used in a liquid state or as powders. Emulsion technologies were used to study the increase of curcumin bioavailability in gastric and intestinal digestion. It was observed that the phenolic compound curcumin exhibited increased stability compared with its native form and was rapidly released in a simulated intestinal medium post encapsulation [126, 127].

Overall, nanodelivery systems are an effective measure and strategy that can be applied to improve and increase the bioavailability and bioaccessibility of polyphenols in any host systems.

16.6 Conclusion

Polyphenols obtained from fruits, vegetables, cereals, and beverages provide significant protection against the development of chronic diseases. Recently, new sources have been explored for the potential use of polyphenols in the food industry and health industry. Efforts in the optimization of the polyphenol preparation and standardization of the quality are being relentlessly pursued. Bioavailability of polyphenols is one of the main factors causing low efficacy of the mode of action of polyphenols. Low absorption of these compounds has played a significant role in hindering studies conducted with respect to these compounds. The pharmacokinetics of phenolic compounds, the factors affecting the bioavailability of polyphenols, certain measures and strategies being used to work around these factors to increase the bioavailability, as well as bioaccessibility of these compounds were discussed. Additionally, the advantages of using nanotechnology to tackle the bioavailability and pharmacokinetic issues were presented.

References

1 Halbwirth, H. (2010). The creation and physiological relevance of divergent hydroxylation patterns in the flavonoid pathway. *International Journal of Molecular Sciences* 11 (2): 595–621.

2 Kris-Etherton, P.M., Hecker, K. D., Bonanome, A. et al. (2002). Bioactive compounds in foods: their role in the prevention of cardiovascular disease and cancer. *The American Journal of Medicine* 113 (9): 71–88.

3 Cheynier, V. (2005). Polyphenols in foods are more complex than often thought. *The American Journal of Clinical Nutrition* 81 (1): 223S–229S.

4 Pandey, K.B. and Rizvi, S.I. (2009). Plant polyphenols as dietary antioxidants in human health and disease. *Oxidative Medicine and Cellular Longevity* 2: 270–278.

5 Wan, M.L.Y., Co, V.A., and El-Nezami, H. (2021). Dietary polyphenol impact on gut health and microbiota. *Critical Reviews in Food Science and Nutrition* 61 (4): 690–711.

6 Farha, A.K., Gan, R-Y., Li, H-B. et al. (2022). The anticancer potential of the dietary polyphenol rutin: current status, challenges, and perspectives. *Critical Reviews in Food Science and Nutrition* 62 (3): 832–859.

7 Aravind, S.M., Wichienchot, A., Tsao, R. et al. (2021). Role of dietary polyphenols on gut microbiota, their metabolites and health benefits. *Food Research International* 142: 110189.

8 Scalbert, A., Manach, C., Morand, C. et al. (2005). Dietary polyphenols and the prevention of diseases. *Critical Reviews in Food Science and Nutrition* 45 (4): 287–306.

9 Perez-Jimenez, J., Fezeu, L., Touvier, M. et al. (2011). Dietary intake of 337 polyphenols in French adults. *The American Journal of Clinical Nutrition* 93 (6): 1220–1228.

10 Stockley, C., Teissedre, P-T., Boban, M. et al. (2012). Bioavailability of wine-derived phenolic compounds in humans: a review. *Food & Function* 3 (10): 995–1007.

11 Kim, K.-H., Tsao, R., Yang, R., and Cui, S.W. (2006). Phenolic acid profiles and antioxidant activities of wheat bran extracts and the effect of hydrolysis conditions. *Food Chemistry* 95 (3): 466–473.

12 Adom, K.K. and Liu, R.H. (2002). Antioxidant activity of grains. *Journal of Agricultural and Food Chemistry* 50 (21): 6182–6187.

13 Chandrasekara, A. and Shahidi, F. (2010). Content of insoluble bound phenolics in millets and their contribution to antioxidant capacity. *Journal of Agricultural and Food Chemistry* 58 (11): 6706–6714.

14 Nursten, H. (1996). Food phenolics: sources, chemistry, effects, applications. *F. Shahidi & M. Naczk. Technomic Publishing Co. Inc, Lancaster, PA, USA, and Basel, 1995. ix+ 331 pp. Price: $125.00. Food Chemistry* 3 (57): 481–482.

15 Fenga, C., Costa, C., Caruso, E. et al. (2016). Current evidence on the protective effect of dietary polyphenols on breast cancer. *Farmacia* 64 (1): 1–12.

16 Tomas-Barberan, F. and Clifford, M. (2000). Dietary hydroxybenzoic acid derivatives and their possible role in health protection. *Journal of the Science of Food and Agriculture* 80: 1024–1032.

17 Clifford, M. and Scalbert, A. (2000). Ellagitannins, occurrence in food, bioavailability and cancer prevention. *Journal of the Science of Food and Agriculture* 80: 1118–1125.

18 Boz, H. (2015). p-Coumaric acid in cereals: presence, antioxidant and antimicrobial effects. *International Journal of Food Science & Technology* 50 (11): 2323–2328.

19 Sun-Waterhouse, D., Zhou, J., Miskelly, G.M. et al. (2011). Stability of encapsulated olive oil in the presence of caffeic acid. *Food Chemistry* 126 (3): 1049–1056.

20 Hong SHHHong, S., Deul, Y.Y., Shinji, K. et al. (2016). Radical copolymerization of ferulic acid derivatives with ethylenic monomers. *Journal of Fiber Science and Technology* 72 (3): 74–79.

21 Chen, C. (2016). Sinapic acid and its derivatives as medicine in oxidative stress-induced diseases and aging. *Oxidative Medicine and Cellular Longevity* 2016.

22 Miglio, C., Chiavarao, E., Visconti, A. et al. (2008). Effects of different cooking methods on nutritional and physicochemical characteristics of selected vegetables. *Journal of Agricultural and Food Chemistry* 56 (1): 139–147.

23 Boudhrioua, N., Bahloul, N., Slimen, I.B., and Kechaou, N. et al. (2009). Comparison on the total phenol contents and the color of fresh and infrared dried olive leaves. *Industrial Crops and Products* 29 (2–3): 412–419.

24 Engels, C., Schieber, A., and Gänzle, M.G. (2012). Sinapic acid derivatives in defatted oriental mustard (Brassica juncea L.) seed meal extracts using UHPLC-DAD-ESI-MS n and identification of compounds with antibacterial activity. *European Food Research and Technology* 234: 535–542.

25 Sosulski, F., Krygier, K., and Hogge, L. (1982). Free, esterified, and insoluble-bound phenolic acids. 3. Composition of phenolic acids in cereal and potato flours. *Journal of Agricultural and Food Chemistry* 30 (2): 337–340.

26 Lempereur, I., Rouau, X., and Abecassis, J. (1997). Genetic and agronomic variation in arabinoxylan and ferulic acid contents of durum wheat (Triticum durumL.) grain and its milling fractions. *Journal of Cereal Science* 25 (2): 103–110.

27 Dyrda-Terniuk, T., Sugajski, M., Pryshchepa, O. et al. (2022). The study of protein–cyclitol interactions. *International Journal of Molecular Sciences* 23 (6): 2940.

28 Buiarelli, F., Coccioli, F., Jasionowska, R. and Terracciano, A. (2010). Identification of hydroxycinnamic acid–tartaric acid esters in wine by HPLC–tandem mass spectrometry. *Food Chemistry* 123 (3): 827–833.

29 Roopa, G.S. and Kasiviswanatham, V. (2013). Extraction of tartaric acid from tamarind pulp and analysis of the acid composition in leaves. *International Journal of Studied Research Technology Management* 1 (5): 478–488.

30 Clifford, M.N. (1999). Chlorogenic acids and other cinnamates–nature, occurrence and dietary burden. *Journal of the Science of Food and Agriculture* 79 (3): 362–372.

31 Macheix, J.-J. (2018). *Fruit Phenolics*. CRC press.

32 D Archivio, M., Filesi, C., Bennedetto, R.D. et al. (2007). Polyphenols, dietary sources and bioavailability. *Annali-Istituto Superiore di Sanita* 43 (4): 348.

33 Andersen, O.M. and Markham, K.R. (2005). *Flavonoids: Chemistry, Biochemistry and Applications*. CRC press.

34 Czeczot, H. (2000). Biological activities of flavonoids-a review. *Polish Journal of Food and Nutrition Sciences* 9 (4): 3–13.

35 Harborne, J.B. (2013). *The flavonoids: advances in research*. https://doi.org/10.1007/978-1-4899-2913-6.

36 Tsao, R. (2010). Chemistry and biochemistry of dietary polyphenols. *Nutrients* 2 (12): 1231–1246.

37 Farha, A.K., Yang, Q-Q., Kim, G. et al. (2020). Tannins as an alternative to antibiotics. *Food Bioscience* 38: 100751.

38 Wang, H.-J. and Murphy, P.A. (1994). Isoflavone content in commercial soybean foods. *Journal of Agricultural and Food Chemistry* 42 (8): 1666–1673.

39 Mazur, W.M., Duke, J.A., Wahala, K. et al. (1998). Isoflavonoids and lignans in legumes: nutritional and health aspects in humans. *The Journal of Nutritional Biochemistry* 9 (4): 193–200.

40 Tsao, R., Papadopoulos, Y., Yang, R. et al. (2006). Isoflavone profiles of red clovers and their distribution in different parts harvested at different growing stages. *Journal of Agricultural and Food Chemistry* 54 (16): 5797–5805.

41 Garazd, M., Garazd, Y.L., and Khilya, V. (2003). Neoflavones. 1. Natural distribution and spectral and biological properties. *Chemistry of Natural Compounds* 39: 54–121.

42 Tsao, R., Yang, R., Young, J.C., and Zhu, H. (2003). Polyphenolic profiles in eight apple cultivars using high-performance liquid chromatography (HPLC). *Journal of Agricultural and Food Chemistry* 51 (21): 6347–6353.

43 Zhao, F., Watanabe, Y., Nozawa, H. et al. (2005). Prenylflavonoids and phloroglucinol derivatives from hops (Humulus lupulus). *Journal of Natural Products* 68 (1): 43–49.

44 Williams, C.A. (2006). Flavone and flavonol O-glycosides. In: *Flavonoids: Chemistry, Biochemistry and Applications* (ed. Ø.M. Andersen and K.R. Markham). CRC Press. ISBN 9781498774826.

45 Grayer, R. and Veitch, N. (2006). Flavanones and dihydroflavonols. In: *Flavonoids: Chemistry, Biochemistry and Application* (ed. Ø.M. Andersen and K.R. Markham). CRC Press. ISBN 9781498774826.

46 Tsao, R. and McCallum, J. (2009). Chemistry of flavonoids. *Fruit and Vegetable Phytochemicals: Chemistry, Nutritional Value and Stability* 5: 131–153.

47 Kawaii, S., Tomono, Y., Katase, E. et al. (1999). Quantitation of flavonoid constituents in citrus fruits. *Journal of Agricultural and Food Chemistry* 47 (9): 3565–3571.

48 Stoupi, S., Williamson, G., Viton, F. et al. (2010). In vivo bioavailability, absorption, excretion, and pharmacokinetics of [14C] procyanidin B2 in male rats. *Drug Metabolism and Disposition* 38 (2): 287–291.

49 Mazza, G. (1993). Miniati e. anthocyanins in fruits, vegetables and grains. In: *Boca Raton, Fla.* USA: CRC Press.

50 Khoo, H.E., Azlan, A., Tang, S.T., and Lim, S.M. (2017). Anthocyanidins and anthocyanins: colored pigments as food, pharmaceutical ingredients, and the potential health benefits. *Food & Nutrition Research* 61 (1): 1361779.

51 Chong, J., Poutaraud, A., and Hugueney, P. (2009). Metabolism and roles of stilbenes in plants. *Plant Science* 177 (3): 143–155.

52 Wahab, A., Gao, K., Jia, C. et al. (2017). Significance of resveratrol in clinical management of chronic diseases. *Molecules* 22 (8): 1329.

53 Bertelli, A., Bertelli, A.A., Gozzini, A.,and Giovannini, L. (1998). Plasma and tissue resveratrol concentrations and pharmacological activity. *Drugs under Experimental and Clinical Research* 24 (3): 133–138.

54 Bhat, K.P. and Pezzuto, J.M. (2002). Cancer chemopreventive activity of resveratrol. *Annals of the New York Academy of Sciences* 957 (1): 210–229.

55 Vitrac, X., Monti, J., Vercauteren, J. et al. (2002). Direct liquid chromatographic analysis of resveratrol derivatives and flavanonols in wines with absorbance and fluorescence detection. *Analytica Chimica Acta* 458 (1): 103–110.

56 Adlercreutz, H. and Mazur, W. (1997). Phyto-oestrogens and Western diseases. *Annals of Medicine* 29 (2): 95–120.

57 Goldberg, D. (2010). Critical reviews in clinical laboratory sciences. *Critical reviews in clinical laboratory sciences* 47 (1): 1–4.

58 Ray, S.K. and Mukherjee, S. (2021). Evolving interplay between dietary polyphenols and gut microbiota—an emerging importance in healthcare. *Frontiers in Nutrition* 8: 634944.

59 Record, I.R. and Lane, J.M. (2001). Simulated intestinal digestion of green and black teas. *Food Chemistry* 73 (4): 481–486.

60 Rios, L.Y., Bennett, R.N., Lazarus, S. A. et al. (2002). Cocoa procyanidins are stable during gastric transit in humans. *The American Journal of Clinical Nutrition* 76 (5): 1106–1110.

61 Galanakis, C.M. (2018). *Polyphenols: Properties, Recovery, and Applications*. Woodhead Publishing.

62 Tagliazucchi, D., Verzelloni, E., Bertolini, D., and Conte, A. (2010). In vitro bio-accessibility and antioxidant activity of grape polyphenols. *Food Chemistry* 120 (2): 599–606.

63 Boyer, J., Brown, D., and Liu, R.H. (2005). In vitro digestion and lactase treatment influence uptake of quercetin and quercetin glucoside by the Caco-2 cell monolayer. *Nutrition Journal* 4: 1–15.

64 Spencer, J.P., Chowrimootoo, G., Choudhury, R., Debnam, E.S. et al. (1999). The small intestine can both absorb and glucuronidate luminal flavonoids. *FEBS Letters* 458 (2): 224–230.

65 Župerl, Š., Fornasaro, S., Novič, M., and Passamonti, S. (2011). Experimental determination and prediction of bilitranslocase transport activity. *Analytica chimica acta* 705 (1–2): 322–333.

66 Karawajczyk, A., Drgan, V., Medic, N. et al. (2007). Properties of flavonoids influencing the binding to bilitranslocase investigated by neural network modelling. *Biochemical Pharmacology* 73 (2): 308–320.

67 Farrell, T.L., Gomez-Juaristi, M., Poquet, L. et al. (2012). Absorption of dimethoxycinnamic acid derivatives in vitro and pharmacokinetic profile in human plasma following coffee consumption. *Molecular Nutrition & Food Research* 56 (9): 1413–1423.

68 Lafay, S., Gil-Izquierdo, A., Manach, C. et al. (2006). Chlorogenic acid is absorbed in its intact form in the stomach of rats. *The Journal of Nutrition* 136 (5): 1192–1197.

69 Konishi, Y., Zhao, Z., and Shimizu, M. (2006). Phenolic acids are absorbed from the rat stomach with different absorption rates. *Journal of Agricultural and Food Chemistry* 54 (20): 7539–7543.

70 Kuhnle, G., Spencer, J.P., Schroeter, H. et al. (2000). Epicatechin and catechin are O-methylated and glucuronidated in the small intestine. *Biochemical and Biophysical Research Communications* 277 (2): 507–512.

71 Weinert, C. H., Wiese, S., Rawel, H.M. et al. (2012). Methylation of catechins and procyanidins by rat and human catechol-O-methyltransferase: metabolite profiling and molecular modeling studies. *Drug Metabolism and Disposition* 40 (2): 353–359.

72 Aura, A.-M. (2008). Microbial metabolism of dietary phenolic compounds in the colon. *Phytochemistry Reviews* 7: 407–429.

73 Roowi, S., Stalmach, A., Mullen, W. et al. (2010). Green tea flavan-3-ols: colonic degradation and urinary excretion of catabolites by humans. *Journal of Agricultural and Food Chemistry* 58 (2): 1296–1304.

74 Crespy, V., Morand, C., Besson, C. et al. (2003). The splanchnic metabolism of flavonoids highly differed according to the nature of the compound. *American Journal of Physiology-Gastrointestinal and Liver Physiology* 284 (6): G980–G988.

75 Crespy, V., Morand, C., Besson, C. et al. (2003). Quantitation of soy-derived phytoestrogens in human breast tissue and biological fluids by high-performance liquid chromatography. *Journal of Chromatography B* 784 (1): 137–144.

76 Sfakianos, J., Coward, L., Kirk, M., and Barnes, S. (1997). Intestinal uptake and biliary excretion of the isoflavone genistein in rats. *The Journal of Nutrition* 127 (7): 1260–1268.

77 Yang, B., Arai, K., and Kusu, F. (2000). Determination of catechins in human urine subsequent to tea ingestion by high-performance liquid chromatography with electrochemical detection. *Analytical Biochemistry* 283 (1): 77–82.

78 Donovan, J.L., Bell, J.R., Kasim-Karakas, S. et al. (1999). Catechin is present as metabolites in human plasma after consumption of red wine. *The Journal of Nutrition* 129 (9): 1662–1668.

79 Baba, S., Osakabe, N., Yasuda, A. et al. (2000). Bioavailability of (-)-epicatechin upon intake of chocolate and cocoa in human volunteers. *Free Radical Research* 33 (5): 635–641.

80 Bourne, L.C. and Rice-Evans, C. (1998). Bioavailability of ferulic acid. *Biochemical and Biophysical Research Communications* 253 (2): 222–227.

81 Cao, G., Muccitelli, H.U., Sánchez-Moreno, C., and Prior, R.L. (2001). Anthocyanins are absorbed in glycated forms in elderly women: a pharmacokinetic study. *The American Journal of Clinical Nutrition* 73 (5): 920–926.

82 Matsumoto, H., Inaba, H., Kishi, M. et al. (2001). Orally administered delphinidin 3-rutinoside and cyanidin 3-rutinoside are directly absorbed in rats and humans and appear in the blood as the intact forms. *Journal of Agricultural and Food Chemistry* 49 (3): 1546–1551.

83 McGhie, T.K. and Walton, M.C. (2007). The bioavailability and absorption of anthocyanins: towards a better understanding. *Molecular Nutrition & Food Research* 51 (6): 702–713.

84 Felgines, C., Talavéra, S., Texier, O. et al. (2003). Strawberry anthocyanins are recovered in urine as glucuro-and sulfoconjugates in humans. *The Journal of Nutrition* 133 (5): 1296–1301.

85 Schümann, K., Classen, H.G., Hages, M. et al. (1997). Bioavailability of oral vitamins, minerals, and trace elements in perspective. *Arzneimittel-forschung* 47 (4): 369–380.

86 Stahl, W., van den Berg, H., Arthur, J. et al. (2002). Bioavailability and metabolism. *Moecularl Aspects Medicine* 23 (1-3): 39–100.

87 Manach, C., Scalbert, A., Morand, C. et al. (2004). Polyphenols: food sources and bioavailability. *The American Journal of Clinical Nutrition* 79 (5): 727–747.

88 Scalbert, A. and Williamson, G. (2000). Dietary intake and bioavailability of polyphenols. *The Journal of Nutrition* 130 (8): 2073S–2085S.

89 Simonetti, P., Gardana, C., and Pietta, P. (2001). Caffeic acid as biomarker of red wine intake. *Methods in Enzymology* 335: 122–130.

90 Nardini, M., Cirillo, E., Natella, F., and Scaccini, C. (2002). Absorption of phenolic acids in humans after coffee consumption. *Journal of Agricultural and Food Chemistry* 50 (20): 5735–5741.

91 Hollman, P.C., van Trijp, J.M., Buysman, M.N. et al. (1997). Relative bioavailability of the antioxidant flavonoid quercetin from various foods in man. *FEBS Letters* 418 (1–2): 152–156.

92 Xu, X., Wang, H., Murphy, P.A. et al. (1994). Daidzein is a more bioavailable soymilk isoflavone than is genistein in adult women. *The Journal of Nutrition* 124 (6): 825–832.

93 Erlund, I., Meririnne, E., Alfthan, G., and Aro, A. (2001). Plasma kinetics and urinary excretion of the flavanones naringenin and hesperetin in humans after ingestion of orange juice and grapefruit juice. *The Journal of Nutrition* 131 (2): 235–241.

94 Vitaglione, P., Donnarumma, G., Napolitano, A. et al. (2007). Protocatechuic acid is the major human metabolite of cyanidin-glucosides. *The Journal of Nutrition* 137 (9): 2043–2048.

95 D'Archivio, M., Filesi, C., Varì, R. et al. (2010). Bioavailability of the polyphenols: status and controversies. *International Journal of Molecular Sciences* 11 (4): 1321–1342.

96 Pérez-Jiménez, J., Serrano, J., Tabernero, M. et al. (2009). Bioavailability of phenolic antioxidants associated with dietary fiber: plasma antioxidant capacity after acute and long-term intake in humans. *Plant Foods for Human Nutrition* 64: 102–107

97 Rubio, L., Macia, A., and Motilva, M.-J. (2014). Impact of various factors on pharmacokinetics of bioactive polyphenols: an overview. *Current Drug Metabolism* 15 (1): 62–76.
98 Manach, C., Williamson, G., Morand, C. et al. (2005). Bioavailability and bioefficacy of polyphenols in humans. I. Review of 97 bioavailability studies. *The American Journal of Clinical Nutrition* 81 (1): 230S–242S.
99 Kardum, N., and Glibetic, M. (2018). Polyphenols and their interactions with other dietary compounds: Implications for human health. Advances in Food and Nutrition Research 103–144. https://doi.org/10.1016/bs.afnr.2017.12.001.
100 Tulipani, S., Martinez Huelamo, M., Rotches Ribalta, M. et al. (2012). Oil matrix effects on plasma exposure and urinary excretion of phenolic compounds from tomato sauces: evidence from a human pilot study. *Food Chemistry* 130 (3): 581–590.
101 Schramm, D.D., Karim, M., Schrader, H.R. et al. (2003). Food effects on the absorption and pharmacokinetics of cocoa flavanols. *Life Sciences* 73 (7): 857–869.
102 Faria, A., Pestana, D., Azevedo, J. et al. (2009). Absorption of anthocyanins through intestinal epithelial cells–Putative involvement of GLUT2. *Molecular Nutrition & Food Research* 53 (11): 1430–1437.
103 Felgines, C., Talavera, S., Texier, O. et al. (2005). Blackberry anthocyanins are mainly recovered from urine as methylated and glucuronidated conjugates in humans. *Journal of Agricultural and Food Chemistry* 53 (20): 7721–7727.
104 Day, A.J., Bao, Y., Morgan, M. R., and Williamson, G. (2000). Conjugation position of quercetin glucuronides and effect on biological activity. *Free Radical Biology and Medicine* 29 (12): 1234–1243.
105 Piskula, M.K. and Terao, J. (1998). Accumulation of (−)-epicatechin metabolites in rat plasma after oral administration and distribution of conjugation enzymes in rat tissues. *The Journal of Nutrition* 128 (7): 1172–1178.
106 Tilgmann, C. and Ulmanen, I. (1996). Purification methods of mammalian catechol-O-methyltransferases. *Journal of Chromatography B: Biomedical Sciences and Applications* 684 (1–2): 147–161.
107 Falany, C.N. (1997). Enzymology of human cytosolic sulfotransferases. *The FASEB Journal* 11 (4): 206–216.
108 Boersma, M.G., Van der Woude, H., Bogaards, J. et al. (2002). Regioselectivity of phase II metabolism of luteolin and quercetin by UDP-glucuronosyl transferases. *Chemical Research in Toxicology* 15 (5): 662–670.
109 Selma, M.V., Espin, J.C., and Tomas-Barberan, F.A. (2009). Interaction between phenolics and gut microbiota: role in human health. *Journal of Agricultural and Food Chemistry* 57 (15): 6485–6501.
110 Liu, Z. and Hu, M. (2007). Natural polyphenol disposition via coupled metabolic pathways. *Expert Opinion on Drug Metabolism & Toxicology* 3 (3): 389–406.
111 Atkinson, C., Frankenfeld, C.L., and Lampe, J.W. (2005). Gut bacterial metabolism of the soy isoflavone daidzein: exploring the relevance to human health. *Experimental Biology and Medicine* 230 (3): 155–170.
112 Bräunlich, M., Slimestad, R., Wangensteen, H. et al. (2013). Extracts, anthocyanins and procyanidins from Aronia melanocarpa as radical scavengers and enzyme inhibitors. *Nutrients* 5 (3): 663–678.
113 Shishikura, Y., Khokhar, S., and Murray, B.S. (2006). Effects of tea polyphenols on emulsification of olive oil in a small intestine model system. *Journal of Agricultural and Food Chemistry* 54 (5): 1906–1913.

114 Petzke, K.J., Schuppe, S., Rohn, S. et al. (2005). Chlorogenic acid moderately decreases the quality of whey proteins in rats. *Journal of Agricultural and Food Chemistry* 53 (9): 3714–3720.

115 Scheepens, A., Tan, K., and Paxton, J.W. (2010). Improving the oral bioavailability of beneficial polyphenols through designed synergies. *Genes & Nutrition* 5: 75–87.

116 Fernstrom, J.D. (1990). Aromatic amino acids and monoamine synthesis in the central nervous system: influence of the diet. *The Journal of Nutritional Biochemistry* 1 (10): 508–517.

117 Fuhr, U., Klittich, K., and Staib, A.H. (1993). Inhibitory effect of grapefruit juice and its bitter principal, naringenin, on CYP1A2 dependent metabolism of caffeine in man. *British Journal of Clinical Pharmacology* 35 (4): 431–436.

118 Gandhi, L., Harding, M.W., Neubauer, M. et al. (2007). *A phase II study of the safety and efficacy of the multidrug resistance inhibitor VX-710 combined with doxorubicin and vincristine in patients with recurrent small cell lung cancer. Cancer Interdisciplinary International Journal of the American Cancer Society* 109 (5): 924–932.

119 Stein, G.E. and Gurwith, M.J. (1984). Amoxicillin-potassium clavulanate, a beta-lactamase-resistant antibiotic combination. *Clinical Pharmacy* 3 (6): 591–599.

120 Youdim, K.A., Qaiser, M., Begley, D.J. et al. (2004). Flavonoid permeability across an in situ model of the blood–brain barrier. Free Radical Biology and Medicine 36 (5): 592–604.

121 Youdim, M.B., Edmondson, D., and Tipton, K.F. (2006). The therapeutic potential of monoamine oxidase inhibitors. *Nature Reviews Neuroscience* 7 (4): 295–309.

122 Rudrapal, M., Mishra, A.K., Rani, L. et al. (2022). Nanodelivery of dietary polyphenols for therapeutic applications. *Molecules* 27 (24): 8706.

123 Xie, X., Tao, Q., Zou, Y. et al. (2011). PLGA nanoparticles improve the oral bioavailability of curcumin in rats: characterizations and mechanisms. *Journal of Agricultural and Food Chemistry* 59 (17): 9280–9289.

124 Kakkar, V., Singh, S., Singla, D., and Kaur, I.P. (2011). Exploring solid lipid nanoparticles to enhance the oral bioavailability of curcumin. *Molecular Nutrition & Food Research* 55 (3): 495–503.

125 Vitaglione, P., Barone Lumaga, R., Ferracane, R. et al. (2012). Curcumin bioavailability from enriched bread: the effect of microencapsulated ingredients. *Journal of Agricultural and Food Chemistry* 60 (13): 3357–3366.

126 Tikekar, R.V., Pan, Y., and Nitin, N. (2013). Fate of curcumin encapsulated in silica nanoparticle stabilized pickering emulsion during storage and simulated digestion. *Food Research International* 51 (1): 370–377.

127 Marín, L., Miguélez, E.M., Villar, C.J., and Lombó, F. (2015). Bioavailability of dietary polyphenols and gut microbiota metabolism: antimicrobial properties. *BioMed Research International* 2015: 1–8.

Index

Page numbers in *italic* refer to figures; page numbers in **bold** refer to tables. US spelling is used in this index.

e

l

m

p

Printed and bound by CPI Group (UK) Ltd, Croydon, CR0 4YY

17/09/2026

14964314-0006